Studies in Systems, Decision and Control

Volume 115

Series editor

Janusz Kacprzyk, Polish Academy of Sciences, Warsaw, Poland
e-mail: kacprzyk@ibspan.waw.pl

About this Series

The series "Studies in Systems, Decision and Control" (SSDC) covers both new developments and advances, as well as the state of the art, in the various areas of broadly perceived systems, decision making and control- quickly, up to date and with a high quality. The intent is to cover the theory, applications, and perspectives on the state of the art and future developments relevant to systems, decision making, control, complex processes and related areas, as embedded in the fields of engineering, computer science, physics, economics, social and life sciences, as well as the paradigms and methodologies behind them. The series contains monographs, textbooks, lecture notes and edited volumes in systems, decision making and control spanning the areas of Cyber-Physical Systems, Autonomous Systems, Sensor Networks, Control Systems, Energy Systems, Automotive Systems, Biological Systems, Vehicular Networking and Connected Vehicles, Aerospace Systems, Automation, Manufacturing, Smart Grids, Nonlinear Systems, Power Systems, Robotics, Social Systems, Economic Systems and other. Of particular value to both the contributors and the readership are the short publication timeframe and the world-wide distribution and exposure which enable both a wide and rapid dissemination of research output.

More information about this series at http://www.springer.com/series/13304

Shihua Li · Xinghuo Yu · Leonid Fridman
Zhihong Man · Xiangyu Wang
Editors

Advances in Variable Structure Systems and Sliding Mode Control—Theory and Applications

Springer

Editors
Shihua Li
School of Automation
Southeast University
Nanjing
China

Xinghuo Yu
Research & Innovation Portfolio
RMIT University
Melbourne, VIC
Australia

Leonid Fridman
Division of Electrical Engineering
National Autonomous University of Mexico
Mexico City, Distrito Federal
Mexico

Zhihong Man
Faculty of Science, Engineering
and Technology
Swinburne University of Technology
Melbourne, VIC
Australia

Xiangyu Wang
School of Automation
Southeast University
Nanjing
China

ISSN 2198-4182 ISSN 2198-4190 (electronic)
Studies in Systems, Decision and Control
ISBN 978-3-319-62895-0 ISBN 978-3-319-62896-7 (eBook)
DOI 10.1007/978-3-319-62896-7

Library of Congress Control Number: 2017946037

Printed on acid-free paper

This Springer imprint is published by Springer Nature
The registered company is Springer International Publishing AG
The registered company address is: Gewerbestrasse 11, 6330 Cham, Switzerland

Preface

Variable structure systems (VSS) and its main mode of operation sliding mode control (SMC) are recognized as one of the most efficient tools to deal with uncertain systems due to their robustness and even insensitivity to perturbations [1–3].

The main advantages of VSS/SMC methodology are:

- theoretical insensitivity with respect to the matched perturbations;
- reduced order of sliding mode dynamics;
- finite-time convergence to zero for sliding mode variables.

However, the development of the VSS/SMC theory has shown their main drawbacks: the chattering phenomenon, namely high-frequency oscillations appearing due to the presence of parasitic dynamics of actuators, sensors, and other non-ideality.

During the last decade, one of the main lines in development of the SMC theory was development of the homogeneous higher-order sliding mode controllers (HOSMC) (see [4–6]). At the first stage, the proof of such algorithm was based on the arguments of homogeneity and geometry.

The main driver of development in recent two years is the new Lyapunov-based approaches for HOSMC design and gain selection [7, 8]. Moreover, the development of Lyapunov function approaches allows to design continuous sliding mode algorithms [9–13].

Different properties of SMC algorithms are investigated, like properties of HOSMC for wider classes of homogeneous systems, as well as properties of SMC for stochastic systems [14] and properties of SMC in frequency domain [15, 16]. Different adaptive algorithms were recently developed [17, 18]. These new algorithms were actively used to both ensure the tracking in different control problems and implement it for control in different real-life systems.

This book is an attempt to reflect the recent developments in VSS/SMC theory and reflect the results which are presented. The book consists of three parts: in the first part (i.e., Chaps. 1–7), new VSS/SMC algorithms are proposed and its properties are analyzed; in the second part (i.e., Chaps. 8–13), the usage of VSS/SMC techniques for solutions of different control problems is given; in the third part (i.e., Chaps. 14–16), applications of VSS/SMC to real-time systems are exhibited.

Part I: New VSS/SMC Algorithms and Their Properties (Chaps. 1–7)

In Chap. 1 "Lyapunov-Based Design of Homogeneous High-Order Sliding Modes" by Prof. Jaime A. Moreno, the author provides a Lyapunov-based design of homogeneous high-order sliding mode (HOSM) control and observation (differentiation) algorithms of arbitrary order for a class of single-input-single-output uncertain nonlinear systems. First, the authors recall the standard problem of HOSM control, which corresponds to the design of a state feedback control and an observer for a particular differential inclusion (DI), which represents a family of dynamic systems including bounded matched perturbations/uncertainties. Next, the author provides a large family of zero-degree homogeneous discontinuous controllers solving the state feedback problem based on a family of explicit and smooth homogeneous Lyapunov functions. The author shows the formal relationship between the control laws and the Lyapunov functions. This also gives a method for the calculation of controller gains ensuring the robust and finite-time stability of the sliding set. The required unmeasured states can be estimated robustly and in finite time by means of an observer or differentiator, originally proposed by Prof. A. Levant. The author gives explicit and smooth Lyapunov functions for the design of gains ensuring the convergence of the estimated states to the actual ones in finite time, despite the non-vanishing bounded perturbations or uncertainties acting on the system. Finally, it is shown that a kind of separation principle is valid for the interconnection of the HOSM controller and observer, and the author illustrates the results by means of a simulation on an electromechanical system.

In Chap. 2 "Robustness of Homogeneous and Homogeneizable Differential Inclusions" by Dr. Emmanuel Bernuau, Prof. Denis Efimov, and Prof. Wilfrid Perruquetti, the authors study the problem of robustness of sliding mode control and estimation algorithms with respect to matched and unmatched disturbances. Using the homogeneous theories and locally homogeneous differential inclusions, two sets of conditions are developed to verify the input-to-state stability property of discontinuous systems. The advantage of the proposed conditions is that they are not based on the Lyapunov function method, but more related to algebraic operations over the right-hand side of the system.

In Chap. 3 "Stochastic Sliding Mode Control and State Estimation" by Prof. Alex S. Poznyak, the author deals with the SMC technique applied to stochastic systems affected by additive as well as multiplicative stochastic white noise. The existence of a strong solution to the corresponding stochastic differential inclusion is discussed. It is shown that this approach is workable with the gain control parameter state-dependent on norms of system states. It is demonstrated that under such modification of the conventional SMC, the exponential convergence of the averaged squared norm of the sliding variable to a zone (around the sliding surface) can be guaranteed, of which the bound is proportional to the diffusion parameter in the model description and inversely depending on the gain parameter.

The behavior of a standard super-twist controller under stochastic perturbations is also studied. For system quadratically stable in the mean-squared sense, a sliding mode observer with the gain parameter linearly depending on the norm of the output estimation error is suggested. It has the same structure as deterministic observer based on "the Equivalent Control Method." The workability of the suggested observer is guaranteed for the group of trajectories with the probabilistic measure closed to one. All theoretical results are supported by numerical simulations.

In Chap. 4 "Practical Stability Phase and Gain Margins Concept" by Prof. Yuri Shtessel, Prof. Leonid Fridman, Dr. Antonio Rosales, and Dr. Chandrasekhara Bharath Panathula, the authors present a new concept of chattering characterization for the systems driven by finite-time convergent controllers (FTCC) in terms of practical stability margins. Unmodeled dynamics of order two or more incite chattering in FTCC-driven systems. In order to analyze the FTCC robustness to unmodeled dynamics, the novel paradigm of tolerance limits (TL) is introduced to characterize the acceptable emerging chattering. Following this paradigm, the authors introduce a new notion of Practical Stability Phase Margin (PSPM) and Practical Stability Gain Margin (PSGM) as a measure of robustness to cascade unmodeled dynamics. Specifically, PSPM and PSGM are defined as the values that have to be added to the phase and gain of dynamically perturbed system driven by FTCC so that the characteristics of the emerging chattering reach TL. For practical calculation of PSPM and PSGM, the harmonic balance (HB) method is employed, and a numerical algorithm to compute describing functions (DFs) for families of FTCC (specifically, for nested, and quasi-continuous higher-order sliding mode (HOSM) controllers) was proposed. A database of adequate DFs was developed. A numerical algorithm for solving HB equation using the Newton–Raphson method is suggested to obtain predicted chattering parameters. Finally, computational algorithms to that identify PSPM and PSGM for the systems driven by FTCC were proposed. The algorithm of a cascade linear compensator design that corrected the FTCC, making the values of PSPM and PSGM to fit the prescribed quantities, is suggested. In order to design the flight-certified FTCC for attitude for the F-16 jet fighter, the proposed technique was employed as a case study. The prescribed robustness to cascade unmodeled actuator dynamics was achieved.

In Chap. 5 "On Inherent Gain Margins of Sliding-Mode Control Systems" by Prof. Igor Boiko, the author defines notion of inherent gain margin of sliding mode control systems. It is demonstrated through analysis and examples that an inherent gain margin depends on the sliding mode control algorithm and not on the plant. This property makes the inherent gain margin a characteristic suitable for comparison of different control algorithms. Analysis of the first-order sliding mode, hysteresis relay control, twisting algorithm, and suboptimal algorithm is presented.

In Chap. 6 "Adaptive Sliding Mode Control Based on the Extended Equivalent Control Concept for Disturbances with Unknown Bounds" by Prof. Tiago Roux Oliveira, Prof. José Paulo V.S. Cunha, and Prof. Liu Hsu, the authors propose an adaptive sliding mode framework based on extended equivalent control to deal with disturbances of unknown bounds. Nonlinear plants are considered with a quite general class of (non)smooth disturbances. The proposed adaptation method is able to make the control gain large when the disturbance grows and decrease it if the latter vanishes, allowing for a minimized chattering occurrence. Global stability of the closed-loop system is demonstrated using the proposed adaptive sliding mode control law. Simulations are presented to show the potential of the new adaptation scheme in this adverse scenario of possibly growing or temporarily large disturbances.

In Chap. 7 “Indirect Adaptive Sliding-Mode Control Using the Certainty-Equivalence Principle” by Dr. Alexander Barth, Prof. Markus Reichhartinger, Prof. Kai Wulff, Prof. Johann Reger, Prof. Stefan Koch, and Prof. Martin Horn, the authors address the design of adaptive sliding mode controllers. The presented controllers compensate uncertainties acting on the input channel of the considered system and are characterized by a possible separation into a structured and an unstructured part. The latter class of uncertainty may affect the system in terms of an external disturbance, whereas a structured uncertainty typically occurs in the case of uncertain plant parameters. The presented controller design methodology enhances standard sliding mode controllers by an additional control action generated from an adaptation mechanism. Applying the certainty equivalence principle, it is possible to systematically handle both classes of uncertainties. The controller design is introduced step by step and demonstrated in detail for systems designated to be controlled by the super-twisting algorithm. The deviation of the adaptive part of the controller is thoroughly demonstrated by deriving three different types of adaptation laws. The requirement to enhance sliding mode controllers by the presented adaptive scheme is underpinned by a simulation scenario demonstrating cascaded feedback loops used for speed and current control of a DC motor. Experimental results obtained by a laboratory test-rig consisting of a motor with unbalanced load demonstrate the applicability of the discussed controller design method.

Part II: The Usage of VSS/SMC Techniques for Solutions of Different Control Problems (Chaps. 8–13)

In Chap. 8 “Variable Structure Observers For Nonlinear Interconnected Systems” by Dr. Mokhtar Mohamed, Prof. Xing-Gang Yan, Prof. Sarah K. Spurgeon, and Prof. Zehui Mao, the authors are concentrated on observer design for nonlinear interconnected systems in the presence of nonlinear interconnections and uncertainties. An approach to deal with nonlinear interconnections is proposed by separating the interconnections to linear and nonlinear parts based on an appropriate transformation. Using the structure property of the interconnected systems, novel variable structure dynamics are designed to observe the state variables of the interconnected systems asymptotically with low conservatism. A simulation example and a case study are presented to demonstrate the effectiveness and the feasibility of the developed results.

In Chap. 9 “A Unified Lyapunov Function for Finite Time Stabilization of Continuous and Variable Structure Systems with Resets” by Dr. Harshal B. Oza, Prof. Yury V. Orlov, and Prof. Sarah K. Spurgeon, the authors present a unified

Lyapunov function for finite-time stabilization of continuous and variable structure systems with resets. This chapter aims to uniformly stabilize a perturbed dynamics of the double integrator in the presence of impacts due to the constraints on the position variable. A non-smooth transformation is proposed to first transform the system into a variable structure system that can be studied within the framework of a conventional discontinuous paradigm. Then, a finite-time stable continuous controller is utilized, and stability of the closed-loop dynamics is proven by identifying a new set of Lyapunov functions. The chapter thus contributes to the VSS and SMC theory by the developing mathematical tools for the finite-time stability analysis of such systems in the presence of impacts.

In Chap. 10 "Robustification of Cooperative Consensus Algorithms in Perturbed Multi-Agents Systems" by Prof. Alessandro Pilloni, Prof. Alessandro Pisano, and Prof. Elio Usai, the authors exploit the integral sliding mode design paradigm in the framework of multi-agent systems. Particularly, it is shown how to redesign standard distributed algorithms for estimating the average value and the median value of the agent's initial conditions in spite of perturbations acting on the agent's dynamics. Constructive Lyapunov-based analysis is presented along with simulation results corroborating the developed treatment.

In Chap. 11 "Finite-Time Consensus for Disturbed Multi-Agent Systems with Unmeasured States via Nonsingular Terminal Sliding-Mode Control" by Dr. Xiangyu Wang and Prof. Shihua Li, the authors study the finite-time output consensus problem for leader–follower higher-order multi-agent systems with mismatched disturbances and unmeasured states. This problem is solved by using a feedforward–feedback composite control method which combines the integral-type non-singular terminal sliding mode control approach and a finite-time observer technique together. The main contributions include three aspects: Firstly, in the presence of mismatched disturbances and unmeasured states, the finite-time output consensus is realized by utilizing the distributed active anti-disturbance control for the first time. Secondly, the results extend the applicable scope of the distributed active anti-disturbance control from state feedback to output feedback. Thirdly, the disturbances considered in this chapter are allowed to be faster time-varying or have higher-order forms, which are not limited to slow time-varying types any more.

In Chap. 12 "Discrete Event-Triggered Sliding Mode Control" by Prof. Abhisek K. Behera and Prof. Bijnan Bandyopadhyay, the authors present a discrete event-triggered SMC strategy for linear systems. Generally, in the event-triggered control, the state is continuously monitored to generate the possible triggering instant, which may incur additional cost and complexity. To overcome this, a discrete event-triggered SMC is proposed which evaluates event periodically and also guarantees the robust performance of the system. The discrete-time SMC is designed considering the triggering rule that ensures the stability with the discrete event-triggering strategy.

In Chap. 13 "Fault Tolerant Control Using Integral Sliding Modes" by Prof. Christopher Edwards, Dr. Halim Alwi, and Dr. Mirza Tariq Hamayun, the authors consider so-called integral sliding modes (ISM) and demonstrate how they can be employed in the context of fault-tolerant control. Two distinct classes of problems are considered: Firstly, a fault-tolerant ISM controller is designed for an over-actuated linear system; secondly, an ISM scheme is retrofitted to an existing feedback control scheme for an over-actuated uncertain linear system with the objective of retaining the preexisting nominal performance in the face of faults and failures. The chapter includes with a case study describing the implementation of an LPV extension of one of the ISM schemes on a motion simulator configured to represent a Boeing 747 aircraft subject to realistic fault scenarios.

Part III: Applications of VSS/SMC to Real-Time Systems (Chaps. 14–16)

In Chap. 14 "Speed Control of Induction Motor Servo Drives Using Terminal Sliding-Mode Controller" by Prof. Yong Feng, Dr. Minghao Zhou, Prof. Fengling Han, and Prof. Xinghuo Yu, the authors apply a non-singular terminal sliding mode control method for the servo system of induction motors. The non-singular terminal sliding mode controllers for speed, flux, and currents are presented, respectively. The switching signals in the controller are softened to generate the continuous output signals of the controllers using the equivalent low-pass filters. Therefore, both the chattering is attenuated and the singularity is eliminated, which means that the controllers can be used in the practical applications.

In Chap. 15 "Sliding Modes Control in Vehicle Longitudinal Dynamics Control" by Prof. Antonella Ferrara and Dr. Gian Paolo Incremona, the authors present recent developments produced at the University of Pavia on application of sliding mode control to the automotive field. Specifically, the chapter focuses on the use of advanced SMC schemes to efficiently solve traction control and vehicle platooning control problems. A slip ratio SMC scheme is described, analyzed, and assessed in simulation. Then, the vehicle platooning control problem is introduced as an extended case of the previously described problem. A vehicle longitudinal dynamics control scheme, based on a suboptimal second-order SMC, is presented and coupled with the slip rate control scheme which allows to generate the correct traction control. The validation in simulation on a realistic scenario of the overall scheme is also discussed.

In Chap. 16 "Sliding Mode Control of Power Converters with Switching Frequency Regulation" by Dr. Víctor Repecho, Dr. Domingo Biel, Dr. Josep M. Olm, and Prof. Enric Fossas, the authors introduce a hysteresis band control loop that provides fixed switched frequency in sliding mode controlled systems while keeping the beneficial properties of sliding motion. The proposal is exemplified in DC-to-DC and DC-to-AC power converters carrying out regulation and tracking tasks, respectively, in the face of load disturbances and input voltage variations.

Nanjing, China — Shihua Li
Melbourne, Australia — Xinghuo Yu
Mexico City, Mexico — Leonid Fridman
Melbourne, Australia — Zhihong Man
Nanjing, China — Xiangyu Wang
March 2017

References

1. Utkin, V.: Sliding Modes in Optimization and Control Problems. Springer Verlag, New York (1992)
2. Edwards, C., Spurgeon, S.K.: Sliding Mode Control: Theory and Applications. Taylor & Francis, London (1998)
3. Shtessel, Y., Edwards, C., Fridman, L., Levant, A.: Sliding Mode Control and Observation. Springer, New York (2014)
4. Levant, A.: Universal SISO sliding-mode controllers with finite-time convergence. IEEE Trans. Autom. Control **46**(9), 1447–1451 (2001)
5. Levant, A.: High-order sliding modes: differentiation and output feedback control. Int. J. Control **76**(9–10), 924–941 (2003)
6. Levant, A.: Quasi-continuous high-order sliding-mode controllers. IEEE Trans. Autom. Control **50**, 1812–1816 (2005)
7. Ding, S., Levant, A., Li, S.: Simple homogeneous sliding-mode controller. Automatica **67**, 22–32 (2016)
8. Fridman, L., Barbot, J.P., Plestan F.: Recent Trends in Sliding Mode Control. IET (2016)
9. Yu, X., Efe, O.: Studies in Systems, Decision and Control 24, Springer, Switzerland, 5–35 (2015)
10. Kamal, S., Moreno, J., Chalanga, A., Bandyopadhyay, B., Fridman, L.: Continuous terminal sliding mode control. Automatica **69**(7), 308–314 (2016)
11. Torres, V., Sanchez, T., Fridman, L., Moreno, J.: Design of continuous twisting algorithm. Automatica **80**, 119–126 (2017)
12. Moreno, J.A.: Discontinuous integral control for mechanical systems. In: Proceedings of the 14th International Workshop on Variable Structure Systems, pp. 142–147 (2016)
13. Edwards, C., Shtessel, Y.: Adaptive continuous higher order sliding mode control. Automatica **65**, 183–190 (2016)
14. Poznyak, A.: Sliding mode control in stochastic continuos-time systems: μ-zone MS-convergence. IEEE Trans. Autom. Control **62**(2), 863–868 (2017)
15. Rosales, A., Shtessel, Y., Fridman, L., Panathula, C.B.: Chattering analysis of HOSM controlled systems: frequency domain approach. IEEE Trans. Autom. Control **62**(8), 4109–4115 (2017)
16. Rosales, A., Shtessel, Y., Fridman, L.: Analysis and design of systems driven by finite-time convergent controllers: practical stability approach. Int. J. Control (2016) doi: 10.1080/00207179.2016.1255354
17. Shtessel, Y., Fridman, L., Plestan, F.: Adaptive sliding mode control and observation. Int. J. Control 89(9), 1743–1746 (2016)
18. Roux Oliveira, T., Fridman, L., Ortega, R.: From adaptive control to variable structure systems-Seeking for harmony. special issue in honor of Professor Liu's 70th Birthday. Int. J. Adapt. Control Signal Processing **30**(8–10), 1074–1079 (2016)

Acknowledgements

Professor Shihua Li gratefully acknowledges the financial support from the National Natural Science Foundation of China under Grants 61473080 and 61633003. Professor Xinghuo Yu gratefully acknowledges the financial support from the Australian Research Council under Grants DP140100544 and DP170102303. Professor Fridman gratefully acknowledges the financial support from Programa de Apoyo a Proyectos de Investigacion e Innovacion Tecnologica of UNAM under Grant 113216 and DGAPA PASPA Program. Doctor Xiangyu Wang gratefully acknowledges the financial support from the National Natural Science Foundation of China under Grant 61503078 and the Natural Science Foundation of Jiangsu Province under Grant BK20150626.

Nanjing, China — Shihua Li
Melbourne, Australia — Xinghuo Yu
Mexico City, Mexico — Leonid Fridman
Melbourne, Australia — Zhihong Man
Nanjing, China — Xiangyu Wang
March 2017

Contents

Part I
New VSS/SMC Algorithms and Their Properties

Chapter 1
Lyapunov-Based Design of Homogeneous High-Order Sliding Modes

Jaime A. Moreno

1.1 Introduction

Sliding Mode (SM) Control (SMC) [65, 66] aims at designing a sliding variable σ and to force it to $\sigma \equiv 0$ in finite time and to keep it in zero for all future times despite uncertainties and perturbations. For this it is required a discontinuous control action. Classical (or First Order (FO)) SMC achieves this objective when the sliding variable has relative degree $\rho = 1$ with respect to the control variable. Higher Order Sliding Mode (HOSM) Control [24, 38, 40, 41, 43, 65] extends these results to sliding variables σ with arbitrary relative degree $\rho > 1$. Since the implementation of a SMC requires the values of the sliding variable and all its derivatives up to order $\rho - 1$, i.e. $\sigma(t), \dot{\sigma}(t), \ldots, \sigma^{(\rho-1)}(t)$, in HOSMC it has been necessary to develop HOSM Differentiators [4, 5, 16, 20, 22, 35, 37, 39, 64] capable of estimating these derivatives of the sliding variable also in finite time and despite of the uncertainties and perturbations present in the system. These Exact Differentiators make also use of discontinuous output injection to achieve this goal, since smooth observers or differentiators are not able to achieve the objective in the presence of non vanishing uncertainties/perturbations. Since the classical FOSMC does not require any derivative of the sliding variable to be implemented, the Exact Differentiators are a particular development of HOSM's.

Due to the uncertainties and perturbations present in the system, the description of the dynamics of the ρ sliding variables $\sigma, \dot{\sigma}, \ldots, \sigma^{(\rho-1)}$ is naturally described not by a differential equation (DE) but by a Differential Inclusion (DI). One of the main tasks of SMC consists in designing an appropriate sliding variable σ. The sliding variable σ is selected in such a way that the reduced dynamics living on the sliding

J.A. Moreno (✉)
Instituto de Ingeniería, Universidad Nacional Autónoma de México (UNAM),
04510 Coyoacán, Ciudad de México, Mexico
e-mail: JMorenoP@ii.unam.mx

S. Li et al. (eds.), *Advances in Variable Structure Systems and Sliding Mode Control—Theory and Applications*, Studies in Systems, Decision and Control 115, DOI 10.1007/978-3-319-62896-7_1

set $\sigma(t) \equiv 0,\ \dot{\sigma}(t) \equiv 0, \ldots, \sigma^{(\rho-1)}(t) \equiv 0$ has the desired behavior, as e.g. it has a robust and asymptotically stable equilibrium point. And thus the main problem of HOSMC reduces to the Finite Time (FT) stabilization of the sliding set for the DI describing its behavior, and this task includes also the FT estimation of the sliding variables.

One of the main ingredients of SMC, the discontinuous control action, becomes also its main disadvantage for the applications: forcing a sliding mode induces a high frequency switching of the control variable and this produces the infamous "chattering" effect, which has undesirable effects as reducing the life of the actuators and exciting high frequency dynamics of the system. HOSM Control helps in mitigating the chattering effect, because introducing extra integrators in the control, and therefore increasing artificially the relative degree of the plant, a continuous (or even smooth) control action can be achieved, at the cost of a higher order sliding set. A further benefit of SMC is the order reduction of the plant's dynamics, since the main design work has to be done on the (reduced) dynamics living on the sliding set. Classical FOSMC allows a reduction of only one dimension (in the Single Input case) while HOSMC permits a reduction of an arbitrary number of degrees up to the order of the system.

(Weighted) Homogenous Differential Equations (HDE) are a very special class of nonlinear systems having very nice and simple properties [2, 3, 10, 28]. For example, for homogeneous systems: (i) local attractivity is equivalent to global asymptotic stability, (ii) internal stability of a system with inputs is equivalent to external stability, (iii) asymptotic stability with negative degree of homogeneity is equivalent to FT stability, etc. These nice properties are also valid for DI's [6, 8, 9, 40, 44]. The FT stabilization and the FT and exact estimation of the sliding variables of the DI describing them requires discontinuous control actions in the controller and discontinuous injection terms in the observer (differentiator). The design and analysis of the robustness, accuracy and convergence properties of the discontinuous controller and observer becomes much simpler if the homogeneity property is imposed on the controlled system and on the estimation error of the observer. This explains that homogeneity has become the main ingredient of HOSMs: essentially all HOSM controllers and observers designed up to now are homogenous.[1]

In fact, the design of discontinuous (and so called quasi-continuous) HOSM controllers and differentiators has been based on geometric methods (which are usually effective for low order or low relative degree systems) [36–39] or, more recently, on the use of Homogeneity and contraction properties of Differential Inclusions [40–43]. It is precisely the homogeneity [3, 8, 40] the property allowing to establishing basic *qualitative* properties of homogeneous HOSM algorithms, as e.g. globality, finite-time convergence, robustness and the type of accuracy.

In contrast, Modern Control Theory is based on the use of Lyapunov or Lyapunov-like Functions (LF) for analysis and design [23, 50]. This is due to the tight connections of this formalism with optimal control, robustness and the diverse internal

[1]For FOSM homogeneity does not play an important role.

and external stability concepts [23, 50]. In particular, for the design of feedback controllers the concept of (Robust) Control Lyapunov Functions (CLFs) has played a major role in the development of control design methods in the last twenty five years [23]. In particular, Classical SMC design is based on the use of Lyapunov functions. One advantage of LF's is that they provide *quantitative* measures which are helpful in the design of controllers and observers. Many of the modern and numerically effective design methods have at their core LF's, as e.g. LMI's for linear and nonlinear systems. It is therefore a natural idea to try to combine homogeneity with a Lyapunov-based design to enhance the modern HOSM control theory, rendering it more quantitative.

In recent years some efforts have been devoted to build explicit smooth and non-smooth, weak and strong Lyapunov functions for some Second Order Sliding Mode (SOSM) Controllers and observers, such as Twisting and Super-Twisting Algorithms [45, 47–49, 54–57]. In [46] a Lyapunov-based design of an output feedback controller, comprising homogeneous SOSM controller and observer, has been obtained. For HOSMC in [27] the authors use the basic idea of the Lyapunov redesign [34] to render a nominal finite-time convergent controller, as e.g. those proposed in [2, 30, 31, 61], and for which a Lyapunov function is already known, robust against matched and bounded perturbations by means of an extra discontinuous control. Unfortunately, the resulting closed loop system is not homogeneous, so that it does not have the nice properties of classical HOSMC [40].

Homogeneous HOSM controllers of arbitrary order based on (explicit) Lyapunov functions were obtained for the first time in [12] (see also [11, 13, 15]) while for arbitrary order HOSM differentiators explicit and smooth Lyapunov Functions were obtained in [14]. Very recently, a new family of so called relay or quasi-continuous polynomial HOSM controllers has been introduced in [18, 19], and a Lyapunov function is obtained for some relay polynomial cases, but no Lyapunov approach is developed for the HOSM differentiator. In [58, 59] Finite-Time convergent controllers have been designed by means of *implicit* Lyapunov functions (ILF), and (quasi-continuous) HOSM controllers can be obtained if some Matrix Inequalities are fulfilled. However, the quasi-continuous controller is also implicitly defined, so that for its implementation the Lyapunov function has to be calculated on-line. The ILF method provides only quasi-continuous controllers, and it has not been yet possible to design exact HOSM differentiators using ILFs.

The main purpose of this chapter is to present some recent advances towards developing a Lyapunov-based approach to the design of homogeneous HOSM control and observation. We develop *explicit* LFs for HOSM controllers and Observers (Differentiators) for the DI describing the dynamics of the sliding variables in HOSM. The use of Lyapunov functions provides a procedure for the gain tuning of the HOSM controllers and observers, it allows the estimation of the convergence time; and it permits the extension to variable-gain discontinuous and quasi-continuous HOSM controllers. Our results are inspired by and constitute a generalization to the discontinuous case of the results for continuous systems [2, 30, 31, 61, 62, 68–70].

The rest of the chapter is organized as follows. In the Sect. 1.2, we give some preliminaries on homogeneous functions and systems. In Sect. 1.3 we formulate the (standard HOSM) problem to be solved. Section 1.4 presents the Lyapunov-based HOSM controllers along with the explicit Lyapunov Functions associated to them. Section 1.5 presents the proofs of the main results of the previous section, and it can be skipped from the first reading without loosing the main track of the ideas. In Sect. 1.6 we show that for discontinuous, homogeneous differentiators there exist smooth LFs for the differentiators for appropriate values of the gains. The proof of this important fact is given in Sect. 1.7. Although this has been shown for the first time in the discontinuous case in [14] we provide here a different Lyapunov function that allows to design the gains of the differentiator independently of the order. In Sect. 1.8 it is shown that the combination of a homogeneous HOSM controller with a homogeneous HOSM observer leads to a globally FT stable output feedback controller, and that a kind of separation principle is available in the global case. Section 1.9 presents some numerical results and in Sect. 1.10 we draw some final comments and conclusions. The Appendix "Some Technical Lemmas on Homogeneous Functions" contains some technical results.

1.2 Preliminaries: Differential Inclusions and Homogeneity

We recall some important concepts about DI's, homogeneity and homogeneous DI's [2, 3, 6–10, 17, 21, 29, 40, 44], which are used in the chapter.

Uncertain or discontinuous systems are more appropriately described by Differential Inclusions (DI) $\dot{x} \in F(t, x)$ than by Differential Equations (DE). A solution of this DI is any function $x(t)$, defined in some interval $I \subseteq [0, \infty)$, which is absolutely continuous on each compact subinterval of I and such that $\dot{x}(t) \in F(t, x(t))$ almost everywhere on I. Thus, for a discontinuous DE $\dot{x} = f(t, x)$ the function $x(t)$ is said to be a generalized solution of the DE if and only if it is a solution of the associated DI $\dot{x} \in F(t, x)$. We will consider the DI $\dot{x} \in F(t, x)$ associated to $\dot{x} = f(t, x)$, as the one given by the approach of A.F. Filippov [3, 21, Sect. 1.2]. So, we refer to such DI as Filippov DI and to its solutions as Filippov solutions.

The multivalued map $F(t, x)$ satisfies the *standard assumptions* if: (H1) $F(t, x)$ is a nonempty, compact, convex subset of $\mathbb{R}^n$, for each $t \geq 0$ and each $x \in \mathbb{R}^n$; (H2) $F(t, x)$ as a set valued map of x, is upper semi-continuous for each $t \geq 0$; (H3) $F(t, x)$ as a set valued map of t, is Lebesgue measurable for each $x \in \mathbb{R}^n$. (H4) $F(t, x)$ is locally bounded. Recall that a set valued map $G : \mathbb{R}^{n_1} \rightrightarrows \mathbb{R}^{n_2}$ with compact values is *upper-semicontinuous* if for each x_0 and for each $\varepsilon > 0$ there exists $\delta > 0$ such that $G(x) \subseteq G(x_0) + B_\varepsilon$, provided that $x \in B_\delta(x_0)$. It is well-known that, see [21] or [3, Theorem 1.4], if the multivalued map $F(t, x)$ satisfies the standard assumptions then for each pair $(t_0, x_0) \in [0, \infty) \times \mathbb{R}^n$ there is an interval I and at least a solution $x(t) : I \to \mathbb{R}^n$ such that $t_0 \in I$ and $x(t_0) = x_0$. A DI $\dot{x} \in F(x)$ (a DE $\dot{x} = f(x)$) is called *globally uniformly finite-time stable* (GUFTS) at 0, if $x(t) = 0$ is a Lyapunov-stable solution and for any $R > 0$ there exists $T > 0$ such that the trajectory starting within the ball $\|x\| < R$ reaches zero in the time T.

Continuous and discontinuous homogeneous functions and systems have a long history [2, 3, 6, 8–10, 25, 28, 40, 44, 51–54, 71]. We recall this important property. For a given vector $x = [x_1, \ldots, x_n]^\top \in \mathbb{R}^n$ and for every $\varepsilon > 0$, the dilation operator is defined as $\Delta_\varepsilon^r x := [\varepsilon^{r_1} x_1, \ldots, \varepsilon^{r_n} x_n]^\top$, where $r_i > 0$ are the weights of the coordinates, and let $\mathbf{r} = [r_1, \ldots, r_n]^\top$ be the vector of weights. A function $V : \mathbb{R}^n \to \mathbb{R}$ (respectively, a vector field $f : \mathbb{R}^n \to \mathbb{R}^n$, or a vector-set field $F(x) \subset \mathbb{R}^n$) is called **r**-homogeneous of degree $l \in \mathbb{R}$ if the identity $V(\Delta_\varepsilon^r x) = \varepsilon^l V(x)$ holds for every $\varepsilon > 0$ (resp., $f(\Delta_\varepsilon^r x) = \varepsilon^l \Delta_\varepsilon^r f(x)$, or $F(\Delta_\varepsilon^r x) = \varepsilon^l \Delta_\varepsilon^r F(x)$). Along this paper we refer to this property as **r**-*homogeneity* or simply *homogeneity*. A system is called homogeneous if its vector field (or vector-set field) is **r**-homogeneous of some degree.

Given a vector **r** and a dilation $\Delta_\varepsilon^r x$, the homogeneous norm is defined by $\|x\|_{\mathbf{r},p} := \left(\sum_{i=1}^n |x_i|^{\frac{p}{r_i}}\right)^{\frac{1}{p}}$, $\forall x \in \mathbb{R}^n$, for any $p \geq 1$, and it is an **r**-homogeneous function of degree 1. The set $S = \{x \in \mathbb{R}^n : \|x\|_{\mathbf{r},p} = 1\}$ is the corresponding homogeneous unit sphere. The following Lemma provides some important properties of homogeneous functions and vector fields (some others are recalled in the Appendix).

Lemma 1.1 ([3, 10, 29]) *For a given family of dilations $\Delta_\varepsilon^r x$, and continuous real-valued functions V_1, V_2 on $\mathbb{R}^n$ (resp., a vector field f) which are* **r**-*homogeneous of degrees $m_1 > 0$ and $m_2 > 0$ (resp., $l \in \mathbb{R}$), we have:*

(i) $V_1 V_2$ is homogeneous of degree $m_1 + m_2$.

(ii) For every $x \in \mathbb{R}^n$ and each positive-definite function V_1, we have $c_1 V_1^{\frac{m_2}{m_1}}(x) \leq V_2(x) \leq c_2 V_1^{\frac{m_2}{m_1}}(x)$, where $c_1 \triangleq \min_{\{z: V_1(z)=1\}} V_2(z)$ and $c_2 \triangleq \max_{\{z: V_1(z)=1\}} V_2(z)$. Moreover, if V_2 is positive definite, there exists $c_1 > 0$.

(iii) $\partial V_1(x) / \partial x_i$ is homogeneous of degree $m_1 - r_i$, with r_i being the weight of x_i.

(iv) The Lie's derivative of $V_1(x)$ along $f(x)$, $L_f V_1(x) := \frac{\partial V_1(x)}{\partial x} \cdot f(x)$, is homogeneous of degree $m_1 + l$.

It is worth to recall that for homogeneous systems the local stability implies global stability and if the homogeneous degree is negative asymptotic stability implies finite-time stability [3, 8, 40, 44]. (Asymptotic) stability of homogeneous systems and homogeneous DI's can be studied by means of homogeneous LFs (HLFs), see for example [3, 6, 8–10, 25, 28, 40, 44, 51, 63, 71]: Assume that the origin of a homogeneous Filippov DI $\dot{x} \in F(x)$ is strongly globally AS. Then, there exists a $\mathscr{C}^\infty$ homogeneous strong LF.

The following robustness result of asymptotically stable homogeneous Filippov Differential Inclusions is of paramount importance for the assertion of the accuracy properties of HOSM algorithms in presence of measurement or discretization noise or also delay and external perturbations. They have been established by Levant [8, 39, 40, 44].

Theorem 1.1 *Let $\dot{x} \in F(x)$ be a globally uniformly finite-time stable homogeneous Filippov inclusion with homogeneity weights $\mathbf{r} = (r_1, \ldots, r_n)$ and degree $l < 0$, and let $\tau > 0$. Suppose that a continuous function $x(t)$ is defined for any $t \geq -\tau^l$ and* *satisfy some initial conditions* $x(t) = \xi(t)$ *for* $t \in [-\tau^l, 0]$. *Then if* $x(t)$ *is a solution of the perturbed differential inclusion*

$$\dot{x}(t) \in F_\tau\left(x\left(t + \left[-\tau^l, 0\right]\right)\right), \ 0 < t < \infty,$$

then the inequalities $|x_i| < \gamma_i \tau^{r_i}$ *are established in finite time with some positive constants* γ_i *independent of* τ *and* ξ.

Along this paper we use the following notation. For a real variable $z \in \mathbb{R}$ and a real number $p \in \mathbb{R}$ the symbol $\lceil z \rfloor^p = |z|^p sign[z]$ is the sign preserving power p of z. According to this $\lceil z \rfloor^0 = \text{sign}[z]$, $\frac{\mathrm{d}}{\mathrm{d}z}\lceil z \rfloor^p = p\,|z|^{p-1}$ and $\frac{\mathrm{d}}{\mathrm{d}z}|z|^p = p\lceil z \rfloor^{p-1}$ almost everywhere for z. Note that $\lceil z \rfloor^2 = |z|^2\text{sign}[z] \neq z^2$, and if p is an odd number then $\lceil z \rfloor^p = z^p$ and $|z|^p = z^p$ for any even integer p. Moreover, $\lceil z \rfloor^p \lceil z \rfloor^q = |z|^{p+q}$, $\lceil z \rfloor^p \lceil z \rfloor^0 = |z|^p$, and $\lceil z \rfloor^0 |z|^p = \lceil z \rfloor^p$. We also use the following notation: For a vector $x \in \mathbb{R}^n$ we denote by $\bar{x}_i \in \mathbb{R}^i$ the vector of the first i components, i.e. $\bar{x}_i = [x_1, \ldots, x_i]^T$. Similarly, we denote by $\underline{x}_i \in \mathbb{R}^{n-(i-1)}$ the vector of the last components, i.e. $\underline{x}_i = [x_i, \ldots, x_n]^T$. Note that $x = \bar{x}_n = \underline{x}_1$ are equivalent.

1.3 SISO Regulation and Tracking Problem

Consider a SISO dynamical system affine in the control

$$\dot{z} = f(t, z) + g(t, z)\,u\,, \ y = h(t, z)\,, \tag{1.1}$$

where $z \in \mathbb{R}^n$ defines the state vector, $u \in \mathbb{R}$ is the control input, $y \in \mathbb{R}$ is the output and $h(t, z) : \mathbb{R} \times \mathbb{R}^n \to \mathbb{R}$ is a smooth output function. A standard problem of control is the output tracking problem [32], consisting in forcing the output y to track a (time-varying) signal $y_R(t)$. Usually this problem has associated a (robust) disturbance decoupling or attenuation property [32, 33]. For our purposes we assume that the functions $f(t, z)$ and $g(t, z)$ are uncertain smooth vector fields on $\mathbb{R}^n$ and the dimension n can also be unknown. The control objective, i.e. the standard HOSM control problem [38, 40, 65], consists in making the output $\sigma = y - y_R$ vanishes in finite time and to keep $\sigma \equiv 0$ exactly by a bounded (discontinuous) feedback control. All differential equations are understood in the Filippov's sense [21].

When the relative degree ρ with respect to σ is known, well defined and constant this is equivalent to designing a controller for the DI

$$\Sigma_{DI}: \begin{cases} \dot{x}_i = x_{i+1}, \ i = 1, \ldots, \rho - 1, \\ \dot{x}_\rho \in [-C, C] + [K_m, K_M]u\,, \end{cases} \tag{1.2}$$

where $x = (x_1, \ldots, x_\rho)^T = (\sigma, \dot{\sigma}, \ldots, \sigma^{(\rho-1)})^T$ and $\sigma^{(i)} = \frac{d^i}{dt^i}h(z, t)$. Note that Σ_{DI} does not depend on the particular properties of the original systems' dynamics and the DI only retains the constants ρ, C, K_m and K_M. Due to the persisting uncertainty/perturbation causing the constant $C > 0$ the stabilization of $x = 0$ requires a control discontinuous at $x = 0$, and therefore the classical nonlinear control techniques, that aim at designing a continuous controller as e.g. [32–34], cannot be applied.

For homogenous HOSM [40] the problem is solved by designing a bounded memoryless feedback **r**-homogeneous control law of degree 0 (called also ρ-sliding homogeneous)

$$u = \varphi\left(x_1, x_2, \ldots, x_\rho\right) = \varphi\left(\varepsilon^{r_1} x_1, \varepsilon^{r_2} x_2, \ldots, \varepsilon^{r_\rho} x_\rho\right), \forall \varepsilon > 0, \tag{1.3}$$

with $\mathbf{r} = (\rho, \rho - 1, \ldots, 1)$, that renders the origin $x = 0$ finite-time stable for Σ_{DI}. The motion on the set $x = 0$, which consists of Filippov trajectories [21], is called an *ρth-order sliding mode*. The function φ is discontinuos at the ρ-sliding set ($x = 0$). The closed-loop inclusion (1.2)–(1.3) is an **r**-homogeneous Filippov DI of degree -1 satisfying *standard assumptions*. In the next Sect. 1.4 we provide some families of homogeneous HOSM controllers that solve the problem for any set of parameters (ρ, C, K_m, K_M). They are similar to the ones proposed by A. Levant but are characterized by the fact that they are obtained by means of explicit smooth (Control) Lyapunov Functions.

Since the implementation of the controller (1.3) requires the values of σ and its derivatives up to $\sigma^{(\rho-1)}$, i.e. the state x of Σ_{DI}, we provide in Sect. 1.6 a homogeneous HOSM observer, able to estimate in finite time and robustly the states of Σ_{DI} for any set of parameters (ρ, C, K_m, K_M). Again this observer corresponds to Levant's robust and exact differentiator [39, 40], but our results are distinguished by the fact that the design is based on explicit homogeneous smooth Lyapunov functions.

Finally we note that if the control enters the system (1.1) non affinely the problem can be reduced to the affine form by introducing an integrator and extending the relative degree to $\rho + 1$.

1.4 Lyapunov Based HOSM Controllers

In a series of (by now) classical works, and using basically geometric tools and homogeneous differential equations, A. Levant has derived some families of homogeneous Second and Higher Order Sliding Mode Controllers [36, 38–43]. Recently, in [18, 19] (see also [11–13, 18, 19]), he has obtained also Lyapunov functions for some "relay polynomial" controllers.

Based on smooth (Control) Lyapunov functions we derive a full family of homogeneous HOSM Controllers, which are different from Levant's families (see [11–13]). Given the relative degree $\rho \geq 2$ we assign the homogeneity weights $r_i = \rho - i + 1$ to the variables x_i, obtaining the vector $\mathbf{r} = (\rho, \rho - 1, \ldots, 1)$, and we define an arbitrary non-decreasing sequence of positive real numbers α_i, so that $0 \leq \alpha_1 \leq \cdots \leq \alpha_{\rho-1} \leq \alpha_\rho$. Furthermore, we define recursively, for $i = 2, \ldots, \rho$, the $\mathscr{C}^1$ **r**-homogeneous functions

$$\sigma_1(\bar{x}_1) = \lceil x_1 \rfloor^{\frac{\rho+\alpha_1}{\rho}}, \ldots, \sigma_i(\bar{x}_i) = \lceil x_i \rfloor^{\frac{\rho+\alpha_i}{\rho-i+1}} + k_{i-1}^{\frac{\rho+\alpha_i}{\rho-i+1}} \lceil \sigma_{i-1}(\bar{x}_{i-1}) \rfloor^{\frac{\rho+\alpha_i}{\rho+\alpha_{i-1}}}, \quad (1.4)$$

with constants $k_i > 0$. Recall that $\bar{x}_i = [x_1, \ldots, x_i]^T$.

For any constant $m \geq \max_{1 \leq i \leq \rho} \{2\rho + 1 + \alpha_{i-1} - i\}$ we also define recursively, for $i = 2, \ldots, \rho$, the $\mathscr{C}^1$ **r**-homogeneous functions

$$V_1(x_1) = \frac{\rho}{m} |x_1|^{\frac{m}{\rho}}, \ldots, V_i(\bar{x}_i) = \gamma_{i-1} V_{i-1}(\bar{x}_{i-1}) + W_i(\bar{x}_i) \quad (1.5)$$

$$W_i(\bar{x}_i) = \frac{r_i}{m} |x_i|^{\frac{m}{r_i}} - \lceil \nu_{i-1}(\bar{x}_{i-1}) \rfloor^{\frac{m-r_i}{r_i}} x_i + \left(1 - \frac{r_i}{m}\right) |\nu_{i-1}(\bar{x}_{i-1})|^{\frac{m}{r_i}}, \quad (1.6)$$

$$\nu_1(x_1) = -k_1 \lceil \sigma_1 \rfloor^{\frac{r_2}{\rho+\alpha_1}} = -k_1 \lceil x_1 \rfloor^{\frac{\rho-1}{\rho}}, \ldots, \nu_i(\bar{x}_i) = -k_i \lceil \sigma_i(\bar{x}_i) \rfloor^{\frac{r_{i+1}}{\rho+\alpha_i}}, \quad (1.7)$$

with (arbitrary) constants $\gamma_i > 0$. $\sigma_i(\bar{x}_i)$, $\nu_i(\bar{x}_i)$ and $V_i(\bar{x}_i)$ are **r**-homogeneous of degrees $\rho + \alpha_i$, r_{i+1} and m, respectively. As it will be shown in Sect. 1.5 $V_c(x) = V_\rho(\bar{x}_\rho)$ is a smooth (Control) Lyapunov Function for the uncertain plant (1.2).

From $V_c(x)$ we can derive different controllers for (1.2). In particular, we obtain the following family of Discontinuous and Quasi-continuous controllers

$$u_D = -k_\rho \varphi_D(x) = -k_\rho \lceil \sigma_\rho(x) \rfloor^0, \quad (1.8)$$

$$u_Q = -k_\rho \varphi_Q(x) = -k_\rho \frac{\sigma_\rho(x)}{M(x)}, \quad (1.9)$$

where $M(x)$ is any continuous **r**-homogeneous positive definite function of degree $\rho + \alpha_\rho$, and (for simplicity) we assume that it is scaled so that $\left|\frac{\sigma_\rho(x)}{M(x)}\right| \leq 1$. The homogeneous controllers (1.8)–(1.9) are derived from $V_c(x) = V_\rho(\bar{x}_\rho)$ by imposing the condition $\frac{\partial V_c(x)}{\partial x_\rho} \varphi(x) > 0$ at all points where $\frac{\partial V_c(x)}{\partial x_\rho} \neq 0$.

The values of k_i, for $i = 1, \ldots, \rho - 1$ can be fixed depending only on ρ and α_i, they are the same for the Discontinuous and the Quasi-continuous controllers, and they are independent of K_m, K_M and C. k_ρ in contrast is selected depending on the values K_m, K_M and C to induce the **r**th order sliding mode, and they are different for the Discontinuous and the Quasi-continuous controllers. Discontinuous controllers are discontinuous not only on the sliding set $\{x = 0\}$ but also when $\sigma_\rho(x) = 0$, while Quasi-Continuous controllers are discontinuous only on the sliding set. Due to this fact they produce less chattering.

Depending on the selection of the free parameters $0 \leq \alpha_1 \leq \cdots \leq \alpha_\rho$ we obtain different families of controllers. We illustrate them presenting the controllers of orders $\rho = 2, 3, 4$:

- Discontinuous Controllers
  - Nested Sliding Controllers: when some of the α_i are different

$$\begin{aligned}
u_{2D} &= -k_2 \left\lceil \lceil x_2 \rfloor^{2+\alpha_2} + k_1^{2+\alpha_2} \lceil x_1 \rfloor^{\frac{2+\alpha_2}{2}} \right\rfloor^0 \\
u_{3D} &= -k_3 \left\lceil \lceil x_3 \rfloor^{3+\alpha_3} + k_2^{3+\alpha_3} \left\lceil \lceil x_2 \rfloor^{\frac{3+\alpha_2}{2}} + k_1^{\frac{3+\alpha_2}{2}} \lceil x_1 \rfloor^{\frac{3+\alpha_2}{3}} \right\rfloor^{\frac{3+\alpha_3}{3+\alpha_2}} \right\rfloor^0 \\
u_{4D} &= -k_4 \left\lceil \lceil x_4 \rfloor^{4+\alpha_4} + k_3^{4+\alpha_4} \left\lceil \lceil x_3 \rfloor^{\frac{4+\alpha_3}{2}} + k_2^{\frac{4+\alpha_3}{2}} \left\lceil \lceil x_2 \rfloor^{\frac{4+\alpha_2}{3}} + \right.\right.\right. \\
&\qquad \left.\left.\left. k_1^{\frac{4+\alpha_2}{3}} \lceil x_1 \rfloor^{\frac{4+\alpha_2}{4}} \right\rfloor^{\frac{4+\alpha_3}{4+\alpha_2}} \right\rfloor^{\frac{4+\alpha_4}{4+\alpha_3}} \right\rfloor^0
\end{aligned} \tag{1.10}$$

  - Relay Polynomial Controllers: when $\alpha_\rho = \alpha_{\rho-1} = \cdots = \alpha_1 = \alpha \geq 0$

$$\begin{aligned}
u_{2R} &= -k_2 \text{sign} \left[\lceil x_2 \rfloor^{2+\alpha} + \bar{k}_1 \lceil x_1 \rfloor^{\frac{2+\alpha}{2}} \right], \\
u_{3R} &= -k_3 \text{sign} \left[\lceil x_3 \rfloor^{3+\alpha} + \bar{k}_2 \lceil x_2 \rfloor^{\frac{3+\alpha}{2}} + \bar{k}_1 \lceil x_1 \rfloor^{\frac{3+\alpha}{3}} \right] \\
u_{4R} &= -k_4 \text{sign} \left[\lceil x_4 \rfloor^{4+\alpha} + \bar{k}_3 \lceil x_3 \rfloor^{\frac{4+\alpha}{2}} + \bar{k}_2 \lceil x_2 \rfloor^{\frac{4+\alpha}{3}} + \bar{k}_1 \lceil x_1 \rfloor^{\frac{4+\alpha}{4}} \right]
\end{aligned} \tag{1.11}$$

where for $\rho = 2, \bar{k}_1 = k_1^{2+\alpha}$; for $\rho = 3, \bar{k}_1 = k_2^{3+\alpha} k_1^{\frac{3+\alpha}{2}}, \bar{k}_2 = k_2^{3+\alpha}$; and for general ρ, $\bar{k}_i = \prod_{j=i}^{\rho-1} k_j^{\frac{\rho+\alpha}{\rho-j}}$, for $i = 1, \ldots, \rho - 1$. Relay Polynomial controllers are specially simple in its form.

- Quasi-Continuous Controllers
  - Nested Sliding Controllers: when some of the α_i are different. The parameters $\beta_i > 0$ are arbitrary

$$\begin{aligned}
u_{2Q} &= -k_2 \frac{\lceil x_2 \rfloor^{2+\alpha_2} + k_1^{2+\alpha_2} \lceil x_1 \rfloor^{\frac{2+\alpha_2}{2}}}{|x_2|^{2+\alpha_2} + \beta_1 |x_1|^{\frac{2+\alpha_2}{2}}}, \\
u_{3Q} &= -k_3 \frac{\lceil x_3 \rfloor^{3+\alpha_3} + k_2^{3+\alpha_3} \left\lceil \lceil x_2 \rfloor^{\frac{3+\alpha_2}{2}} + k_1^{\frac{3+\alpha_2}{2}} \lceil x_1 \rfloor^{\frac{3+\alpha_2}{3}} \right\rfloor^{\frac{3+\alpha_3}{3+\alpha_2}}}{|x_3|^{3+\alpha_3} + \beta_2 |x_2|^{\frac{3+\alpha_3}{2}} + \beta_1 |x_1|^{\frac{3+\alpha_3}{3}}} \\
u_{4Q} &= -k_4 \frac{\lceil x_4 \rfloor^{4+\alpha_4} + k_3^{4+\alpha_4} \left\lceil \lceil x_3 \rfloor^{\frac{4+\alpha_3}{2}} + k_2^{\frac{4+\alpha_3}{2}} \left\lceil \lceil x_2 \rfloor^{\frac{4+\alpha_2}{3}} + k_1^{\frac{4+\alpha_2}{3}} \lceil x_1 \rfloor^{\frac{4+\alpha_2}{4}} \right\rfloor^{\frac{4+\alpha_3}{4+\alpha_2}} \right\rfloor^{\frac{4+\alpha_4}{4+\alpha_3}}}{|x_4|^{4+\alpha_4} + \beta_3 |x_3|^{\frac{4+\alpha_4}{2}} + \beta_2 |x_2|^{\frac{4+\alpha_4}{3}} + \beta_1 |x_1|^{\frac{4+\alpha_4}{4}}}
\end{aligned} \tag{1.12}$$

– Relay Polynomial Controllers: when $\alpha_i = \alpha \geq 0$

$$\begin{aligned}
u_{2QR} &= -k_2 \frac{\lceil x_2 \rfloor^{2+\alpha} + k_1^{2+\alpha} \lceil x_1 \rfloor^{\frac{2+\alpha}{2}}}{|x_2|^{2+\alpha} + \beta_1 |x_1|^{\frac{2+\alpha}{2}}}, \\
u_{3QR} &= -k_3 \frac{\lceil x_3 \rfloor^{3+\alpha} + \bar{k}_2 \lceil x_2 \rfloor^{\frac{3+\alpha}{2}} + \bar{k}_1 \lceil x_1 \rfloor^{\frac{3+\alpha}{3}}}{|x_3|^{3+\alpha} + \beta_2 |x_2|^{\frac{3+\alpha}{2}} + \beta_1 |x_1|^{\frac{3+\alpha}{3}}} \\
u_{4QR} &= -k_4 \frac{\lceil x_4 \rfloor^{4+\alpha} + \bar{k}_3 \lceil x_3 \rfloor^{\frac{4+\alpha}{2}} + \bar{k}_2 \lceil x_2 \rfloor^{\frac{4+\alpha}{3}} + \bar{k}_1 \lceil x_1 \rfloor^{\frac{4+\alpha}{4}}}{|x_4|^{4+\alpha} + \beta_3 |x_3|^{\frac{4+\alpha}{2}} + \beta_2 |x_2|^{\frac{4+\alpha}{3}} + \beta_1 |x_1|^{\frac{4+\alpha}{4}}}
\end{aligned} \tag{1.13}$$

All these controllers solve the problem posed in the previous Sect. 1.3.

Theorem 1.2 *For any $\rho \geq 2$ each controller of the families of Discontinuous or Quasi-continuous controllers in* (1.8)–(1.9)*, with arbitrary parameters $0 \leq \alpha_1 \leq \cdots \leq \alpha_\rho$, is ρ-sliding homogeneous, and for k_ρ sufficiently large the ρth order sliding mode $x = 0$ is established in Finite-Time for the uncertain system* (1.2) *for properly chosen gains $k_1, \ldots, k_{\rho-1}$ and $\beta_1, \ldots, \beta_{\rho-1}$.*

In the case of measurement noise and/or perturbations we obtain (from the homogeneity [39, 40, 44]) the following accuracy properties (see Theorem 1.1).

Theorem 1.3 *Consider the uncertain plant* (1.2) *with any of the (state) feedback controllers* (1.8) *or* (1.9) *and suppose that conditions of Theorem 1.2 are satisfied. Suppose that the control is realized with a sampling interval τ. In this case the state x reaches after a finite time a neighborhood of the origin characterized by*

$$|x_1(t)| \leq \delta_1 \tau^\rho,\ |x_2(t)| \leq \delta_2 \tau^{\rho-1}, \ldots, |x_i(t)| \leq \delta_i \tau^{\rho-(i-1)},\ |x_\rho(t)| \leq \delta_\rho \tau,$$

and x stays in this vicinity of zero for all future times. Here $\delta_1, \ldots, \delta_\rho > 0$ are some (positive) constants depending only on the chosen controller, the parameters (C, K_m, K_M, ρ) and the gains, but they are independent on τ and the initial conditions.

Using the CLF (1.5) we show that the convergence time is a bounded function of the initial states [8, Theorems 5.6, 5.7].

Proposition 1.1 *Controllers* (1.8)–(1.9)*, in closed-loop with system* (1.2)*, enforce the state trajectories, starting at initial state $x_0 = x(0) \in \mathbb{R}^n$, to reach $x = 0$ in a finite time smaller than*

$$T(x_0) \leq m\eta_\rho V_\rho^{\frac{1}{m}}(x_0), \tag{1.14}$$

where η_ρ is a function of the gains $(k_1, \ldots, k_\rho)$, K_m and C.

When the bound of the perturbation $[-C, C]$ is time-varying, it is possible to design the following variable-gain controller, with a slight variation of the proof of Theorem 1.2.

Theorem 1.4 *Consider that in* (1.2) $C = \bar{C} + \Theta(t, z)$, *where the function* $\Theta(t, z) \geq 0$ *is known. Then the discontinuous and quasi-continuous controllers* (1.8)–(1.9), *with* k_ρ *replaced by the variable-gain* $\left(K(t, z) + k_\rho\right)$, *stabilize the origin* $x = 0$ *in Finite-Time if the gain* k_ρ *is chosen large enough and* $K_m K(t, z) \geq \Theta(t, z)$.

By making a linear change of variables $\zeta = Lx$, $L \geq 1$, it is easy to show that if the vector of gains $\mathbf{k} = (k_1, \ldots, k_\rho)$ is stabilizing, then so is the scaled vector of gains $\mathbf{k}_L = (L^{\frac{1}{\rho}} k_1, \ldots, L^{\frac{1}{\rho+1-i}} k_i, \ldots, Lk_\rho)$ for any α_j. For the Relay Polynomial Controllers and the relative degree ρ, the gains $\bar{k}_i = \prod_{j=i}^{\rho-1} k_j^{\frac{\rho+\alpha}{\rho-j}}$, for $i = 1, \ldots, \rho - 1$ are scaled as $\bar{k}_i \to L^{\frac{(\rho-i)(\rho+\alpha)}{\rho-i+1}} \bar{k}_i$. Moreover, the convergence will be accelerated for $L > 1$, or the size of the allowable perturbation C will be incremented to LC. Note that the gains obtained by means of the LF can be very large for practical applications, so that a simulation-based gain design is eventually necessary (see [13]).

In next Sect. 1.5 it will be shown that the gains k_i can be calculated recursively as

$$k_1 > 0, \ldots, k_{i+1} > G_{i+1}(k_1, \ldots, k_i), \ldots, k_\rho > \frac{1}{K_m}\left(G_\rho\left(k_1, \ldots, k_{\rho-1}\right) + C\right),$$

where the functions G_i are obtained from the LF $V_c(x)$ and they depend on ρ, γ_i and α_i. We can parametrize the gains in terms of k_1 as

$$k_1 > 0, \ldots, k_i = \mu_i k_1^{\frac{\rho}{\rho-(i-1)}}, \; k_\rho > \frac{1}{K_m}\left(\mu_\rho k_1^\rho + C\right), \tag{1.15}$$

for some positive constants μ_i depending on ρ, γ_i and α_i. Some values, calculated numerically for u_D in (1.11), for $\alpha = 0$, are: $\rho = 2$, $\mu_2 = 1.62$; $\rho = 3$, ($\mu_2 = 1.5$, $\mu_3 = 3.25$); $\rho = 4$, ($\mu_2 = 2$, $\mu_3 = 8.45$, $\mu_4 = 30$). We note that this parametrization can be used for all controllers, except the value of k_ρ, which is different for the discontinuous and the quasi-continuous controllers. The values can also be used with the variable gain controller.

Remark 1.1 We note that using the family of CLFs (1.5) we obtain a large family of HOSM controllers, related to the ones proposed by A. Levant in his works. However, not all Levant's controllers have been provided with a Lyapunov function. For example, [19] derive Polynomial controllers for arbitrary $\alpha > -\rho$, while the Lyapunov functions proposed here (and also in [19]) are only valid for $\alpha \geq 0$. The construction of Lyapunov functions for the controllers for the values of α in the interval $-\rho < \alpha < 0$ is an open problem.

Remark 1.2 It is easy to see that controllers (1.8) and (1.9) can be modified without changing their properties: suppose that $\varsigma_\rho(x)$ is a continuous $\mathbf{r}$-homogeneous function of degree $\rho + \alpha_\rho$ such that (i) $\left\{x \in \mathbb{R}^\rho \mid \varsigma_\rho(x) = 0\right\} = \left\{x \in \mathbb{R}^\rho \mid \sigma_\rho(x) = 0\right\}$, i.e. σ_ρ and ς_ρ vanish at the same points, and (ii) $\varsigma_\rho(x)\sigma_\rho(x) \geq 0$. In this case the controllers

$$u_D = -k_\rho \lceil \varsigma_\rho (x) \rfloor^0 ,$$
$$u_Q = -k_\rho \frac{\varsigma_\rho (x)}{M(x)} ,$$

have the same properties as (1.8) and (1.9), respectively. For example, instead of the controller u_{3D} in (1.10) we can also implement the controller

$$u_{3D} = -k_3 \left\lceil \left\lceil \lceil x_3 \rfloor^{3+\alpha_2} + k_2^{3+\alpha_2} \lceil x_2 \rfloor^{\frac{3+\alpha_2}{2}} \right\rfloor^{\frac{3+\alpha_3}{3+\alpha_2}} + k_1^{\frac{3+\alpha_3}{2}} k_2^{3+\alpha_3} \lceil x_1 \rfloor^{\frac{3+\alpha_3}{3}} \right\rfloor^0 .$$

1.5 The Lyapunov Function for the HOSM Controllers

This section can be skipped in a fast reading. We show that the continuously differentiable function $V_\rho (x)$ (1.5) is an **r**-homogeneous Lyapunov Function of degree m for the uncertain system (1.2) for all $\rho \geq 2$, all $\gamma_j > 0$ and sufficiently large values of $k_j > 0$, $j = 1, 2, \ldots, \rho - 1$.

1.5.1 Proof by the Lyapunov Approach

In this subsection we establish the relationship between the proposed family of controllers and the Control Lyapunov functions. First we present some preliminary results.

We define recursively the auxiliary variables

$$s_1 = x_1, \ldots, s_i = x_i - \nu_{i-1} (\bar{x}_{i-1}) ,$$

$$s_{1,d} = \lceil x_1 \rfloor^{\frac{m-r_1}{r_1}} , \ldots, s_{i,d} = \lceil x_i \rfloor^{\frac{m-r_i}{r_i}} - \lceil \nu_{i-1} (\bar{x}_{i-1}) \rfloor^{\frac{m-r_i}{r_i}} .$$

Lemma 1.2 *For $\alpha > 1$ and $\beta > 0$ consider the function of the two real variables $x, y \in \mathbb{R}$*

$$F (x, y) = \frac{1}{\alpha} |x|^\alpha - x \lceil y \rfloor^\beta + (1-\tfrac{1}{\alpha}) |y|^{\beta \frac{\alpha}{\alpha-1}} . \quad (1.16)$$

Then $F (x, y) \geq 0$ and $F (x, y) = 0$ if and only if $\lceil x \rfloor^\alpha = \lceil y \rfloor^{\beta \frac{\alpha}{\alpha-1}}$.

Proof The conclusion follows immediately from Young's inequality (see Lemma 1.4), since it implies

$$x \lceil y \rfloor^\beta \leq \frac{1}{\alpha} |x|^\alpha + \left(1 - \frac{1}{\alpha}\right) |y|^{\beta \frac{\alpha}{\alpha-1}} .$$

□

From Lemma 1.2 it follows that $W_i(\bar{x}_i) \geq 0$ and $W_i(\bar{x}_i) = 0$ iff $x_i = \nu_{i-1}(\bar{x}_{i-1})$, so that $V_i(\bar{x}_i)$ is positive definite for any $\gamma_{i-1} > 0$. The following relations will be used in the sequel: for $1 \leq j \leq i-1$

$$\frac{\partial W_i(\bar{x}_i)}{\partial x_j} = -\frac{m-r_i}{\rho+\alpha_{i-1}} k_{i-1}^{\frac{\rho+\alpha_{i-1}}{r_i}} |\nu_{i-1}|^{\frac{m-r_i-\rho-\alpha_{i-1}}{r_i}} s_i \frac{\partial \sigma_{i-1}(\bar{x}_{i-1})}{\partial x_j}, \tag{1.17}$$

$$\frac{\partial W_i(\bar{x}_i)}{\partial x_i} = \lceil x_i \rfloor^{\frac{m-r_i}{r_i}} - \lceil \nu_{i-1}(\bar{x}_{i-1}) \rfloor^{\frac{m-r_i}{r_i}} = s_{i,d}. \tag{1.18}$$

Notice also that $s_i = 0 \Leftrightarrow \sigma_i = 0 \Leftrightarrow s_{i,d} = 0$, and that they have the same sign, i.e. $s_i\sigma_i > 0$, $s_i s_{i,d} > 0$ and $\sigma_i s_{i,d} > 0$ when $s_i \neq 0$, $\sigma_i \neq 0$, $s_{i,d} \neq 0$.

For $i = 1, \ldots, \rho-1$ we introduce also the functions

$$Z_1(\bar{x}_1) = \nu_1(\bar{x}_1) \frac{dV_1(x_1)}{dx_1}, \; Z_i(\bar{x}_i) \triangleq \sum_{j=1}^{i-1} x_{j+1} \frac{\partial V_i(\bar{x}_i)}{\partial x_j} + \nu_i(\bar{x}_i) \frac{\partial V_i(\bar{x}_i)}{\partial x_i}. \tag{1.19}$$

Using (1.5) in (1.19), and with (1.17), (1.18) we obtain

$$\begin{aligned} Z_i(\bar{x}_i) =& \gamma_{i-1} \left(\sum_{j=1}^{i-2} x_{j+1} \frac{\partial V_{i-1}(\bar{x}_{i-1})}{\partial x_j} + x_i \frac{\partial V_{i-1}(\bar{x}_{i-1})}{\partial x_{i-1}} \right) \\ & - \frac{m-r_i}{\rho+\alpha_{i-1}} k_{i-1}^{\frac{\rho+\alpha_{i-1}}{r_i}} |\nu_{i-1}|^{\frac{m-r_i-\rho-\alpha_{i-1}}{r_i}} s_i \sum_{j=1}^{i-1} x_{j+1} \frac{\partial \sigma_{i-1}(\bar{x}_{i-1})}{\partial x_j} + s_{i,d}\nu_i. \end{aligned}$$

Using $x_i = \nu_{i-1} + s_i$ in the first line we can obtain the recursive expression

$$\begin{aligned} Z_i(\bar{x}_i) &= \gamma_{i-1} Z_{i-1}(\bar{x}_{i-1}) + s_i \Psi_{i-1}(\bar{x}_i) + s_{i,d}\nu_i \\ \Psi_{i-1} &\triangleq \gamma_{i-1} s_{i-1,d} - \tfrac{m-r_i}{\rho+\alpha_{i-1}} k_{i-1}^{\frac{\rho+\alpha_{i-1}}{r_i}} |\nu_{i-1}|^{\frac{m-r_i-\rho-\alpha_{i-1}}{r_i}} \sum_{j=1}^{i-1} x_{j+1} \frac{\partial \sigma_{i-1}(\bar{x}_{i-1})}{\partial x_j} \end{aligned} \tag{1.20}$$

Note that if in (1.19) we set $\frac{\partial V_i(\bar{x}_i)}{\partial x_i} = s_{i-1,d} = s_i = 0$ it follows from (1.20) that

$$\sum_{j=1}^{i-1} x_{j+1} \frac{\partial V_i(\bar{x}_i)}{\partial x_j} = \gamma_{i-1} Z_{i-1}(\bar{x}_{i-1}). \tag{1.21}$$

The Main Argument

The proof can be divided in two parts. We take the derivative of $V_\rho(x)$ along the trajectories of (1.2)

$$\dot{V}_{\rho}(x) \in \sum_{j=1}^{\rho-1} \frac{\partial V_{\rho}(x)}{\partial x_j} x_{j+1} + \frac{\partial V_{\rho}(x)}{\partial x_{\rho}} ([-C, C] + [K_m, K_M]u) ,$$

with the controller (1.8) or (1.9). Note that

$$\frac{\partial V_{\rho}(x)}{\partial x_{\rho}} = \frac{\partial W_{\rho}(x)}{\partial x_{\rho}} = s_{\rho,d}(x) = \lceil x_{\rho} \rfloor^{\frac{m-r_{\rho}}{r_{\rho}}} - \lceil \nu_{\rho-1}(\bar{x}_{\rho-1}) \rfloor^{\frac{m-r_{\rho}}{r_{\rho}}}$$

and

$$\sigma_{\rho}(x) = \lceil x_{\rho} \rfloor^{\frac{\rho+\alpha_{\rho}}{r_{\rho}}} - \lceil \nu_{\rho-1} \rfloor^{\frac{\rho+\alpha_{\rho}}{r_{\rho}}} = \lceil x_{\rho} \rfloor^{\frac{\rho+\alpha_{\rho}}{r_{\rho}}} + k_{\rho-1}^{\frac{\rho+\alpha_{\rho}}{r_{\rho}}} \lceil \sigma_{\rho-1} \rfloor^{\frac{\rho+\alpha_{\rho}}{\rho+\alpha_{\rho-1}}} .$$

Moreover, $s_{\rho,d}(x)\,\sigma_{\rho}(x) \geq 0$ and it is zero only if $s_{\rho,d}(x) = \sigma_{\rho}(x) = 0$.

We consider here only the Discontinuous controller (1.8), the proof for the Quasi-Continuous one follows the same path. For (1.8) the multivalued function $\psi(u) = [-C, C] - k_{\rho}[K_m, K_M]u$, when evaluated at (1.8), i.e.

$$\psi(\sigma_{\rho}(x)) = [-C, C] - k_{\rho}[K_m, K_M] \lceil \sigma_{\rho}(x) \rfloor^{0} ,$$

can be represented as

$$\psi(\sigma_{\rho}(x)) = \begin{cases} -\left[(C + k_{\rho}K_m), (k_{\rho}K_M - C)\right] & \text{if } \sigma_{\rho}(x) > 0 \\ \left[-(C + k_{\rho}K_M), (C + k_{\rho}K_M)\right] & \text{if } \sigma_{\rho}(x) = 0 \\ \left[(k_{\rho}K_m - C), (C + k_{\rho}K_M)\right] & \text{if } \sigma_{\rho}(x) < 0 \end{cases}$$

If we assume that $\frac{C}{K_m} < k_{\rho}$ then we conclude that

$$\psi(\sigma_{\rho}(x))\,\sigma_{\rho}(x) \leq 0, \text{ and } \psi(\sigma_{\rho}(x))\,s_{\rho,d}(x) \leq 0,$$

and they are zero only if $\sigma_{\rho}(x) = 0$.

Using these results we conclude that

$$\dot{V}_{\rho}(x) \leq \sum_{j=1}^{\rho-1} \frac{\partial V_{\rho}(x)}{\partial x_j} x_{j+1} - k_{\rho}\left(K_m - \frac{C}{k_{\rho}}\right) \left|s_{\rho,d}(x)\right| .$$

If we assume that

$$\forall x \in \left\{x \in \mathbb{R}^{\rho} | s_{\rho,d}(x) = 0\right\} = \left\{x \in \mathbb{R}^{\rho} | \sigma_{\rho}(x) = 0\right\} \Rightarrow \sum_{j=1}^{\rho-1} \frac{\partial V_{\rho}(x)}{\partial x_j} x_{j+1} < 0 \tag{1.22}$$

then, due to homogeneity (Lemma 1.5), $\dot{V}_\rho(x) < 0$ for a sufficiently large value of k_ρ. This value can be calculated from the previous inequality as

$$k_\rho > \frac{1}{K_m}\left(\frac{\sum_{j=1}^{\rho-1}\frac{\partial V_\rho(x)}{\partial x_j}x_{j+1}}{\left|s_{\rho,d}(x)\right|} + C\right). \tag{1.23}$$

The fulfilling of the hypothesis (1.22) can be shown for the function (1.5) in a recursive manner. Note that since

$$\left\{x \in \mathbb{R}^\rho | \sigma_\rho(x) = 0\right\} = \left\{x \in \mathbb{R}^\rho | \left\lceil x_\rho \right\rfloor^{\frac{\rho+\alpha_\rho}{r_\rho}} = \left\lceil \nu_{\rho-1} \right\rfloor^{\frac{\rho+\alpha_\rho}{r_\rho}} = -k_{\rho-1}^{\frac{\rho+\alpha_\rho}{r_\rho}} \left\lceil \sigma_{\rho-1} \right\rfloor^{\frac{\rho+\alpha_\rho}{\rho+\alpha_{\rho-1}}}\right\},$$

we can rewrite (1.22) as

$$\sum_{j=1}^{\rho-1}\frac{\partial V_\rho(x)}{\partial x_j}x_{j+1} = \sum_{j=1}^{\rho-2}\frac{\partial V_\rho(x)}{\partial x_j}x_{j+1} + \frac{\partial V_\rho(x)}{\partial x_{\rho-1}}\nu_{\rho-1} = Z_{\rho-1}\left(\bar{x}_{\rho-1}\right) < 0, \tag{1.24}$$

with Z_i defined in (1.19).

The Induction Argument

We show recursively, using (1.20), that (1.24) is fulfilled for appropriate selected gains k_i.

First consider $i = 1$, i.e. $Z_1(\bar{x}_1)$. Since $Z_1(\bar{x}_1) = \nu_1(\bar{x}_1)\frac{dV_1(z_1)}{dx_1} = -k_1 |x_1|^{\frac{m-1}{\rho}} < 0$ it is negative definite for any $k_1 > 0$.

From (1.20) we obtain in case $i = 2$ for $Z_2(\bar{x}_2)$

$$Z_2(\bar{x}_2) = -k_1\gamma_1 |x_1|^{\frac{m-1}{\rho}} + s_2\left(\gamma_1 \left\lceil x_1 \right\rfloor^{\frac{m-\rho}{\rho}} - \frac{m-\rho+1}{\rho}k_1^{\frac{m+1-\rho}{\rho-1}} |x_1|^{\frac{m-2\rho+1}{\rho}} x_2\right) \tag{1.25}$$
$$- k_2 s_{2,d} \left\lceil \sigma_2 \right\rfloor^{\frac{\rho-2}{\rho-\alpha_2}}.$$

Note that the last term in $Z_2(\bar{x}_2)$ is negative, i.e. $-k_2 s_{2,d}\left\lceil \sigma_2 \right\rfloor^{\frac{\rho-2}{\rho+\alpha_2}} < 0$ and it is zero only if $\sigma_2 = 0$. Since when $\sigma_2 = 0$ we obtain that

$$\sigma_2 = 0 \Rightarrow Z_2(\bar{x}_2) = -k_1\gamma_1 |x_1|^{\frac{m-1}{\rho}} < 0$$

we conclude from Lemma 1.5 that $Z_2(\bar{x}_2) < 0$ everywhere for $k_2 > 0$ sufficiently large.

By induction it is easy to see that from (1.20) we obtain in case i for $Z_i(\bar{x}_i)$

$$Z_i(\bar{x}_i) = \gamma_{i-1} Z_{i-1}(\bar{x}_{i-1}) + s_i \Psi_{i-1}(\bar{x}_i) - k_i s_{i,d} \lceil \sigma_i \rfloor^{\frac{r_{i+1}}{\rho+\alpha_i}} \tag{1.26}$$

$$\Psi_{i-1}(\bar{x}_i) \triangleq \gamma_{i-1} s_{i-1,d} - \frac{m-r_i}{\rho+\alpha_{i-1}} k_{i-1}^{\frac{\rho+\alpha_{i-1}}{r_i}} |\nu_{i-1}|^{\frac{m-r_i-\rho-\alpha_{i-1}}{r_i}} \sum_{j=1}^{i-1} x_{j+1} \frac{\partial \sigma_{i-1}(\bar{x}_{i-1})}{\partial x_j}$$

Note that the last term in $Z_i(\bar{x}_i)$ is negative, i.e. $-k_i s_{i,d} \lceil \sigma_i \rfloor^{\frac{r_{i+1}}{\rho+\alpha_i}} < 0$ and it is zero only if $\sigma_i = 0$. Since when $\sigma_i = 0$ we obtain that

$$\sigma_i = 0 \Rightarrow Z_i(\bar{x}_i) = \gamma_{i-1} Z_{i-1}(\bar{x}_{i-1}) .$$

Since by the induction hypothesis $Z_{i-1}(\bar{x}_{i-1}) < 0$ we conclude from Lemma 1.5 that $Z_i(\bar{x}_i) < 0$ everywhere for $k_i > 0$ sufficiently large. By induction we can therefore conclude that $Z_\rho(x) < 0$. □

In the last step of the proof it is shown that $Z_{i-1} < 0$. This can be interpreted as rendering the origin of the system $\dot{x}_1 = x_2$, $\dot{x}_{i-1} = x_i$ GFTS with the virtual controller $x_i = \nu_{i-1}(\bar{x}_{i-1})$. This is the typical Backstepping method [34], also used in [30, 31] for continuous controllers. However, since controller $\nu_{i-1}(\bar{x}_{i-1})$ is not differentiable, the standard Backstepping procedure fails. To overcome this problem we use the desingularization method proposed in [60]. For this, instead of using the function $s_i = x_i - \nu_{i-1}$, a desingularization function $s_{i,d} = \lceil x_i \rfloor^{\frac{m-r_i}{r_i}} - \lceil \nu_{i-1} \rfloor^{\frac{m-r_i}{r_i}}$ is introduced, which vanishes at the same set as s_i.

1.5.2 Gain Calculation

The controller gains k_i, for $i = 1, \ldots, \rho - 1$, have to be calculated such that V_ρ is a CLF, while k_ρ is selected large enough so that inequality (1.23) holds. After fixing the parameters m, ρ, α_i and γ_i, the values of the required k_i can be obtained recursively from the condition that Z_i in (1.26) has to be negative definite. We require $k_1 > 0$ and for $i = 2, \ldots, \rho - 1$ we can calculate k_i as

$$k_i > \max_{\bar{x}_i \in S_i} \left\{ \frac{\gamma_{i-1} Z_{i-1}(\bar{x}_{i-1}) + s_i \Psi_{i-1}(\bar{x}_i)}{s_{i,d}(\bar{x}_i) \lceil \sigma_i(\bar{x}_i) \rfloor^{\frac{r_{i+1}}{\rho+\alpha_i}}} \right\} =: G_i(k_1, \ldots, k_{i-1}) , \tag{1.27}$$

where $S_i = \{\bar{x}_i \in \mathbb{R}^i : \|\bar{x}_i\|_{\mathbf{r},p} = 1\}$ is the unit homogeneous sphere, which is a compact set. The maximizations are feasible since the function to be maximized has the following properties: (i) It is $\mathbf{r}$-homogeneous of degree 0, so that it achieves all its values on the unit sphere S_i, and (ii) it is upper-semicontinuous, since it is continuous at the points where the denominator does not vanish, and when the denominator vanishes the numerator is negative, as shown in the previous proof. It is well-known that an upper-semicontinuous function has a maximum on a compact set (see Lemma 1.6).

1.5.3 Analytical Gain Calculation

Instead of the maximization approach presented above, it is possible to provide for any order an analytical estimation of the values of the gains using classical inequalities to verify the condition that Z_i in (1.26) has to be negative definite (or (1.23) holds for k_ρ). Since the general derivation and expressions are lengthy, we illustrate the procedure with the simplest case for $Z_2(\bar{x}_2)$, that provides an explicit parametrization of the set of gains (k_1, k_2) for $\rho = 3$, i.e. u_{3D}, u_{3R} and u_{3Q}.

Lemma 1.3 *For any values of $\alpha_3 \geq \alpha_2 \geq 0$, $m \geq r_2 + \rho + \alpha_2$, $\gamma_1 > 0$, $0 < \eta < 1$ and $k_1 > 0$, Z_2 in (1.25) can be rendered negative definite for any k_2 such that*

$$k_2 > \frac{2^{\frac{m-2r_2}{\rho+\alpha_2}} r_2}{m-1} \left(\frac{2^{\frac{\rho+\alpha_2-r_2}{\rho+\alpha_2}} (m-r_1)}{m-1} \right)^{\frac{m-r_1}{r_2}} \frac{\left(\gamma_1 + \frac{m-r_2}{r_1} k_1^{\frac{m}{r_2}} \right)^{\frac{m-1}{r_2}}}{(\eta \gamma_1 k_1)^{\frac{m-r_1}{r_2}}} . \tag{1.28}$$

Proof Recalling from (1.4) that $\sigma_1(\bar{x}_1) = \lceil x_1 \rfloor^{\frac{\rho+\alpha_1}{\rho}}$, and using $x_2 = -k_1 \lceil \sigma_1 \rfloor^{\frac{r_2}{\rho+\alpha_1}} + s_2$ in (1.25) we can write

$$Z_2(\bar{x}_2) = -k_1\gamma_1 |\sigma_1|^{\frac{m-1}{\rho+\alpha_1}} - \frac{m-r_2}{r_1} k_1^{\frac{m-r_2}{r_2}} |\sigma_1|^{\frac{m-r_2-r_1}{\rho+\alpha_1}} |s_2|^2 +$$
$$\left(\gamma_1 + \frac{m-r_2}{r_1} k_1^{\frac{m}{r_2}} \right) \lceil \sigma_1 \rfloor^{\frac{m-r_1}{\rho+\alpha_1}} s_2 - k_2 |\sigma_2|^{\frac{r_2-1}{\rho+\alpha_2}} \left| s_{2,d} \right| .$$

Recall the fact (see e.g. [19]) that for any two reals x_1, x_2 and any positive real numbers $0 < p \leq q$

$$\left| \lceil x_1 \rfloor^p - \lceil x_2 \rfloor^p \right|^{\frac{1}{p}} \leq 2^{\frac{1}{p}-\frac{1}{q}} \left| \lceil x_1 \rfloor^q - \lceil x_2 \rfloor^q \right|^{\frac{1}{q}} .$$

From it follows that

$$|s_i| \leq 2^{\frac{\rho+\alpha_i-r_i}{\rho+\alpha_i}} |\sigma_i|^{\frac{r_i}{\rho+\alpha_i}} \leq 2^{\frac{m-2r_i}{m-r_i}} \left| s_{i,d} \right|^{\frac{r_i}{m-r_i}}$$

As a consequence we get that

$$\left| s_{i,d} \right| \geq 2^{-\frac{m-r_i-\rho-\alpha_i}{\rho+\alpha_i}} |\sigma_i|^{\frac{m-r_i}{\rho+\alpha_i}}$$

and

$$\lceil \sigma_1 \rfloor^{\frac{m-r_1}{\rho+\alpha_1}} s_2 \leq 2^{\frac{\rho+\alpha_2-r_2}{\rho+\alpha_2}} |\sigma_1|^{\frac{m-r_1}{\rho+\alpha_1}} |\sigma_2|^{\frac{r_2}{\rho+\alpha_2}} .$$

These results and the classical Young's inequality (Lemma 1.4) imply that for any $c > 0$

$$Z_2 \leq -\left[k_1\gamma_1 - 2^{\frac{\rho+\alpha_2-r_2}{\rho+\alpha_2}}\left(\gamma_1 + \frac{m-r_2}{r_1}k_1^{\frac{m}{r_2}}\right)\frac{m-r_1}{m-1}c^{\frac{m-1}{m-r_1}}\right]|\sigma_1|^{\frac{m-1}{\rho+\alpha_1}}$$
$$-\left[\frac{k_2}{2^{\frac{m-r_2-\rho-\alpha_2}{\rho+\alpha_2}}} - 2^{\frac{\rho+\alpha_2-r_2}{\rho+\alpha_2}}\left(\gamma_1 + \frac{m-r_2}{r_1}k_1^{\frac{m}{r_2}}\right)\frac{r_2 c^{-\frac{m-1}{r_2}}}{m-1}\right]|\sigma_2|^{\frac{m-1}{\rho+\alpha_2}}$$

For any k_2 fulfilling (1.28) it is possible to select c such that Z_2 is negative definite. □

1.6 The Arbitrary-Order Exact Differentiator

Let the input signal $f(t)$ to the differentiator be a Lebesgue-measurable function defined on $[0, \infty)$. The signal $f(t)$ is assumed to be decomposed as $f(t) = f_0(t) + \nu(t)$. The first term is the unknown base signal $f_0(t)$, to be differentiated, and belonging to the class $\mathscr{S}_L^n$ of signals which are $(n-1)$-times differentiable and with a $(n-1)$th time derivative having a known Lipschitz constant $L > 0$, i.e. $\left|f_0^{(n)}(t)\right| \leq L$. The second term $\nu(t)$ corresponds to a Lebesgue-measurable uniformly bounded noise signal $|\nu(t)| \leq \varepsilon$, $\forall t \geq 0$. Defining the variables $\varsigma_1 = f_0(t)$, $\varsigma_2 = \dot{f}_0(t), \ldots,$ $\varsigma_n = f_0^{(n-1)}(t)$, where $f_0^{(i)}(t) = \frac{d^i}{dt^i} f_0(t)$, for all $i = 1, \ldots, n$, a state representation of the base signal is given by $\dot{\varsigma}_i = \varsigma_{i+1}$, $i = 1, \ldots, n-1$, $\dot{\varsigma}_n = f_0^{(n)}(t)$. To estimate the derivatives of the base signal, consider Levant's $(n-1)$th order homogeneous discontinuous exact differentiator [39, 40] $(i = 1, \ldots, n-1)$

$$\begin{aligned} \dot{x}_i &= -\lambda_i L^{\frac{i}{n}} \lceil x_1 - f \rfloor^{\frac{n-i}{n}} + x_{i+1} , \\ \dot{x}_n &= -\lambda_n L \lceil x_1 - f \rfloor^0 . \end{aligned} \tag{1.29}$$

Since (1.29) is discontinuous, its solutions are understood in the sense of Filippov [21].

Defining the differentiation error as $e_i \triangleq x_i - f_0^{(i-1)}$ and performing the state transformation

$$z_1 = \frac{e_1}{L}, \ldots, z_i = \frac{e_i}{\lambda_{i-1} L}, \; i = 1, \ldots, n,$$

the error dynamics is (for $i = 1, \ldots, n-1$)

$$\dot{z}_i = -\tilde{\lambda}_i \left(\lceil z_1 + \tilde{\nu}(t) \rfloor^{\frac{n-i}{n}} - z_{i+1} \right) , \tag{1.30}$$
$$\dot{z}_n \in -\tilde{\lambda}_n \left(\lceil z_1 + \tilde{\nu}(t) \rfloor^{0} + \frac{1}{\lambda_n} [-1,\ 1] \right) ,$$

where $\tilde{\nu}(t) = \frac{\nu(t)}{L} \in [-1,\ 1]\frac{\varepsilon}{L}$ and

$$\lambda_0 = 1,\ \tilde{\lambda}_i = \frac{\lambda_i}{\lambda_{i-1}},\ i = 1, \ldots, n .$$

For $\varepsilon = 0$ system (1.30) is a homogeneous DI with homogeneity degree $d = -1$ and weights $\mathbf{r} = [r_1, \ldots, r_n]$. where $r_i = n + 1 - i$ for $i = 1, \ldots, n$. For $n \geq 2$ fix p_i such that $p_i > r_i + r_{i+1} = 2(n-i) + 1$ and $p = p_1 > r_1 + r_2 = 2n - 1 > 1$. Note that $p_i \geq p_{i+1}$ for all $i = 1, \ldots, n$.

Consider as Lyapunov Function candidate for (1.30) the $\mathbf{r}$-homogeneous of degree p_1 and continuously differentiable function $V(z) = V_1(z)$, defined (backwards) recursively for $i = n, \ldots, 1$ as (we define the vector of the last $(n-i)$ components of z as $\underline{z}_i \triangleq (z_i, \ldots, z_n)$)

$$V_i\left(\underline{z}_i\right) = \beta_i Z_i\left(z_i,\ z_{i+1}\right) + V_{i+1}^{\frac{p_i}{p_{i+1}}}\left(\underline{z}_{i+1}\right) \tag{1.31}$$

where the recursion starts with $i = n$

$$V_n\left(z_n\right) = \beta_n \frac{1}{p_n} \left|z_n\right|^{p_n}$$

and with $\beta_i > 0$, $i = 1, \ldots, n$ and

$$Z_i\left(z_i,\ z_{i+1}\right) = \frac{n+1-i}{p_i} \left|z_i\right|^{\frac{p_i}{n+1-i}} - z_i \left\lceil z_{i+1} \right\rfloor^{\frac{p_i - n - 1 + i}{n-i}} + \left(\frac{p_i - n - 1 + i}{p_i}\right) \left|z_{i+1}\right|^{\frac{p_i}{n-i}} . \tag{1.32}$$

which for $i = 1, \ldots, n-1$ are continuously differentiable $\mathbf{r}$-homogeneous of degree p_i and positive semidefinite functions.

The main result of this section is the following

Theorem 1.5 *Under the stated assumptions for the signal $f(t)$, in the absence of noise ($\nu(t) \equiv 0$) and for properly chosen gains $\lambda_i > 0$, $i = 1, \ldots, n$, the origin $z = 0$ of the differentiation error dynamics* (1.30) *is Finite-Time stable and $V(z)$ in* (1.31) *is a smooth Lyapunov function for every $p \geq 2n - 1$ and $\beta_i > 0$. Moreover, $V(z)$ satisfies, for some $\kappa > 0$, the differential inequality*

$$\dot{V} \leq -\kappa V(z)^{\frac{p-1}{p}} . \tag{1.33}$$

A convergence time estimation is given by

$$T(z_0) \leq \frac{p}{\kappa} V^{\frac{1}{p}}(z_0) . \tag{1.34}$$

The proof of this Theorem and of the following Proposition 1.2 are given in Sect. 1.7. The result of Theorem 1.5 has been already obtained in [14], but we provide a different Lyapunov function allowing the design of gains independently of the order of the differentiator. From the Lyapunov Function $V(z)$ (1.31) we can obtain not only an estimation of the convergence time, but also a method of calculating gains λ_i assuring the convergence of the differentiator. This is of paramount importance for applications. Until now this has been only possible (in general) by means of simulations [39, 40]. The following proposition provides a method to calculate gains for differentiators of any order.

Proposition 1.2 *For a given order* $n-1>1$ *of the differentiator select* p_i *such that* $p_i > r_i + r_{i+1} = 2(n-i)+1$ *and* $p = p_1 > r_1 + r_2 = 2n-1 > 1$*:*

(1) The sequence of (positive) gains $\tilde{\lambda}_1, \ldots, \tilde{\lambda}_n$ *can be calculated backwards as follows: (a) Select* $\lambda_n > 1$*, and* $\tilde{\lambda}_n > 0$*. (b) For* $i = n, n-1, \ldots, 2$ *select* $\tilde{\lambda}_{i-1}$ *such that*

$$\tilde{\lambda}_{i-1} > \omega_{i-1}(\tilde{\lambda}_i, \ldots, \tilde{\lambda}_n)$$

where ω_{i-1} *is a function obtained from the Lyapunov Function and depending only on* $(\tilde{\lambda}_i, \ldots, \tilde{\lambda}_n)$*,* p_i *and* $\beta_{i-1}, \ldots \beta_n$*.*

(2) For any $j = 1, \ldots, n-1$ *the gains* $(\tilde{\lambda}_j, \ldots \tilde{\lambda}_n)$ *are appropriate for the differentiator of order* $n-j$*.*

A method to calculate the gains $\tilde{\lambda}_i$ using the Lyapunov function is described in Sect. 1.7.2 below, where functions ω_i are derived from the Lyapunov function $V(z)$. We see from the previous Proposition that the gains can be parametrized in terms of the last one $\tilde{k}_n$.

Item (2) in Proposition 1.2 corresponds to the affirmation in [37, 39, 40] that the gains of the differentiator can be designed as an infinite sequence independent of the order of differentiation.

As a consequence of homogeneity (see Theorem 1.1) we obtain

Proposition 1.3 *For a uniformly bounded noise* ($|\nu(t)| \leq \varepsilon$) *and appropriate gains for the differentiation error* $x_i(t) - f_0^{(i-1)}(t)$*,* $i = 1, \ldots, n$*, the following inequalities are achieved in finite time*

$$\left| x_i(t) - f_0^{(i-1)}(t) \right| \leq \vartheta_i L^{\frac{i-1}{n}} |\varepsilon|^{\frac{n-i+1}{n}} . \tag{1.35}$$

ϑ_i *depend on the gains* λ_i *and* $\vartheta_1 > 1$*.*

The proof of this Proposition has been given in [40, 44]. Alternatively, the results of [8] show that the differentiation error is ISS with respect to the noise signal $\nu(t)$, and homogeneity implies the inequalities (1.35). The advantage of having an explicit Lyapunov Function, is that it is possible to estimate the constants ϑ_i appearing in (1.35). The calculation can be done in the following form: For $L = 1$ and $\varepsilon = 1$ find the infimum value c such that for the level set $\mathscr{L}_c = \{z \in \mathbb{R}^n | V(z) \leq c\}$ of the Lyapunov function the inequality $\dot{V} < 0$ is fulfilled on the border of $\mathscr{L}_c$. This implies that the trajectories of the differentiator subject to measurement noise will reach $\mathscr{L}_c$ in finite time and will remain there for all future times. The constants λ_i are then given by

$$\vartheta_i = \max_{\mathscr{L}_c} |z_i| \,. \tag{1.36}$$

We obtain the accuracy with respect to noise (1.35) reported in [37] for the first order and in [39] for the arbitrary order differentiator. And the orders obtained with respect to ε and L are the optimal ones, according to the discussion in [37].

Note that the due to the discontinuous term Levant's differentiator (1.29) has an astonishing distinguishing feature: in the absence of noise it is *exact* (i.e. it converges in finite time) to a much larger class of signals $\mathscr{S}_L^n$ than its continuous counterparts, which are only exact for the much thinner set $\mathscr{S}_p^n = \left\{\left|f^{(n)}(t)\right| = 0\right\} \subset \mathscr{S}_L^n$. For a more detailed discussion see [36, 37, 67].

1.7 Proofs of Main Results on Differentiators

In this section we provide the proofs of Theorem 1.5 and Proposition 1.2. We make use of some properties of continuous homogeneous functions, Lemmas 1.4–1.6, contained in the Appendix "Some Technical Lemmas on Homogeneous Functions".

1.7.1 Proof of Theorem 1.5

Due to Lemma 1.2 for $i = 1, \ldots, n-1$ functions (1.32) are positive semidefinite, i.e. $Z_i(z_i, z_{i+1}) \geq 0$ and $Z_i(z_i, z_{i+1}) = 0$ if and only if $\lceil z_i \rfloor^{\frac{p_i}{n+1-i}} = \lceil z_{i+1} \rfloor^{\frac{p_i}{n-i}}$.

For simplicity we introduce the variables

$$\begin{aligned}
\sigma_i &= \lceil z_i \rfloor^{\frac{p_i-(n+1-i)}{n+1-i}} - \lceil z_{i+1} \rfloor^{\frac{p_i-(n+1-i)}{n-i}} \\
s_i &= z_i - \lceil z_{i+1} \rfloor^{\frac{n+1-i}{n-i}} \,, \\
\varphi_i &= -\tfrac{p_i-(n+1-i)}{n-i} \left| z_{i+1} \right|^{\frac{p_i-1-2(n-i)}{n-i}} s_i \,.
\end{aligned}$$

Note that Z_i, σ_i, s_i and φ_i vanish on the same set. The partial derivatives of Z_i are

$$\frac{\partial Z_i(z_i, z_{i+1})}{\partial z_i} = \sigma_i \,, \quad \frac{\partial Z_i(z_i, z_{i+1})}{\partial z_{i+1}} = \varphi_i \,,$$

which are continuous and become zero at the points where Z_i achieves its minimum. $V(z)$ in (1.31) is non negative, since it is the sum of non negative terms, and it is positive definite since $V(z) = 0$ implies $z = 0$. Due to homogeneity it is radially unbounded [10].

Its derivative is

$$\dot{V}(z) = \beta_1 (\sigma_1 \dot{z}_1 + \varphi_1 \dot{z}_2) + \sum_{j=2}^{n-1} \beta_j \frac{p_1}{p_j} \prod_{k=2}^{j} V_k^{\frac{p_{k-1}-p_k}{p_k}} (\underline{z}_k) (\sigma_j \dot{z}_j + \varphi_j \dot{z}_{j+1})$$
$$+ \beta_n \frac{p_1}{p_n} \prod_{k=2}^{n} V_k^{\frac{p_{k-1}-p_k}{p_k}} (\underline{z}_k) |z_n|^{p_n - 1} \dot{z}_n$$

or taking terms together we get

$$\dot{V}(z) = \sum_{j=0}^{n-2} \upsilon_{j+1} \dot{z}_{j+1} + \upsilon_n \dot{z}_n$$

where for $j = 1, \ldots, n-2$

$$\upsilon_1 = \beta_1 \sigma_1$$
$$\upsilon_2 = \beta_1 \varphi_1 + \beta_2 \frac{p_1}{p_2} \prod_{k=2}^{2} V_k^{\frac{p_{k-1}-p_k}{p_k}} (\underline{z}_k) \sigma_2$$
$$\upsilon_{j+1} = \beta_j \frac{p_1}{p_j} \prod_{k=2}^{j} V_k^{\frac{p_{k-1}-p_k}{p_k}} (\underline{z}_k) \varphi_j + \beta_{j+1} \frac{p_1}{p_{j+1}} \prod_{k=2}^{j+1} V_k^{\frac{p_{k-1}-p_k}{p_k}} (\underline{z}_k) \sigma_{j+1}$$
$$\upsilon_n = \beta_{n-1} \frac{p_1}{p_{n-1}} \prod_{k=2}^{n-1} V_k^{\frac{p_{k-1}-p_k}{p_k}} (\underline{z}_k) \varphi_{n-1} + \beta_n \frac{p_1}{p_n} \prod_{k=2}^{n} V_k^{\frac{p_{k-1}-p_k}{p_k}} (\underline{z}_k) |z_n|^{p_n - 1} .$$

Using the dynamics (1.30), in the absence of noise ($\nu(t) \equiv 0$), we obtain

$$\dot{V} \in -\sum_{j=0}^{n-2} \tilde{\lambda}_{j+1} \upsilon_{j+1} \left(\lceil z_1 \rfloor^{\frac{n-j-1}{n}} - z_{j+2} \right) - \tilde{\lambda}_n \upsilon_n \left(\lceil z_1 \rfloor^0 + \frac{1}{\lambda_n} [-1, 1] \right) .$$

Note that since System (1.30) is a DI $\dot{V}$ is a multivalued map. Lyapunov's Theorem of asymptotic stability for DI (see e.g. [3, Theorem 4.1]) requires that $\forall z \neq 0$, $\dot{V} < 0$ for all possible values of the right hand side. To simplify this analysis we find a single-valued and continuous upper bound for $\dot{V}$. For this it is enough to consider the multivalued last term, since all other terms in $\dot{V}$ are single-valued and continuous. When $\lambda_n > 1$ the multivalued map can be written as

$$\lceil z_1 \rfloor^0 + \frac{1}{\lambda_n}[-1,\ 1] = \lceil z_1 \rfloor^0 \left[1 - \frac{1}{\lambda_n},\ 1 + \frac{1}{\lambda_n}\right]$$

From this expression we easily find the following single-valued and continuous upper bounds

$$- \lceil z_1 \rfloor^{\frac{p_1-1}{n}} \left(\lceil z_1 \rfloor^0 + \frac{1}{\lambda_n}[-1,\ 1]\right) \leq \left(\frac{1}{\lambda_n} - 1\right) |z_1|^{\frac{p_1-1}{n}} ,$$
$$\left|\lceil z_1 \rfloor^0 + \frac{1}{\lambda_n}[-1,\ 1]\right| \leq \left(1 + \frac{1}{\lambda_n}\right) .$$

From them and adding and subtracting $\beta_n \lceil z_1 \rfloor^{\frac{p_1-1}{n}}$ to υ_n in $\dot{V}$ we get the following single-valued and continuous upper bound for the last term of $\dot{V}$

$$-\tilde{\lambda}_n \upsilon_n \left(\lceil z_1 \rfloor^0 + \frac{1}{\lambda_n}[-1,\ 1]\right) \leq \tilde{\lambda}_n \theta_n \left(1 + \frac{1}{\lambda_n}\right) - \tilde{\lambda}_n \beta_n \left(1 - \frac{1}{\lambda_n}\right) |z_1|^{\frac{p_1-1}{n}} ,$$

$$\theta_n = \beta_{n-1} \frac{p_1}{p_{n-1}} \prod_{k=2}^{n-1} V_k^{\frac{p_{k-1}-p_k}{p_k}} \left(\underline{z}_k\right) |\varphi_{n-1}| +$$
$$\beta_n \left| \frac{p_1}{p_n} \prod_{k=2}^{n} V_k^{\frac{p_{k-1}-p_k}{p_k}} \left(\underline{z}_k\right) |z_n|^{p_n-1} - \lceil z_1 \rfloor^{\frac{p_1-1}{n}} \right| \geq 0 .$$

We can therefore write

$$\dot{V}(z) \leq -W(z) , \tag{1.37}$$

with the single-valued, continuous and homogeneous function W

$$W(z) = \sum_{j=0}^{n-2} \tilde{\lambda}_{j+1} \upsilon_{j+1} \left(\lceil z_1 \rfloor^{\frac{n-j-1}{n}} - z_{j+2}\right)$$
$$- \tilde{\lambda}_n \theta_n \left(1 + \frac{1}{\lambda_n}\right) + \tilde{\lambda}_n \beta_n \left(1 - \frac{1}{\lambda_n}\right) |z_1|^{\frac{p_1-1}{n}} . \tag{1.38}$$

We show that there exist values of $\tilde{\lambda}_i > 0$ such that $W(z) > 0$. For this consider the values of W restricted to some surfaces: for $i = 1, \ldots, n-1$

$$\mathscr{Z}_1 = \left\{\lceil z_1 \rfloor^{\frac{n-1}{n}} = z_2\right\} ,$$
$$\mathscr{Z}_i = \left\{\lceil z_1 \rfloor^{\frac{n-1}{n}} = z_2 \wedge \cdots \wedge \lceil z_1 \rfloor^{\frac{n-i}{n}} = z_{i+1}\right\} .$$

These sets are related as $\mathscr{Z}_{n-1} \subset \cdots \subset \mathscr{Z}_1$.

We have that $\sigma_i = s_i = \varphi_i = 0$ on $\mathcal{Z}_j$, for every $j \geq i$. Let $W_i = W_{\mathcal{Z}_i}$ represent the function $W(z)$ restricted to the manifold $\mathcal{Z}_i$. We can obtain the value of W_1 by replacing in $W(z)$ the variable z_1 by $z_1 = \lceil z_2 \rfloor^{\frac{n}{n-1}}$, so that W_1 becomes a function of $(z_2, \ldots, z_n)$. In general, we obtain the value of W_i, for $i = 1, \ldots, n-1$, by replacing in $W(z)$ the variables $(z_1, \ldots, z_i)$ by their values in terms of z_{i+1}, i.e. $z_1 = \lceil z_{i+1} \rfloor^{\frac{n}{n-i}}, \ldots, z_i = \lceil z_{i+1} \rfloor^{\frac{n+1-i}{n-i}}$, so that W_i becomes a function of $\underline{z}_{i+1} \triangleq (z_{i+1}, \ldots, z_n)$. For example, the value of the expression $\left(\lceil z_1 \rfloor^{\frac{n-i-1}{n}} - z_{i+2}\right)$ on $\mathcal{Z}_i$ is $\left(\lceil z_{i+1} \rfloor^{\frac{n-i-1}{n-i}} - z_{i+2}\right)$.

Our strategy to prove that there exist gains $\tilde{k}_i$ such that $W(z) > 0$ is as follows: we show (recursively) that if $W_i > 0$ then, using homogeneity, we can render $W_{i-1} > 0$ by selecting $\tilde{\lambda}_i$ sufficiently large. To start we see from (1.38) that we can write $W(z)$ as

$$W(z) = \tilde{\lambda}_1 \eta_1(z_1, z_2) + \mu_1(z),$$
$$\eta_1(z_1, z_2) = \upsilon_1\left(\lceil z_1 \rfloor^{\frac{n-1}{n}} - z_2\right) = \beta_1\left(\lceil z_1 \rfloor^{\frac{p_1-n}{n}} - \lceil z_2 \rfloor^{\frac{p_1-n}{n-1}}\right)\left(\lceil z_1 \rfloor^{\frac{n-1}{n}} - z_2\right)$$

$$\mu_1(z) = \sum_{j=1}^{n-2} \tilde{\lambda}_{j+1} \upsilon_{j+1}\left(\lceil z_1 \rfloor^{\frac{n-j-1}{n}} - z_{j+2}\right) - \tilde{\lambda}_n \theta_n\left(1 + \frac{1}{\lambda_n}\right) + \tilde{\lambda}_n \beta_n\left(1 - \frac{1}{\lambda_n}\right)|z_1|^{\frac{p_1-1}{n}}.$$

$\eta_1(z_1, z_2)$ is positive everywhere except on the set $\mathcal{Z}_1$, where it vanishes, i.e. $\eta_1(z_1, z_2) = 0$ for $z \in \mathcal{Z}_1$. Note also that μ_1 does not depend on the gain $\tilde{\lambda}_1$. According to Lemma 1.5 there exists a sufficiently large positive value of $\tilde{\lambda}_1$ such that $W(z) > 0$ if the value of μ_1 restricted to $\mathcal{Z}_1$, i.e. $\mu_{1\mathcal{Z}_1} = W_1(\underline{z}_2)$, is positive.

$W_1(\underline{z}_2)$ can be written as

$$W_1(\underline{z}_2) = \tilde{\lambda}_2 \eta_2(z_2, z_3) + \mu_2(\underline{z}_2),$$

where

$$\eta_2(z_2, z_3) = \beta_2 \frac{p_1}{p_2} \prod_{k=2}^{2} V_k^{\frac{p_{k-1}-p_k}{p_k}}(\underline{z}_k)\left(\lceil z_2 \rfloor^{\frac{p_2-n+1}{n-1}} - \lceil z_3 \rfloor^{\frac{p_2-n+1}{n-2}}\right)\left(\lceil z_2 \rfloor^{\frac{n-2}{n-1}} - z_3\right)$$

$$\mu_2(\underline{z}_2) = \sum_{j=2}^{n-2} \tilde{\lambda}_{j+1} \upsilon_{j+1}\left(\lceil z_2 \rfloor^{\frac{n-j-1}{n-1}} - z_{j+2}\right) - \tilde{\lambda}_n \theta_n\left(1 + \frac{1}{\lambda_n}\right) + \tilde{\lambda}_n \beta_n\left(1 - \frac{1}{\lambda_n}\right)|z_2|^{\frac{p-1}{n-1}},$$

where everywhere we set $z_1 = \lceil z_2 \rfloor^{\frac{n}{n-1}}$. $\eta_2(z_2, z_3)$ is positive on $\mathscr{Z}_1$ except on the set $\mathscr{Z}_2$, where it vanishes, and $\mu_2(\underline{z}_2)$ does not depend on the gain $\tilde{\lambda}_2$. According to Lemma 1.5 there exists a sufficiently large positive value of $\tilde{\lambda}_2$ such that $W_1(\underline{z}_2)$ is positive definite on $\mathscr{Z}_1$ if $\mu_{2\mathscr{Z}_2} = W_2(\underline{z}_3)$ is positive.

Doing this recursively, we see that $W_i(\underline{z}_{i+1})$, for $i = 1, \ldots, n-2$, can be written as

$$W_i(\underline{z}_{i+1}) = \tilde{\lambda}_{i+1}\eta_{i+1}(z_{i+1}, z_{i+2}) + \mu_{i+1}(\underline{z}_{i+1}), \tag{1.39}$$

$$\eta_{i+1}(\cdot) = \beta_{i+1}\frac{p_1}{p_{i+1}}\prod_{k=2}^{i-1} V_k^{\frac{p_{k-1}-p_k}{p_k}}(\underline{z}_k)\left(\lceil z_{i+1}\rfloor^{\frac{p_{i+1}-n+i}{n-i}} - \lceil z_{i+2}\rfloor^{\frac{p_{i+1}-n+i}{n-i-1}}\right)\times \left(\lceil z_{i+1}\rfloor^{\frac{n-i-1}{n-i}} - z_{i+2}\right) \tag{1.40}$$

$$\mu_{i+1}(\underline{z}_{i+1}) = \sum_{j=i+1}^{n-2}\tilde{\lambda}_{j+1}\upsilon_{j+1}\left(\lceil z_{i+1}\rfloor^{\frac{n-j-1}{n-i}} - z_{j+2}\right) - \tilde{\lambda}_n\theta_n\left(1+\frac{1}{\lambda_n}\right) - \tilde{\lambda}_n\beta_n\left(1-\frac{1}{\lambda_n}\right)|z_{i+1}|^{\frac{p_1-1}{n-i}}, \tag{1.41}$$

where everywhere we set $z_1 = \lceil z_{i+1}\rfloor^{\frac{n}{n-i}}, \ldots, z_i = \lceil z_{i+1}\rfloor^{\frac{n+1-i}{n-i}}$. Since $\eta_{i+1}(z_{i+1}, z_{i+2})$ is positive on $\mathscr{Z}_i$ except on the set $\mathscr{Z}_{i+1}$, where it vanishes, and $\mu_{i+1}(\underline{z}_{i+1})$ does not depend on the gain $\tilde{\lambda}_{i+1}$, Lemma 1.5 assures the existence of a sufficiently large positive value of $\tilde{\lambda}_{i+1}$ such that $W_i(\underline{z}_{i+1})$ is positive definite on $\mathscr{Z}_i$ if $\mu_{i+1,\mathscr{Z}_{i+1}} = W_{i+1}(\underline{z}_{i+2})$ is positive.

Finally, since

$$W_{n-1}(z_n) = \tilde{\lambda}_n\beta_n\left(1-\frac{1}{\lambda_n}\right)|z_n|^{p_1-1},$$

which is positive on $\mathscr{Z}_{n-1}$ for all positive values of $\tilde{\lambda}_n$, β_n, and $\lambda_n > 1$, we conclude that there exist positive values of $\left(\tilde{\lambda}_1, \ldots, \tilde{\lambda}_n\right)$ such that $W(z) > 0$. Lyapunov's theorem for DI [3] implies the stated stability property.

Moreover, Lemma 1.6 implies that $W(z) \geq \kappa V(z)^{\frac{p-1}{p}}$ for some positive κ. Applying this in inequality (1.37) leads to (1.33). Since the solution of the differential equation $\dot{v}(t) = -\kappa v(t)^{\frac{p-1}{p}}$, $v(0) = v_0 > 0$ is given by $t = p/\kappa\left(v_0^{\frac{1}{p}} - v(t)^{\frac{1}{p}}\right)$ we obtain that the time T such that $v(T) = 0$ is $T = p/\kappa v_0^{\frac{1}{p}}$. Using the comparison principle for (1.33) we conclude (1.34).

1.7.2 Selection of the Gains. Proof of Proposition 1.2

From the proof of Theorem 1.5 the gains $\tilde{\lambda}_i$ should be selected backwards, i.e. from $\tilde{\lambda}_n$ to $\tilde{\lambda}_1$. First select $\lambda_n > 1$, in order to compensate the effect of $f_0^{(n)}$, and $\tilde{\lambda}_n > 0$, so that $W_{n-1}(z_n) > 0$. Given $\lambda_n > 1$ select $\tilde{\lambda}_{n-1}$ large enough such that $W_{n-2}\left(\underline{z}_{n-1}\right) > 0$, i.e. such that

$$\tilde{\lambda}_{n-1} > \omega_{n-1}(\tilde{\lambda}_n) := \max\left\{\frac{-\mu_{n-1}\left(\underline{z}_{n-1}\right)}{\eta_{n-1}\left(\underline{z}_{n-1}\right)}\right\}.$$

Recursively, for $i = n, n-1, \ldots, 2$ and given the gains $(\tilde{\lambda}_i, \ldots, \tilde{\lambda}_n)$ select $\tilde{\lambda}_{i-1}$ large enough such that $W_i\left(\underline{z}_{i-1}\right) > 0$ (see (1.39)), i.e. such that

$$\tilde{\lambda}_{i-1} > \omega_{i-1}(\tilde{\lambda}_i, \ldots, \tilde{\lambda}_n) := \max\left\{\frac{-\mu_{i-1}\left(\underline{z}_{i-1}\right)}{\eta_{i-1}\left(\underline{z}_{i-1}\right)}\right\},$$

where η_i and μ_i are given by (1.40) and (1.41), respectively.

Functions $\mu_{i-1}\left(\underline{z}_{i-1}\right)/\eta_{i-1}\left(\underline{z}_{i-1}\right)$ are homogeneous of degree 0, so that they achieve their maximum value on the unit sphere S.

Item (2) of the Proposition follows immediately from the following observation: Setting $z_1 = \lceil z_2 \rfloor^{\frac{n}{n-1}}$ in the error system (1.30) of differentiator of order $n-1$ one obtains the error system for the differentiator of order $n-2$. The same is true for the Lyapunov function and its derivative. Note also that when $z_1 = \lceil z_2 \rfloor^{\frac{n}{n-1}}$ the Lyapunov function $V_1(z)$ reduces to $V_2^{\frac{p_1}{p_2}}\left(\underline{z}_2\right)$, which is a Lyapunov Function exactly when $V_2\left(\underline{z}_2\right)$ is. And this is obviously valid recursively.

1.8 Output Feedback HOSM Control

Given the state feeback HOSM controllers and the differentiators it is possible to construct an output feedback HOSM control. Consider the system (1.2) with the feedback (1.8) or (1.9). Suppose that the states x are estimated by means of the differentiator (1.29), so that we obtain the following output feedback

$$\Sigma_{ID} : \begin{cases} \dot{x}_i = x_{i+1}, \ i = 1, \ldots, \rho - 1, \\ \dot{x}_\rho \in [-C, C] + [K_m, K_M]u, \end{cases}$$

$$u = -k_\rho \varphi\left(\hat{x}\right) = \begin{cases} -k_\rho \left\lceil \sigma_\rho\left(\hat{x}\right)\right\rfloor^0, \\ -k_\rho \frac{\sigma_\rho(\hat{x})}{M(\hat{x})}, \end{cases}$$

$$\dot{\hat{x}}_i = -k_i L^{\frac{i}{\rho}} \left\lceil \hat{x}_1 - y \right\rfloor^{\frac{\rho-i}{\rho}} + \hat{x}_{i+1} ,$$
$$\dot{\hat{x}}_\rho = -k_\rho L \left\lceil \hat{x}_1 - y \right\rfloor^{0} .$$

Defining the estimation errors as $e_i \triangleq \hat{x}_i - x_i$ and performing the state transformation

$$z_1 = \frac{e_1}{L}, \ldots, z_i = \frac{e_i}{\lambda_{i-1} L}, \; i = 1, \ldots, \rho,$$

it is possible to write the dynamics of the closed loop in the state variables (x, z) (see (1.30) for the estimation error dynamics)

$$\begin{aligned} \dot{x}_i &= x_{i+1}, \\ \dot{x}_\rho &\in [-C, C] - k_\rho [K_m, K_M] \varphi(x+e) , \\ \dot{z}_i &= -\tilde{\lambda}_i \left(\lceil z_1 \rfloor^{\frac{\rho-i}{\rho}} - z_{i+1} \right) , \\ \dot{z}_\rho &\in -\tilde{\lambda}_\rho \left(\lceil z_1 \rfloor^{0} + \frac{C + K_M k_\rho}{L \lambda_\rho} [-1, 1] \right) . \end{aligned} \tag{1.42}$$

We prove the convergence of the closed output feedback

Theorem 1.6 *Consider the Output Feedback system* (1.42)*, composed of the uncertain plant* (1.2) *and the observer (differentiator)* (1.29)*, where $\varphi(x)$ is any discontinuous* (1.8) *or Quasi-Continuous* (1.9) *controller. Suppose that the gains of the controller $(k_1, \ldots, k_\rho)$ are selected large enough so that Theorem 1.2 is satisfied and the observer gains $(\lambda_1, \ldots, \lambda_\rho)$ and $L > C + k_\rho K_M$ are such that Theorem 1.5 is fulfilled. Under these conditions the ρth order sliding mode $(x, e) = 0$ is established in Finite-Time.*

Proof In the absence of noise under the stated conditions Theorem 1.5 assures that $z(t) \to 0$ in finite time, i.e. for $t \geq T_o(z_0)$, and so will also $e(t) \to 0$. Therefore, for $t \geq T_o(z_0)$ it happens that $e(t) \equiv 0$. It is easy to see that for any Lebesgue-measurable uniformly bounded control $u(t)$ the solutions of (1.2) exist for all future times, i.e. there is no escape to infinity in finite time. Since the controller $\varphi(x)$ is homogeneous of degree zero and is locally bounded, it is also globally bounded, and therefore the solutions of (1.2) with the controller exist for all future times, even in the case when $e(t) \neq 0$. During the time interval $t \in [0, T_o(z_0)]$ when the observer has not converged, the trajectories of the plant will exist and its state will reach a finite value $x(T_o(z_0))$. And then Theorem 1.2 assures the convergence of $x(t)$ to zero in finite time, i.e. for $t \geq T_c(x(T_o(z_0)))$ it happens that $x(t) \equiv 0$ (in the absence of noise). □

This result is a kind of separation principle for HOSM control, since any stabilizing HOSM controller and any robust HOSM observer (satisfying additionally the gain condition $L > C + k_\rho K_M$) leads to a globally stable closed loop. It is important to remark that for this separation result the model (1.2) has to be valid globally, otherwise the separation is not valid anymore.

In the case of measurement noise and/or perturbations we obtain from the homogeneity [39, 40, 44] the following accuracy properties.

Theorem 1.7 *Consider the Output Feedback controlled system* (1.42) *satisfying the conditions of Theorem 1.6. Suppose that the measurement is realized without noise but it is sampled with a sampling interval* τ. *In this case the states* x *reach after a finite time a neighborhood of the origin and stay there for all future times*

$$|x_1(t)| \leq \delta_1 \tau^{\rho},\ |x_2(t)| \leq \delta_2 \tau^{\rho-1}, \ldots, |x_i(t)| \leq \delta_i \tau^{\rho-(i-1)},\ \left|x_\rho(t)\right| \leq \delta_\rho \tau,$$

where $\delta_1, \ldots, \delta_\rho > 0$ *are some (positive) constants. If the measurement is done continuously but with a noise of maximal magnitude* $\varepsilon > 0$ *then the obtained accuracies are*

$$|x_1(t)| \leq \tilde{\delta}_1 \varepsilon,\ |x_2(t)| \leq \tilde{\delta}_2 \varepsilon^{\frac{\rho-1}{\rho}}, \ldots, |x_i(t)| \leq \tilde{\delta}_i \varepsilon^{\frac{\rho-i+1}{\rho}},\ \left|x_\rho(t)\right| \leq \tilde{\delta}_\rho \varepsilon^{\frac{1}{\rho}},$$

where $\tilde{\delta}_1, \ldots, \tilde{\delta}_\rho > 0$ *are some (positive) constants. In all cases the constants depend only on the chosen controller, the parameters* $(C,\ K_m,\ K_M,\ \rho)$ *and the controller and observer gains, but they are independent of* $\tau,\ \varepsilon$ *and the initial conditions.*

1.9 Application Example: The Magnetic Levitation System

To illustrate the results consider the dynamics of a magnetic levitation system (see [34])

$$\begin{aligned}
\dot{x}_1 &= x_2 \\
\dot{x}_2 &= -\frac{k}{m}x_2 - \frac{aL_0}{2m}\frac{x_3^2}{(a+x_1)^2} + g \\
\dot{x}_3 &= \frac{1}{L(x_1)}\left(-Rx_3 + aL_0\frac{x_2 x_3}{(a+x_1)^2} + u\right)
\end{aligned}$$

where $x_1 = y \in \mathbb{R}_+$ is the the vertical (downward) distance of the ball measured from the coil (when the ball is next to the coil $x_1 = 0$), $x_2 = \dot{y}$ is the velocity, m is the mass of the ball, g is the gravity acceleration, k is a viscous friction coefficient, $L(x_1) = L_1 + \frac{aL_0}{a+x_1}$ is the inductance of the electromagnet, which changes with the position of the ball, $x_3 = i \in \mathbb{R}_+$ is the electric current, R is the electric resistance on the circuit and the control u is the voltage applied. All parameters $k,\ m,\ a,\ L_0,\ L_1,\ g,\ R$ are positive constants. We assume that the only measured (output) signal is the position x_1.

Our control objective is that the ball's position x_1 tracks a desired signal $r(t)$, i.e. to render $\sigma(t) = x_1(t) - r(t) \equiv 0$ robustly and in finite time. σ has full relative degree with respect to the control u, i.e. $\rho = 3$, so that there is no zero dynamics, and the system is brought to the normal form (1.2) by using as state variables $(z_1, z_2, z_3) = (\sigma, \dot{\sigma}, \ddot{\sigma})$, i.e.

$$z_1 = x_1 - r,\ z_2 = x_2 - \dot{r},\ z_3 = -\frac{k}{m}x_2 - \frac{aL_0}{2m}\frac{x_3^2}{(a+x_1)^2} + g - \ddot{r}$$

and therefore the inverse transformation is

$$x_1 = z_1 + r,\ x_2 = z_2 + \dot{r},\ x_3^2 = \frac{2m}{aL_0}\left(g - z_3 - \frac{k}{m}z_2 - \left(\ddot{r} + \frac{k}{m}\dot{r}\right)\right)(a+x_1)^2 .$$

Note that the quadratic term x_3^2 in the dynamics imposes the restriction that

$$g - z_3 - \frac{k}{m}z_2 - \left(\ddot{r} + \frac{k}{m}\dot{r}\right) \geq 0 .$$

The transformation is well defined under the restrictions $x_1 \geq 0$ and $x_3 > 0$. The dynamics becomes

$$\begin{aligned}
\dot{z}_1 &= z_2 \\
\dot{z}_2 &= z_3 \\
\dot{z}_3 &= f(t, z) - \gamma(t, z)\, u
\end{aligned}$$

where

$$\begin{aligned}
f(t, z) &= -\frac{k}{m}z_3 + \frac{aL_0}{m}\frac{x_2 x_3^2}{(a+x_1)^3} - \left(\dddot{r} + \frac{k}{m}\ddot{r}\right) - \gamma(t, z)\left(-R + aL_0\frac{x_2}{(a+x_1)^2}\right)x_3 \\
&= -\frac{k}{m}z_3 + \frac{aL_0}{m}\frac{(z_2+\dot{r})\, h^2(t, z)}{(a+z_1+r)^3} - \gamma(t, z)\, h(t, z)\left(-R + aL_0\frac{z_2+\dot{r}}{(a+z_1+r)^2}\right) + \\
&\quad - \left(\dddot{r} + \frac{k}{m}\ddot{r}\right) \\
\gamma(t, z) &= \frac{aL_0}{m}\frac{x_3}{(a+x_1)^2}\frac{1}{L(x_1)} = \frac{aL_0}{m}\frac{h(t, z)}{(a+z_1+r)^2 L(z_1+r)} \\
h(t, z) &= x_3 = \sqrt{\frac{2m}{aL_0}\left(g - z_3 - \frac{k}{m}z_2 - \left(\ddot{r} + \frac{k}{m}\dot{r}\right)\right)}\,(a+z_1+r)
\end{aligned}$$

Note that for the tracking $(z(t) \equiv 0)$ the control variable is required to be

$$\begin{aligned}
u_{tk}(t) = \frac{f(t, 0)}{g(t, 0)} = \frac{\sqrt{\frac{m}{2aL_0}}\,(a+r)\, L(r)}{\sqrt{g - \left(\ddot{r} + \frac{k}{m}\dot{r}\right)}}\Bigg[2\left(g - \left(\ddot{r} + \frac{k}{m}\dot{r}\right)\right)\times \\
\left\{\frac{\dot{r}}{a+r} - \frac{1}{L(r)}\left(-R + \frac{aL_0\dot{r}}{(a+r)^2}\right)\right\} - \left(\dddot{r} + \frac{k}{m}\ddot{r}\right)\Bigg]
\end{aligned}$$

and for regulation ($\dot{r} = \ddot{r} = \dddot{r} = 0$) this reduces to

$$u = R\sqrt{\frac{2mg}{aL_0}}\,(a + r)\ .$$

Note that (1.2) is not a global model of the actual plant, and thus the control and observation results are only valid locally. In particular, the separation principle for the controller/observer is not valid globally. As HOSM controller we use a discontinuous relay polynomial of order $\rho = 3$ with $\alpha = 0$ (see (1.11)). The controller is therefore given by

$$u = u_{tk}\,(t) + k_3 L \operatorname{sign}\left(\lceil x_3 \rfloor^3 + k_2 L^{\frac{3}{2}} \lceil x_2 \rfloor^{\frac{3}{2}} + k_1 L^2 x_1\right)\ .$$

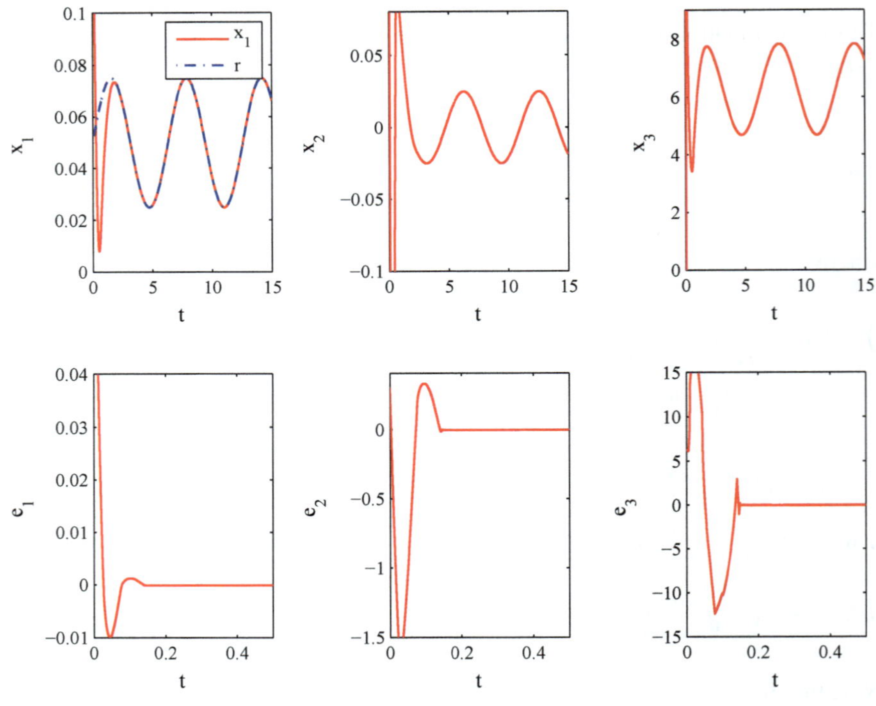

Fig. 1.1 *Upper graph* time evolution of the plant's states $(x_1,\ x_2,\ x_3)$. The ball's position x_1 tracks perfectly after ca. 1 s the reference signal $r\,(t) = 0.05 + 0.025\sin\,(t)$. *Lower graph* behavior of the estimation errors $e = \hat{x} - x$

The observer is given by

$$
\begin{aligned}
\dot{\hat{z}}_1 &= -\lambda_1 l^{\frac{1}{3}} \left\lceil \hat{z}_1 - z_1 \right\rfloor^{\frac{2}{3}} + \hat{z}_2 \\
\dot{\hat{z}}_2 &= -\lambda_2 l^{\frac{2}{3}} \left\lceil \hat{z}_1 - z_1 \right\rfloor^{\frac{1}{3}} + \hat{z}_3 \\
\dot{\hat{z}}_3 &= -\lambda_3 l \left\lceil \hat{z}_1 - z_1 \right\rfloor^{0} + f\left(t, \hat{z}\right) - g\left(t, \hat{z}\right) u \,.
\end{aligned}
$$

For the simulations we used as parameters of the plant: $L_1 = 0.02$ [H]; $L_0 = 0.01$ [H], $a = 0.05$ [m], $k = 0.001$ [N/m/s], $m = 0.1$ [Kg], $g = 9.81$ [m/s^2], $R = 1[\Omega]$. The ball should track a reference signal $r(t) = 0.05 + 0.025 \sin(t)$. Following the design rules presented in the previous sections, we have selected the controller gains as $k_1 = 1.5$, $k_2 = 5$, $k_3 = 25$, $L = 4$, while the observer (differentiator) gains were set to $\lambda_1 = 3$, $\lambda_2 = 1.5\sqrt{3}$, $\lambda_3 = 1.1$, and $l = 700$. The initial conditions for the plant were $z_0 = [0.1,\ 0,\ 0.3]$ and for the observer $\hat{z}_0 = [0.2,\ 0.3,\ 0]$. We consider the parameters $k,\ m,\ R$ uncertain, so that for the observer their values were set 10% above their nominal ones. The simulation was performed with a fixed step Euler algorithm, with a step size of 10^{-5}.

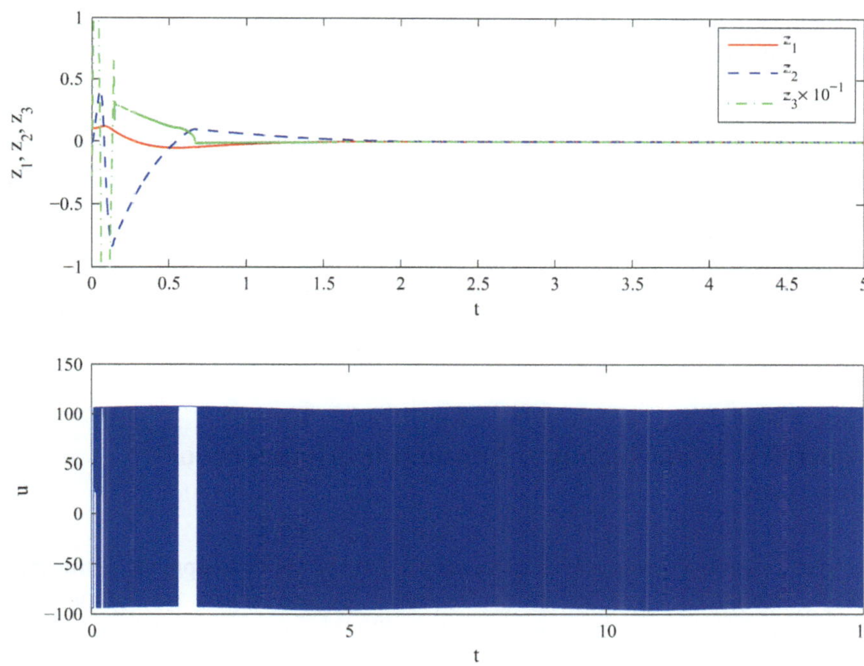

Fig. 1.2 *Upper Graph* time evolution of the sliding variables z. The sliding set $z \equiv 0$ is reached in finite time. *Lower graph* behavior of the control signal u

Figure 1.1 shows a simulation result of the output feedback control, i.e. the discontinuous polynomial controller with the observer. The upper graphs present the three states of the plant and, for the output $x_1 = z_1$ the value of the reference signal $r(t) = 0.05 + 0.025 \sin(t)$. We see that the ball's position x_1 tracks perfectly the reference $r(t)$ after a short period of time (about 1 s) despite of the uncertainties on the model of the system, which was the control objective. The lower graphs show the observer's estimation errors for the original states $x(t)$ of the plant. The observer converges exactly after a finite time smaller than 0.2 s. This is critical for this system, since otherwise the ball can go out of the physical zone if the control does not act appropriately. The upper graph in Fig. 1.2 shows the time evolution of the sliding variables $z(t)$. The sliding set $z \equiv 0$ is reached in less than (2) two seconds. The lower graph in Fig. 1.2 shows the control variable u, which shows the characteristic chattering of the SM control.

1.10 Concluding Remarks

In this chapter we have presented an explicit Lyapunov-based design of homogeneous HOSM controllers and observers (differentiators) for the standard HOSM problem. This allows to put the HOSM design in the same framework as the modern nonlinear control methods, which are pretty much based on Lyapunov approach. In fact, this has been achieved as an extension of previous results in control and observer design for continuous homogeneous systems to discontinuous ones [1, 2, 61, 62, 68–70]. This opens an avenue to find gains for HOSM controllers and observers to meet not only stability and robustness specifications but also other optimal performance requirements.

Acknowledgements The author would like to thank the financial support from PAPIIT-UNAM (Programa de Apoyo a Proyectos de Investigación e Innovación Tecnológica), projects IN113614 and IN113617; Fondo de Colaboración II-FI UNAM, Project IISGBAS-100-2015; CONACyT (Consejo Nacional de Ciencia y Tecnología), project 241171.

Appendix: Some Technical Lemmas on Homogeneous Functions

We recall or prove some useful Lemmas needed for the development of our main results. Lemmas 1.5 and 1.6 are extensions of classical results for homogeneous continuous functions to semicontinuous ones [29, Theorems 4.1 and 4.4].

Lemma 1.4 ([26]) (Young's inequality) *For any positive real numbers $a > 0, b > 0, c > 0, p > 1$ and $q > 1$, with $\frac{1}{p} + \frac{1}{q} = 1$, the following inequality is always satisfied*

$$ab \leq \frac{c^p}{p} a^p + \frac{c^{-q}}{q} b^q,$$

and equality holds if and only if $a^p = b^q$.

Lemma 1.5 *Let* $\eta : \mathbb{R}^n \to \mathbb{R}$ *and* $\gamma : \mathbb{R}^n \to \mathbb{R}_+$ *be two lower (upper) semicontinuous single-valued* **r**-*homogeneous functions of degree* $m > 0$. *Suppose that* $\gamma(x) \geq 0$ $(\gamma(x) \leq 0)$ *on* $\mathbb{R}^n$. *If* $\eta(x) > 0$ $(\eta(x) < 0)$ *for all* $x \neq 0$ *such that* $\gamma(x) = 0$, *then there is a constant* $\lambda^* \in \mathbb{R}$ *and a constant* $c > 0$ *such that for all* $\lambda \geq \lambda^*$ *and for all* $x \in \mathbb{R}^n \setminus \{0\}$,

$$\eta(x) + \lambda\gamma(x) \geq c \|x\|_{\mathbf{r},p}^m$$

$$\left(\eta(x) + \lambda\gamma(x) \leq -c \|x\|_{\mathbf{r},p}^m\right).$$

Proof By virtue of the homogeneity of η and γ, it is sufficient to establish the result on the unit sphere $S = \{x \in \mathbb{R}^n : \|x\|_{\mathbf{r},p} = 1\}$. Suppose that this relation is not valid. Then for every integer q there is a point x_q in S such that

$$\eta(x_q) + q\gamma(x_q) < \frac{1}{q}. \tag{1.43}$$

The sequence $\{x_q\}$, being bounded, has a subsequence converging to a point x_0, and we can accordingly suppose that $\{x_q\}$ converges to x_0. Since $\gamma(x) \geq 0$ on S, it follows from (1.43) and the lower semicontinuity of η and γ, i.e. $\liminf_{x=x_0} \eta(x) \geq \eta(x_0)$, $\liminf_{x=x_0} \gamma(x) \geq \gamma(x_0)$, that $\eta(x_0) \leq 0$, $\gamma(x_0) = 0$. This contradicts our hypothesis, and the first part of the Lemma is established. The second part is proved in the same manner since if $\eta(x)$ is lower semicontinuous then $-\eta(x)$ is upper semicontinuous. □

Lemma 1.6 *Let* $\eta : \mathbb{R}^n \to \mathbb{R}$ *be an upper semicontinuous, single-valued* **r**-*homogeneous function, with weights* $\mathbf{r} = [r_1, \ldots, r_n]^\top$ *and degree* $m > 0$. *Then there is a point* x_2 *in the unit homogeneous sphere* $S = \{x \in \mathbb{R}^n : \|x\|_{\mathbf{r},p} = 1\}$ *such that the following inequality holds for all* $x \in \mathbb{R}^n$

$$\eta(x) \leq \eta(x_2) \|x\|_{\mathbf{r},p}^m. \tag{1.44}$$

Under the same conditions, if η *is lower semicontinuous, there is a point* x_1 *in the unit homogeneous sphere* S *such that the following inequality holds for all* $x \in \mathbb{R}^n$

$$\eta(x_1) \|x\|_{\mathbf{r},p}^m \leq \eta(x). \tag{1.45}$$

Proof By virtue of the homogeneity of η, it is sufficient to establish the inequality (1.44) on the unit homogeneous sphere $S = \{x \in \mathbb{R}^n : \|x\|_{\mathbf{r},p} = 1\}$, i.e. $\eta(x) \leq \eta(x_2)$. Since S is compact and non empty, the latter inequality is a consequence of the fact that an upper semicontinuous function has a finite maximum value on a compact set and it achieves it at some point x_2 [29, Theorem 3.2]. The second part of the Lemma, i.e. inequality (1.45), is obtained by applying the same arguments to $-\eta(x)$, which is upper semicontinuous. If η is continuous, then we obtain item (ii) in Lemma 1.1. □

References

1. Andrieu, V., Praly, L.: A unifying point of view on output feedback designs for global asymptotic stabilization. Automatica **45**(8), 1789–1798 (2009)
2. Andrieu, V., Praly, L., Astolfi, A.: Homogeneous approximation, recursive observer design and output feedback. SIAM J. Control Optim. **47**(4), 1814–1850 (2008)
3. Bacciotti, A., Rosier, L.: Liapunov Functions and Stability in Control Theory, 2nd edn. Springer, New York (2005)
4. Bartolini, G., Pisano, A., Usai, E.: First and second derivative estimation by sliding mode technique. J. Signal Process. **4**(2), 167–176 (2000)
5. Bejarano, F., Fridman, L.: High order sliding mode observer for linear systems with unbounded unknown inputs. Int. J. Control **83**(9), 1920–1929 (2010)
6. Bernuau, E., Efimov, D., Perruquetti, W., Polyakov, A.: On an extension of homogeneity notion for differential inclusions. In: Proceedings of the European Control Conference (2013)
7. Bernuau, E., Polyakov, A., Efimov, D., Perruquetti, W.: Verification of ISS, iISS and IOSS properties applying weighted homogeneity. Syst. Control Lett. **62**(12), 1159–1167 (2013)
8. Bernuau, E., Efimov, D., Perruquetti, W., Polyakov, A.: On homogeneity and its application in sliding mode control. J. Frankl. Inst. **351**(4), 1816–1901 (2014)
9. Bernuau, E., Efimov, D., Perruquetti, W., Polyakov, A.: Homogeneity of differential inclusions. Control, Robotics and Sensors, pp. 103–118. Institution of Engineering and Technology (2016). doi:10.1049/PBCE102E_ch2.1
10. Bhat, S.P., Bernstein, D.S.: Geometric homogeneity with applications to finite-time stability. Math. Control Signals Syst. **17**(2), 101–127 (2005)
11. Cruz-Zavala, E., Moreno, J.A.: Improved convergence rate of discontinuous finite-time controllers. In: Proceedings of the 19th IFAC World Congress, pp. 8636–8641 (2014)
12. Cruz-Zavala, E., Moreno, J.A.: A new class of fast finite-time discontinuous controllers. In: Proceedings of the 13th IEEE Workshop on Variable Structure Systems, pp. 1–6 (2014)
13. Cruz-Zavala, E., Moreno, J.A.: Lyapunov approach to higher-order sliding mode design. Control, Robotics and Sensors, pp. 3–28. Institution of Engineering and Technology (2016). doi:10.1049/PBCE102E_ch1.1
14. Cruz-Zavala, E., Moreno, J.A.: Lyapunov functions for continuous and discontinuous differentiators. In: Proceedings of the 10th IFAC Symposium on Nonlinear Control Systems (2016)
15. Cruz-Zavala, E., Moreno, J.A.: Homogeneous high order sliding mode design: a Lyapunov approach. Automatica. **80**, 232–238 (2017). doi:10.1016/j.automatica.2017.02.039
16. Davila, J., Fridman, L., Levant, A.: Second-order sliding-mode observer for mechanical systems. IEEE Trans. Autom. Control **50**(11), 1785–1789 (2005)
17. Deimling, K.: Multivalued Differential Equations. Walter de gruyter, Berlin (1992)
18. Ding, S., Levant, A., Li, S.: New families of high-order sliding-mode controllers. In: Proceedings of the 54th IEEE Conference on Decision and Control, pp. 4752–4757 (2015)
19. Ding, S., Levant, A., Li, S.: Simple homogeneous sliding-mode controller. Automatica **67**(5), 22–32 (2016)
20. Efimov, D., Fridman, L.: A hybrid robust non-homogeneous finite-time differentiator. IEEE Trans. Autom. Control **56**, 1213–1219 (2011)
21. Filippov, A.: Differential Equations with Discontinuous Righthand Side. Kluwer, Dordrecht (1988)
22. Floquet, T., Barbot, J.P.: Super twisting algorithm-based step-by-step sliding mode observers for nonlinear systems with unknown inputs. Int. J. Syst. Sci. **38**(10), 803–815 (2007)

23. Freeman, R., Kokotovic, P.: Robust Nonlinear Control Design: State Space and Lyapunov Techniques. Modern Birkhäuser Classics, Boston (1996)
24. Fridman, L., Levant, A.: Sliding Mode Control in Engineering. Marcel Dekker Ink., New York (2002)
25. Hahn, W.: Stability of Motion. Springer, Berlin (1967)
26. Hardy, G.H., Littlewood, J.E., Polya, G.: Inequalities. Cambridge University Press, London (1951)
27. Harmouche, M., Laghrouche, S., Chitour, Y.: Robust and adaptive higher order sliding mode. In: Proceedings of the IEEE 51st Annual Conference on Decision and Control (2012)
28. Hermes, H.: Homogeneus coordinates and continuous asymptotically stabilizing feedback controls. In: Elaydi, S. (ed.) Differential Equations, Stability and Control. LNPAM, vol. 127, pp. 249–260. Marcel Dekker Inc., New York (1991)
29. Hestenes, M.R.: Calculus of Variations and Optimal Control Theory. Wiley, New York (1966)
30. Hong, Y.: Finite-time stabilization and stabilizability of a class of controllable systems. Syst. Control Lett. **46**, 231–236 (2002)
31. Huang, X., Lin, W.: Global finite-time stabilization of a class of uncertain nonlinear systems. Automatica **41**, 881–888 (2005)
32. Isidori, A.: Nonlinear Control Systems. Springer, Berlin (1995)
33. Isidori, A.: Nonlinear Control Systems II. Springer, London (1999)
34. Khalil, H.: Nonlinear Systems, 3rd edn. Prentice-Hall, New Jersey (2002)
35. Kobayashi, S., Suzuki, S., Furuta, K.: Frequency characteristics of Levant's differentiator and adaptive sliding mode differentiator. Int. J. Syst. Sci. **38**(10), 825–832 (2007)
36. Levant, A.: Sliding order and sliding accuracy in sliding mode Control. Int. J. Control **58**(6), 1247–1263 (1993)
37. Levant, A.: Robust exact differentiation via sliding mode technique. Automatica **34**(3), 379–384 (1998)
38. Levant, A.: Universal single-input single-output (siso) sliding-mode controllers with finite-time. IEEE Trans. Autom. Control **46**(9), 1447–1451 (2001)
39. Levant, A.: High-order sliding modes: differentiation and output-feedback control. Int. J. Control **76**(9), 924–941 (2003)
40. Levant, A.: Homogeneity approach to high-order sliding mode design. Automatica **41**, 823–830 (2005)
41. Levant, A.: Quasi-continuous high-order sliding-mode controllers. IEEE Trans. Autom. Control **50**(11), 1812–1816 (2005)
42. Levant, A.: Principles of 2-sliding mode design. Automatica **43**(4), 576–586 (2007)
43. Levant, A.: Adjustment of high-order sliding-mode controllers. Int. J. Robust Nonlinear Control **19**, 1657–1672 (2009)
44. Levant, A., Livne, M.: Weighted homogeneity and robustness of sliding mode control. Automatica **72**, 186–193 (2016)
45. Moreno, J.: Lyapunov approach for analysis and design of second order sliding mode algorithms. In: Fridman, L., Moreno, J., Iriarte, R. (eds.) Sliding Modes after the first decade of the 21st Century. LNCIS, vol. 412, pp. 113–150. Springer, Berlin (2011)
46. Moreno, J.: A Lyapunov approach to output feedback control using second-order sliding modes. IMA J. Math. Control Inf. **29**(3), 291–308 (2012)
47. Moreno, J.: On discontinuous observers for second order systems: properties, analysis and design. In: Bandyopadhyay, B., Janardhanan, S., Spurgeon, S.K. (eds.) Advances in Sliding Mode Control-Concepts, Theory and Implementation. LNCIS, vol. 440, pp. 243–265. Springer, Berlin (2013)
48. Moreno, J., Osorio, M.: A Lyapunov approach to second-order sliding mode controllers and observers. In: Proceedings of the 47th IEEE Conference on Decision and Control, pp. 0191–2216 (2008)
49. Moreno, J., Osorio, M.: Strict Lyapunov functions for the super-twisting algorithm. IEEE Trans. Autom. Control **57**(4), 1035–1040 (2012)

50. Moulay, E., Perruquetti, W.: Finite time stability and stabilization of a class of continuous systems. J. Math. Anal. Appl. **323**(2), 1430–1443 (2006)
51. Nakamura, H., Yamashita, Y., Nishitani, H.: Smooth Lyapunov functions for homogeneous differential inclusions. In: Proceedings of the 41st SICE Annual Conference, pp. 1974–1979 (2002)
52. Nakamura, N., Nakamura, H., Yamashita, Y., Nishitani, H.: Homogeneous stabilization for input affine homogeneous systems. IEEE Trans. Autom. Control **54**(9), 2271–2275 (2009)
53. Orlov, Y.: Finite time stability of homogeneous switched systems. In: Proceedings of the 42nd IEEE Conference on Decision and Control, pp. 4271–4276 (2003)
54. Orlov, Y.V.: Discontinuous Systems: Lyapunov Analysis and Robust Synthesis under Uncertainty Conditions. Springer, Berlin (2009)
55. Polyakov, A., Poznyak, A.: Lyapunov function design for finite-time convergence analysis: "twisting" controller for second-order sliding mode realization. Automatica **45**(2), 444–448 (2009)
56. Polyakov, A., Poznyak, A.: Reaching time estimation for "super-twisting" second-order sliding mode controller via Lyapunov function designing. IEEE Trans. Autom. Control **54**(8), 1951–1955 (2009)
57. Polyakov, A., Poznyak, A.: Unified Lyapunov function for a finite-time stability analysis of relay second-order sliding mode control systems. IMA J. Math. Control Inf. **29**(4), 529–550 (2012)
58. Polyakov, A., Efimov, D., Perruquetti, W.: Finite-time and fixed-time stabilization: implicit Lyapunov function method. Automatica **51**(1), 332–340 (2015)
59. Polyakov, A., Efimov, D., Perruquetti, W.: Robust stabilization of MIMO systems in finite/fixed time. Int. J. Robust Nonlinear Control **26**(1), 69–90 (2016)
60. Praly, L., Andrea-Novel, B., Coron, J.: Lyapunov design of stabilizing controllers for cascaded systems. IEEE Trans. Autom. Control **36**, 1177–1181 (1991)
61. Qian, C., Lin, W.: A continuous feedback approach to global strong stabilization of nonlinear systems. IEEE Trans. Autom. Control **46**(7), 1061–1079 (2001)
62. Qian, C., Lin, W.: Recursive observer design, homogeneous approximation, and nonsmooth output feedback stabilization of nonlinear systems. IEEE Trans. Autom. Control **51**(9), 1457–1471 (2006)
63. Rosier, L.: Homogeneous Lyapunov function for homogeneous continuous vector field. Syst. Control Lett. **19**, 467–473 (1992)
64. Shtessel, Y.B., Shkolnikov, I.A.: Aeronautical and space vehicle control in dynamic sliding manifolds. Int. J. Control **76**(9/10), 1000–1017 (2003)
65. Shtessel, Y., Edwards, C., Fridman, L., Levant, A.: Sliding Mode Control and Observation. Birkhauser, Springer, New York (2014)
66. Utkin, V., Guldner, J., Shi, J.: Sliding Mode Control in Electro-Mechanical Systems, 2nd edn. CRC Press, Taylor & Francis, London (2009)
67. Vasiljevic, L.K., Khalil, H.K.: Error bounds in differentiation of noisy signals by high-gain observers. Syst. Control Lett. **57**, 856–862 (2008)
68. Yang, B., Lin, W.: Output feedback stabilization of a class of homogeneous and high-order nonlinear systems. In: Proceedings of the 42nd IEEE Conference on Decision and Control, pp. 37–42 (2003)
69. Yang, B., Lin, W.: Homogeneous observers, iterative design, and global stabilization of high-order nonlinear systems by smooth output feedback. IEEE Trans. Autom. Control **49**(7), 1069–1080 (2004)
70. Yang, B., Lin, W.: Correction to "homogeneous observers, iterative design, and global stabilization of high-order nonlinear systems by smooth output feedback". IEEE Trans. Autom. Control **50**(11), 1916–1916 (2005)
71. Zubov, V.I.: Methods of A. M. Lyapunov and their Applications. Groningen: P. Noordho: Limited, Netherlands (1964)

Chapter 2
Robustness of Homogeneous and Homogeneizable Differential Inclusions

Emmanuel Bernuau, Denis Efimov and Wilfrid Perruquetti

2.1 Introduction

The problem of robustness and stability analysis with respect to external inputs (like exogenous disturbances or measurement noises) for dynamical systems is in the center of attention of many research works [8, 13, 22, 24, 26, 28]. One of the most popular theories, which can be used for this robustness analysis of nonlinear systems, was originated more than 20 years ago [23] and it is based on the Input-to-State Stability (ISS) property and many related notions. The advantages of ISS theory include a complete list of necessary and sufficient conditions, existence of the Lyapunov method extension, a rich variety of stability concepts adopted for different control and estimation problems.

The main tool to check the ISS property for a nonlinear system consists in a Lyapunov function design satisfying sufficient conditions. As usual, there is no generic approach to select a Lyapunov function for nonlinear systems. Therefore, computationally tractable approaches for ISS verification for particular classes of

---

E. Bernuau (✉)
AgroParisTech and INRA, UMR1145 Ingénierie Procédés Aliments Aliments,
1 Avenue des Olympiades, 91300 Massy, France
e-mail: emmanuel.bernuau@agroparistech.fr

D. Efimov · W. Perruquetti
Inria, Parc Scientifique de la Haute Borne, 40 av. Halley,
59650 Villeneuve-d'Ascq, France
e-mail: denis.efimov@inria.fr

D. Efimov · W. Perruquetti
CRIStAL (UMR-CNRS 9189), Ecole Centrale de Lille, BP 48,
Cité Scientifique, 59651 Villeneuve-dAscq, France
e-mail: wilfrid.perruquetti@inria.fr

S. Li et al. (eds.), *Advances in Variable Structure Systems and Sliding Mode Control—Theory and Applications*, Studies in Systems, Decision and Control 115, DOI 10.1007/978-3-319-62896-7_2

nonlinear systems are of great importance, and they are highly demanded in applications. In this chapter we are going to propose and extend such techniques for checking ISS and Input-to-State Practical Stability (ISpS) for a class of homogeneous and homogenizable discontinuous systems.

The *homogeneity* is an intrinsic property of an object on which the flow of a particular vector field, called Euler vector field, operates as a scaling. This property entails a lot of qualitative results for a homogeneous object, and is of particular interest in view of stability purposes. The notion of homogeneity was found useful by many authors [1, 6, 7, 11, 12, 15, 20, 29]. The main feature of this property is that any local property of the system is in fact global. Obviously, some systems are not homogeneous. The *homogenization* notion has been proposed in [2, 9, 29] and is to homogeneous systems what linearization is to linear system: it allows a system to be locally approximated by a homogeneous one. In the literature, this property is sometimes called "local homogeneity" but we will prefer the terminology "homogenization" to highlight the parallel with linearization. Qualitative properties of the homogeneous approximation are shown to persist locally for the starting system.

The ISS properties of homogeneous or homogenizable *continuous* systems have been studied in [2, 14, 21]. But continuity assumption is not always verified. For instance, mechanical systems with friction or systems controlled by a Sliding Mode Control (SMC) induce a discontinuous vector field. In this chapter, ISS and ISpS properties for discontinuous systems and systems which dynamics are given by a Differential Inclusion (DI) are provided.

Numerous frameworks have been given to deal with discontinuous systems. We will focus here on the Filippov's solution [10]. Filippov's idea is to replace a (discontinuous) vector field by a set-valued map, mapping a point to a set of admissible velocities. The solutions are then absolute continuous curves which derivative belongs to this set of admissible velocities, leading hence to a DI. Different notions of homogeneity for DI have been proposed [4, 10, 17, 18]. In the last paper, a converse homogeneous Lyapunov theorem was proved, on which we shall rely to prove ISS properties. This result was already used to get ISS properties for DI in [5].

In this chapter, our objective is twofold. First, we shall generalize the notion of homogenization to differential inclusions and second we shall formulate conditions of ISS and ISpS properties of discontinuous systems using homogeneity and homogenization. We will present these results using geometric homogeneity to have the most generic formulation.

The outline of the chapter is as follows. Section 2.2 is devoted to the introduction of notations and results that will be used in the sequel. Section 2.3 presents the new framework of homogenization of a DI and the associated stability results. Section 2.4 gives the ISS and ISpS results obtained using homogeneity techniques. Finally, a conclusion will sum up the chapter and will give some perspectives.

2.2 Preliminaries

2.2.1 Notations

We denote n a positive integer and we will be interested in systems defined on $\mathbb{R}^n$. We endow $\mathbb{R}^n$ with the Lebesgue measure and denote $\mathscr{N}$ the set of all zero-measure subsets of $\mathbb{R}^n$. For $x \in \mathbb{R}^n$ and $\varepsilon > 0$, we denote $\mathbb{B}(x, \varepsilon) = \{x\} \dot{+} \mathbb{B}(\varepsilon)$ the open ball centered in x and of radius ε. If $g : \mathbb{R}^n \to \mathbb{R}^p$ is a differentiable mapping, we denote $d_x g$ the value at point $x \in \mathbb{R}^n$ of the differential of g; hence, $d_x g$ is a linear form on $\mathbb{R}^p$.

We shall consider here locally essentially bounded vector fields. The set of locally essentially bounded vector fields is denoted by $\mathscr{L}^\infty_{\text{loc}}(\mathbb{R}^n, \mathbb{R}^n)$.

Definition 2.1 The set of nonempty compact subsets of $\mathbb{R}^n$ is denoted by $\mathfrak{H}(\mathbb{R}^n)$. The *Hausdorff distance* between $X, Y \in \mathfrak{H}(\mathbb{R}^n)$ is defined by:

$$\mathfrak{d}(X, Y) = \max\left(\sup_{x \in X} \mathsf{d}(x, Y), \sup_{y \in Y} \mathsf{d}(y, X) \right). \tag{2.1}$$

where the distance between a point and a compact set is defined by

$$\mathsf{d}(x, Y) = \inf_{y \in Y} \|x - y\|.$$

Proposition 2.1 *The Hausdorff distance defines a distance on $\mathfrak{H}(\mathbb{R}^n)$. Endowed with this distance, $(\mathfrak{H}(\mathbb{R}^n), \mathfrak{d})$ is a complete metric space. Moreover, for all $\lambda \in \mathbb{R}$, $\mathfrak{d}(\lambda X, \lambda Y) = |\lambda| \mathfrak{d}(X, Y)$.*

We will denote by ∂A the boundary of a bounded set A and $\|A\| = \sup_{a \in A} \|a\|$. If A is compact, $\|A\| = \mathfrak{d}(A, \{0\})$.

Definition 2.2 Let (E, d) be a metric space, and let $u_k : \mathbb{R}^n \to E$ be a sequence of mappings. We say that this sequence *converges uniformly on compact sets to* $u : \mathbb{R}^n \to E$, denoted $u_k \xrightarrow[k \to +\infty]{CUC} u$, iff for any compact set $K \subset \mathbb{R}^n$ and for all $\varepsilon > 0$ there exists a $k_0 > 0$ such that for all $k > k_0$, $\sup_{x \in K} \mathsf{d}(u_n(x), u(x)) < \varepsilon$.

2.2.2 Differential Inclusions

We refer to [3, 10] for the basic definitions and the technical material on set-valued maps and DI. In this section, we will only recall the definitions and results that will be used hereafter, without any proof.

The Filippov's regularization procedure consists in the construction of a set-valued map F starting with a vector field $f \in \mathscr{L}^\infty_{\text{loc}}(\mathbb{R}^n, \mathbb{R}^n)$:

$$\mathscr{F}[f](x) = \bigcap_{\varepsilon>0}\bigcap_{\mathrm{N}\in\mathscr{N}} \overline{\mathrm{conv}}(f(\mathbb{B}(x,\varepsilon)\setminus \mathrm{N})). \tag{2.2}$$

By construction, for all $x \in \mathbb{R}^n$, the set $\mathscr{F}[f](x)$ is compact and convex. Moreover, the set-valued map $\mathscr{F}[f]$ is upper semicontinuous.

In many applications, the DI is given by the set-valued map coming from the Filippov's procedure. We will therefore focus on set-valued map with the properties inherited by this procedure.

Definition 2.3 Let F be a set-valued map. We say that F verifies the *standard assumptions* (SA) if F is upper semicontinuous and if for any $x \in \mathbb{R}^n$, $F(x)$ is a nonempty compact convex set.

2.2.3 Homogeneity

To introduce the notion of geometric homogeneity, the class of Euler vector fields has to be defined.

Definition 2.4 [16] A vector field $\nu \in \mathscr{C}^1(\mathbb{R}^n, \mathbb{R}^n)$ is said to be *Euler* if it is complete and if the origin is a GAS equilibrium of $-\nu$.

We will always write Φ the flow of ν, that is $\Phi^s(x)$ is the current state at time s of the trajectory of ν starting from x at $s = 0$. We also denote $d_x\Phi^s$ the differential of the diffeomorphism Φ^s at a fixed $s \in \mathbb{R}$, taken at $x \in \mathbb{R}^n$. We are now able to state the classical definitions of geometric homogeneity.

Definition 2.5 Let ν be an Euler vector field.

- A function $V : \mathbb{R}^n \to \mathbb{R}$ is ν-homogeneous of degree $\kappa \in \mathbb{R}$ if:

$$V(\Phi^s(x)) = e^{\kappa s}V(x) \qquad \forall x \in \mathbb{R}^n, \forall s \in \mathbb{R}.$$

- A vector field $f : \mathbb{R}^n \to \mathbb{R}^n$ is ν-homogeneous of degree $\kappa \in \mathbb{R}$ if:

$$f(\Phi^s(x)) = e^{\kappa s}d_x\Phi^s f(x) \qquad \forall x \in \mathbb{R}^n, \forall s \in \mathbb{R}. \tag{2.3}$$

The relation (2.3) can be recast under a more compact form $\mathscr{H}_\kappa^s(f) = f$, where the vector field $\mathscr{H}_\kappa^s(f)$ is defined by:

$$\mathscr{H}_\kappa^s(f) : x \mapsto e^{-\kappa s}\left(d_x\Phi^s\right)^{-1} f(\Phi^s(x)). \tag{2.4}$$

2.2.4 Homogeneous Differential Inclusions

In this subsection, we recall some definitions and results obtained in [4] that we will need in the sequel.

Definition 2.6 [4] Let ν be an Euler vector field. A set-valued map $F : \mathbb{R}^n \rightrightarrows \mathbb{R}^n$ is ν-homogeneous of degree $\kappa \in \mathbb{R}$ if for all $s \in \mathbb{R}$ we have:

$$\mathscr{H}_\kappa^s(F) = F,$$

where we extend the operator $\mathscr{H}_\kappa^s$ defined in (2.4) by:

$$\mathscr{H}_\kappa^s(F) : x \mapsto e^{-\kappa s} \left(d_x \Phi^s\right)^{-1} \cdot F(\Phi^s(x)).$$

Proposition 2.2 *Let $f \in \mathscr{L}_{loc}^\infty(\mathbb{R}^n, \mathbb{R}^n)$ be a vector field. Then for all $s \in \mathbb{R}$ and all $\kappa \in \mathbb{R}$ we have:*

$$\mathscr{H}_\kappa^s(\mathscr{F}[f]) = \mathscr{F}[\mathscr{H}_\kappa^s(f)].$$

Proof Since for all $\varepsilon > 0$ there exist $\varepsilon_- > 0$ and $\varepsilon_+ > 0$ such that $\Phi^s(\mathbb{B}(x, \varepsilon_-)) \subset \mathbb{B}(\Phi^s(x), \varepsilon) \subset \Phi^s(\mathbb{B}(x, \varepsilon_+))$ we have

$$\begin{aligned}\mathscr{F}[f](\Phi^s(x)) &= \bigcap_{\varepsilon>0} \bigcap_{N\in\mathscr{N}} \overline{\text{conv}}(f(y), y \in \mathbb{B}(\Phi^s(x), \varepsilon) \setminus N)\\ &= \bigcap_{\varepsilon>0} \bigcap_{N\in\mathscr{N}} \overline{\text{conv}}(f(y), y \in \Phi^s(\mathbb{B}(x, \varepsilon)) \setminus N)\\ &= \bigcap_{\varepsilon>0} \bigcap_{N\in\mathscr{N}} \overline{\text{conv}}(f(\Phi^s(z)), z \in \mathbb{B}(x, \varepsilon) \setminus N)\end{aligned}$$

Hence we find that

$$\mathscr{H}_\kappa^s(\mathscr{F}[f])(x) = \bigcap_{\varepsilon>0} \bigcap_{N\in\mathscr{N}} \overline{\text{conv}}((d_x\Phi^s)^{-1} d_z\Phi^s \mathscr{H}_\kappa^s(f)(z), z \in \mathbb{B}(x, \varepsilon) \setminus N).$$

Let us denote by $\sigma_{max}((d_x\Phi^s)^{-1}d_z\Phi^s)$ the biggest singular value of the linear mapping $(d_x\Phi^s)^{-1}d_z\Phi^s$. The function $\phi : z \mapsto |\sigma_{max}((d_x\Phi^s)^{-1}d_z\Phi^s) - 1|$ is continuous and therefore bounded on $\mathbb{B}(x, \varepsilon)$ and moreover vanishes at $z = x$. For all $z \in \mathbb{B}(x, \varepsilon)$ we have:

$$\|(d_x\Phi^s)^{-1} d_z\Phi^s \mathscr{H}_\kappa^s(f)(z) - \mathscr{H}_\kappa^s(f)(z)\| \le M(\varepsilon),$$

where

$$M(\varepsilon) = \sup_{\mathbb{B}(x,\varepsilon)} \phi \text{ ess sup}_{\mathbb{B}(x,\varepsilon)} \|\mathscr{H}_\kappa^s(f)\|.$$

The function M is continuous at zero and $M(0) = 0$. We have proved that

$$(d_x \Phi^s)^{-1} d_z \Phi^s \mathscr{H}_\kappa^s(f)(z) \in \mathscr{H}_\kappa^s(f)(z) \dot{+} \mathbb{B}(0, M(\varepsilon)).$$

It follows that

$$\begin{aligned}
\mathscr{H}_\kappa^s(\mathscr{F}[f])(x) &= \bigcap_{\varepsilon>0} \bigcap_{N\in\mathscr{N}} \overline{\text{conv}}((d_x \Phi^s)^{-1} d_z \Phi^s \mathscr{H}_\kappa^s(f)(z), z \in \mathbb{B}(x, \varepsilon) \setminus N) \\
&\subset \bigcap_{\varepsilon>0} \bigcap_{N\in\mathscr{N}} \overline{\text{conv}}(\mathscr{H}_\kappa^s(f)(z) + \mathbb{B}(0, M(\varepsilon)), z \in \mathbb{B}(x, \varepsilon) \setminus N) \\
&= \bigcap_{\varepsilon>0} \left[\left(\bigcap_{N\in\mathscr{N}} \overline{\text{conv}}(\mathscr{H}_\kappa^s(f)(z), z \in \mathbb{B}(x, \varepsilon) \setminus N) \right) + \mathbb{B}(0, M(\varepsilon)) \right] \\
&= \bigcap_{\varepsilon>0} \bigcap_{N\in\mathscr{N}} \overline{\text{conv}}(\mathscr{H}_\kappa^s(f)(z), z \in \mathbb{B}(x, \varepsilon) \setminus N) \\
&= \mathscr{F}[\mathscr{H}_\kappa^s(f)](x).
\end{aligned}$$

The proof of the converse inclusion is similar. □

Corollary 2.1 *Let $f \in \mathscr{L}^\infty_{loc}(\mathbb{R}^n, \mathbb{R}^n)$ be a vector field. Suppose f is ν-homogeneous of degree κ. Then $\mathscr{F}[f]$ is ν-homogeneous of degree κ.*

Proof Since f is ν-homogeneous of degree κ, we have $\mathscr{H}_\kappa^s(f) = f$. Hence $\mathscr{F}[f] = \mathscr{F}[\mathscr{H}_\kappa^s(f)] = \mathscr{H}_\kappa^s(\mathscr{F}[f])$ by Proposition 2.2 and therefore $\mathscr{F}[f]$ is ν-homogeneous of degree κ. □

The following theorem asserts that a strongly globally asymptotically stable homogeneous differential inclusion admits a homogeneous Lyapunov function. This result is a generalization of the theorem proved for ODE in [20].

Theorem 2.1 [4] *Let F be a ν-homogeneous set-valued map of degree κ, satisfying the SA. Then the following statements are equivalent:*

- *The origin is (strongly) GAS for the system $\dot{x} \in F(x)$.*
- *For all $\mu > \max(-\kappa, 0)$, there exists a pair (V, W) of continuous functions, such that:*

1. *$V \in \mathscr{C}^\infty(\mathbb{R}^n, \mathbb{R})$, V is positive definite and ν-homogeneous of degree μ;*
2. *$W \in \mathscr{C}^\infty(\mathbb{R}^n \setminus \{0\}, \mathbb{R})$, $W(x) > 0$ for all $x \neq 0$ and W is ν-homogeneous of degree $\mu + \kappa$;*
3. *$\max_{v\in F(x)} d_x V v \leq -W(x)$ for all $x \neq 0$.*

2.3 Homogenization of a Differential Inclusion

The following definition extends the notion of homogenization to DI.

Definition 2.7 Let F be a set-valued map and ν be an Euler vector field.

- The set-valued map $H : \mathbb{R}^n \to \mathfrak{H}(\mathbb{R}^n)$ is the ν-homogenization of F at the origin if $H \neq \{0\}$ and if there exists $\kappa \in \mathbb{R}$ such that:

$$\mathscr{H}_\kappa^s(F)(x) \underset{s\to-\infty}{\overset{CUC}{\longrightarrow}} H(x) \qquad \forall x \in \mathbb{R}^n. \tag{2.5}$$

- The set-valued map $H : \mathbb{R}^n \to \mathfrak{H}(\mathbb{R}^n)$ is the ν-homogenization of F at infinity if $H \neq \{0\}$ and there exists $\kappa \in \mathbb{R}$ such that:

$$\mathscr{H}_\kappa^s(F)(x) \underset{s\to+\infty}{\overset{CUC}{\longrightarrow}} H(x) \qquad \forall x \in \mathbb{R}^n. \tag{2.6}$$

Proposition 2.3 *Let F be a set-valued map, ν be an Euler vector field and H be the ν-homogenization of F at the origin (resp. at infinity). The following properties hold:*

1. *H is unique;*
2. *H is ν-homogeneous;*
3. *If the standard assumptions hold for F, they hold for H.*

Proof We will only give the proofs in the case of homogenization at the origin, the case of homogenization at infinity being similar.

By uniqueness of the limit, for a given $\kappa \in \mathbb{R}$, the possible limit of $\mathscr{H}_\kappa^s(F)$ is unique. Assume now that there exists a degree $\mu \neq \kappa$ such that $\mathscr{H}_\mu^s(F)$ converges to $\tilde{H}$. We will consider two cases.

If $\mu > \kappa$, we have:

$$\begin{aligned}\mathfrak{d}(\mathscr{H}_\mu^s(F)(x), \{0\}) &= e^{(\kappa-\mu)s}\mathfrak{d}(\mathscr{H}_\kappa^s(F)(x), \{0\}) \\ &\leq e^{(\kappa-\mu)s}[\mathfrak{d}(\mathscr{H}_\kappa^s(F)(x), H(x)) \\ &\quad +\mathfrak{d}(H(x), \{0\})],\end{aligned}$$

and therefore for all compact set $X \subset \mathbb{R}^n$:

$$\begin{aligned}\sup_{x\in X}\mathfrak{d}(\mathscr{H}_\mu^s(F)(x), \{0\}) \leq e^{(\kappa-\mu)s}[&\sup_{x\in X}\mathfrak{d}(\mathscr{H}_\kappa^s(F)(x), H(x)) \\ &+\sup_{x\in X}\mathfrak{d}(H(x), \{0\})].\end{aligned}$$

Since $\sup_{x\in X}\mathfrak{d}(H(x), \{0\})$ is finite and $\sup_{x\in X}\mathfrak{d}(\mathscr{H}_\kappa^s(F)(x), H(x))$ tends to zero when $s \to -\infty$, we conclude that $\sup_{x\in X}\mathfrak{d}(\mathscr{H}_\mu^s(F)(x), \{0\}) \to 0$, that is $\tilde{H} = \{0\}$, which is a contradiction.

If $\kappa > \mu$, consider $z \in \mathbb{R}^n$ such that $H(z) \neq \{0\}$. The application $X \in \mathfrak{H}(\mathbb{R}^n) \mapsto \sup_{x\in X} \|x\|$ is continuous, hence $\sup_{v\in\mathscr{H}_\kappa^s(F)(z)} \|v\| \to \alpha > 0$ when $s \to -\infty$ and therefore $\sup_{v\in\mathscr{H}_\mu^s(F)(z)} \|v\| = e^{(\mu-\kappa)s} \sup_{v\in\mathscr{H}_\kappa^s(F)(z)} \|v\| \to +\infty$ when $s \to -\infty$, but $\sup_{v\in\mathscr{H}_\mu^s(F)(z)} \|v\|$ converges to $\sup_{v\in\tilde{H}(z)} \|v\|$ as well and thus $\tilde{H}(z)$ is not bounded, which is a contradiction. This proves the first point.

The homogeneity of H is a consequence of the following computation:

$$\begin{aligned} H(\Phi^\sigma(x)) &= \lim_{s\to-\infty} \mathscr{H}_\kappa^s(F)(\Phi^\sigma(x)) \\ &= \lim_{s\to-\infty} e^{-\kappa s}\left(d_{\Phi^\sigma(x)}\Phi^s\right)^{-1} \cdot F(\Phi^{s+\sigma(x)}) \\ &= \lim_{s\to-\infty} e^{\kappa\sigma} d_x\Phi^\sigma e^{-\kappa(s+\sigma)}\left(d_x\Phi^{s+\sigma}\right)^{-1} \cdot F(\Phi^{s+\sigma(x)}(x)) \\ &= \lim_{u\to-\infty} e^{\kappa\sigma} d_x\Phi^\sigma e^{-\kappa(u)}\left(d_x\Phi^{u}\right)^{-1} \cdot F(\Phi^{u}(x)) \\ &= e^{\kappa\sigma} d_x\Phi^\sigma \cdot \lim_{u\to-\infty} \mathscr{H}_\kappa^u(F)(x) \\ &= e^{\kappa\sigma} d_x\Phi^\sigma \cdot H(x) \end{aligned}$$

Finally, $H(x)$ is a nonempty compact set by construction. It is well known that the convexity is preserved at the limit by the Hausdorff distance (see for instance [27]), so $H(x)$ is convex. Only the USC remains to prove.

Consider $\mathscr{V}$ an open neighborhood of $H(x)$. We can assume that $\mathscr{V}$ is bounded; if not, we replace it by $\mathscr{V} \cap \mathbb{B}(r)$ for $r > 0$ such that $H(x) \subset \mathbb{B}(r)$. Denote

$$\alpha = \inf\{d(h, v), h \in H(x), v \in \partial\mathscr{V}\} > 0.$$

We have

$$H(x)\dot{+}\mathbb{B}(\alpha) \subset \mathscr{V}.$$

By the uniform convergence, there exists s such that for all $y \in \mathbb{B}(x, 1)$, we have

$$\mathfrak{d}(H_\kappa^s(F)(y), H(y)) < \varepsilon/3.$$

In particular,

$$H(y) \subset \mathscr{H}_\kappa^s(F)(y)\dot{+}\mathbb{B}(\varepsilon/3)$$

and

$$\mathscr{H}_\kappa^s(F)(x) \subset H(x)\dot{+}\mathbb{B}(\varepsilon/3).$$

By USC of $\mathscr{H}_\kappa^s(F)$, there exists a neighborhood of x, $\mathscr{U} \subset \mathbb{B}(x, 1)$, such that for all $y \in \mathscr{U}$,

$$\mathscr{H}_\kappa^s(F)(y) \subset \mathscr{H}_\kappa^s(F)(x)\dot{+}\mathbb{B}(\varepsilon/3).$$

Hence, for all $y \in \mathscr{U}$,

$$H(y) \subset H_\kappa^s(F)(y)\dot{+}\mathbb{B}(\varepsilon/3) \subset H_\kappa^s(F)(x)\dot{+}\mathbb{B}(2\varepsilon/3) \subset H(x)\dot{+}\mathbb{B}(\varepsilon) \subset \mathscr{V}.$$

□

The Definition 2.7 allows us to build a local approximation of a given set-valued map. But we can also apply this procedure to a vector field. Denoting f a locally essentially bounded vector field with a ν-homogenization h, F the regularization of f via the Filippov's procedure and H the ν-homogenization of F, then we can naturally wonder whether $\mathscr{F}(h) = H$. The following proposition answers positively to this question.

Proposition 2.4 *Consider a locally essentially bounded vector field f with a ν-homogenization h of degree κ. Then $\mathscr{F}[f]$ admits a ν-homogenization H of degree κ and moreover $H = \mathscr{F}[h]$.*

Proof Consider a sequence of locally essentially bounded vector fields (f_k) converging to f uniformly on compact sets. Let us prove that $\mathscr{F}(f_n)$ converges to $\mathscr{F}[f]$ uniformly on compact sets.

For every compact set Y, for all $\varepsilon > 0$, there exists $N(Y) > 0$ such that for all $k \geq N(Y)$, $\sup_{y\in Y} \|f_n(y) - f(y)\| \leq \varepsilon$, that is $f_n(y) \in f(y)\dot{+}\mathbb{B}(\varepsilon)$ and $f(y) \in f_n(y)\dot{+}\mathbb{B}(\varepsilon)$.

Consider a compact set X and fix $\varepsilon > 0$. Denote $Y = X + \bar{\mathbb{B}}(1)$. For all $x \in X$ and all $\delta < 1$ we have $\mathbb{B}(x, \delta) \subset Y$. Thus for $n \geq N(Y)$:

$$\begin{aligned}
\mathscr{F}[f_n](x) &= \bigcap_{\delta>0} \bigcap_{\mathrm{N}\in\mathscr{N}} \overline{\mathrm{conv}}(f_n(y), y \in \mathbb{B}(x, \delta) \setminus \mathrm{N}) \\
&\subset \bigcap_{\delta>0} \bigcap_{\mathrm{N}\in\mathscr{N}} \overline{\mathrm{conv}}(f(y)\dot{+}\mathbb{B}(\varepsilon), y \in \mathbb{B}(x, \delta) \setminus \mathrm{N}) \\
&\subset \bigcap_{\delta>0} \bigcap_{\mathrm{N}\in\mathscr{N}} \overline{\mathrm{conv}}(f(y), y \in \mathbb{B}(x, \delta) \setminus \mathrm{N}) + \mathbb{B}(\varepsilon) \\
&\subset \mathscr{F}[f](x)\dot{+}\mathbb{B}(\varepsilon).
\end{aligned}$$

The converse inclusion $\mathscr{F}[f](x) \subset \mathscr{F}[f_n](x) + \mathbb{B}(\varepsilon)$ is obtained similarly. Finally, for $n \geq N(Y)$, for all $x \in X$, $\mathfrak{d}(\mathscr{F}[f](x), \mathscr{F}[f_n](x)) < \varepsilon$ and we get the uniform convergence.

Now, by Proposition 2.2, for all $s \in \mathbb{R}$ we have $\mathscr{H}_\kappa^s(\mathscr{F}[f]) = \mathscr{F}[\mathscr{H}_\kappa^s(f)]$. Since $\mathscr{H}_\kappa^s(f)$ is converging uniformly on compact sets to h, $\mathscr{F}[\mathscr{H}_\kappa^s(f)]$ converges to $\mathscr{F}[h]$ and hence $\mathscr{H}_\kappa^s(\mathscr{F}[f])$ converges to $\mathscr{F}[h]$. Since h is ν-homogeneous of degree κ, so is $\mathscr{F}[h]$ and then by definition $\mathscr{F}[h]$ is the ν-homogenization of $\mathscr{F}[f]$, that is $\mathscr{F}[h] = H$. □

Theorem 2.2 *Let F be a set-valued map for which the standard assumptions hold and H be its homogenization at the origin. If the origin is a GAS equilibrium of H, it is a LAS equilibrium of F. If moreover the degree of H is negative, the origin is a locally finite-time stable equilibrium of F.*

Proof Let (V, W) be a ν-homogeneous Lyapunov pair for H, with V of degree $\mu > \max\{0, -\kappa\}$. Let us denote $\mathbb{S} = \{V = 1\}$ and fix $x \in \mathbb{S}$ and $s \in \mathbb{R}$. For the homogenization of F at the origin H, we have:

$$\forall \varepsilon > 0\, \exists g(\varepsilon) \in \mathbb{R},\ \forall s \leq g(\varepsilon)\, \forall x \in \mathbb{S}$$
$$\eth(\mathscr{H}_\kappa^s(F)(x), H(x)) < \varepsilon.$$

Hence, denoting $a = \inf_{\mathbb{S}} W$ and $b = \sup_{\mathbb{S}} \|d_x V\|$, for all $s \leq g(\frac{a}{2b})$ and all $v \in F(\Phi^s(x))$, there exists $w \in H(x)$ such that $\|e^{-\kappa s}\,(d_x\Phi^s)^{-1}\,v - w\| < \frac{a}{2b}$. Therefore, for all $s \leq g(\frac{a}{2b})$ and all $v \in F(\Phi^s(x))$:

$$\begin{aligned}
d_{\Phi^s(x)}Vv &= d_{\Phi^s(x)}V(v - e^{\kappa s}d_x\Phi^s w) + d_{\Phi^s(x)}V(e^{\kappa s}d_x\Phi^s w) \\
&= e^{(\kappa+\mu)s}\Big[e^{-\mu s}d_{\Phi^s(x)}Vd_x\Phi^s\left(e^{-\kappa s}\left(d_x\Phi^s\right)^{-1}v - w\right) \\
&\quad + e^{-\mu s}d_{\Phi^s(x)}Vd_x\Phi^s w\Big] \\
&= e^{(\kappa+\mu)s}\left[d_xV\left(e^{-\kappa s}\left(d_x\Phi^s\right)^{-1}v - w\right) + d_xVw\right] \\
&\leq e^{(\kappa+\mu)s}\left[b\|e^{-\kappa s}\left(d_x\Phi^s\right)^{-1}v - w\| - a\right] \\
&\leq -\frac{a}{2}e^{(\kappa+\mu)s} = -\frac{a}{2}V(\Phi^s(x))^{\frac{\kappa+\mu}{\mu}}.
\end{aligned}$$

Thus, for all $y \neq 0$ such that $V(y) \leq e^{\kappa g(\frac{a}{2})}$, we find that for all $v \in F(y)$:

$$d_yVv \leq -\frac{a}{2}V(y)^{\frac{\kappa+\mu}{\mu}}. \tag{2.7}$$

The relation (2.7) proves that V is a local Lyapunov function for F, and then the origin is a LAS equilibrium of F. Moreover, if $\kappa < 0$ then $0 < \frac{\kappa+\mu}{\mu} < 1$. Classical techniques then show that the convergence to the origin is performed in a finite time. □

Example 2.1 Consider the following system from [19] (with the particular choice of $\varepsilon = 1/2$):

$$\begin{cases} \dot{e}_1 = e_2 - k_1 e_1^{[1/2]} - k_2 e_1 \\ \dot{e}_2 = -k_3\mathrm{sign}[e_1] - k_4 e_1 \end{cases}.$$

Taking $\nu = 2x_1 \frac{\partial}{\partial x_1} + x_2 \frac{\partial}{\partial x_2}$, we can compute the ν-homogenization of the system at the origin. A direct computation yields the following ν-homogenization of degree -1:

$$\begin{cases} \dot{e}_1 = e_2 - k_1 e_1^{[1/2]} \\ \dot{e}_2 = -k_3 \text{sign}[e_1] \end{cases}.$$

The origin is known to be globally finite-time stable for this system, and we conclude by the Theorem 2.2 that the origin is a locally finite-time stable equilibrium of the initial system.

Corollary 2.2 *Let F be a set-valued map for which the standard assumptions hold and H be its homogenization at the origin. If the origin is a GAS equilibrium of both F and H and if the degree of H is negative, then the origin is a globally finite-time stable equilibrium of F.*

Proof Theorem 2.2 yields that the origin is a locally finite-time stable equilibrium of F. Given that we also assumed the origin to be asymptotically stable, the origin is therefore globally finite-time stable.

2.4 Robustness of Homogeneous and Homogenizable Systems

In this section we consider a measurable set-valued map $F : \mathbb{R}^n \times \mathbb{R}^m \rightrightarrows \mathbb{R}^n$. We denote $F_\Delta(x) = F(x, \Delta)$. We will be interested in proving robustness properties of the system defined by:

$$\dot{x} \in F(x, \Delta), \quad x \in \mathbb{R}^n, \quad \Delta \in \mathscr{L}^\infty_{loc}(\mathbb{R}, \mathbb{R}^m). \tag{2.8}$$

2.4.1 ISS Definitions and Properties

In this chapter, we will be interested in the following stability properties [23, 25].

Definition 2.8 The system (2.8) is called *input-to-state practically stable (ISpS)*, if for any input $\Delta \in \mathscr{L}^\infty_{loc}(\mathbb{R}, \mathbb{R}^m)$ and any $x_0 \in \mathbb{R}^n$ there are some functions $\beta \in \mathscr{KL}$, $\gamma \in \mathscr{K}$ and $c \geq 0$ such that for any solution x of (2.8):

$$\|x(t)\| \leq \beta(\|x_0\|, t) + \gamma(\sup_{\tau\in[0,t]} \|\Delta(\tau)\|) + c \quad \forall t \geq 0.$$

The function γ is called *nonlinear asymptotic gain*. The system is called *ISS* if $c = 0$.

These properties have the following Lyapunov function characterizations.

Definition 2.9 A smooth function $V : \mathbb{R}^n \to \mathbb{R}_+$ is called *ISpS Lyapunov function* for the system (2.8) if for all $x \in \mathbb{R}^n$, $\Delta \in \mathbb{R}^m$ and some $r \geq 0, \alpha_1, \alpha_2, \alpha_3 \in \mathscr{K}_\infty$ and $\theta \in \mathscr{K}$:

$$\alpha_1(\|x\|) \leq V(x) \leq \alpha_2(\|x\|),$$
$$\sup_{v \in F(x,\Delta)} d_x V v \leq r + \theta(\|\Delta\|) - \alpha_3(\|x\|).$$

Such a function V is called *ISS Lyapunov function* if $r = 0$.

Note that an ISpS Lyapunov function can also satisfy the following equivalent condition for some $\alpha_4 \in \mathscr{K}_\infty$, $\chi \in \mathscr{K}$ and $\rho \geq 0$:

$$\|x\| > \chi(\|\Delta\|) + \rho \Rightarrow \sup_{v \in F(x,\Delta)} d_x V v \leq -\alpha_4(\|x\|).$$

Proposition 2.5 *If there exists an ISpS (resp. ISS) Lyapunov function for the system (2.8), then the system is ISpS (resp. ISS).*

2.4.2 ISS of Homogeneous Differential Inclusions

In the following results, we will need some assumptions on F.

Assumption 2.1 For all $\Delta \in \mathbb{R}^m$ the set-valued map F_Δ verifies the SA.

This assumption ensures that solutions of the system (2.8) exist.

Assumption 2.2 There exists a ν-homogeneous set-valued map H of degree κ verifying the SA such that

A. the origin is a GAS equilibrium of H. We denote (V, W) a ν-homogeneous Lyapunov pair for H given by Theorem 2.1.
B. for all $\varepsilon > 0$ and for all $D \geq 0$ there exists $\eta > 0$ such that for all $s \geq \eta$, for all $x \in \mathbb{S} = \{V = 1\}$ and for all $\|\Delta\| \leq D$, we have $\mathscr{H}_\kappa^s(F_\Delta)(x) \subset H(x) + \mathbb{B}(\varepsilon)$.

Following the notations used in Sect. 2.3, we denote

$$a = \inf_{\mathbb{S}} W \quad \text{and} \quad b = \sup_{\mathbb{S}} \|d_x V\|.$$

We also denote

$$h(D) = \inf \left\{ \eta \in \mathbb{R} : \forall s \geq \eta \; \forall \|\Delta\| \leq D, \; \mathscr{H}_\kappa^s(F_\Delta)(x) \subset H(x) + \mathbb{B}(\frac{a}{2b}) \right\}.$$

By Assumption 2.2 B, $h(D) < +\infty$. We allow $h(D) = -\infty$, denote $\ell = \lim_{D \to 0^+} h(D)$.

Theorem 2.3 *Under Assumptions 2.1 and 2.2, the system (2.8) is:*

ISS if $\ell = -\infty$,
ISpS if $\ell \neq -\infty$.

Remark 2.1 The following hint for a selection of H can be proposed. When F_0 is ν-homogeneous of degree μ, Assumption 2.2B gives, for $x \in \mathbb{S}$:

$$\lim_{s \to +\infty} e^{(\mu-\kappa)s} F_0(x) \subset H(x).$$

If $\mu > \kappa$, $e^{(\mu-\kappa)s} F_0(x)$ diverges when $s \to +\infty$, and if $\mu < \kappa$, we get that $0 \in H(x)$, which is a contradiction to the global asymptotic stability of H (Assumption 2.2A). Thus $\mu = \kappa$ and $F_0 \subset H$. Similarly, if F_0 admits a ν-homogeneization H_0 of degree μ, we find that $\mu = \kappa$ and $H_0 \subset H$. This remark gives us a candidate for H in some situations and it will be used in Theorem 2.4.

To prove the Theorem 2.3, we need some technical lemmas.

Lemma 2.1 *Let* $\sigma : \mathbb{R}_+ \to \mathbb{R}$ *be an increasing function such that* $\lim_{x \to 0^+} \sigma(x) = 0$. *Then there exists a class* $\mathcal{K}$ *function* $\bar{\sigma}$ *such that for all* $x \in \mathbb{R}_+$, $\sigma(x) \leq \bar{\sigma}(x)$.

Proof Let us first remark that $\sigma(x) \geq 0$ for all $x > 0$.

- For all $n \in \mathbb{N}^*$ and all $x \in [n, n+1[$, let us define:

$$\bar{\sigma}(x) = (\sigma(n+2) - \sigma(n+1) + 1)(x - n) + \sigma(n+1) + n.$$

  We find $\bar{\sigma}(n) = \sigma(n+1) + n$ and

$$\lim_{x \to n+1; x < n+1} \bar{\sigma}(x) = \sigma(n+2) + n + 1 = \bar{\sigma}(n+1).$$

  Hence $\bar{\sigma}$ is continuous on $[1, +\infty[$ and clearly strictly increasing. Moreover, for all $n \in \mathbb{N}^*$ and all $x \in [n, n+1[$:

$$\bar{\sigma}(x) \geq \bar{\sigma}(n) = \sigma(n+1) + n > \sigma(n+1) \geq \sigma(x).$$

- For all $x \in [\frac{1}{2}, 1[$, we set $\bar{\sigma}(x) = (2x-1)(\sigma(2) - \sigma(1) + \frac{1}{2}) + \sigma(1) + \frac{1}{2}$. We easily check that $\bar{\sigma}(x) \to \bar{\sigma}(1)$ when $x \to 1, x < 1$, and we find $\bar{\sigma}(\frac{1}{2}) = \sigma(1) + \frac{1}{2}$. This construction proves that $\bar{\sigma}$ is continuous and strictly increasing on $[\frac{1}{2}, +\infty[$. For $x \in [\frac{1}{2}, 1[$, we have $\bar{\sigma}(x) \geq \bar{\sigma}(\frac{1}{2}) = \sigma(1) + \frac{1}{2} > \sigma(x)$.
- For all $n \in \mathbb{N}$, $n \geq 2$, and all $x \in [\frac{1}{n+1}, \frac{1}{n}[$ we set:

$$\bar{\sigma}(x) = n(n+1)\left(x - \frac{1}{n+1}\right)$$

$$\times\left[\sigma\left(\frac{1}{n-1}\right)-\sigma\left(\frac{1}{n}\right)+\frac{1}{n(n+1)}\right]$$
$$+\sigma\left(\frac{1}{n}\right)+\frac{1}{n+1}.$$

We find $\bar{\sigma}\left(\frac{1}{n+1}\right)=\sigma\left(\frac{1}{n}\right)+\frac{1}{n+1}$ and $\bar{\sigma}(x)\to\sigma\left(\frac{1}{n-1}\right)+\frac{1}{n}$ when $x\to\frac{1}{n}$, $x<\frac{1}{n}$. Thus $\bar{\sigma}$ is continuous and strictly increasing on $]0,+\infty[$. For $x\in[\frac{1}{n+1},\frac{1}{n}[$, we have $\bar{\sigma}(x)\geq\bar{\sigma}(\frac{1}{n+1})=\sigma\left(\frac{1}{n}\right)+\frac{1}{n+1}>\sigma\left(\frac{1}{n}\right)\geq\sigma(x)$.
- The function $\bar{\sigma}$ being increasing on $]0,+\infty[$, the limit of $\bar{\sigma}$ at 0 is the limit of $\bar{\sigma}\left(\frac{1}{n+1}\right)$ when $n\to+\infty$. But $\bar{\sigma}\left(\frac{1}{n+1}\right)=\sigma\left(\frac{1}{n}\right)+\frac{1}{n+1}$ converges to zero when $n\to+\infty$.

Finally setting $\bar{\sigma}(0)=0$ proves that $\bar{\sigma}$ is a class $\mathscr{K}$ function such that $\bar{\sigma}(x)\geq\sigma(x)$ for all $x\geq 0$. □

Lemma 2.2 *Let $V:\mathbb{R}^n\to\mathbb{R}$ be a continuous positive definite ν-homogeneous function of degree $\kappa>0$. There exist σ_- and σ_+ two functions of class $\mathscr{K}$ such that for all $x\in\mathbb{R}^n$:*

$$\sigma_-(\|x\|)\leq V(x)\leq\sigma_+(\|x\|).$$

Proof Denote $\sigma_+(r)=\sup_{\|x\|\leq r}V(x)$. The function σ_+ is clearly continuous, increasing and verifies $\sigma_+(0)=0$. Let us show that σ_+ is strictly increasing. It is enough to prove that for any x_0 such that $\|x_0\|\leq r$ and $V(x_0)=\sigma_+(r)$ verifies $\|x_0\|=r$. Assume by contradiction that $\|x_0\|<r$. By continuity of the flow Φ, there exists $\varepsilon>0$ such that for all $s\in[0,\varepsilon[$, we have $\|\Phi^s(x_0)\|<r$ and thus $V(\Phi^s(x_0))\leq\sigma_+(r)=V(x_0)$. But $V(\Phi^s(x_0))=e^{\kappa s}V(x_0)>V(x_0)$ for $s>0$, which is a contradiction.

The function σ_- is defined by $\sigma_-^{-1}(r)=\sup_{V(x)\leq r}\|x\|$. We similarly prove that $\sigma_-^{-1}\in\mathscr{K}$.

Finally,

$$V(x)\leq\sup_{\|y\|\leq\|x\|}V(y)=\sigma_+(\|x\|)$$

and

$$\|x\|\leq\sup_{V(y)\leq V(x)}\|y\|=\sigma_-^{-1}(V(x)),$$

that is $V(x)\geq\sigma_-(\|x\|)$. □

We can now prove Theorem 2.3.

Proof (*of Theorem* 2.3) Let $x\in\mathbb{S}$, $s\in\mathbb{R}$ and $v\in F_\Delta(\Phi^s(x))$ be fixed. We have $e^{-\kappa s}(d_x\Phi^s)^{-1}v\in\mathscr{H}_\kappa^s(F_\Delta)(x)$, thus for $s\geq h(\|\Delta\|)$ and by Assumption 2.2 there exists $w\in H(x)$ such that:

$$\|e^{-\kappa s}\left(d_x\Phi^s\right)^{-1}v - w\| < \frac{a}{2b}.$$

Hence for $s \geq h(\|\Delta\|)$:

$$\begin{aligned}
d_{\Phi^s(x)}Vv &= e^{(\kappa+\mu)s}d_xVe^{-\kappa s}\left(d_x\Phi^s\right)^{-1}v \\
&= e^{(\kappa+\mu)s}\left[d_xVw + d_xV\left(e^{-\kappa s}\left(d_x\Phi^s\right)^{-1}v - w\right)\right] \\
&\leq e^{(\kappa+\mu)s}\left[-a + b\|e^{-\kappa s}\left(d_x\Phi^s\right)^{-1}v - w\|\right] \\
&\leq -\frac{a}{2}e^{(\kappa+\mu)s} = -\frac{a}{2}V(\Phi^s(x))^{\frac{\kappa+\mu}{\mu}}.
\end{aligned}$$

Finally, for all $y \in \mathbb{R}^n$ such that $V(y) \geq e^{\mu h(\|\Delta\|)}$ we have:

$$\sup_{v\in F_d(y)} d_yVv \leq -\frac{a}{2}V(y)^{\frac{\kappa+\mu}{\mu}}. \tag{2.9}$$

Let us define $\varpi = e^{\mu\ell}$ if $\ell \neq -\infty$, and $\varpi = 0$ else, and then $\sigma(D) = e^{\mu h(D)} - \varpi$. Since h is clearly increasing and $\mu > 0$, σ is increasing and positive. Moreover $\lim_{D\to 0^+}\sigma(D) = 0$. By Lemma 2.1, there exists $\bar{\sigma} \in \mathcal{K}$ such that $\bar{\sigma}(D) \geq \sigma(D)$. Therefore, for all $y \in \mathbb{R}^n$ such that $V(y) \geq \bar{\sigma}(\|\Delta\|) + \varpi$, in equation (2.9) holds. Denoting now $\gamma(s) = \sigma_-^{-1}(\bar{\sigma}(s) + \varpi) - \sigma_-^{-1}(\varpi)$, where σ_- is the class $\mathcal{K}$ function given by Lemma 2.2, we have that if $\|y\| \geq \gamma(\|\Delta\|) + \sigma_-^{-1}(\varpi)$ then inequality (2.9) holds. We conclude by standard arguments on ISS/ISpS Lyapunov functions. □

Inequality (2.9) gives one straightforward Corollary.

Corollary 2.3 *Under Assumptions 2.1 and 2.2, if $\kappa > 0$, then the system (2.8) is ISS. If moreover we denote γ the asymptotic gain, then any trajectory of the system converges to the ball of radius $\gamma(\sup_{t\geq 0}\|\Delta(t)\|)$ in a* uniform finite-time.

The following Corollary of Theorem 2.3 shows how to use it when dealing with more concrete systems.

Corollary 2.4 *Consider the system $\dot{x} \in F_0(x)\dot{+}\mathbb{B}(\|\Delta\|)$. Assume that:*

1. *F_0 verifies the SA;*
2. *the origin is a GAS equilibrium of F_0;*
3. *there exists a* linear *Euler vector field $\nu(x) = Ax$ such that F_0 is ν-homogeneous of degree κ;*
4. *if ρ denotes the smallest real part of the eigenvalues of A, $\kappa + \rho > 0$.*

Then the system $\dot{x} \in F_0(x)\dot{+}\mathbb{B}(\|\Delta\|)$ is ISS.

Proof Assumption 1 and 2A are clearly verified, taking $H = F_0$. Noting that $\mathcal{H}_\kappa^s(F_\Delta)(x) = F_0(x) + \exp((-\kappa I - A)s)\cdot\mathbb{B}(\|\Delta\|)$, we see that Assumption 2B holds because the matrix $-\kappa I - A$ is Hurwitz ($\rho + \kappa > 0$). Finally, $\lim_{D\to 0^+} h(D) = -\infty$ follows from classical matrices considerations.

Theorem 2.4 *Consider a set-valued map* $F : \mathbb{R}^n \times \mathbb{R}^m \rightrightarrows \mathbb{R}^n$ *verifying Assumption 2.1. Assume moreover that the following hypothesis hold:*

1. *The origin is a GAS equilibrium of* F_0. *We denote* (V, W) *a* ν*-homogeneous Lyapunov pair for* F_0 *given by Theorem 2.1.*
2. *There exists an Euler vector field* $\tilde{\nu}$ *on* $\mathbb{R}^m$, *which flow is denoted* $\tilde{\Phi}$, *such that:*

$$F(\Phi^s(x), \tilde{\Phi}^s(\Delta)) = e^{\kappa s} d_x \Phi^s F(x, \Delta).$$

3. *There exists a function* $\sigma \in \mathcal{K}$ *such that for all* $x \in \mathbb{S} = \{V = 1\}$ *we have* $F_\Delta(x) \subset F_0(x) \dot{+} \mathbb{B}(\sigma(\|\Delta\|))$.

Then the system (2.8) is ISS.

Proof Let us show first that Assumption 2.2 holds for $H = F_0$. The point Assumption 2.2A is given by the point 1 of the hypothesis of this theorem.

By the point 2 of the hypothesis, $\mathcal{H}_\kappa^s(F_\Delta)(x) = F_{\tilde{\Phi}^{-s}(\Delta)}(x)$. Consider N a continuous positive definite $\tilde{\nu}$-homogeneous function of degree 1 on $\mathbb{R}^m$ and denote $\eta(\varepsilon, D) = \ln\left(\frac{\sigma_+(D)}{\sigma_- \circ \sigma^{-1}(\varepsilon)}\right)$, where the functions σ_- and σ_+ are given by Lemma 2.2 with respect to the function N. Consider $s \geq \eta(\varepsilon, D)$. Then $e^{-s}\sigma_+(D) \leq \sigma_- \circ \sigma^{-1}(\varepsilon)$ and for all $\|\Delta\| \leq D$ we have

$$\|\tilde{\Phi}^{-s}(\Delta)\| \leq \sigma_-^{-1}(N(\tilde{\Phi}^{-s}(\Delta))) \leq \sigma_-^{-1}(e^{-s}N(\Delta)) \leq \sigma_-^{-1}(e^{-s}\sigma_+(\|\Delta\|)) \leq \sigma^{-1}(\varepsilon).$$

Hence for all $s \geq \eta(\varepsilon, D)$, for all $x \in \mathbb{S}$ and all $\|\Delta\| \leq D$ we have

$$\mathcal{H}_\kappa^s(F_\Delta)(x) = F_{\tilde{\Phi}^{-s}(\Delta)}(x) \subset F_0(x) \dot{+} \mathbb{B}(\varepsilon)$$

by hypothesis 3. Therefore Assumption 2.2 holds for $H = F_0$. Since $h(D) \leq \eta(\frac{a}{2b}, D)$ and $\lim_{D \to 0^+} \eta(\frac{a}{2b}, D) = -\infty$, we conclude by Theorem 2.3.

Example 2.2 Consider the following disturbed system:

$$\begin{cases} \dot{x}_1 = & x_2 + \Delta \\ \dot{x}_2 = & -k_1 \text{sign}[x_1] - k_2 \text{sign}[x_2] \end{cases}$$

where $k_1 > k_2 > 0$ are fixed gains. When $d = 0$, it is well-known that the system is ν-homogeneous of degree $\kappa = -1$ with $\nu = 2x_1 \frac{\partial}{\partial x_1} + x_2 \frac{\partial}{\partial x_2}$ and GAS, that is hypothesis 1 holds. Taking $\tilde{\nu} = \Delta \frac{\partial}{\partial \Delta}$, we see that hypothesis 2 holds. Finally the hypothesis 3 also holds with $\sigma(D) = D$ and the system is ISS by Theorem 2.4.

Corollary 2.5 *Let* $f : \mathbb{R}^n \times \mathbb{R}^m \to \mathbb{R}^n$ *be a continuous vector field. Assume that there exists an Euler vector field* $\tilde{\nu}$ *on* $\mathbb{R}^m$, *which flow is denoted* $\tilde{\Phi}$, *such that:*

$$f(\Phi^s(x), \tilde{\Phi}^s(\Delta)) = e^{\kappa s} d_x \Phi^s f(x, \Delta),$$

and assume moreover that the origin is a GAS equilibrium of f_0, *where* $f_0(x) = f(x, 0)$. *Then the system* $\dot{x} = f(x, \Delta)$ *is ISS.*

Proof We take $F_\Delta(x) = \{f_\Delta(x)\}$. The hypothesis 1 and 2 of Theorem 2.4 are clearly satisfied. The continuity of f and the compactness of $\mathbb{S}$ give that the function $\sigma(\Delta) = \sup_{x\in\mathbb{S}, \|\Delta\|\leq D} \|f_\Delta(x) - f_0(x)\|$ belongs to class $\mathcal{K}$, which gives in turn hypothesis 3 and concludes the proof.

2.5 Conclusion

In this chapter, we achieved two objectives. First, we introduced homogenization for DI. We proved that this notion is consistent with the Filippov's procedure and that the local stability is inherited by a system which has a GAS homogenization. Second, we applied homogeneity and homogenization techniques to prove ISS and ISpS properties of systems defined by DI. All these results were presented using geometric homogeneity.

In the future, we plan to use these results and techniques for designing SMC and getting a good understanding of the associated robustness properties of such systems. In particular, controlling the asymptotic gain could give a way of reducing the chattering effect.

Acknowledgements This chapter is supported by ANR Finite4SoS (ANR 15 CE23 0007).

References

1. Aleksandrov, A.Y., Kosov, A.A., Platonov, A.V.: On the asymptotic stability of switched homogeneous systems. Syst. Control Lett. **61**(1), 127–133 (2012)
2. Andrieu, V., Praly, L., Astolfi, A.: Homogeneous approximation, recursive observer design, and output feedback. SIAM J. Control Optim. **47**(4), 1814–1850 (2008)
3. Aubin, J., Cellina, A.: Differential Inclusions. Grundlehren der Math. Wissenschaften 264. Springer, Berlin (1984)
4. Bernuau, E., Efimov, D., Perruquetti, W., Polyakov, A.: On an extension of homogeneity notion for differential inclusions. In: Proceedings of the European Control Conference (ECC), pp. 2204–2209 (2013)
5. Bernuau, E., Polyakov, A., Efimov, D., Perruquetti, W.: Robustness of finite-time stability property for sliding modes. In: Proceedings of 5th IFAC Symposium on System Structure and Control, pp. 391–396 (2013)
6. Bhat, S.P., Bernstein, D.S.: Geometric homogeneity with applications to finite-time stability. Math. Control Signals Syst. **17**, 101–127 (2005)
7. Bokharaie, V.S., Mason, O., Verwoerd, M.: D-stability and delay-independent stability of homogeneous cooperative systems. IEEE Trans. Aut. Contr. **55**(12), 2882–2885 (2010)
8. Doyle, J., Francis, B., Tannenbaum, A.: Feedback Control Systems. Mac Millan Publishing Co. (1992)
9. Efimov, D., Perruquetti, W.: Oscillations conditions in homogenous systems. In: Proceedings of IFAC NOLCOS Symposium, pp. 1379–1384 (2010)

10. Filippov, A.F.: Differential Equations with Discontinuous Righthand Sides. Kluwer Academic (1988)
11. Grüne, L.: Homogeneous state feedback stabilization of homogeneous systems. SIAM J. Control Optim. **38**(4), 1288–1314 (2000)
12. Hermes, H.: Homogeneous feedback controls for homogeneous systems. Syst. Control Lett. **24**, 7–11 (1995)
13. Hill, D.J., Moylan, P.J.: Dissipative dynamical systems: basic input-output and state properties. J. Frankl. Inst. **309**(5), 327–357 (1980)
14. Hong, Y.: H_∞ control, stabilization, and input-output stability of nonlinear systems with homogeneous properties. Automatica **37**(7), 819–829 (2001)
15. Kawski, M.: Nilpotent Lie algebras of vectorfields. J. Reine Sngew. Math. **388**, 1–17 (1988)
16. Kawski, M.: Families of dilations and asymptotic stability. Birkhäuser, Boston **8**, 285–294 (1991)
17. Levant, A.: Homogeneity approach to high-order sliding mode design. Automatica **41**(5), 823–830 (2005)
18. Orlov, Y.: Finite time stability and robust control synthesis of uncertain switched systems. SIAM J. Control Optim. **43**(4), 1253–1271 (2005)
19. Orlov, Y., Aoustin, Y., Chevallereau, C.: Finite time stabilization of a perturbed double integrator - Part I: continuous sliding mode-based output feedback synthesis. IEEE Trans. Aut. Contr. **56**(3), 614–618 (2011)
20. Rosier, L.: Homogeneous Lyapunov function for homogeneous continuous vector field. Syst. Control Lett. **19**, 467–473 (1992)
21. Ryan, E.P.: Universal stabilization of a class of nonlinear systems with homogeneous vector fields. Syst. Control Lett. **26**, 177–184 (1995)
22. Schaft, A.: L_2-gain and Passivity Techniques in Nonlinear Control. Springer, London Ltd (1996)
23. Sontag, E.D.: Smooth stabilization implies coprime factorization. IEEE Trans. Aut. Control. **34**(4), 435–443 (1989)
24. Sontag, E.D.: The ISS Philosophy as a Unifying Framework for Stability-like Behavior. Springer, London (2001)
25. Sontag, E.D., Wang, Y.: On characterizations of the input-to-state stability property. Syst. Control Lett. **24**(5), 351–359 (1995)
26. Vidyasagar, M.: Input-output Analysis of Large-scale Interconnected Systems. Springer, Berlin (1981)
27. Wijsman, R.: Convergence of sequences of convex sets, cones and functions. II. Trans. Amer. Math. Soc. **123**, 32–45 (1966)
28. Willems, J.C.: Dissipative dynamical systems. I. General theory. Arch. Rational Mech. Anal. **45**, 321–351 (1972)
29. Zubov, V.I.: On ordinary differential equations with generalized homogeneous right-hand sides. Izvestiya vuzov, Matematika. **1**(2), 80–88 (1958)

Chapter 3
Stochastic Sliding Mode Control and State Estimation

Alex S. Poznyak

3.1 Stochastic Sliding Mide Control: Basic Features and Notions

3.1.1 Brief Review

The classical publications on SMC (see [1, 2]) have been devotes to the stability analysis of a class of nonlinear discontinuous feedback systems containing *bounded deterministic uncertainties*.

On our opinion, the paper [3] (which has also been discussed later in Chap. 14 (4) of [2]) may be considered as the first publication dealing with *stochastic* SMC, more exactly, with the filtering problem of linear systems with time-varying parameters. Instead of the residual term (the difference between the measured output and its estimate) as in the Kalman filter this paper suggested to use the *sign*-function of this residual multiplied by the scalar gain matrix proportional to ε^{-1} (ε is a small positive parameter). It was shown that for $\varepsilon \to 0$, that is, increasing the gain matrix, the behavior of such sliding-mode observer became close (in probability) to one of the Kalman filter which is optimal for this problem in mean-square sense.

In [4] the SMC design for a class of nonlinear Itô-type stochastic systems with actuator nonlinearities and possible *time-delay terms* is considered. The control input may contain both *sector nonlinearities* and *dead-zones*. Unknown nonlinear functions are not required to satisfy the matching condition. An integral-type sliding surface is constructed, and, the reachability of the specified sliding surface is ensured by the proposed SMC law. Sufficient conditions for the globally asymptotic stability

A.S. Poznyak (✉)
Automatic Control Department, CINVESTAV-IPN, Mexico, D.F., Mexico
e-mail: apoznyak@ctrl.cinvestav.mx

S. Li et al. (eds.), *Advances in Variable Structure Systems and Sliding Mode Control—Theory and Applications*, Studies in Systems, Decision and Control 115, DOI 10.1007/978-3-319-62896-7_3

(in probability) of the sliding motion are derived via linear matrix inequality. The paper [5] investigates the design of reliable SMC for uncertain stochastic systems with possible occurrence of actuator faults. A SMC law with a fixed gain is firstly designed. Then, both the reachability of the specified sliding surface and the stability of sliding mode dynamics are analyzed in the presence of actuator faults. Because the fixed reliable control may bring more conservativeness, an *adaptive reliable SMC law* is further proposed in this work. By means of the on-line estimation of effectivenessloss values of faulty actuators, SMC gain is adoptively updated to compensate the effects of actuator faults. In both papers the external noise is *multiplicative* one.

The paper [6] addresses the problem of H^{∞} *control for a class of uncertain stochastic systems with Markovian switching and time-varying delays*. The considered model is subject to time-varying norm-bounded parameter and unknown nonlinear function in the state. An integral sliding surface corresponding to every mode is first constructed, and the given sliding mode controller concerning the transition rates of modes can deal with the effect of Markovian switching. The synthesized sliding mode control law ensures the reachability of the sliding surface for corresponding subsystems and the global stochastic stability of the sliding mode dynamics.

In [7] there is studied the sliding mode control (SMC) of nonlinear singular stochastic systems with Markovian switching. An integral sliding surface function is designed, and the resulting sliding mode dynamics is a full-order Markovian jump singular stochastic system. By introducing some specified matrices, a new sufficient condition is proposed in terms of strict linear matrix inequality (LMI), which guarantees the stochastic stability of the sliding mode dynamics. Then, a SMC law is synthesized for reaching motion. Moreover, when there exists an external disturbance, the $\mathscr{L}_2$ disturbance attenuation performance is analyzed for the sliding mode dynamics. Some related sufficient conditions are also established. Here again the external white noise is supposed to be multiplicative.

The paper [8] addresses the optimal controller problem for a *polynomial system over linear observations* with respect to different Bolza–Meyer criteria, where the integral control and state energy terms are quadratic and the non-integral term is of the first degree, or the control energy term is quadratic and the state energy terms are of the first degree. The optimal solutions are obtained as sliding mode controllers, each consisting of a sliding mode filter and a sliding mode regulator, whereas the conventional feedback polynomial-quadratic controller fails to provide a causal solution.

In this section

- we consider the additive stochastic noise effect;
- it is shown that the special design of the sliding mode gain parameters (in fact, linearly depending on the norm of the function $s(x)$, defining the sliding surface) may guarantee the exponential convergence of the averaged squared-norm of the state vector to a μ-zone (i.e., μ-neighborhood) around the sliding surface $s(x) = 0$; such convergence is suggested to be referred to as μMS-convergence;

- this μ-zone is exactly defined by the diffusion parameter σ in the model description: smaller σ - smaller μ-zone of the means square convergence; this zone is also inversely depending on the gain parameter.

3.1.2 SMC for a Simplest Scalar Itô Dynamic Model

3.1.2.1 Model Description

- Consider a filtered probability space $(\Omega, F, \{H_t\}_{t\geq 0}, P)$ where σ − algebra H_0 contains all the P-null sets from F, the filtration $\{H_t\}_{t\geq 0}$ is right continuous, that is, $H_{t+} := \cap_{s>t} H_s = H_t$. Let $\{(W_t, F_t)\}_{t\geq 0}$, be scalar standard Brownian motion, $F_t \subset F$. $H_t \subset F$ stands for the smallest σ − algebra containing σ −algebras $F_t := \sigma(\mathscr{F}_s : 0 \leq s < t)$.
- In this probability space consider the simplest scalar controllable stochastic model given in the Itô form:

$$dx_t \overset{a.s.}{=} [u_t + f(x_t, t)]\, dt + \sigma(x_t, t)\, dW_t, \ \ x_0 \overset{a.s.}{=} \bar{x} \tag{3.1}$$

- $x_t = x_t(\omega) \in R$ $(\omega \in \Omega)$ is the system state at time $t \geq 0$,
- $u_t \in R$ is a control ($\mathscr{F}_t$-measurable random variable),
- $f(x_t, t)$ is $\mathscr{F}_t$-measurable (but may be uncertain (unavailable on-line) term satisfying $|f(x_t, t)| \overset{a.s}{\leq} L$,
- W_t - a standard Wiener process, $\sigma(x_t, t) > 0$ is also $\mathscr{F}_t$-measurable.

3.1.2.2 Discontinuous SM-Controller

Select the sliding surface $s(x)$ as

$$s(x) = x = 0$$

and the control as

$$\left.\begin{aligned} u_t &= -Kx_t - k(x_t, t)\ \operatorname{sign}(s(x_t)) \\ &= -Kx_t - k(x_t, t)\ \operatorname{sign}(x_t) \\ K \geq 0,\ &k(x, t) > 0 \text{ for all } t \geq 0 \text{ and } x \in \mathbb{R} \end{aligned}\right\} \tag{3.2}$$

In the classical formulation (see [2]) the gain parameter $k(x, t)$ is selected as a positive constant, that is,

$$k(x, t) = k > 0$$

and

$$K = 0$$

Here

$$\operatorname{sign}(x) := \begin{cases} -1 & \text{if } x < 0 \\ \in [-1, 1] & \text{if } x = 0 \\ 1 & \text{if } x > 0 \end{cases} \tag{3.3}$$

3.1 Our aim is to design $k(x, t)$ in (3.2) which guarantees the stabilization of the system (3.1) in a neighborhood of the origin in some probabilistic sense.

3.1.2.3 Close-Loop Dynamics

Model of the Process as a Stochastic Differential Inclusion

Take $K = 0$. Substitution (3.2) in (3.1) leads to the following description of the closed-loop dynamics:

$$\mathrm{d}x_t = \left[-k(x, t)\ \operatorname{sign}(x_t) + f(x_t, t)\right] \mathrm{d}t + \sigma(x_t, t)\, \mathrm{d}W_t,\ x_0 \overset{a.s.}{=} \bar{x} \tag{3.4}$$

Here $\left[-k(x, t)\ \operatorname{sign}(x_t) + f(x_t, t)\right]$ is the *drift set-valued function* and $\sigma = \sigma(x_t, t)$ is the *diffusion parameter*. In fact, (3.4) is an *stochastic differential inclusion* [9] since the function $\operatorname{sign}(x_t)$ (see Fig. 3.1) is a set-valued *upper semicontinuos mapping*.

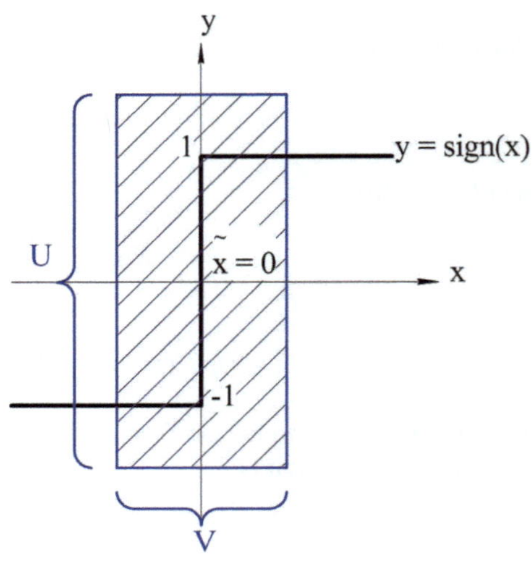

Fig. 3.1 The upper semicontinuos function sign(x)

More exactly, the set-value function $F : \mathbb{R}^n \to \mathbb{R}$ is said to be *upper semi-continuos* at the point $\tilde{x} \in \mathbb{R}^n$ if for every set $U \subset \mathbb{R}$ such that $F(\tilde{x}) \subset U$, there exists a set $V \subset \mathbb{R}^n$ such that $\tilde{x} \in V$ and $F(x) \subset U$ for every $x \in V$.

The relation (3.4) may be represented in the integral form, as

$$x_t - x_s \overset{a.s.}{\in} \int_{\tau=s}^{t} F(x_\tau, \tau)\, d\tau + \int_{\tau=0}^{t} G(x_\tau, \tau)\, dW_\tau \tag{3.5}$$

with

$$\begin{aligned} F(x_\tau, \tau) &:= -k(x_\tau, \tau)\ \mathrm{sign}(x_\tau) + f(x_\tau, \tau) \\ G(x_\tau, \tau) &= \sigma(x_\tau, \tau) \end{aligned} \tag{3.6}$$

Here $\int_{\tau=s}^{t} F(x_\tau, \tau)\, d\tau$ and $\int_{\tau=0}^{t} G(x_\tau, \tau)\, dW_\tau$ denote the Aumann [10] and Itô set-valued integrals [11] of the set-valued processes $F \circ x := (F(x_\tau, \tau))_{\tau \ge 0}$ and $G \circ x := (G(x_\tau, \tau))_{\tau \ge 0}$, respectively.

3.1.2.4 On the Extension of the Filippov's Theorem to the Class of Stochastic Differential Inclusions

The conditions of *existence and uniqueness* of the strong *(or weak)* solution x_t of (3.5) may be found in [9, 11–13]. To define exactly what we mean when talk about a solution of a stochastic differential inclusion we need the following definitions.

Strong Solution of the Differential Inclusion

Definition 3.1 (*Strong solution of* (3.5) [12]) A stochastic process $X_t(\omega)$, defined on $(\Omega, \mathcal{F}, \{\mathcal{H}_t\}_{t\ge 0}, \mathrm{P})$, is referred to as a **strong solution** of the differential inclusion (3.5) if

(1) there exist $\mathcal{F}_t$-measurable (predictable) stochastic processes $f_t(\omega)$ and $g_t(\omega)$, also defined on $(\Omega, \mathcal{F}, \{\mathcal{H}_t\}_{t\ge 0}, \mathrm{P})$, such that for all $t \in [0, T]$

$$f_t(\omega) \overset{a.s.}{\in} F(x_t(\omega), t), \ g_t(\omega) \overset{a.s.}{\in} G(x_t(\omega), t)$$

$$P\omega \in \Omega : \int_{\tau=0}^{t} \left\{ \|F(x_\tau(\omega), \tau)\|^2 + \|G(x_\tau(\omega), \tau)\|^2 \right\} d\tau < \infty = 1$$

(2) for all $t \in [0, T]$

$$X_t(\omega) \overset{a.s.}{=} \bar{x}(\omega) + \int_{\tau=0}^{t} f_\tau(\omega)\,\mathrm{d}\tau + \int_{\tau=0}^{t} g_\tau(\omega)\,\mathrm{d}W_\tau \tag{3.7}$$

(3) $X_t(\omega)$ is continuos for almost all trajectories $\omega \in \Omega$.[1]

Remark that the stochastic processes

$$\{X_t(\omega)\}_{t\geq 0}, \{f_t(\omega)\}_{t\geq 0} \text{ and } \{g_t(\omega)\}_{t\geq 0}$$

are uniquely determined by (3.7) if take into account that two stochastic processes $\{X_t(\omega)\}_{t\geq 0}$ and $\{Y_t(\omega)\}_{t\geq 0}$ are indistinguishable if

$$\mathrm{P}\{X_t(\omega) = Y_t(\omega)\} = 1$$

for all $t \in [0, T]$.

Remark 3.1 The comparisons between weak and strong solutions may be found in [14, 15].

$\mathcal{H}$-Conditions

Introduce some notions which will be useful in the following considerations.

Definition 3.2 Following [10], we say the set-valued mappings $F : \mathbb{R}^n \times [0, T] \to \mathrm{Cl}(\mathbb{R})$ and $G : \mathbb{R}^n \times [0, T] \to \mathrm{Cl}(\mathbb{R})$ (where $\mathrm{Cl}(\mathbb{R})$ is the space of all nonempty closed subsets of $\mathbb{R}^n$ and $T < \infty$) satisfy the $\mathcal{H}$**-conditions** if

(i) F and G are $\mathcal{F}_t$-measurable (predictable),
(ii) F and G are uniformely square integrably bounded, namely, for any $0 \leq t < T$

$$P\left\{\omega \in \Omega : \int_{\tau=0}^{t} \left(\|F(x_\tau, \tau)\|^2 + \|G(x_\tau, \tau)\|^2\right) \mathrm{d}\tau < \infty\right\} = 1$$

Definition 3.3 (*Hausdorff–Lipschitz continuity*) A set-value mapping $F(\cdot, t)$ is said to be $\mathcal{HL}$**-continuos** for almost all fixed $t \in [0, T]$ if there exists $l \in \mathbb{L}^2([0, T], \mathbb{R}^+)$ such that

$$h(F(x,t), F(z,t)) \leq l(t)\|x - z\| \tag{3.8}$$

for almost all $t \in [0, T]$ and all $x, z \in \mathbb{R}^n$. Here the **Hausdorff distance** between two sets A and B is defined as

---

[1]By the Kolmogorov's theorem (see, for example, [11]) the continuos modification of a stochastic process $\{X_t(\omega)\}_{t\geq 0}$ always exists (under the corresponding bounded moments) if $E\left\{\|X_t - X_s\|^\alpha \leq \gamma |t-s|^{1+\beta}\right\}$ for all $s, t \in [0, T]$ and some positive α, β and γ. The process (3.7) obviously satisfies this condition.

$$h(A,B) := \inf_{\varepsilon>0} \{\varepsilon : A \subset V_\varepsilon(B) \text{ and} B \subset V_\varepsilon(A)\}$$

$$V_\varepsilon(C) := \{x \in \mathbb{R}^n : \text{dist}(x,C) \le \varepsilon\} \tag{3.9}$$

denotes the ε-neighborhood of $C \subset \mathbb{R}^n$

Obviously that a Lipschitz continuity implies $\mathscr{H}\mathscr{L}$- continuity, but not inverse.
The Existance of a Strong Solution

The following theorem state the sufficient conditions of the existance of a strong solution of (3.5).

Theorem 3.1 (Chap. 4, Theorem 1.1 in [11]) *If F and G in (3.5) satisfy $\mathscr{H}$-conditions (see Appendix) and $F(\cdot,t)$ and $G(\cdot,t)$ are $\mathscr{H}\mathscr{L}$* - **continuos** *with the Lipschitz constant $l \in \mathbb{L}^2([0,T],\mathbb{R}^+)$ such that*

$$\left(\sqrt{T}+1\right)\sqrt{\int_{t=0}^{T} l^2(t)\,\mathrm{d}t} < 1$$

then the set $X_{\bar{x}}(F,G,W,)$ of strong solutions of the differential inclusion (3.5) on $[0,T]$ with the initial condition $\bar{x}(\omega)$ is non-empty, i.e., $X_{\bar{x}}(F,G,W,) \ne \emptyset$.

Remark 3.2 It is important to note that the set-valued mapping

$$F(x,t) := -k(x,t)\ \text{sign}(x) + f(x,t)$$

is $\mathscr{H}\mathscr{L}$-continuos if

- $f(x,t)$ is $\mathscr{H}\mathscr{L}$-continuos,
- $k(x,t)\ \text{sign}(x)$ is $\mathscr{H}\mathscr{L}$-continuos.

In spite of the fact that the set-valued mapping $\text{sign}(x)$ (3.3) is not $\mathscr{H}\mathscr{L}$-continuos, the product $k(x,t)\ \text{sign}(x)$ may satisfy this condition by the adequate selection of the gain parameter $k(x,t)$ which should be obligatory $\mathscr{H}\mathscr{L}$-continuos. So, in some sense, $k(x,t)$ may be treated as a "**stochastic corrector**" providing the existence of a strong solution for the considered differential inclusion (3.5).

3.1.2.5 Deterministic Dynamics

In the deterministic case ($\sigma(x,t)=0$) for $k(x,t)=k=\text{const}>0$ the Eq. (3.4) becomes

$$\dot{x}_t = -k\text{sign}(x_t) + f(x_t,t)$$

Its solution is intended in the Filippov's sense [16]. Then for the Lyapunov function $V_t := \frac{1}{2}x_t^2$ we have

$$\dot{V}_t = x_t \dot{x}_t = x_t \left[-k\text{sign}(x_t) + f(x_t, t) \right] =$$

$$-k|x_t| + x_t f(x_t, t) \leq -k|x_t| + |x_t| L =$$

$$-\sqrt{2}(k - L)\sqrt{V_t}$$

implying under $k = L + \rho$, $\rho > 0$

$$\frac{dV_t}{\sqrt{V_t}} \leq -\sqrt{2}\rho \mathrm{d}t, \ 0 \leq \sqrt{V_t} \leq \sqrt{V_0} - \frac{\rho}{\sqrt{2}} t$$

$$V_t = 0 \text{ for all } t \geq t_{reach} = \frac{\sqrt{V_0}}{\rho} = \frac{1}{k - L}|x_0|$$

3.1.2.6 Stochastic Dynamics and the Itô Formula

In the stochastic case ($\sigma(x, t) := \sigma = \text{const} > 0$), corresponding the additive noise affect, for any twice differentiable function $V : \mathbb{R} \to \mathbb{R}$ the following Itô formula takes place

$$\mathrm{d}V(x_t) = V'(x_t)\,\mathrm{d}x_t + \underbrace{\frac{1}{2}\sigma^2 V''(x_t)\,\mathrm{d}t}_{\text{the Itô term}}$$

Particularly, for $V(x) := \frac{1}{2}x^2$ on the trajectories of (3.4) we have

$$\mathrm{d}V(x_t) = \left(x_t \left[-k(x, t)\ \text{sign}(x_t) + f(x_t, t) \right] + \frac{1}{2}\sigma^2 \right) \mathrm{d}t + \sigma x_t \mathrm{d}W_t$$

or, in the integral form,

$$V(x_{t+\Delta t}) - V(x_t) \overset{a.s.}{=}$$

$$\int_{\tau=t}^{t+\Delta t} \left(x_\tau \left[-k(x, \tau)\ \text{sign}(x_\tau) + f(x_\tau, \tau) \right] + \tfrac{1}{2}\sigma^2 \right) \mathrm{d}\tau \tag{3.10}$$

$$+\sigma \int_{\tau=t}^{t+\Delta t} x_\tau \mathrm{d}W_\tau$$

3.1.2.7 Storage Function and Its Evaluation

By the Itô integral property

$$E\left\{\int_{\tau=t}^{t+\Delta t} x_\tau \mathrm{d}W_\tau\right\} = 0$$

and, applying the mathematical expectation operator $E\{\cdot\}$ to both sides of (3.10), for $V_t := E\{V(x_t)\}$ we obtain

$$V_{t+\Delta t} - V_t = \int_{\tau=t}^{t+\Delta t} E\left\{x_\tau \left[-k(x,\tau)\ \mathrm{sign}(x_\tau) + f(x_\tau,\tau)\right] + \frac{1}{2}\sigma^2\right\} \mathrm{d}\tau \tag{3.11}$$

Dividing both sides in (3.11) by Δt and taking $\Delta t \to +0$, we finally get

$$\begin{aligned}\dot{V}_t &= E\left\{x_t\left[-k(x,t)\ \mathrm{sign}(x_t) + f(x_t,t)\right]\right\} + \frac{1}{2}\sigma^2 \\ &\leq E\{[-k(x,t)|x_t| + |x_t||f(x_t,t)|]\} + \frac{1}{2}\sigma^2 \\ &\leq E\{[-k(x,t)|x_t| + |x_t|L]\} + \frac{1}{2}\sigma^2\end{aligned} \tag{3.12}$$

Now take

$$\begin{aligned}k(x,t) &= L\varphi_\varepsilon(x) + \tilde{k}(x,t) \\ \varphi_\varepsilon(x) &:= \begin{cases} 1 & \text{if } |x| > \varepsilon \\ \varepsilon^{-1}|x| & \text{if } |x| \leq \varepsilon \end{cases}, \quad \varepsilon > 0\end{aligned} \tag{3.13}$$

which transforms (3.12) in to

$$\begin{aligned}\dot{V}_t &\leq E\left\{\left[L|x_t|[1-\varphi_\varepsilon(x)] - \tilde{k}(x,t)|x_t|\right]\right\} + \frac{1}{2}\sigma^2 \\ &\leq E\left\{\left[L|x_t|\chi(|x| \leq \varepsilon) - \tilde{k}(x,t)|x_t|\right]\right\} + \frac{1}{2}\sigma^2 \\ &\leq E\left\{-\tilde{k}(x,t)\sqrt{2V(x_t)} + \tfrac{1}{2}\sigma^2 + L\varepsilon\right\}\end{aligned} \tag{3.14}$$

3.1.2.8 The Lyapunov's Zone Function Analysis

Introduce the Lyapunov's zone function

$$\bar{V}_t := \left[V_t - \frac{\mu}{2}\right]_+^{1+\varepsilon}, \quad \mu, \varepsilon > 0$$
$$[z]_+ := \begin{cases} z & \text{if } z \geq 0 \\ 0 & \text{if } z < 0 \end{cases} \tag{3.15}$$

The inequality (3.14) leads to

$$\frac{\mathrm{d}}{\mathrm{d}t}\bar{V}_t = (1+\varepsilon)\left[V_t - \mu\right]_+^{\varepsilon} \dot{V}_t \leq$$
$$(1+\varepsilon)\left[V_t - \mu\right]_+^{\varepsilon} E\left\{-\tilde{k}_t\sqrt{2V(x_t)} + \tfrac{1}{2}\sigma^2 + L\varepsilon\right\} \tag{3.16}$$

Select

$$\tilde{k}(x,t) := k_0\sqrt{\frac{V(x_t)}{2}}, \, \mu = \frac{1}{k_0}\left(\frac{\sigma^2}{2} + L\varepsilon\right), \; k_0 > 0 \tag{3.17}$$

Substitution (3.17) in (3.16) gives

$$\frac{\mathrm{d}}{\mathrm{d}t}\bar{V}_t \leq (1+\varepsilon)\left[V_t - \mu\right]_+^{\varepsilon} E\left\{-\tilde{k}(x,t)\sqrt{2V(x_t)} + \frac{1}{2}\sigma^2 + L\varepsilon\right\}$$
$$= -(1+\varepsilon)\left[V_t - \mu\right]_+^{\varepsilon}\left[k_0 V_t - \frac{1}{2}\sigma^2 - L\varepsilon\right]$$
$$= -(1+\varepsilon)k_0\left[V_t - \frac{1}{k_0}\left(\frac{\sigma^2}{2} + L\varepsilon\right)\right]_+^{1+\varepsilon}$$

Corollary 3.1 *Selection*

$$k(x,t) = L\varphi_\varepsilon(x) + k_0\sqrt{\frac{V(x_t)}{2}} = L\varphi_\varepsilon(x) + \frac{k_0}{2}|x_t| \tag{3.18}$$

guarantees the exponential μ-zone convergence for the 2-nd moment of the state, i.e., for $\mu = \frac{1}{k_0}\left(\frac{\sigma^2}{2} + L\varepsilon\right)$.

$$\left[E\left\{x_t^2\right\}-\mu\right]_+=$$

$$2\bar{V}_t^{1/(1+\varepsilon)}\le\left(2\bar{V}_0\right)^{1/(1+\varepsilon)}e^{-t}\to 0\ when\,t\to\infty$$

The zone μ may be done smaller taking k_0 bigger.

3.1.3 SMC for a Conventional Multi-dimensional System

3.1.3.1 Model Description

Now let us consider the following Itô model

$$\left.\begin{aligned}\mathrm{d}x_{1,t}&=x_{2,t}\mathrm{d}t\\ \mathrm{d}x_{2,t}&=\left(u+f\left(x_t,t\right)\right)\mathrm{d}t+\sigma\mathrm{d}W_t\end{aligned}\right\}\tag{3.19}$$

where

$$x_t:=\left(x_{1,t}^{\mathsf{T}},x_{2,t}^{\mathsf{T}}\right)^{\mathsf{T}}\in\mathbb{R}^{2n},\ \|f\left(x_t,t\right)\|\le L$$

and $\sigma\in R^{n\times n}$ is a diffusion matrix of the standard n-dimensional Wiener process, i.e.,

$$E\left\{W_t\mid\mathscr{F}_{t-0}\right\}\overset{a.s.}{=}0,\ E\left\{W_tW_t^{\mathsf{T}}\mid\mathscr{F}_{t-0}\right\}\overset{a.s.}{=}tI_{n\times n}$$

3.1.3.2 Sliding Surface, the Discontinuous Control Structure and Problem Formulation

Define the following **sliding surface** $s\left(x\right)$:

$$s\left(x\right):=x_2+x_1=\dot{x}_1+x_1=0\tag{3.20}$$

and the **sliding mode control** as

$$\begin{aligned}&u_t=-Kx_t-k\left(x,t\right)\ \mathrm{SIGN}\left(s\left(x_t\right)\right),\ 0<k\left(x_t,t\right)\in\mathbb{R}^1\\&\mathrm{SIGN}\left(s\right):=\left(\mathrm{sign}(s_1),\ldots,\ \mathrm{sign}(s_n)\right)^{\mathsf{T}},\ 0<K\in\mathbb{R}^{n\times 2n}\end{aligned}\tag{3.21}$$

Problem 3.1 Design K and $k\left(x,t\right)$ in (3.21) which guarantee the stabilization of the system (3.19) in a μ-neighborhood of the sliding surface $s\left(x\right)=0$ (3.20) in some probabilistic sense.

3.1.3.3 The Deterministic Case

For $\sigma = 0$ the model (3.19) becomes

$$\left.\begin{aligned} \dot{x}_{1,t} &= x_{2,t} \\ \dot{x}_{2,t} &= -Kx_t - k(x_t, t)\ \text{SIGN}(s(x_t)) + f(x_t, t) \end{aligned}\right\} \tag{3.22}$$

Such model describes behavior of a wide class of mechanical systems and has a great spectrum of practical applications. Analogously, for $V(s(x_t)) := \frac{1}{2}\|s(x_t)\|^2$ using the inequality

$$\sum_{i=1}^{n} |s_i(x_t)| \geq \|s(x_t)\|$$

for

$$k_t = k = L + \rho,\ \rho = \text{const} > 0$$

and

$$K = \begin{bmatrix} 0 & I \end{bmatrix} \text{ so that } Kx_t = x_{2,t}$$

one has

$$\begin{aligned} \dot{V}_t &= s^{\top}(x_t)\,\dot{s}(x_t) = \\ & s^{\top}(x_t)\left[x_{2,t} - Kx_t - k\text{SIGN}(s(x_t)) + f(x_t, t)\right] \\ &= -k\sum_{i=1}^{n}|s_i(x_t)| + s^{\top}(x_t)\,f(x_t, t) \\ &\leq -k\|s(x_t)\| + \|s(x_t)\|\,L = \\ &= -\rho\|s(x_t)\| = -\sqrt{2}\rho\sqrt{V_t} \end{aligned}$$

implying $V_t = 0$ for all

$$t \geq t_{reach} = \sqrt{2}\frac{\sqrt{V_0}}{\rho} = \frac{1}{(k-L)}\|s(x_0)\|$$

3.1.3.4 The Zone Lyapunov Function and Its Evaluation in the Stochastic Case

Again, by the Itô formula

$$\mathrm{d}V\left(s\left(x_t\right)\right)=\nabla^{\top} V\left(s\left(x_t\right)\right) \mathrm{d}s+$$

$$\frac{1}{2}\operatorname{tr}\left\{\sigma\sigma^{\top}\nabla^{2} V\left(s\left(x_t\right)\right)\right\}\mathrm{d}t=s^{\top}\left(x_t\right)\left[\mathrm{d}x_{1,t}+\mathrm{d}x_{2,t}\right]+$$

$$\frac{1}{2}\operatorname{tr}\left\{\sigma\sigma^{\top}\right\}\mathrm{d}t=s^{\top}\left(x_t\right)\left(x_{2,t}dt+\left[u+f\left(x_t,t\right)\right]\mathrm{d}t+\sigma\mathrm{d}W_t\right)$$

$$+\frac{1}{2}\operatorname{tr}\left\{\sigma\sigma^{\top}\right\}\mathrm{d}t=$$

$$\left[s^{\top}\left(x_t\right)\left(x_{2,t}+u+f\left(x_t,t\right)\right)+\frac{1}{2}\operatorname{tr}\left\{\sigma\sigma^{\top}\right\}\right]\mathrm{d}t$$

$$+s^{\top}\left(x_t\right)\sigma\mathrm{d}W_t$$

Repeating the same manipulations as in the scalar case for $V_t:=E\left\{V\left(s\left(x_t\right)\right)\right\}$ we obtain

$$\begin{aligned}\dot{V}_t&=E\left\{s^{\top}\left(x_t\right)\left[-k\left(x,t\right)\ \mathrm{SIGN}\left(s\left(x_t\right)\right)+f\left(x_t,t\right)\right]\right\}+\frac{1}{2}\operatorname{tr}\left\{\sigma\sigma^{\top}\right\}\\&\leq E\left\{\left\|s\left(x_t\right)\right\|\left(-k\left(x,t\right)+L\right)\right\}+\frac{1}{2}\operatorname{tr}\left\{\sigma\sigma^{\top}\right\}\end{aligned}\tag{3.23}$$

Taking in (3.23)

$$\begin{gathered}k\left(x,t\right)=L\varphi_{\varepsilon}\left(\|s\|\right)+\tilde{k}\left(x,t\right)\\\varphi_{\varepsilon}\left(\|s\|\right):=\begin{cases}1 & \text{if } \|s\|>\varepsilon\\ \varepsilon^{-1}\|s\| & \text{if } \|s\|\leq\varepsilon\end{cases},\ \varepsilon>0\end{gathered}\tag{3.24}$$

we get

$$\dot{V}_t\leq E\left\{\left[L\left\|s\left(x_t\right)\right\|\left[1-\varphi_{\varepsilon}\left(\|s\|\right)\right]-\tilde{k}\left(x,t\right)\left\|s\left(x_t\right)\right\|\right]\right\}$$

$$+\frac{1}{2}\operatorname{tr}\left\{\sigma\sigma^{\top}\right\}\leq E\left\{\left[L\left\|s\left(x_t\right)\right\|\chi\left(\|s\|\leq\varepsilon\right)-\tilde{k}\left(x,t\right)\left|x_t\right|\right]\right\}$$

$$+\frac{1}{2}\operatorname{tr}\left\{\sigma\sigma^{\top}\right\}\leq E\left\{-\tilde{k}\left(x,t\right)\sqrt{2V\left(x_t\right)}+\frac{1}{2}\operatorname{tr}\left\{\sigma\sigma^{\top}\right\}+L\varepsilon\right\}$$

and with

$$\tilde{k}(x_t, t) := k_0 \frac{\|s(x_t)\|}{2} = k_0 \frac{\|x_{1,t} + x_{2,t}\|}{2}$$

it follows

$$\begin{aligned}\dot{V}_t &\leq E\left\{- \|s(x_t)\| \tilde{k}(x, t) + \tfrac{1}{2}\mathrm{tr}\{\sigma\sigma^{\top}\} + L\varepsilon\right\} \\ &= -k_0 E\{V(s(x_t))\} + \tfrac{1}{2}\mathrm{tr}\{\sigma\sigma^{\top}\} + L\varepsilon = \\ &- \left(k_0 V_t - \left[\frac{1}{2}\mathrm{tr}\{\sigma\sigma^{\top}\} + L\varepsilon\right]\right)\end{aligned} \tag{3.25}$$

Substitution (3.25) in (3.16)

$$\bar{V}_t := [V_t - \mu]_+^{1+\varepsilon}, \quad \mu, \varepsilon > 0$$

with

$$\mu = k_0^{-1}\left[\frac{1}{2}\mathrm{tr}\{\sigma\sigma^{\top}\} + L\varepsilon\right] \tag{3.26}$$

gives

$$\begin{aligned}\frac{\mathrm{d}}{\mathrm{d}t}\bar{V}_t &\leq -(1+\varepsilon)[V_t - \mu]_+^{\varepsilon}\left(k_0 V_t - \left[\tfrac{1}{2}\mathrm{tr}\{\sigma\sigma^{\top}\} + L\varepsilon\right]\right) \\ &= -(1+\varepsilon)k_0[V_t - \mu]_+^{1+\varepsilon} = -(1+\varepsilon)k_0\bar{V}_t\end{aligned}$$

and, as the result,

$$\bar{V}_t \leq \bar{V}_0 e^{-(1+\varepsilon)k_0 t}$$

Corollary 3.2 *Selection*

$$\begin{aligned}u_t &= -x_{2,t} - k(x_t, t)\ SIGN(s(x_t)) \\ k(x_t, t) &= L\varphi_\varepsilon\left(\|x_{1,t} + x_{2,t}\|\right) + \frac{k_0}{2}\|x_{1,t} + x_{2,t}\|\end{aligned} \tag{3.27}$$

guarantees the exponential μ-zone convergence for the 2-nd moment of the state, i.e.,

$$\begin{aligned}&\left[E\{\|s(x_t)\|^2\} - k_0^{-1}\mathrm{tr}\{\sigma\sigma^{\top}\}\right]_+ = \\ &2\bar{V}_t^{1/(1+\varepsilon)} \leq 2\bar{V}_0^{1/(1+\varepsilon)} e^{-k_0 t} \to 0 \ when\, t \to \infty\end{aligned}$$

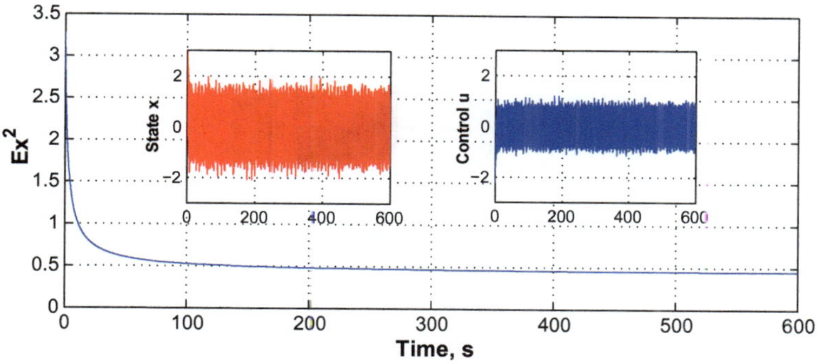

Fig. 3.2 The state $x(t)$, the control $u(t)$ and the estimate of $E\{x_t^2\}$ for the stochastic process (1.4)

3.1.4 *Numerical Examples*

3.1.4.1 Example 1

Figure 3.2 presents the results of the numerical simulation of the process (3.4), obtained by the Euler–Maruyama method [17]) with $k(x, t)$ given by (3.13) where

$$L = 0.3, \sigma = 0.5, x_0 = 3, k_0 = 1, \varepsilon = 0.5$$

and

$$f(x, t) = Lp(t)\frac{\sqrt{|x|}}{1 + |x|}$$

with $p(t) = \text{triangle}(0.1, 90)$ which is the triangle signal of the amplitude *1*, the frequency 0.1 and the phase 90^0. The "*white noise*" is approximated as $(W_t - W_{t-h})\, h^{-1}$ with $h = 10^{-3}$ and W_t as the standard Gaussian noise with zero-mean and variance equal to 1. The mean squared value of the process is estimated as

$$E\left\{x_t^2\right\} \simeq t^{-1} \int\limits_{\tau=0}^{t} x_\tau^2 d\tau$$

One can see that the suggested control

$$u_t = -k(x, t)\ \text{sign}(x_t)$$

with $k(x, t)$ (3.13) provides the convergence of the mean squared sliding variable $s(x_t)$ in the predefined μ-zone.

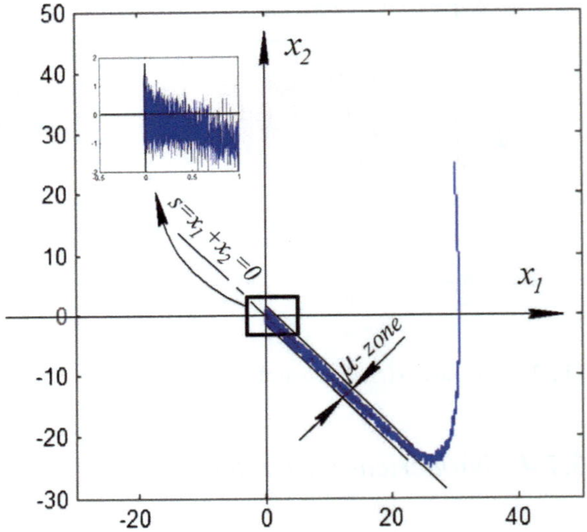

Fig. 3.3 The states $x_{1,t}$ and $x_{2,t}$ behavior in time and their zoom

3.1.4.2 Example 2

Here in Fig. 3.3 the states $x_{1,t}$ and $x_{2,t}$ and the sliding surface $s(x_t)$ for the stochastic process (3.19) are depicted. All parameters are as in Example 1 except k_0 which here is equal to 20. The initial values are $x_{1,0} = 30$ and $x_{2,0} = 25$.

3.2 Stochastic Super-Twist Sliding Mode Controller

3.2.1 Structure of Super-Twist Controller

During the last decade many publications have been dedicated to the stability analysis of the, so-called, *Super-Twist Controller* (STC) which was suggested in [18]. Many important application of this controller can be found in ([19]). Its dynamic is governed by the following scalar ODE

$$\dot{x}(t) = u\,(x(t), t) + F(x(t), t),\;\; x(0) = x_0 \in \mathbb{R}$$

where $F(x, t)$ is uncertain dynamics with bounded derivative, namely,

$$\left|\dot{F}(x, t)\right| \leq L \tag{3.28}$$

and the controller $u(x,t)$ is continuos one and has the structure

$$u(x,t) = -\alpha\sqrt{|x(t)|}\operatorname{sign}(x(t)) - \beta \int_{\tau=0}^{t} \operatorname{sign}(x(\tau))d\tau \tag{3.29}$$

α, β - positive constants

This system may be represented equivalently as a 2-nd order dynamics, that is,

$$\begin{cases} \dot{x}(t) = -\alpha\sqrt{|x(t)|}\operatorname{sign}(x(t)) + y(t) \\ \dot{y}(t) = f(x(t),t) - \beta \operatorname{sign}(x(t)) \end{cases} \tag{3.30}$$

$$f(x(t),t) := \dot{F}(x,t), \ |f(x(t),t)| \le L, \ \alpha > 0, \ \beta > 0$$

Stability analysis of this dynamics, based on Lyapunov function approach, was presented in several publications (see, for example, [20–23]). In these papers there is shown that for special parameters selection α, β such system is finite-time stable. The main assumption, used in this analysis, is that the uncertain term $f(x(t),t)$ is bounded. In the presence of stochastic unbounded (normally distributed) noise this assumptions can not be accepted and therefore the stability analysis, applied for a deterministic case, falls and new technique should be designed.

Below we show one of possible approaches for solving this problem. It is based on the ideas of the analysis developed in previous sections (see also [24]) for standard first-order Sliding Mode Controllers subject additive stochastic perturbations.

3.2.2 Stochastic Super-Twist Model

The stochastic analogue of the deterministic system (3.30) may be expressed as follows

$$\begin{cases} dx(t) = \left[-\alpha\sqrt{|x(t)|}\operatorname{sign}(x(t)) + y(t)\right]dt \\ dy(t) = \left[f(x(t),t) - \beta \operatorname{sign}(x(t))\right]dt + \sigma dW(t) \end{cases}$$

$$|f(x(t),t)| \overset{a.s.}{\le} L, \ \sigma > 0 \tag{3.31}$$

$$\alpha = \alpha(x,y,t), \ \beta = \beta(x,y,t)$$

Stochastic processes $x(t) = x_t(\omega) \in R$ and $y(t) = y_t(\omega) \in R$ are the system states at time $t \ge 0$ with a random realization $\omega \in \Omega$. They are defined on a filtered probability space $(\Omega, \mathcal{F}, \{H_t\}_{t\ge 0}, P)$ defined before. The functions $f(x(t),t)$, $\alpha(x(t), y(t), t)$, $\alpha(x(t), y(t), t)$ are assumed to be $\mathcal{F}_t$-measurable.

The "parameters" $\alpha = \alpha(x, y, t)$, $\beta = \beta(x, y, t)$ should be designed to provide the stability property of the system (3.31) in some probabilistic sense.

3.2.3 Lyapunov Functions Approach

3.2.3.1 Lyapunov Functions for Deterministic ST ($\sigma = 0$)

There are known several different functions $V(x, y)$ which in the deterministic case ($\sigma = 0$) may serve as Lyapunov functions for the stability analysis of (see, for example, [20–23]). All of them posses the property

$$\dot{V}(x, y) \le -qV^{\rho}(x, y), \; q > 0, \; \rho \in [0, 1) \tag{3.32}$$

3.2.3.2 Analysis of Lyapunov Functions Designed for Deterministic Systems on the Trajectories of Stochastic Dynamics

By the Itô formula the differential of a function $V(x, y)$ on the trajectories of the stochastic system (3.31) is as follows:

$$\begin{aligned} dV &= \tfrac{\partial V}{\partial x}dx + \tfrac{\partial V}{\partial y}dy + \\ &\underbrace{\frac{1}{2}\operatorname{tr}\left\{\begin{pmatrix}0\\ \sigma\end{pmatrix}\begin{pmatrix}0\\ \sigma\end{pmatrix}^{\intercal}\nabla^2 V\right\}dt}_{\text{the Itô term}} \\ &= \varphi dt + \frac{\partial V}{\partial y}\sigma dW \end{aligned} \tag{3.33}$$

where

$$\varphi = \varphi_0 + \tfrac{1}{2}\sigma^2\frac{\partial^2}{\partial y^2}V$$

$$\varphi_0 := \frac{\partial V}{\partial x}\left[-\alpha\sqrt{|x|}\operatorname{sign}(x) + y\right] + \frac{\partial V}{\partial y}\left[f - \beta\operatorname{sign}(x)\right]$$

Here the term φ_0 corresponds to the regular part of the stochastic dynamics and we will suppose that it satisfies the inequality

$$\varphi_0 \le -k\phi_t, \; \phi_t \ge 0 \tag{3.34}$$

for some function ϕ_t. Particularly, with $k = q$ and $\phi_t = V^{\rho}(x(t), y(t))$ we obtain the property (3.32).

Defining then $V_t := V(x(t), y(t))$, the Eq. (3.33) may be represented in the integral form as

$$V_{t+\Delta t} - V_t \overset{a.s.}{=} \int_{\tau=t}^{t+\Delta t} \varphi(\tau)\,\mathrm{d}\tau + \int_{\tau=t}^{t+\Delta t} \frac{\partial V_\tau}{\partial y}\sigma \mathrm{d}W(\tau)$$

Taking the mathematical expectation of both sides of this equation and denoting $\bar{V}_t = E\{V_t\}$, in view of the Itô integral property

$$E\left\{\int_{\tau=t}^{t+\Delta t} \frac{\partial V_\tau}{\partial y}\sigma \mathrm{d}W(\tau)\right\} = 0$$

we get

$$\bar{V}_{t+\Delta t} - \bar{V}_t = \int_{\tau=t}^{t+\Delta t} E\{\varphi(\tau)\}\,\mathrm{d}\tau$$

Dividing both parts by Δt and taking $\Delta t \to 0$, in view of (3.34) we finally obtain (the Dynkin's formula)

$$\begin{aligned}\frac{\mathrm{d}}{\mathrm{d}t}\bar{V}_t &= E\{\varphi(t)\} = E\{\varphi_0\} + E\left\{\frac{1}{2}\sigma^2\frac{\partial^2}{\partial y^2}V_t\right\} \\ &\le E\{-k\phi_t\} + E\left\{\frac{1}{2}\sigma^2\frac{\partial^2}{\partial y^2}V_t\right\}\end{aligned} \tag{3.35}$$

Take

$$\begin{aligned}&\alpha = \alpha_0,\ \beta = \beta_0\varphi_\varepsilon(x) + \beta_{ad}(x, y, t) \\ &\alpha_0 = \text{const} > 0,\ \beta_0 = \text{const} > 0 \\ &\varphi_\varepsilon(|x|) := \begin{cases} 1 & \text{if } |x| > \varepsilon \\ \varepsilon^{-1}|x| & \text{if } |x| \le \varepsilon \end{cases},\ \varepsilon > 0\end{aligned} \tag{3.36}$$

where α_0 and β_0 are positive functions making the function $V(x, y)$ the Lyapunov function for the deterministic version ($\sigma = 0$) of the supert-twist algorithm (3.31) and providing a finite-time convergence of this function to zero, namely, guarantying the property

$$\dot{V} \overset{\sigma=0}{=} \varphi_0 = \frac{\partial V}{\partial x}\left[-\alpha_0\sqrt{|x|}\mathrm{sign}(x) + y\right] +$$
$$\frac{\partial V}{\partial y}\left[f - \beta_0 \mathrm{sign}(x)\right] \leq -k\phi_t = -qV^{\rho} \;\; (\rho \in [0, 1)) \tag{3.37}$$

with $k = q > 0$ and $\phi_t = V^{\rho}$. The function $\varphi_{\varepsilon}(|x|)$ is a regularization term providing the $\mathscr{HL}$ *(Hausdorff–Lipschitz)*-property of the term $\varphi_{\varepsilon}(x)\,\mathrm{sign}(x)$ and guarantying the existence of the solution of stochastic differential inclusion (3.31). In view of that modification instead of (3.37) we have

$$\dot{V} \overset{\sigma=0}{=} \varphi_0 = \frac{\partial V}{\partial x}\left[-\alpha_0\sqrt{|x|}\mathrm{sign}(x) + y\right] +$$
$$\frac{\partial V}{\partial y}\left[f - \beta_0\varphi_{\varepsilon}(|x|)\,\mathrm{sign}(x)\right] =$$
$$\frac{\partial V}{\partial x}\left[-\alpha_0\sqrt{|x|}\mathrm{sign}(x) + y\right] + \frac{\partial V}{\partial y} f_{\varepsilon}\mathrm{sign}(x) \leq -k\phi_t$$

with the new constraint to the modified uncertainty

$$f_{\varepsilon} := f - \beta_0\varphi_{\varepsilon}(|x|)\,\mathrm{sign}(x)$$

fulfilling $|f_{\varepsilon}| \leq L + 2\beta_0$. Thus the Eq. (3.35) becomes

$$\frac{\mathrm{d}}{\mathrm{d}t}\bar{V}_t \leq -E\left\{k\phi_t + \frac{\partial V_t}{\partial y}\beta_{ad}\mathrm{sign}(x_t)\right\} + E\left\{\frac{1}{2}\sigma^2\frac{\partial^2}{\partial y^2}V_t\right\} \tag{3.38}$$

In (3.36) the gain parameter $\beta_{ad} = \beta_{ad}(x, y, t)$ should be adapted to provide the system (3.31) with desired properties.

3.2.4 State-Dependent Gain Parameter

If additionally, there exists a nonnegative definite function $v_t = v(x_t, y_t) \geq 0$ such that

$$\frac{1}{2}\frac{\partial^2}{\partial y^2}V_t \leq c_0 + c_1 v_t \tag{3.39}$$

then the right-hand side in (3.38) may be estimated as

$$\frac{\mathrm{d}}{\mathrm{d}t}\bar{V}_t \leq -E\left\{k\phi_t + \frac{\partial V_t}{\partial y}\beta_{ad}\mathrm{sign}(x_t) - c_1\sigma^2 v_t\right\} + c_0\sigma^2 \tag{3.40}$$

Select β_{ad} as

$$\beta_{ad} = -\frac{s_t}{\left|\frac{\partial V_t}{\partial y}\right| + \varepsilon_t} \operatorname{sign}\left(\frac{\partial V_t}{\partial y}\right) \operatorname{sign}(x_t) \tag{3.41}$$

$$s_t := k\phi_t - \theta V_t - c_1\sigma^2 v_t,\ \theta > 0,\ \varepsilon_t > 0$$

Substitution β_{ad} (3.41) into the first term in (3.40) leads to the following relation

$$k\phi_t + \frac{\partial V_t}{\partial y}\beta_{ad}\operatorname{sign}(x_t) - c_1\sigma^2 v_t = k\phi_t - \left|\frac{\partial V_t}{\partial y}\right| \frac{s_t}{\left|\frac{\partial V}{\partial y}\right| + \varepsilon_t}$$

$$-c_1\sigma^2 v_t = \theta V_t + \varepsilon_t \frac{s_t}{\left|\frac{\partial V_t}{\partial y}\right| + \varepsilon_t}$$

and (3.40) becomes

$$\frac{d}{dt}\bar{V}_t \leq -E\left\{\theta V_t + \varepsilon_t \frac{s_t}{\left|\frac{\partial V_t}{\partial y}\right| + \varepsilon_t}\right\} + c_0\sigma^2$$

Select now ε_t in such a way that for a prespecified $\varepsilon > 0$ the following inequality holds:

$$-\varepsilon_t \frac{s_t}{\left|\frac{\partial V_t}{\partial y}\right| + \varepsilon_t} \leq \varepsilon$$

or, equivalently,

$$-\varepsilon_t (s_t + \varepsilon) \leq \varepsilon \left|\frac{\partial V_t}{\partial y}\right|$$

This always may be fulfilled if

$$\varepsilon_t = \begin{cases} \text{any positive value if } s_t + \varepsilon \geq 0 \\ \varepsilon \left|\frac{\partial V}{\partial y}\right| / |s_t + \varepsilon| \quad \text{if } s_t + \varepsilon < 0 \end{cases} \tag{3.42}$$

Therefore, finally for any $\theta > 0$ we obtain

$$\frac{d}{dt}\bar{V}_t \leq -\theta \bar{V}_t + \varepsilon + c_0\sigma^2 \tag{3.43}$$

Now we are ready to formulate the main result.

3.2.5 Main Theorem on μ-MS Convergence

Theorem 3.2 (Main result) *Selecting in (3.31)*

$$\alpha = \alpha_0, \ \beta = \beta_0 \varphi_\varepsilon(x) + \beta_{ad}$$

with α_0, β_0 *providing the property (3.37) and*

$$\beta_{ad} = -\frac{s_t}{\left|\frac{\partial V_t}{\partial y}\right| + \varepsilon_t} \operatorname{sign}\left(\frac{\partial V_t}{\partial y}\right) \operatorname{sign}(x) \tag{3.44}$$

where s_t *and* ε_t *are defined by (3.41) and (3.42), we may guarantee the means square (MS) exponential convergence of* V_t *in the prespecified zone* $\mu = \dfrac{c_0\sigma^2 + \varepsilon}{\theta}$, *that is,*

$$\left[\bar{V}_t - \mu\right]_+^2 = O\left(e^{-2\theta t}\right) \underset{t\to\infty}{\to} 0, \ [z]_+ := \begin{cases} z \ \textit{if}\ z \geq 0 \\ 0 \ \textit{if}\ z < 0 \end{cases} \tag{3.45}$$

Proof For $W_t := \frac{1}{2}\left[\bar{V}_t - \mu\right]_+^2$ we have

$$\dot{W}_t = \left[\bar{V}_t - \mu\right]_+ \frac{\mathrm{d}}{\mathrm{d}t}\bar{V}_t \leq$$
$$\left[\bar{V}_t - \mu\right]_+ \left(-\theta\bar{V}_t + \varepsilon + c_0\sigma^2\right) = -\theta\left[\bar{V}_t - \mu\right]_+\left[\bar{V}_t - \mu\right]$$
$$= -\theta\left[\bar{V}_t - \mu\right]_+^2 = -2\theta W_t$$

which leads to (3.45).

3.2.6 State-Dependence of Gain Parameter for Different Lyapunov Functions (LF)

3.2.6.1 The Polyakov–Poznyak's LF

To analyze the finite-time convergence effect for Super-Twisting second order sliding mode controller in [21] there was suggested to use the following Lyapunov Function

$$V_{PP}(x, y) =$$
$$\begin{cases} \frac{k^2}{4}\left(\frac{y}{\gamma}\mathrm{sign}\,(x) + k_0 e^{m(x,y)}\sqrt{s(x, y)}\right)^2 & \text{if } xy \neq 0 \\ 2\alpha^{-2}\bar{k}^2 y^2 & \text{if } x = 0 \\ |x|/2 & \text{if } y = 0 \end{cases} \tag{3.46}$$

$$\beta > 5L,\ \alpha^2 \in (32L, 8[\beta - L]),\ \gamma \geq \beta + L$$

where $0 < k_0$ (a piece-wise constant function), $\bar{k}$ is large enough and $s(x, y)$ and $m(x, y)$ are as follows

$$s(x, y) = y^2 - \alpha\sqrt{|x|}y\ \mathrm{sign}\,(x) + 2\gamma\,|x|$$

$$m(x, y) = \frac{\arctan\left(\frac{\alpha g\sqrt{|x|}\ \mathrm{sign}(x)}{2\sqrt{g-1}}\frac{1}{y} - \frac{1}{\sqrt{g-1}}\right)}{\sqrt{g-1}},\ g > 1 \tag{3.47}$$

Remark 3.3 The inequality (3.32) is fulfilled with $\rho = \frac{1}{2}$ and some positive $q = q^*$ (see details in [21]).

For this function

- if $xy \neq 0$

$$\frac{\partial V_{PP}}{\partial y} = \frac{k^2}{2}\left(\gamma^{-1}y\ \mathrm{sign}\,(x) + k_0 e^{m(x,y)}\sqrt{s(x, y)}\right)\left[\frac{\mathrm{sign}(x)}{\gamma} + \right.$$
$$\left. k_0 e^{m(x,y)}\left(\sqrt{s(x, y)}\frac{\partial m(x,y)}{\partial y} + \frac{1}{\sqrt{s(x, y)}}\frac{\partial s(x,y)}{\partial y}\right)\right]$$

- if $x = 0$

$$\frac{\partial V_{PP}}{\partial y} = 4\alpha^{-2}\bar{k}^2 y$$

- if $y = 0$

$$\frac{\partial V_{PP}}{\partial y} = 0$$

In this presentation

$$\frac{\partial m(x,y)}{\partial y} = -\frac{c(x)\sqrt{g-1}}{gy^2 - 2c(x)\sqrt{g-1}y + c^2(x)(g-1)}$$

$$c(x) := \frac{\alpha g\sqrt{|x|}\ \mathrm{sign}\ (x)}{2\sqrt{g-1}}$$

$$\frac{\partial s(x,y)}{\partial y} = -\alpha\sqrt{|x|}\ \mathrm{sign}\,(x) + 2y$$

and in the case when $xy \neq 0$ we have

$$\frac{\partial^2 V_{PP}}{\partial y^2} = \tfrac{k^2}{2}\left[\gamma^{-1} y\ \mathrm{sign}\,(x) + \right.$$

$$\left. k_0 e^{m(x,y)}\left(\sqrt{s(x,y)}\tfrac{\partial m(x,y)}{\partial y} + \tfrac{\frac{\partial s(x,y)}{\partial y}}{\sqrt{s(x,y)}}\right)\right]^2 +$$

$$\tfrac{k^2 k_0}{2}\left(\gamma^{-1} y\ \mathrm{sign}\,(x) + k_0 e^{m(x,y)}\sqrt{s(x,y)}\right) e^{m(x,y)} \times$$

$$\left[\left(\sqrt{s(x,y)}\tfrac{\partial m(x,y)}{\partial y} + \tfrac{1}{\sqrt{s(x,y)}}\tfrac{\partial s(x,y)}{\partial y}\right)\tfrac{\partial m(x,y)}{\partial y} + \right.$$

$$\left.\tfrac{\frac{\partial s(x,y)}{\partial y}\frac{\partial m(x,y)}{\partial y}}{\sqrt{s(x,y)}} + \tfrac{\frac{\partial^2 m(x,y)}{\partial y^2}}{\sqrt{s(x,y)}}\left(\left(\tfrac{\partial s(x,y)}{\partial y}\right)^2 + \tfrac{\partial^2 s(x,y)}{\partial y^2}\right)\right]$$

$\dfrac{\partial^2 V_{PP}}{\partial y^2} = 4\alpha^{-2}\bar{k}^2$ if $x = 0$, and $\dfrac{\partial^2 V_{PP}}{\partial y^2} = 0$ if $y = 0$ with the properties

$$|m(x,y)| \le \tfrac{\pi}{2\sqrt{g-1}}, \quad \left|\tfrac{\partial m(x,y)}{\partial y}\right| \le \tfrac{\alpha g}{2\sqrt{g-1}}\sqrt{|x|}$$

$$\tfrac{\partial^2 m(x,y)}{\partial y^2} = 0, \ \sqrt{s(x,y)} \le \sqrt{2\gamma\,|x|} + |y| \ \text{ if } \gamma \ge \alpha^2/8$$

$$\left|\tfrac{\partial s(x,y)}{\partial y}\right| \le \left|-\alpha\sqrt{|x|}\ \mathrm{sign}\,(x) + 2y\right| \le \alpha\sqrt{|x|} + 2\,|y|$$

$$\tfrac{\partial^2 s(x,y)}{\partial y^2} = 2, \quad \left|\tfrac{1}{\sqrt{s(x,y)}}\tfrac{\partial s(x,y)}{\partial y}\right| \le 2$$

and

$$\left|\frac{\partial^2 V}{\partial y^2}\right| \leq$$

$$\frac{k^2}{2}\left[\gamma^{-1}+k_0 e^{\frac{\pi}{2\sqrt{g-1}}}\left(\frac{\alpha g}{2\sqrt{g-1}}\sqrt{s(x,y)}\sqrt{|x|}+2\right)\right]^2$$
$$+\frac{k^2 k_0}{2}\left(\gamma^{-1}|y|+k_0 e^{\frac{\pi}{2\sqrt{g-1}}}\sqrt{s(x,y)}\right)e^{\frac{\pi}{2\sqrt{g-1}}}\times$$
$$\left[\sqrt{s(x,y)}\left(\frac{\alpha g}{2\sqrt{g-1}}\right)^2|x|+\frac{2\alpha g}{\sqrt{g-1}}\sqrt{|x|}+\frac{2}{\sqrt{s(x,y)}}\right]$$
$$\leq\left\{\text{by } ax+\frac{b}{x}\leq 2\sqrt{ab} \text{ for any } x>0 \text{ and } a,b\geq 0\right\}$$
$$\leq\frac{2k^2k_0\alpha g}{\sqrt{g-1}}e^{\frac{\pi}{2\sqrt{g-1}}}\left[\left(\gamma^{-1}+2k_0 e^{\frac{\pi}{2\sqrt{g-1}}}\right)|y|\sqrt{|x|}\right.$$
$$\left.+k_0 e^{\frac{\pi}{2\sqrt{g-1}}}\alpha|x|\right]=\theta_0\left(\theta_1|y|\sqrt{|x|}+\theta_2|x|\right)$$

with

$$\vartheta_0:=2k^2k_0\frac{\alpha g}{\sqrt{g-1}}e^{\frac{\pi}{2\sqrt{g-1}}},\ \theta_1:=\gamma^{-1}+2k_0 e^{\frac{\pi}{2\sqrt{g-1}}}$$

$$\theta_2:=k_0\alpha e^{\frac{\pi}{2\sqrt{g-1}}}$$

Lemma 3.1 *If $\gamma\geq\alpha^2/8$ then when $xy\neq 0$ we have*

$$\left|\frac{\partial^2 V_{PP}}{\partial y^2}\right|\leq\theta_0\left(\theta_1|y|\sqrt{|x|}+\theta_2|x|\right) \tag{3.48}$$

and when $xy=0$

$$\left|\frac{\partial^2 V_{PP}}{\partial y^2}\right|=\begin{cases}4\alpha^{-2}\bar{k}^2 & if\ x=0\\ 0 & if\ y=0\end{cases} \tag{3.49}$$

Finally, we have the fulfilling (3.39) with

$$c_0=2\alpha^{-2}\bar{k}^2,\ c_1=\frac{1}{2}\ \textit{and}\ v_t=\theta_0\left(\theta_1|y|\sqrt{|x|}+\theta_2|x|\right)$$

For the Lyapunov function V_{PP} the adaptation parameter β_{ad} is as follows

$$\boxed{\begin{gathered}\beta_{ad}=-\frac{k\sqrt{V_{PP}}-\theta V_{PP}-c_1\sigma^2 v_t}{\left|\frac{\partial V_{PP}}{\partial y}\right|+\varepsilon_t}\operatorname{sign}\left(\frac{\partial V_{PP}}{\partial y}\right)\operatorname{sign}(x)\\ v_t=\theta_0\left(\theta_1|y|\sqrt{|x|}+\theta_2|x|\right)\end{gathered}}$$

3.2.6.2 The Moreno-Osorio's LF

In [20] for the deterministic case ($\sigma = 0$) there was suggested the following Lyapunov function

$$V_{MO}(x, y) = \begin{pmatrix} \sqrt{|x|}\text{sign}(x) \\ y \end{pmatrix}^{\mathsf{T}} P \begin{pmatrix} \sqrt{|x|}\text{sign}(x) \\ y \end{pmatrix}$$

$$P = \frac{1}{2}\begin{bmatrix} 4\beta + \alpha^2 & -\alpha \\ -\alpha & 2 \end{bmatrix}$$

with

$$\frac{\partial V_{MO}(x, y)}{\partial y} = -2\alpha\sqrt{|x|}\text{sign}(x) + 2y$$

$$\frac{\partial^2 V_{MO}(x, y)}{\partial y^2} = 2$$

which in (3.39) leads to

$$c_0 = 1,\ c_1 = 0$$

If

$$\beta > 3L + 2L^2\alpha_1^{-2}$$

then the inequality (3.32) is fulfilled with

$$\rho = \frac{1}{2},\ q = \frac{\sqrt{\lambda_{\min}(P)}}{\lambda_{\max}(P)}\lambda_{\min}(Q)$$

where the matrix Q is defined as

$$Q = \frac{1}{2}\begin{bmatrix} 2\beta + \alpha_1^2 - L & -\alpha - \dfrac{2}{\alpha} \\ -\alpha - \dfrac{2}{\alpha} & 1 \end{bmatrix} > 0$$

For the Lyapunov function V_{MO} the adaptation parameter β_{ad} is as follows

$$\beta_{ad} = -\frac{k\sqrt{V_{MO}} - \theta V_{MO}}{\left|\dfrac{\partial V_{MO}(x, y)}{\partial y}\right| + \varepsilon_t}\text{sign}\left(\frac{\partial V_{MO}}{\partial y}\right)\text{sign}(x)$$

$$\frac{\partial V_{MO}(x, y)}{\partial y} = -2\alpha\sqrt{|x|}\ \text{sign}(x) + 2y$$

3.2.6.3 The Orlov–Aoustin–Chevellereau's LF

To analyze the stability of the system (3.31) without stochastic noise the authors of [22] suggested to use the following Lyapunov function candidate

$$V_{OACh}(x,y) = 2\beta|x| + \frac{1}{2}y^2 + \frac{1}{2}\left[y - \alpha\sqrt{|x|}\mathrm{sign}(x)\right]^2$$

with

$$\frac{\partial V_{OACh}(x,y)}{\partial y} = 2y - \alpha\sqrt{|x|}\mathrm{sign}(x)$$

$$\frac{\partial^2}{\partial y^2}V_{OACh}(x,y) = 2$$

which in (3.39) leads to

$$c_0 = 1,\ c_1 = 0$$

If

$$\min\left\{\frac{\alpha}{2}, \frac{\alpha\beta}{1+\alpha}\right\} > L$$

then the inequality (3.32) is fulfilled with

$$\rho = \frac{1}{2},\ q = \sqrt{2\beta}\min\left\{\frac{2(\alpha\beta - L - L\alpha)}{3\alpha^2 + 4\beta}, \frac{\alpha - 4L}{1+\alpha}\right\}$$

For the Lyapunov function V_{OACh} the adaptation parameter β_{ad} is as follows

$$\boxed{\begin{array}{c} \beta_{ad} = -\dfrac{k\sqrt{V_{OACh}} - \theta V_{OACh}}{\left|\dfrac{\partial V_{OACh}}{\partial y}\right| + \varepsilon_t}\mathrm{sign}\left(\dfrac{\partial V_{OACh}}{\partial y}\right)\mathrm{sign}(x) \\ \dfrac{\partial V_{OACh}}{\partial y} = 2y - \alpha\sqrt{|x|}\ \mathrm{sign}(x) \end{array}}$$

3.2.6.4 The Utkin's LF

In the book [23] one of the authors suggested the following Lyapunov function for the deterministic version of the system (3.31)

$$V_U(x,y) = 2\sqrt{\frac{1}{2}y^2 + M_0|x|},\ M_0 = \text{const} > 0$$

Its the first and second derivatives, participating in the adaptation procedure (3.44) are

$$\frac{\partial}{\partial y}V_U(x,y) = \frac{y}{\sqrt{\frac{1}{2}y^2 + M_0|x|}}$$

$$\frac{\partial^2}{\partial y^2}V_U(x,y) = \frac{\sqrt{2}}{2}\frac{y\left[V^2 - 2y\right]}{V^3} =$$

$$\frac{\sqrt{2}}{2}\frac{y}{V} - \sqrt{2}\frac{y^2}{V^3} \le \frac{\sqrt{2}}{2}\frac{|y|}{V} \le 1$$

which in (3.39) leads to

$$c_0 = \frac{1}{2},\ c_1 = 0$$

and some $q > 0$. For the Lyapunov function V_U the adaptation parameter β_{ad} is

$$\boxed{\begin{gathered}\beta_{ad} = -\frac{k\sqrt{V_U} - \theta V_U}{\left|\frac{\partial V_U}{\partial y}\right| + \varepsilon_t}\operatorname{sign}\left(\frac{\partial V_U}{\partial y}\right)\operatorname{sign}(x)\\ \frac{\partial}{\partial y}V_U(x,y) = \frac{y}{\sqrt{\frac{1}{2}y^2 + M_0|x|}} = 2\frac{y}{V_U(x,y)}\end{gathered}}$$

3.2.7 Simulation Results

All simulations presented bellow have been completed based on the *Euler–Maruyama Method* which detail description may be found in ([17]).

The results of simulation are depicted at the Figs. 3.4, 3.5, 3.6 and 3.7 where the right-hand graphics of each figure illustrates the behavior of the corresponding Lyapunov function V_t and the approximation of its mathematical expectation (the dashed line) V_{av}.

$$\bar{V}_t = E\{V_t\} \simeq V_{av} = \frac{1}{t + 0.01}\int_{\tau=0}^{t} V_\tau d\tau$$

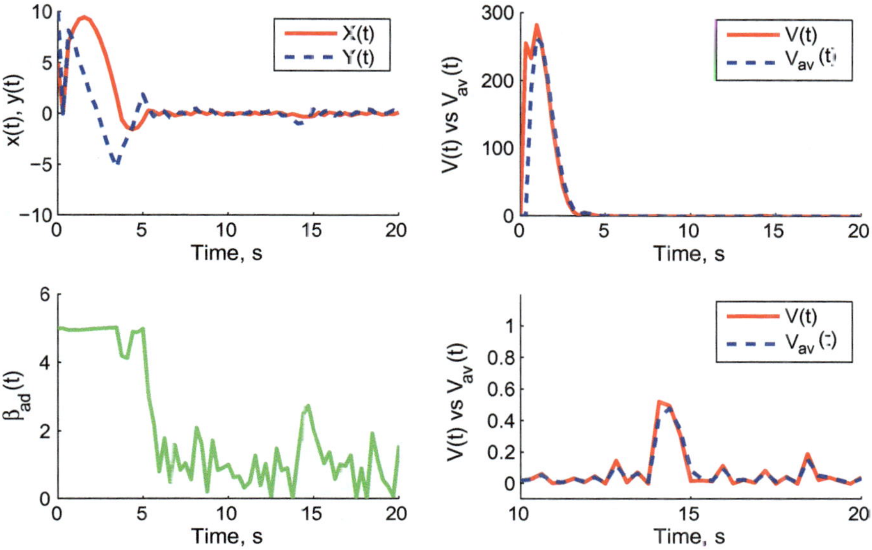

Fig. 3.4 Behavior of the system with the controller based on the Polyakov–Poznyak's LF

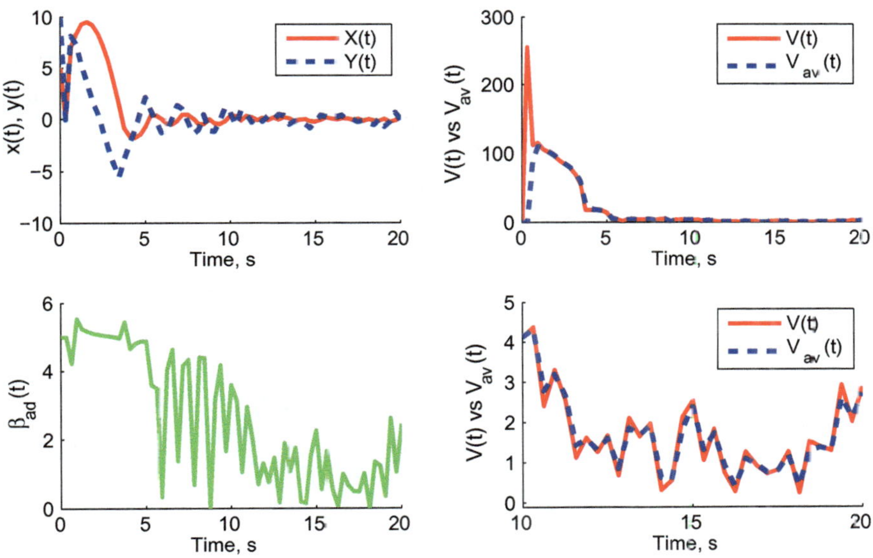

Fig. 3.5 The behavior of the system with the controller based on the Moreno-Osorio's LF

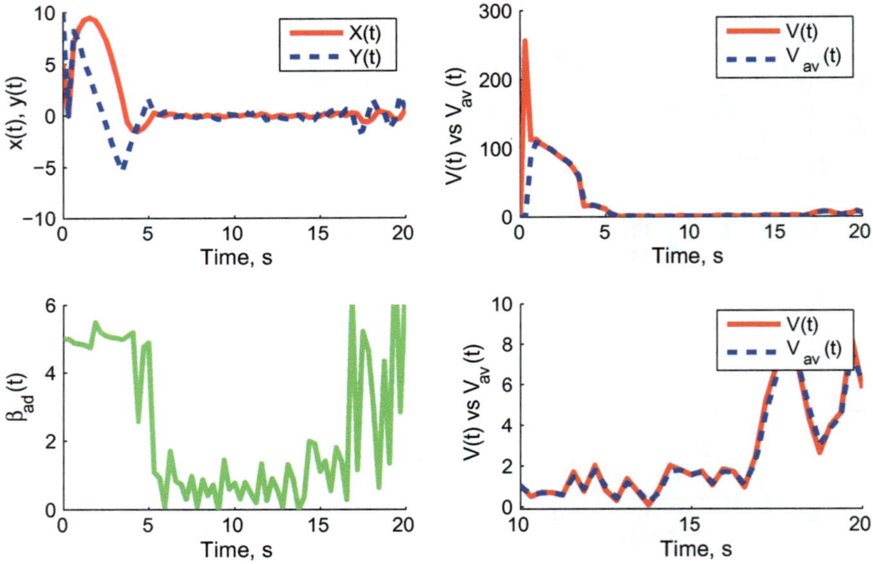

Fig. 3.6 The behavior of the system with the controller based on the Orlov–Aoustin–Chevellereau's LF

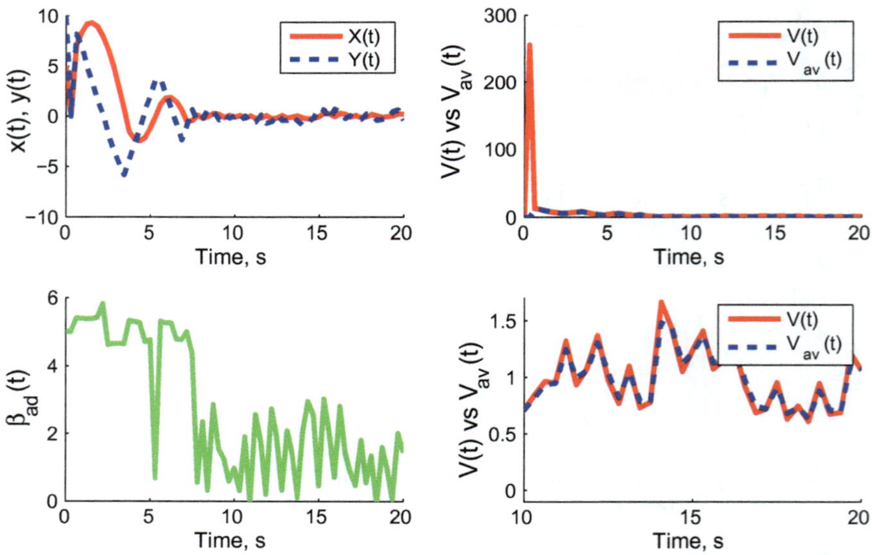

Fig. 3.7 The behavior of the system with the controller based on the Utkin's LF

The following parameters of the stochastic controller (3.29), (3.41) were selected for all tested versions:

$k = 0.5$	$\theta = 0.1$	$\varepsilon = 0.5$
$\sigma = 0.3$	$\alpha = 1.5$	$\beta_0 = 5$

The upper bounds of $|x(t)|$ and $|y(t)|$ for $t \geq 20\,s$. are given in the table below

Lyapunov function	$\max\limits_{t\in[20,50]s} \|x(t)\|$	$\max\limits_{t\in[20,50]s} \|y(t)\|$
V_{PP}	0.3	0.6
V_{MO}	0.3	0.7
V_{OACh}	0.6	1.6
V_U	0.3	0.6

The corresponding μ-zone values were obtained for super-twist controllers with different adaptive gain parameters:

$V:$	V_{PP}	V_{MO}	V_{OACh}	V_U
μ	0.55	4.3	8.1	1.6

One can see that μ-zone under the adaptive gain β_{ad} (3.44), corresponding the Lyapunov function V_{PP}, is less than other ones. This effect may be explained by the fact that the Lyapunov function V_{PP} "*dominates*" other ones, namely, possesses the property

$$V_{FP}(x, y) > V_s(x, y)$$

$$\text{for all } x, y \text{ and } s = MO, OACh, U$$

since it is designed directly by the application of the "Characteristic Function Method" (see [21, 25]) satisfying the inequality (3.32) with more precise upper bound.

3.3 Sliding Mode Observer for Simplest Uncertain Model

3.3.1 Briefly on State Observation Problem

The classical state estimation (filtering) problem for stochastic differential models with complete knowledge of all functions, participating in their descriptions, was solved in 60-es and the obtained state-observer is known as the filter of Dunkan–Mortensen–Zakai. The solution of this problem is given by the Zakai equation [26] which is a bilinear stochastic partial differential equation for the un-normalized density of a hidden state. In contrast, the Kushner equation [27] gives a non-linear stochastic partial differential equation for the normalized density of the hidden state. In principle either approach allows one to estimate a quantity function (the state of a Dynamic system) from noisy measurements, even when the system is non-linear, generalizing the earlier results of Wiener and Kalman for linear systems, and solving a central problem in estimation theory. The application of this approach to a specific engineering situation may be problematic since these equations are quite complex. When the complete information on the dynamics of the system is unavailable robust control methods are required for state estimation designing. One of possible approaches to its realization is the Sliding Mode (SM) approach [2].

The most advanced results have been obtained for *deterministic uncertain systems* [28]. Sliding mode observers have unique properties, in that the ability to generate a sliding motion on the error between the measured plant output and the output of the observer ensures that a sliding mode observer produces a set of state estimates that are precisely commensurate with the actual output of the plant. It is also the case that analysis of the average value of the applied observer injection signal, the so-called equivalent injection signal, contains useful information about the mismatch between the model used to define the observer and the actual plant. These unique properties, coupled with the fact that the discontinuous injection signals which were perceived as problematic for many control applications have no disadvantages for software-based observer frameworks, have generated a ground swell of interest in sliding mode observer methods in recent years. So, in [29] the equivalent control concept, applied for a wide class of nonlinear systems, makes it possible to develop finite-time observers. In [30] there is considered the problem of designing an observer for a linear system subject to unknown inputs. This section shows how the relative degree condition can be weakened if a classical sliding mode observer is combined with sliding mode exact differentiators to essentially generate additional independent output signals from the available measurements. In the paper [31], sliding mode observer design principles based on the equivalent control approach are discussed for a linear time invariant system both in continuous and discrete time. For the continuous case, the observer is designed using a recursive procedure; however, the observer is eventually expressed as a replica of the original system with an additional auxiliary input with a certain nested structure. A direct discrete time counterpart of the sliding mode realization of a reduced order asymptotic observer using the discrete time equivalent control is also developed. The super-twisting second-order sliding-mode algorithm is modified in order to design a velocity observer for uncertain mechanical systems [32]. The finite time convergence of the observer is proved. Thus, the observer can be designed independently of the controller. A discrete version of the observer is considered and the corresponding accuracy is estimated.

As it was mentioned before, the paper [3] (which has also been discussed later in Chap. 14 (4) of [2] may be considered as the first publication dealing with *stochastic* SM approach. The paper [33] investigates the problem of fault estimation and fault-tolerant control against sensor failures for a class of nonlinear Itô stochastic systems with simultaneous input and output disturbances. By using the descriptor sliding mode approach, an accurate estimation of the system states, fault vector and disturbances can be obtained simultaneously. The paper [34] deals with the output feedback sliding mode control for Itô stochastic time-delay systems. The system states are unmeasured, and the uncertainties are unmatched. A sliding mode control scheme is proposed based on the state estimates. By utilizing a special switching function, the derivative of the switching function is ensured to be finite variation. It is shown that the sliding mode in the estimation space can be attained in finite time. The sufficient condition for the asymptotic stability (in probability) of the overall closed-loop stochastic system is derived.

Here we intend to extend the ideas of *the equivalent control method* for the class of mechanical models subjected the stochastic external perturbations of both additive and multiplicative types. The motivation of this study is based on the following remarks:

- Under stochastic perturbations, which are unbounded by its nature, *it is impossible to talk about sliding mode surfaces* as well as on finite-time convergence to it: only the approachment (in some probabilistic sense) to some neighborhood to a sliding surface may be discussed.
- The classical *equivalent control method also can not be applied directly* because of the main reason: any random trajectory can not be differentiable and therefore to talk about its values equal to zero on the sliding surface (which also does not exist) no makes any sense.

We suggest some ideas how to extend the traditional deterministic notions, used in deterministic Sliding Mode theory, to the class of stochastic models where the half of coordinates are not measurable and should be estimated on-line.

3.3.2 Model Description and Problem Formulation

3.3.2.1 Stochastic Model

Consider a filtered probability space $(\Omega, \mathscr{F}, \{H_t\}_{t\geq 0}, P)$ and let $\{(W_t, \mathscr{F}_t)\}_{t\geq 0}$, be vector standard Brownian motion. In this probability space consider the simplest stochastic model given in the Itô form:

$$\left.\begin{aligned} dx_{1,t} &\overset{a.s.}{=} x_{2,t}dt, \\ dx_{2,t} &\overset{a.s.}{=} f(x_t)\,dt + \Xi(x,t)\,dW_t \\ y_t &\overset{a.s.}{=} x_{1,t} \\ x_C = \left(x_{1,0}, x_{2,0}\right) &\overset{a.s.}{=} \bar{x} \end{aligned}\right\} \tag{3.50}$$

- $x_{1,t} = x_{1,t}(\omega) \in \mathbb{R}^n$, $x_{2,t} = x_{2,t}(\omega) \in \mathbb{R}^n$ $(\omega \in \Omega)$ are the coordinates at time $t \geq 0$; the components of the vector $x_{1,t}$ will be referred to as *the states of the system* and the components of the vector $x_{2,t}$ as *their velocities* that corresponds to the standard mechanical interpretation,

- W_t - a standard Wiener process, that is,

$$\mathrm{E}\{W_t/\mathscr{F}_t\} \overset{a.s.}{=} 0,\ \mathrm{E}\left\{W_t W_t^{\mathsf{T}}/\mathscr{F}_t\right\} \overset{a.s.}{=} tI_{n\times n} \tag{3.51}$$

3.3.2.2 Main Assumptions

- The mappings $\Xi : \mathbb{R}^{n\times n} \times \mathbb{R}_+ \to \mathbb{R}^{n\times n}$ and $f : \mathbb{R}^{n\times n} \times \mathbb{R}_+ \to \mathbb{R}^n$ satisfy for all $x \in \mathbb{R}^n$ and $t \geq 0$ the inequalities

$$\|\Xi(x, t)\|^2 \leq d_0 + d_1 \|x\|^2 \tag{3.52}$$

- The model (3.50) is assumed to be *globally quadratically stable*, i.e.,

$$\mathrm{E}\left\{\|x_t\|^2\right\} \leq X_+ < \infty \tag{3.53}$$

(the upper bound X_+ is supposed to be known).

Remark 3.4 Notice that if $d_0 = 0$ we deal with **multiplicative noise**, and if $d_1 = 0$ this case corresponds with the **additive noise** effect.

3.3.2.3 Problem Formulation

The problem of state estimation consists in the designing a vector function $\hat{x}_{2,t}$ such that the error state estimates $e_{2,t} = x_{2,t} - \hat{x}_{2,t}$, obtained by the designed observer, would be as less as possible in some probabilistic sense sense.

3.3.3 SM Observer for Stochastic Models

3.3.3.1 Structure of the Observer

Consider the following observer structure

$$\left.\begin{array}{c}
\mathrm{d}\hat{x}_{1,t} = v_t \mathrm{d}t \\
v_t = -\rho_t \operatorname{Sign}\left(e_{1,t}\right),\ \rho > 0 \\
\\
\rho_t = \rho\left\|e_{1,t}\right\|,\ \rho > 0 \\
e_{1,t} := \hat{x}_{1,t} - x_{1,t} \\
\operatorname{Sign}\left(e_{1,t}\right) := \left(\operatorname{sign}(e_{11,t}), \ldots, \operatorname{sign}(e_{1n,t})\right)^{\mathsf{T}} \\
\\
\hat{x}_{2,t} = v_t
\end{array}\right\} \tag{3.54}$$

3.3.3.2 Main Result

Theorem 3.3 *Under the assumption above the observer (3.54) for large enough*

$$\rho \geq \sqrt{\frac{X_+}{2\varepsilon}},\ \varepsilon > 0$$

provides the "small" error of the state estimate $e_t = \begin{pmatrix} e_{1,t} \\ e_{2,t} = v_t - x_{2,t} \end{pmatrix}$ *fulfilling*

- $$\mathrm{E}\left\{\frac{1}{2}\left\|e_{1,t}\right\|^2\right\} \leq \varepsilon + O\left(e^{-\rho t}\right)$$

- $$\left\|x_{2,t} - v_t\right\| \leq 2\varepsilon^{\alpha(1-\beta)}$$

valid for any $\alpha \leq (0, 1/2)$, $\beta \in (0, 1)$ *and large enough* $t > 0$ *with probability* $1 - 2\varepsilon^{1-2\alpha}$.

Proof For $e_{1,t}$ we have

$$\mathrm{d}e_{1,t} = \left(v_t - x_{2,t}\right)\mathrm{d}t$$

Notice that in this equation the diffusion term is absent. Defining

$$V\left(e_{1,t}\right) = \frac{1}{2}\left\|e_{1,t}\right\|^2$$

we have

$$dV\left(e_{1,t}\right) = e_{1,t}^{\mathsf{T}} de_{1,t} = e_{1,t}^{\mathsf{T}}\left(v_t - x_{2,t}\right)\mathrm{d}t$$

which implies

$$V\left(e_{1,t+\Delta t}\right) - V\left(e_{1,t}\right) = \int_{\tau=t}^{t+\Delta t} e_{1,\tau}^{\mathsf{T}}\left(v_\tau - x_{2,\tau}\right)\mathrm{d}\tau \tag{3.55}$$

Applying the mathematical expectation operator too both sides of (3.55), then deviding by Δt tending to zero, we get

$$\begin{aligned}
&\frac{\mathrm{d}}{\mathrm{d}t}\mathrm{E}\left\{V\left(e_{1,t}\right)\right\} = \mathrm{E}\left\{e_{1,t}^{\mathsf{T}}\left(v_t - x_{2,t}\right)\right\} \\
&= \mathrm{E}\left\{-\rho\left\|e_{1,t}\right\|\sum_{i=1}^{n}\left|e_{1i,t}\right| - e_{1,t}^{\mathsf{T}} x_{2,t}\right\}
\end{aligned}$$

In view of the inequality

$$\sum_{i=1}^{n} \left| e_{1i,t} \right| \geq \left\| e_{1,t} \right\|$$

and applying the Cauchy–Schwartz inequality, for

$$\bar{V}_t := \mathrm{E}\left\{ V\left(e_{1,t}\right)\right\}$$

we obtain

$$\frac{\mathrm{d}}{\mathrm{d}t}\bar{V}_t = \mathrm{E}\left\{ -\rho \left\| e_{1,t} \right\| \sum_{i=1}^{n} \left| e_{1i,t} \right| - e_{1,t}^{\mathsf{T}} x_{2,t} \right\} \leq$$

$$-\mathrm{E}\left\{ \rho \left\| e_{1,t} \right\|^2 \right\} + \sqrt{\mathrm{E}\left\{ \left\| e_{1,t} \right\|^2 \right\}} \sqrt{\mathrm{E}\left\{ \left\| x_{2,t} \right\|^2 \right\}}$$

$$\leq -\mathrm{E}\left\{ \rho \left\| e_{1,t} \right\|^2 \right\} + X_+^{1/2} \sqrt{\mathrm{E}\left\{ \left\| e_{1,t} \right\|^2 \right\}}$$

or, equivalently,

$$\frac{\mathrm{d}}{\mathrm{d}t}\bar{V}_t \leq -2\rho \bar{V}_t + \sqrt{2X_+}\sqrt{\bar{V}_t}$$

$$= 2\rho\sqrt{\bar{V}_t}\left(\frac{\sqrt{X_+/2}}{\rho} - \sqrt{\bar{V}_t} \right)$$

If $\sqrt{\bar{V}_t} > \dfrac{\sqrt{X_+/2}}{\rho}$ the right-hand side of this inequality is negative. Therefore the attraction region is

$$\limsup_{t\to\infty} \sqrt{\bar{V}_t} \leq \frac{\sqrt{X_+/2}}{\rho}$$

and moreover, by the integrating by parts we may conclude that $\bar{V}_t$ exponentially fulfills

$$\bar{V}_t \leq \frac{X_+}{2\rho^2} + ce^{-\rho t}, \ c > 0$$

So, for large enough ρ we can make $\bar{V}_t$ less than any desired ε, namely

$$\bar{V}_t = \mathrm{E}\left\{ V\left(e_{1,t}\right)\right\} \leq \varepsilon + O\left(e^{-\rho t}\right) \tag{3.56}$$

if

$$\rho \geq \sqrt{\frac{X_{+}}{2\varepsilon}} \tag{3.57}$$

By the Chebyshev inequality

$$\mathrm{P}\left\{\left\|e_{1,t}\right\| > \varepsilon\right\} \leq \frac{1}{\varepsilon^2}\mathrm{E}\left\{\left\|e_{1,t}\right\|^2\right\} = \frac{2}{\varepsilon^2}\left(\varepsilon + O\left(e^{-\rho t}\right)\right)$$

For $\varepsilon := \varepsilon^{\alpha}$ ($\alpha \in (0, 1/2)$) we have

$$\mathrm{P}\left\{\left\|e_{1,t}\right| > \varepsilon\right\} \leq O\left(\varepsilon^{1-2\alpha}\right) \underset{\varepsilon\to 0}{\to} 0 \tag{3.58}$$

Consider now the set $\Omega_0 \subset \Omega$ of the realizations $\omega \in \Omega$ containing only the trajectores satisfying $\left\|e_{1,t}(\omega)\right\| \leq \varepsilon$, that is,

$$\Omega_0 := \left\{\omega \in \Omega \mid \left\|e_{1,t}(\omega)\right\| \leq \varepsilon\right\}$$

By (3.58)

$$\mathrm{P}\left\{\Omega_0\right\} = 1 - \mathrm{P}\left\{\left\|e_{1,t}\right\| > \varepsilon\right\} \geq 1 - O\left(\varepsilon^{1-2\alpha}\right) \tag{3.59}$$

and hence, the right-hand side of (3.59) may be done closed to one by the special selection of ρ, satisfying (3.57). In view of that for almost all $\omega \in \Omega_0$

$$\begin{aligned} \mathrm{d}e_{1,t}(\omega) &= \mathrm{d}\hat{x}_{1,t}(\omega) - \mathrm{d}x_{1,t}(\omega) \\ &= \left(v_t(\omega) - x_{2,t}(\omega)\right)\mathrm{d}t \end{aligned}$$

or, in the integral format for any continuous trajectory

$$\begin{aligned} e_{1,t+\Delta t}(\omega) - e_{1,t}(\omega) &= \int_{\tau=t}^{t+\Delta t} \left(v_\tau(\omega) - x_{2,\tau}(\omega)\right)\mathrm{d}\tau \\ &= \left(v_{\tau'}(\omega) - x_{2,\tau'}(\omega)\right)\Delta t, \ \tau' \in (t, t+\Delta t) \end{aligned}$$

This is equivalent to

$$\frac{1}{\Delta t}\left(e_{1,t+\Delta t}(\omega) - e_{1,t}(\omega)\right) = v_{\tau'}(\omega) - x_{2,\tau'}(\omega)$$

which leads for small enough $\Delta t = \varepsilon^{\beta}$ ($\beta \in (0, 1)$) to the following relation

$$\left\|v_{\tau'}(\omega) - x_{2,\tau'}(\omega)\right\| =$$

$$\frac{1}{\Delta t}\left\|e_{1,t+\Delta t}(\omega) - e_{1,t}(\omega)\right\| \leq \frac{2\varepsilon}{\Delta t}$$

$$= 2\varepsilon^{1-\beta} = 2\varepsilon^{\alpha(1-\beta)} = o\left(\varepsilon^{1/2}\right)$$

This means that $x_{2,t}(\omega)$ may be estimated as

$$\hat{x}_{2,t}(\omega) = v_t(\omega) + o\left(\varepsilon^{1/2}\right) \simeq v_t(\omega)$$

for almost all $\omega \in \Omega_0$. According to (3.59) measure of such relation is closed to 1 for large enough ρ (3.57). Theorem is proven.

Remark 3.5 One may consider also the, so-called, **smooth version** of the estimate $\hat{x}_{2,t}(\omega)$ obtained as

$$\left.\begin{aligned} \hat{x}_{2sm,t} &= \hat{v}_{sm,t} \\ \tau \tfrac{d}{dt}\hat{v}_{sm,t} &= -\hat{v}_{sm} - v_t, \; 0 < \tau \ll 1 \end{aligned}\right\} \tag{3.60}$$

Remark 3.6 Notice that when we have no stochastic noise ($d_0 = d_1 = 0$) the suggested state estimation method exactly coincides with the classical equivalent control method with adjusted gain parameter ($\rho_t = \rho\left\|e_{1,t}\right\|$) and in this case $v_t = v_{eq,t}$, where $v_{eq,t}$ is the equivalent control for deterministic case, as well as $\hat{v}_{sm,t} = \tilde{v}_{eq,t}$ where $\tilde{v}_{eq,t}$ is the smooth realization of $v_{eq,t}$ obtained as the output of the first-order low-pass filter (3.60).

3.3.4 Numerical Simulation

3.3.4.1 Single Dimensional State Component

Now for simulation purposes, consider the simple plant with $n = 1$, i.e., one state $x_{1,t}$ and its velocity $x_{2,t}$:

$$\begin{aligned} dx_{1,t} &\overset{a.s.}{=} x_{2,t}dt \\ dx_{2,t} &= -0.5x_2\left[\tan^{-1}(x_1)\right]dt + [16 + x_2]\,dW_t \\ y_t &= x_{1,t} \end{aligned}$$

The stochastic noise is assumed to be both additive and multiplicative nature. In the suggested observer with $\hat{x}_{2,t} = \hat{v}_{sm,t}$ we took $\rho = 90$, and in the low-pass filter

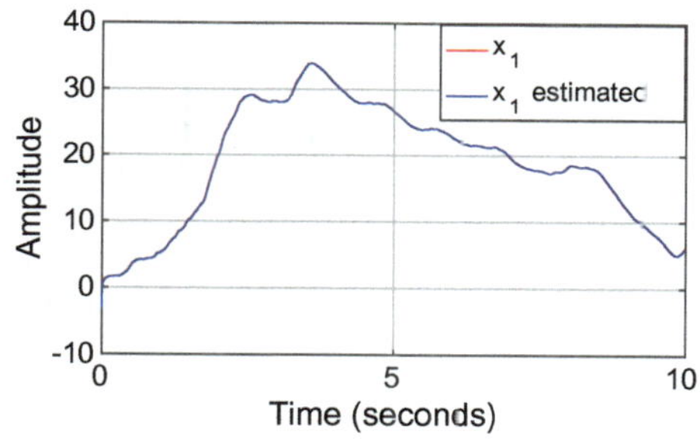

Fig. 3.8 The state $x_{1,t}$ and its estimate $\hat{x}_{1,t}$

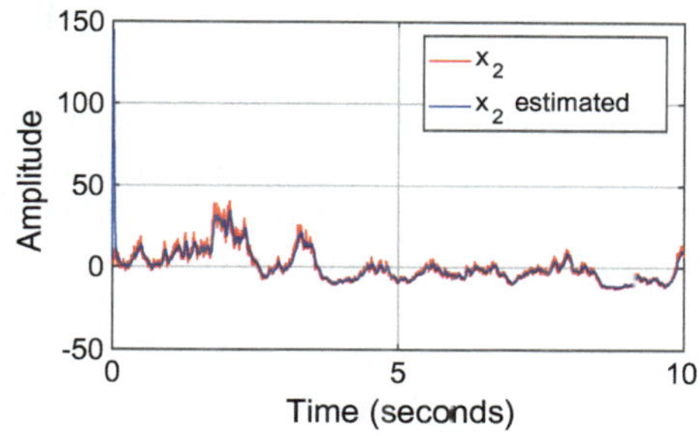

Fig. 3.9 The state $x_{2,t}$ and its estimate $\hat{x}_{2,t}$

the parameter is taken as $\tau = 0.001$. The results of the numerical simulation are shown in Figs. 3.8 and 3.9.

One can see that the estimates $\hat{x}_{1,t}$ of the state $x_{1,t}$ are practically ideal from the first moments of the process. As for the estimates of the velocity component $x_{2,t}$, subjected to stochastic perturbation, we can see the sufficiently good estimation process excepting the first moments in the beginning of the process that may be related to the initial value selection $\hat{v}_{sm,0}$ in the low-pass filter.

3.3.4.2 Two Dimensional State Component Vector

Consider now more complex model containing 2 coordinates $x_{1,t} = \begin{pmatrix} x_{11,t} \\ x_{12,t} \end{pmatrix}$ and the velocities $x_{2,t} = \begin{pmatrix} x_{21,t} \\ x_{22,t} \end{pmatrix}$ connected by the following relation

$$\begin{pmatrix} \mathrm{d}x_{11,t} \\ \\ \mathrm{d}x_{12,t} \end{pmatrix} \stackrel{a.s.}{=} \begin{pmatrix} x_{21,t} \\ \\ x_{22,t} \end{pmatrix} \mathrm{d}t$$

$$
\begin{matrix} \mathrm{d}x_{21,t} \\ \mathrm{d}x_{4,t} \end{matrix} = \begin{pmatrix} -0.5x_{21}\left[\tan^{-1}\left(x_{11,t}\right)\right] \\ \dfrac{\left|x_{21,t}\right| + \left|x_{22,t}\right|}{1 + \left|x_{21,t}\right| + \left|x_{22,t}\right|} \end{pmatrix} \mathrm{d}t
$$

$$
+ \begin{pmatrix} \left[16 + x_{21,t}\right] \\ 0.5x_{12,t} \end{pmatrix} \mathrm{d}W_t
$$

and

$$
y_t = \begin{pmatrix} x_{11,t} \\ x_{21,t} \end{pmatrix}
$$

The parameters of the filter are taken as in the previous example. We obtained the following behavior of the plant and observer coordinates:

The presented Figs. 3.10, 3.11, 3.12 and 3.13 clearly demonstrate a good workability of the suggested observer (3.54).

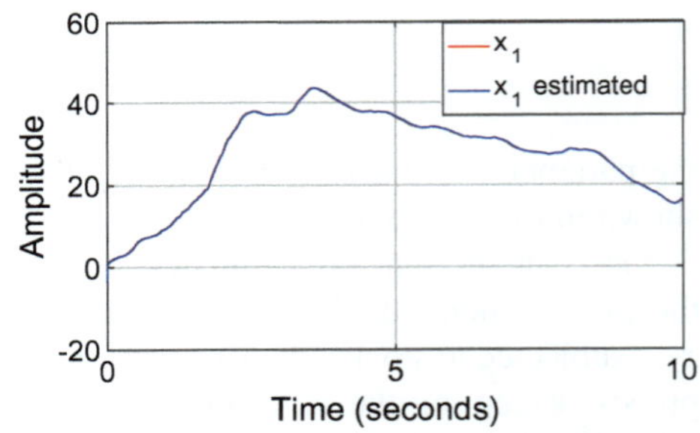

Fig. 3.10 The state $x_{11,t}$ and its estimate $\hat{x}_{11,t}$

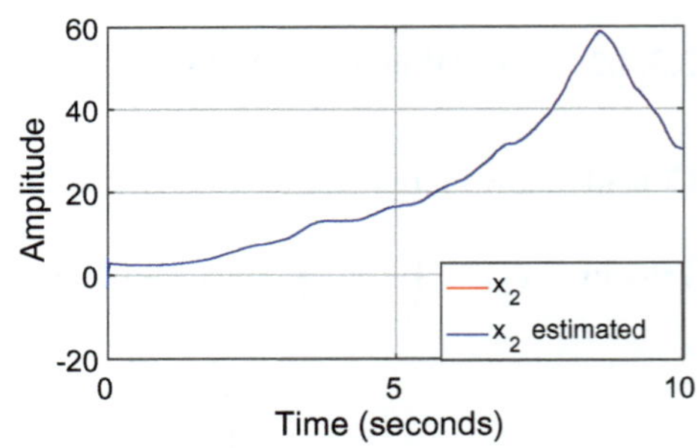

Fig. 3.11 The state $x_{12,t}$ and its estimate $\hat{x}_{12,t}$

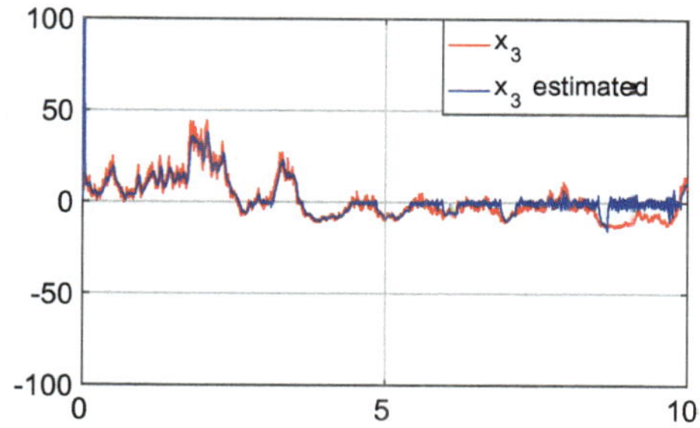

Fig. 3.12 The state $x_{21,t}$ and its estimate $\hat{x}_{21,t}$

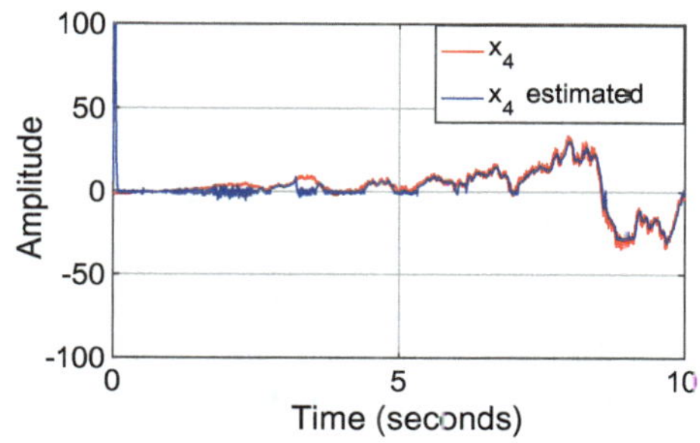

Fig. 3.13 The state $x_{22,t}$ and its estimate $\hat{x}_{22,t}$

3.3.5 Conclusion

Here we demostrate that

- in the case of stochastic systems, governed by the stochastic differential equations of the Itô type, which contain stochastic unbounded and deterministic bounded perturbations, *we can not speak on SMC in the traditional deterministic sense*. Sliding Mode dynamics on the sliding surface $s(x) = 0$ may be understood only as the asymptotic convergence of the mean squared value $E\left\{ |s(x_t)\|^2 \right\}$ to the μ-zone suggested to be referred to as μMS-convergence;
- to realize this convergence the gain parameter of discontinuous controller should be state-dependent on norms $\|s(x_t)\|$ (if we deal with multi sliding surfaces) considered on the system trajectories;
- the exponential convergence of the square-norm of the averaged squared norm of the sliding variable to μ-zone (around the sliding surface) is guaranteed. It can be done small enough by higher gain parameter selection k_0;
- for stability analysis of *the stochastic version of super-twist controller* we suggest to apply a new technique, based on the Lyapunov functions suggested for the deterministic versions of the same system. It is shown that the special selection of one of gain-parameters of such controller provides it with an "*adaptivity property*" which guarantees the mean-square convergence of the applied Lyapunov function

to the prespecified zone depending on the diffusion parameter and parameters of the controller;

- we suggest the sliding mode observer with the gain parameter linearly depending on the norm of the output estimation error which is available during the process. It is shown that for any desired $\varepsilon > 0$ the selection of the large enough gain parameter $\rho \geq \sqrt{\frac{X_+}{2\varepsilon}}$ may guarantee for large enough $t > 0$ the quality of the second vector-component

$$\left\| x_{2,t} - v_t \right\| \leq 2\varepsilon^{\alpha(1-\beta)}, \ \alpha \leq (0, 1/2), \beta \in (0, 1)$$

with the probability

$$P\left\{\omega \in \Omega \mid \left\| x_{2,t}(\omega) - v_t(\omega) \right\| \leq 2\varepsilon^{\alpha(1-\beta)}\right\} \geq 1 - 2\varepsilon^{1-2\alpha}$$

closed to one.

The numerical simulations show a good workability of the suggested technique.

Stochastic Sliding Mode Control and State Estimation Approach, discussed in this chapter, may be considered and spread in the following directions: the Equivalent Control Method for stochastic SM, Stochastic Twist, Super-Twist and Nested controllers, Stochastic Integral SM, Stochastic SM observers, Output based SM with stochastic perturbations, Adaptive stochastic SM and etc. The deterministic versions of some problems mentioned above can be found in [2, 21, 35, 36].

Acknowledgements The author would like to express his appreciation to Prof. V. Utkin for his critical comments and suggestions concerning the first part of this chapter and to my MS-student Edgar Pelaez for his help with the simulations of the stochastic observer examples.

References

1. Edwards, C., Spurgeon, S.: Sliding Mode Control: Theory and Applications. Taylor Francis, New York (1998)
2. Utkin, V.: Sliding Modes in Control Optimization. Springer, Berlin (1992)
3. Drakunov, S.: On adaptive quasioptimal filter with discontinuous parameters. Autom. Remote Control **44**(9), 1167–1175 (1983)
4. Niu, Y., Ho, D.: Design of sliding mode control for nonlinear stochastic systems subject to actuator nonlinearity. IEE Proc. Control Theory Appl. **153**(6), 737–744 (2006)
5. Niu, Y., Liu, Y., Jia, T.: Reliable control of stochastic systems via sliding mode technique. Optim. Control. Appl. Methods **34**(6), 712–727 (2013)
6. Gao, L., Wu, Y.: Control for stochastic systems with Markovian switching and time-varying delay via sliding mode design. Mathematical Problems in Engineering (2013, in press). doi:10.1155/2013/898172
7. Wu, L., Daniel, W.: Sliding mode control of singular stochastic hybrid systems. Automatica **46**(4), 779–783 (2010)
8. Basin, M., Rodriguez-Ramirez, P.: Sliding mode controller design for stochastic polynomial systems with unmeasured states. IEEE Trans. Ind. Electron. **61**(1), 387–396 (2013)

9. Aubin, J., Prato, G., Frankowska, H.: Stochastic invariance for differential inclusions. Stoch. Anal. Appl. **8**, 181–201 (2000)
10. Aumann, R.: Integrals of set-valued functions. J. Math. Anal. Appl. **12**, 1–12 (1965)
11. Kisielewicz, M.: Stochastic Differential Inclusions and Applications. Springer Optimization and Its Applications. Springer, New York (2013)
12. Pratro, G., Frankowska, H.: A stochastic Filippov theorem. Stoch. Anal. Appl. **12**(4), 409–426 (1994)
13. Veretennikov, A.: On strong solutions and explicit formula for solution of stochastic integral equations. Math. USSR Sbornik **39**(3), 387–403 (1981)
14. Jacod, J., Memin, J.: Weak and strong solutions of stochastic differential equations: existence and stability. Stochastic Integrals. LNM, vol. 851, pp. 169–213. Springer, Berlin (1980)
15. Øksendal, B.K.: Stochastic Differential Equations: An Introduction with Applications. Springer, Berlin (2003)
16. Filippov, A.: Differential equations with discontinuous right-hand sides. Mat. Sb. (N.S.) **51**(93), 1, 99–128 (1960)
17. Higham, D.J.: An algorithmic introduction to numerical simulation of stochastic differential equations. SIAM Rev. **43**(3), 525–546 (2001)
18. Levant, A.: Sliding order and sliding accuracy in sliding mode control. Int. J. Control **58**(6), 1247–1263 (1993)
19. Shtessel, Y., Edwards, C., Fridman, L., Levant, A.: Sliding Mode Control and Observation. Birkhauser, New York (2014)
20. Moreno, J., Osorio, M.: Strict Lyapunov functions for the supertwisting algorithm. IEEE Trans. Autom. Control **57**(4), 1035–1040 (2012)
21. Polyakov, A., Poznyak, A.: Reaching time estimation for super-twisting second order sliding mode controller via Lyapunov function designing. IEEE Trans. Autom. Control **54**(8), 1951–1955 (2009)
22. Orlov, Y., Aoustin, Y., Chevellereau, C.: Finite time stabilization of a perturbed double integrator - part i: continuous sliding modebased output feedback synthesis. IEEE Trans. Autom. Control **56**(3), 1035–1040 (2010)
23. Utkin, V., Guldner, J., Shi, J.: Sliding Mode Control in Electro-Mechanical Systems. CRC Press, Boca Raton (2009)
24. Poznyak, A.: Sliding mode control in stochastic continuos-time systems: μ-zone MS-convergence. IEEE Trans. Autom. Control (2016). doi:10.1109/TAC.2016 2557759
25. Poznyak, A.: The method of characteristics for the Lyapunov function design in the finite-time stability analysis of sliding mode controlled systems with uncertainties. In: Preprints of the 18th IFAC World Congress, Milano, pp. 768–773 (2011)
26. Zakai, M.: On the optimal filtering of diffusion processes. Zeitschrift für Wahrscheinlichkeitstheorie und Verwandte Gebiete **11**(3), 230–243 (1969)
27. Kushner, H.: On the differential equations satisfied by conditional probability densities of Markov processes with applications. J. SIAM Control Ser. A **2**(1), 106–119 (1964)
28. Spurgeon, S.: Sliding mode observers: a survey. Int. J. Syst. Sci. **39**(8), 751–764 (2008)
29. Drakunov, S., Utkin, V.: Sliding mode observers. Tutorial. In: Proceedings of the 34th IEEE Conference on Decision and Control, pp. 3376–3378 (1995)
30. Floquet, T., Edwards, C., Spurgeon, S.: On sliding mode observers for systems with unknown inputs. Int. J. Adapt. Control Signal Process. **21**(8–9), 638–656 (2007)
31. Haskara, I.: On sliding mode observers via equivalent control approach. Int. J. Control **71**(6), 1051–1067 (1998)
32. Davila, J., Fridman, L., Levant, A.: Second-order sliding-mode observer for mechanical systems. IEEE Trans. Autom. Control **50**(11), 1785–1789 (2005)
33. Liu, M., Shi, P.: Sensor fault estimation and tolerant control for Itô stochastic systems with a descriptor sliding mode approach. Automatica **49**(5), 1242–1250 (2013)
34. Niu, Y., Ho, D.: Robust observer design for Itô stochastic time-delay systems via sliding mode control. Syst. Control Lett. **55**(10), 781–793 (2006)

35. Plestan, F., Shtessel, Y., Bré geault, V., Poznyak, A.: New methodologies for adaptive sliding mode control. Int. J. Control **83**(9), 1907–1919 (2010)
36. Utkin, V., Poznyak, A.: Adaptive sliding mode control with application to super-twist algorithm: equivalent control method. Automatica **49**, 39–47 (2013)
37. Poznyak, A.: Advanced Mathematical Tools for Automatic Control Engineers: Stochastic Technique, vol. 2. Elsevier, Amsterdam (2009)

Chapter 4
Practical Stability Phase and Gain Margins Concept

Yuri Shtessel, Leonid Fridman, Antonio Rosales and Chandrasekhara Bharath Panathula

4.1 Introduction

Finite-Time Convergent Controllers (FTCC), specifically, Sliding Mode/Higher Order Sliding Mode (SM/HOSM) controllers provide finite-time convergence for sliding set (sliding variable and consecutive time sliding variable derivatives) of any arbitrary relative degree system to the origin in the presence of matched and bounded perturbations, which significantly affects sliding variable dynamics [1, 2]. The advantages of FTCC have been investigated and reported by many works, some of which are [1–5].

A compelling issue is with the certification of FTCC robustness to unmodeled dynamics for practical application. In classical control theory, robustness of linear controllers to unmodeled dynamics is characterized by Phase Margin (*PM*) and Gain

L. Fridman gratefully acknowledges the financial support from by Programa de Apoyo a Proyectos de Investigacion e Innovacion Tecnologica (UNAM) 113216, and DGAPA PASPA Program.

Y. Shtessel · C.B. Panathula
The Department of Electrical and Computer Engineering,
University of Alabama in Huntsville, Huntsville, AL, USA
e-mail: shtessy@uah.edu

C.B. Panathula
e-mail: cpanathula@gmail.com

L. Fridman (✉)
The Department of Control Engineering and Robotics, Universidad Nacional Autonoma de Mexico, 04510 Mexico City, CDMX, Mexico
e-mail: lfridman@unam.mx

A. Rosales
Tecnologico de Monterrey, Campus Ciudad de Mexico, MX, Mexico
e-mail: jarm@ieee.org

S. Li et al. (eds.), *Advances in Variable Structure Systems and Sliding Mode Control—Theory and Applications*, Studies in Systems, Decision and Control 115, DOI 10.1007/978-3-319-62896-7_4

Margin (GM). It is to be worth noting that controller must achieve prescribed values on PM and GM for obtaining deployment certification [6]. For example, according to Goddard space flight center [6], required $PM = 30°$ and $GM = 12\,\text{dB}$ in order for controller to be eligible for flight space control. Unfortunately, these classical robustness metrics (PM and GM) are hard to justify for nonlinear FTCC.

A side effect of finite-time convergence provided by FTCC is chattering/limit-cycle (self-sustained oscillations exhibited by sliding variable with non-zero amplitude and high, but finite, frequency) that may occur due to the presence of unmodeled dynamics, which are exists in any mathematical model used for control design [7], and due to theoretical infinite gain at the origin. Chattering in systems controlled by FTCC due to unmodeled dynamics of physical actuators, such as, hydraulic and pneumatic actuators, etc., was investigated by several authors, including but not limited to [3, 8–17].

The predicted chattering in systems controlled by FTCC due to the presence of unmodeled dynamics can be analyzed using time domain techniques like Poincare maps [11, 18, 19]. Note that such techniques are often used for analysis of second order systems, and are hard to determine chattering/limit-cycle parameters (amplitude and frequency of main harmonic). On the other hand, frequency domain techniques like Tsypkin Locus [20], LPRS [3], the method of solving Harmonic Balance equation using Describing Function of FTCC (DF-HB) [21, 22] can be used to determine predicted chattering parameters. However, Tsypkin Locus [20] and LPRS [3] methods are mostly limited to relay controlled system.

DF-HB method is the only engineering method that is heavily used for predicting limit-cycle parameters (amplitude and frequency of main harmonic) in nonlinear systems [21, 22]. The exactness of DF-HB method to dynamically perturbed systems controlled by FTCC in predicting and analyzing chattering parameters was confirmed by many works, such as [3, 9].

DFs for second order FTCC, such as, twisting and super-twisting SM controllers (see [23, 24], for example), were realized analytically by representing the controllers with known nonlinearities, whose DFs are available in DF tables [14, 21, 22, 25], since the complexity of controllers is less. Analytical computation of DFs for higher order FTCC, specifically, nested [26] and quasi-continuous [27] HOSM controllers is very hard due to the complexity of control algorithms.

A certain level of chattering/limit-cycle parameters (maximum value of amplitude and minimum value of frequency for main harmonic), which are acceptable in practical control system in the sense of practical stability, is defined as Tolerance Limits (TL). Then, based on TL, acceptable chattering is characterized in terms of Practical Stability Phase Margin ($PSPM$) and Practical Stability Gain Margin ($PSGM$) as maximum additional phase lag and maximum additional gain that can be added to FTC-controlled system, while sliding variable exhibits chattering/limit-cycle satisfying TL, respectively. These robustness metrics ($PSPM$ and $PSGM$) acts as a tool for FTCC to present robustness to unmodeled dynamics to obtain deployment certification.

The main contributions of this work are as follows:

1. A new concept of chattering characterization for the systems driven by finite-time convergent controllers (FTCC) in terms of practical stability margins is presented. Specifically,
    a. The notion of Tolerance Limits (TL) is proposed as the admissible level of chattering exhibited by sliding variable due to the presence of unmodeled dynamics in systems controlled by FTCC.
    b. The metrics of robustness, Practical Stability Phase Margins ($PSPM$) and Practical Stability Gain Margin ($PSGM$), are proposed for quantifying FTCC robustness to unmodeled dynamics.
    c. It is proposed to cascade FTCC with linear compensator for achieving prescribed values of $PSPM$ and $PSGM$.
2. The numerical algorithms for identifying proposed robustness metrics ($PSPM$ and $PSGM$) are developed and proposed using DF-HB method.
    a. A numerical algorithm for the identification of DF of FTCC is proposed. A database of DFs for FTCC, specifically, N-HOSM and Q-HOSM controllers is developed.
    b. A numerical algorithm for obtained chattering parameters, amplitude and frequency, by solving HB equation is proposed.
    c. Numerical algorithms for the identification of proposed robustness metrics ($PSPM$ and $PSGM$) are proposed.
3. It is shown how the concept of practical stability margin can close the gap of certification for FTCC robustness to unmodeled dynamics. Proposed approach is validated on a case study of controlling attitude of F-16 aircraft.

4.2 Problem Formulation

Consider SISO continuous-time dynamic system

$$\dot{x}(t) = a(t, x) + b(t, x)v, \quad \sigma(t) = \sigma(t, x), \tag{4.1}$$

where $x(t) \in \mathbb{R}^n$ is a vector of system states, $a(t, x)$, $b(t, x) \in \mathbb{R}^n$ are partially known smooth enough and known Lipschitz vector fields, respectively, $\sigma(t) \in \mathbb{R}$ is a sliding variable or a system output, and $v \in \mathbb{R}$ is a small constant characterizing unmodeled dynamics. Let r_d be relative degree of system (4.1).

The input-output dynamics of system (4.1) are derived as

$$\sigma^{(r_d)} = g(t, x) + h(t, x)v, \quad r_d \leq n, \tag{4.2}$$

where $g(t,x) = \sigma^{(r_d)}|_{v=0}$ is unknown smooth function, and $h(t,x) = \frac{\partial}{\partial v}(\sigma^{(r_d)}) \neq 0$ is a known smooth function. Assuming that $g(t,x)$ does not depend on x, $g(t,x) = \zeta(t)$ and $\dot{\zeta}(t)$ is bounded, i.e., $|\dot{\zeta}(t)| \leq L$, $L > 0$, $h(t,x)v = u(t)$, and internal/zero dynamics are stable, input-output dynamics in Eq. (4.2) are rewritten as

$$\sigma^{(r_d)} = u(t) + \zeta(t). \tag{4.3}$$

Suppose system (4.1) is actuated by unmodeled dynamics

$$\mu \dot{z} = f(z,u);\ v = v(z), \tag{4.4}$$

where $z \in \mathbb{R}^m$, $u \in \mathbb{R}$ is a FTCC, $\mu > 0$ is a small time constant, output of $v(z)$ is a continuous function, and $f(z,u)$ is a Lipschitz function. From [4, 26, 27], it is known that FTCC of order $r = r_d$, being applied to system (4.1), drives sliding set $\{\sigma, \dot{\sigma}, \ddot{\sigma}, \ldots, \sigma^{(r-1)}\}$ to zero in finite time in the case of SM/HOSM control, and to a small vicinity of the origin in the other FTCC cases. However, FTCC u that is applied to dynamically perturbed system (4.1), (4.4) may not drive the set $\{\sigma, \dot{\sigma}, \ddot{\sigma}, \ldots, \sigma^{(r-1)}\} \to 0$, and chattering (limit cycle with non-zero amplitude and finite frequency) may occur due to theoretical infinite gain at the origin.

Dynamically perturbed system (4.1), (4.4) is linearized in the vicinity of the origin,

$$\begin{aligned} \dot{x} &= Ax + Bu + \xi, \\ \sigma &= Cx, \end{aligned} \tag{4.5}$$

where $x \in \mathbb{R}^{n+m}$, $u \in \mathbb{R}$, matrix A and vectors B, C have appropriate dimensions, $\sigma \in \mathbb{R}$, and $\xi \in \mathbb{R}^{n+m}$ represents a vector of uncertainties and perturbations, including the difference between original and linearized systems. Usually, FTCC is able to compensate the perturbations/uncertainties (SM/HOSM controllers compensate them exactly, while the other FTCC attenuate them). A side effect of this control/compensation is chattering/limit-cycle (self-sustained oscillations exhibit by sliding-variable/system-output σ) that may occur due to cascade unmodeled dynamics. In this chapter, the robustness of FTCC to unmodeled dynamics is quantified, while chattering in dynamically perturbed FTC-controlled systems is analyzed assuming that the uncertainty/perturbation ξ is compensated completely by FTCC.

The following example illustrates the side effect (chattering) of finite-time convergence provided by FTCC to system (4.1) output in the presence of cascade unmodeled dynamics.

Example 4.1 Consider a linearized system (4.5) as

$$\dot{x}_1 = x_2,\ \dot{x}_2 = -8x_1 - 3x_2 + u_1,\ \sigma = x_1. \tag{4.6}$$

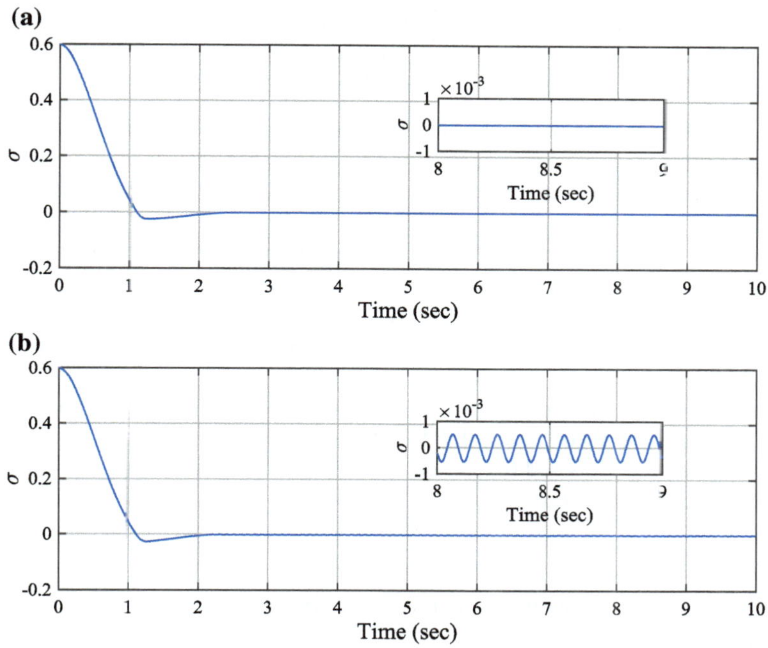

Fig. 4.1 σ in the **a** dynamically unperturbed (4.6) and **b** dynamically perturbed (4.6), (4.8) system controlled by FTCC (4.7)

The dynamically unperturbed system (4.6) is controlled by the FTCC [4]

$$u = -\mu_1|\sigma|^{\frac{\alpha}{2-\alpha}}\text{sign}[\sigma] - \mu_2|\dot{\sigma}|^{\alpha}\text{sign}[\dot{\sigma}], \tag{4.7}$$

with $\alpha = 0.5$ and $\mu_1, \mu_2 = 1$, i.e., $u_1 = u$.

Figure 4.1a shows that the output σ is driven to zero in finite-time, when order of control (4.7) is equal to relative degree of system (4.6), i.e., $r = r_d$.

Next, cascade unmodeled dynamics are considered as

$$\dot{u}_1 = u_2, \ \dot{u}_2 = -\frac{1}{T^2}u_1 - \frac{1}{T}u_2 + \frac{1}{T^2}u. \tag{4.8}$$

Then, the same control (4.7), including same control gains, is given to the dynamically perturbed system (4.6), (4.8) with $T = 0.01$.

Figure 4.1b shows that the output σ of dynamically perturbed system (4.6), (4.8) controlled by FTCC (4.7) exhibits chattering with non-zero amplitude and high, but finite, frequency due to infinite gain at the origin.

In this chapter, the robustness of FTCC to cascade unmodeled dynamics is characterized in terms of chattering parameters (amplitude and frequency of main harmonic) using DF-HB technique. The outlined problem is as follows:

(a) To introduce the notion of Tolerance Limits in terms of the amplitude $A_c > 0$ and the frequency $\omega_c > 0$ that correspond to acceptable self-sustained oscillations of sliding variable σ in dynamically perturbed system (4.1), (4.4) controlled by FTCC.
(b) To introduce the notions of Practical Stability Margins ($PSPM$ and $PSGM$) that characterize the practical stability of closed loop system (4.1), (4.4) controlled by FTCC to additional dynamic perturbations.
(c) To develop the algorithms for computing $PSPM$ and $PSGM$ in the system (4.1), (4.4) controlled by FTCC.
(d) To develop the design algorithm of cascade linear compensator to FTCC for enforcing prescribed values of $PSPM$ and $PSGM$.
(e) To validate the proposed theoretical developments on a case study of controlling attitude of F-16 aircraft.

4.3 Frequency Domain Analysis of Systems Controlled by FTCC

DF-HB technique, a frequency domain method, allows computing the predicted chattering/limit-cycle parameters (amplitude A and the frequency ω of fundamental harmonic) in dynamically perturbed FTC-controlled system (4.1), (4.4) by solving HB equation [3, 9, 21, 22]:

$$W(j\omega) = -N^{-1}(A, \omega), \tag{4.9}$$

where $N(A, \omega)$ is DF of FTCC, and $W(j\omega) = C(sI - A)^{-1}B$. It is assumed that the computed $W(j\omega)$ meets the following assumptions:

A1. $W(j\omega)$ has the low-pass filter properties, i.e., $|W(j\omega)| >> |W(jn\omega)|$ for $n = 2, 3, \ldots$
A2. Amplitude and phase frequency characteristics of $W(j\omega)$ are monotonously decreasing functions, i.e., for $\omega_1 < \omega_2$, $|W(j\omega_1)| > |W(j\omega_2)|$ and arg $W(j\omega_1) >$ arg $W(j\omega_2)$.

4.3.1 *Tolerance Limits in FTC-controlled Systems*

Definition 4.1 The frequency $0 < \omega_c < \infty$ and the amplitude $A_c > 0$ are said to be the Tolerance Limits (TL) for the limit cycle of sliding variable σ, if the amplitude $A \le A_c$ and the frequency $\omega \ge \omega_c$ of the limit cycle correspond to the acceptable performance of the FTC-controlled dynamically perturbed closed-loop system.

Note that σ-limit cycle may not satisfy TL in all the cases loosing practical stability due to the presence of unmodeled dynamics, which are inescapable in any mathematical model used for control design. Therefore, it is proposed to quantify FTCC robustness to unmodeled dynamics by using practical stability Margins ($PSPM$ and $PSGM$), whose definitions are introduced in the next section, like in classical linear control systems.

4.3.2 *Practical Stability Margins in FTC-controlled Systems*

Definition 4.2 (*Practical Stability Gain Margin*) The $PSGM$ in the closed loop system controlled by FTCC is the maximum additional gain added to the frequency characteristic of the linear (linearized) plant $W(j\omega)$, while the output σ exhibits a limit cycle with marginally reached amplitude $A = A_c$ and/ frequency $\omega = \omega_c$ whichever comes first.

Definition 4.3 (*Practical Stability Phase Margin*) The $PSPM$ in the closed loop system controlled by FTCC is the maximal additional phase shift that can be added to the frequency characteristic of the linear (linearized) plant $W(j\omega)$, while the output σ exhibits a limit cycle with marginally reached amplitude $A = A_c$ and/ frequency $\omega = \omega_c$ whichever comes first.

The metrics ($PSPM$ and $PSGM$) are used to quantify FTCC robustness to unmodeled dynamics, and are used as a tool for obtaining control certification for practical implementation. The algorithms for the identification of robustness metrics ($PSPM$ and $PSGM$) are developed using DF-HB method in the next section.

4.3.3 *Development of Algorithms for the Identification of Practical Stability Margins*

DF-HB method is used to develop algorithms for the identification of practical stability margins ($PSPM$ and $PSGM$), while assuming A1 and A2.

4.3.3.1 Numerical Algorithm for DF Identification

It is very hard to compute analytical DFs for higher order ($r \geq 2$) FTCC, since the complexity of control algorithms will be increased. So, it is proposed to use the following numerical algorithm for the identification of DF for FTCC [28]:

1. Isolate the FTCC from the frequency characteristics of the linearized dynamically perturbed system (4.5), i.e., $W(j\omega)$.
2. Excite the FTCC by a set of sinusoidal inputs $E_{k,i} = A_k \sin(\omega_i t)$ of the known amplitude A_k and the known frequency ω_i, where $k, i = 1, 2, 3, \ldots$
3. Compute the output $u_{k,i}$ of the FTCC.
4. Using the Fourier series technique, compute the DF of the FTCC as follows [21, 22],

$$N_{k,i}(A_k, \omega_i) = \frac{2}{TA_k}\int_0^T f_{k,i}\left(A_k \sin(\omega_i t), t\right)\left(\sin(\omega_i t) + j\cos(\omega_i t)\right) dt, \tag{4.10}$$

where $T = \frac{2\pi}{\omega_i}$ and $f_{k,i}(A_k \sin(\omega_i t), t) = u_{k,i}$.

Negative reciprocals of DF, $-N_{i,k}^{-1}(A_k, \omega_i)$, for FTCC is represented in the form of a look-up Table 4.1, where each element $\mathrm{Re}(-N^{-1}(A_k, \omega_i)) + j\mathrm{Im}(-N^{-1}(A_k, \omega_i))$ corresponds to the sinusoidal input $E_{k,i} = A_k \sin(\omega_i t)$.

A database of DFs for higher order FTCC, specifically, Nested [26] (N-HOSM) and Quasi-continuous [27] (Q-HOSM) controllers (upto $r \leq 4$) is developed using the proposed numerical algorithm as follows:

The family of N-HOSM control algorithms for any arbitrary value of r is given by [26]

$$u = -\lambda \Psi_{r-1,r}\left(\sigma, \dot{\sigma}, .., \sigma^{(r-1)}\right), \tag{4.11}$$

where

$$E_{i,r} = \left(|\sigma|^{\kappa/r} + |\dot{\sigma}|^{\kappa/(r-1)} + \ldots + |\sigma^{(i-1)}|^{\kappa/(r-i+1)}\right),$$
$$\Psi_{0,r} = \mathrm{sign}[\sigma[, \quad \Psi_{i,r} = \mathrm{sign}[\sigma^{(i)} + \chi_i E_{i,r} \Psi_{i-1,r}],$$

$\lambda > 0$ is a controller gain, $\kappa > 1$ and $\chi_i > 0$ $(i = 1, 2, .., r-1)$ represent the control parameters.

Table 4.1 $-N^{-1}(A_k, \omega_i)$ look-up table representation

ω	A		
	A_1	$\cdots$	A_k
ω_1	$\mathrm{Re}(-N^{-1}(A_1, \omega_1)) +$ $j\mathrm{Im}(-N^{-1}(A_1, \omega_1))$	$\cdots$	$\mathrm{Re}(-N^{-1}(A_k, \omega_1)) +$ $j\mathrm{Im}(-N^{-1}(A_k, \omega_1))$
ω_2	$\mathrm{Re}(-N^{-1}(A_1, \omega_2)) +$ $j\mathrm{Im}(-N^{-1}(A_1, \omega_2))$	$\cdots$	$\mathrm{Re}(-N^{-1}(A_k, \omega_2)) +$ $j\mathrm{Im}(-N^{-1}(A_k, \omega_2))$
$\vdots$	$\vdots$	$\vdots$	$\vdots$
ω_i	$\mathrm{Re}(-N^{-1}(A_1, \omega_i)) +$ $j\mathrm{Im}(-N^{-1}(A_1, \omega_i))$	$\cdots$	$\mathrm{Re}(-N^{-1}(A_k, \omega_i)) +$ $j\mathrm{Im}(-N^{-1}(A_k, \omega_i))$

For instance, the N-HOSM control algorithms for $r \leq 4$ can be written as [26]

$$\begin{aligned}
r = 1;\quad u &= -\lambda \text{sign}[\sigma],\\
r = 2;\quad u &= -\lambda \text{sign}\left[\dot{\sigma} + |\sigma|^{1/2}\text{sign}[\sigma]\right],\\
r = 3;\quad u &= -\lambda \text{sign}\left[\ddot{\sigma} + 2\,(|\dot{\sigma}|^3 + |\sigma|^2)^{1/6}\text{sign}\,[\dot{\sigma} + |\sigma|^{2/3}\text{sign}[\sigma]]\right],\\
r = 4;\quad u &= -\lambda \text{sign}\left[\dddot{\sigma} + 3\,(\ddot{\sigma}^6 + \dot{\sigma}^4 + |\sigma|^3]^{1/12}\text{sign}\left[\ddot{\sigma} + (\dot{\sigma}^4 + |\sigma|^3)^{1/6}\times\right.\right.\\
&\qquad \left.\left.\text{sign}\left[\dot{\sigma} + 0.5|\sigma|^{3/4}\text{sign}[\sigma]\right]\right]\right],
\end{aligned} \tag{4.12}$$

where $\lambda > 0$ is a controller gain.

The family of Q-HOSM control algorithms for any arbitrary value of r is given by [27]

$$u = -\lambda\Psi_{r-1,r}(\sigma, \dot{\sigma}, .., \sigma^{(r-1)}), \tag{4.13}$$

where

$$\begin{aligned}
E_{0,r} &= \sigma,\ D_{0,r} = |\sigma|,\ \Psi_{0,r} = \frac{E_{0,r}}{D_{0,r}} = \text{sign}[\sigma],\ \Psi_{i,r} = \frac{E_{i,r}}{D_{i,r}},\\
E_{i,r} &= \sigma^{(i)} + \chi_i D_{i-1,r}^{(r-i)/(r-i+1)}\Psi_{i-1,r};\ D_{i,r} = |\sigma^{(i)}| + \chi_i D_{i-1,r}^{(r-i)/(r-i+1)},
\end{aligned}$$

$\lambda > 0$ is the controller gain, and $\chi_i > 0$ $(i = 1, 2, .., r-1)$ represents the control parameters.

For instance, the Q-HOSM control algorithms for $r \leq 4$ can be written as [27]

$$\begin{aligned}
r = 1;\quad u &= -\lambda \text{sign}[\sigma],\\
r = 2;\quad u &= -\lambda\left(\dot{\sigma} + |\sigma|^{1/2}\text{sign}[\sigma]\right) / \left(|\dot{\sigma}| + |\sigma|^{1/2}\right),\\
r = 3;\quad u &= -\lambda\left(\ddot{\sigma} + 2\,(|\dot{\sigma}| + |\sigma|^{2/3})^{-1/2}(\dot{\sigma} + |\sigma|^{2/3}\text{sign}[\sigma])\right) /\\
&\qquad \left(|\ddot{\sigma}| + 2\,(|\dot{\sigma}| + |\sigma|^{2/3})^{1/2}\right),\\
r = 4;\quad u &= -\lambda E_{3,4}/D_{3,4},\\
E_{3,4} &= \dddot{\sigma} + 3\left(\ddot{\sigma} + (|\dot{\sigma}| + 0.5|\sigma|^{3/4})^{-1/3}(\dot{\sigma} + 0.5|\sigma|^{3/4}\text{sign}[\sigma])\right)\times\\
&\qquad \left(|\ddot{\sigma}| + (|\dot{\sigma}| + 0.5|\sigma|^{3/4})^{2/3}\right)^{1/2},\\
D_{3,4} &= |\dddot{\sigma}| + 3\left(\ddot{\sigma} + (|\dot{\sigma}| + 0.5|\sigma|^{3/4})^{2/3}\right)^{1/2},
\end{aligned} \tag{4.14}$$

where $\lambda > 0$ is a controller gain.

DFs for N-HOSM (4.12) and Q-HOSM (4.14) controllers with $r = 2, 3, 4$ and $\lambda = \chi = 1$ are computed using the proposed numerical procedure. The controllers are excited by a set of sinusoidal inputs $E_{k,i} = A_k \sin(\omega_i t)$ with $A_k \in [0.0001, 0.01]$

Table 4.2 $-N^{-1}(A_k, \omega_i)$ look-up table for N-HOSM ($r = 2$)

ω	A			
	0.0001	$\cdots$	0.00998	0.01
30	$-7.821 \times 10^{-5} + j7.046 \times 10^{-6}$	$\cdots$	$-2.555 \times 10^{-3} + j7.406 \times 10^{-3}$	$-0.002.560 \times 10^{-3} + j7.421 \times 10^{-3}$
30.541	$-7.819 \times 10^{-5} + j7.339 \times 10^{-6}$	$\cdots$	$-2.510 \times 10^{-3} + j7.421 \times 10^{-3}$	$-0.002.515 \times 10^{-3} + j7.436 \times 10^{-3}$
$\vdots$	$\vdots$	$\vdots$	$\vdots$	$\vdots$
298.917	$-2.667 \times 10^{-5} + j7.333 \times 10^{-5}$	$\cdots$	$-0.445 \times 10^{-3} + j7.772 \times 10^{-3}$	$-0.000.446 \times 10^{-3} + j7.787 \times 10^{-3}$
299.458	$-2.736 \times 10^{-5} + j7.343 \times 10^{-5}$	$\cdots$	$-0.380 \times 10^{-3} + j7.811 \times 10^{-3}$	$-0.000.381 \times 10^{-3} + j7.826 \times 10^{-3}$
300	$-2.586 \times 10^{-5} + j7.382 \times 10^{-5}$	$\cdots$	$-0.298 \times 10^{-3} + j7.799 \times 10^{-3}$	$-0.000.298 \times 10^{-3} + j7.815 \times 10^{-3}$

Table 4.3 $-N^{-1}(A_k, \omega_i)$ look-up table for N-HOSM ($r = 3$)

ω	A			
	0.0001	$\cdots$	0.00998	0.01
30	$-4.826 \times 10^{-5} + j6.192 \times 10^{-5}$	$\cdots$	$7.777 \times 10^{-3} + j0.974 \times 10^{-3}$	$7.792 \times 10^{-3} + j0.976 \times 10^{-3}$
30.541	$-3.867 \times 10^{-5} + j7.989 \times 10^{-5}$	$\cdots$	$7.779 \times 10^{-3} + j0.951 \times 10^{-3}$	$7.795 \times 10^{-3} + j0.953 \times 10^{-3}$
$\vdots$	$\vdots$	$\vdots$	$\vdots$	$\vdots$
298.917	$7.8247 \times 10^{-5} + j6.243 \times 10^{-6}$	$\cdots$	$7.825 \times 10^{-3} + j0.393 \times 10^{-3}$	$7.841 \times 10^{-3} + j0.394 \times 10^{-3}$
299.458	$7.8280 \times 10^{-5} + j6.136 \times 10^{-6}$	$\cdots$	$7.833 \times 10^{-3} + j0.267 \times 10^{-3}$	$7.848 \times 10^{-3} + j0.267 \times 10^{-3}$
300	$7.8235 \times 10^{-5} + j6.539 \times 10^{-6}$	$\cdots$	$7.831 \times 10^{-3} + j0.300 \times 10^{-3}$	$7.846 \times 10^{-3} + j0.301 \times 10^{-3}$

and $\omega_i \in [30, 300]$, and the corresponding negative inverse of DFs are tabulated in Tables 4.2, 4.3, 4.4, 4.5, 4.6 and 4.7.

4.3.3.2 Numerical Algorithm for Computing Predicted Chattering Parameters

The parameters of predicted chattering in dynamically perturbed FTC-controlled system (4.1), (4.4) are computed by solving DF-HB eq. numerically as follows.

Table 4.4 $-N^{-1}(A_k, \omega_i)$ look-up table for N-HOSM ($r = 4$)

ω	A			
	0.0001	⋯	0.00998	0.01
30	$2.597 \times 10^{-5} - j7.432 \times 10^{-5}$	⋯	$0.302 \times 10^{-3} - j7.852 \times 10^{-3}$	$0.303 \times 10^{-3} - j7.868 \times 10^{-3}$
30.541	$2.507 \times 10^{-5} - j7.462 \times 10^{-5}$	⋯	$0.297 \times 10^{-3} - j7.852 \times 10^{-3}$	$0.285 \times 10^{-3} - j7.868 \times 10^{-3}$
⋮	⋮	⋮	⋮	⋮
298.917	$4.698 \times 10^{-6} - j8.020 \times 10^{-5}$	⋯	$0.468 \times 10^{-3} - j8.005 \times 10^{-3}$	$0.469 \times 10^{-3} - j8.020 \times 10^{-3}$
299.458	$4.011 \times 10^{-6} - j8.062 \times 10^{-5}$	⋯	$0.278 \times 10^{-3} - j8.051 \times 10^{-3}$	$0.279 \times 10^{-3} - j8.067 \times 10^{-3}$
300	$3.137 \times 10^{-6} - j8.055 \times 10^{-5}$	⋯	$0.313 \times 10^{-3} - j8.039 \times 10^{-3}$	$0.279 \times 10^{-3} - j8.067 \times 10^{-3}$

Table 4.5 $-N^{-1}(A_k, \omega_i)$ look-up table for Q-HOSM ($r = 2$)

ω	A			
	0.0001	⋯	0.00998	0.01
30	$-8.289 \times 10^{-5} + j2.837 \times 10^{-5}$	⋯	$-4.453 \times 10^{-3} + j7.774 \times 10^{-3}$	$-4.460 \times 10^{-3} + j7.791 \times 10^{-3}$
30.541	$-8.285 \times 10^{-5} + j2.876 \times 10^{-5}$	⋯	$-4.408 \times 10^{-3} + j7.790 \times 10^{-3}$	$-4.415 \times 10^{-3} + j7.806 \times 10^{-3}$
⋮	⋮	⋮	⋮	⋮
298.917	$-4.527 \times 10^{-5} + j7.672 \times 10^{-5}$	⋯	$-1.038 \times 10^{-3} + j7.956 \times 10^{-3}$	$-1.040 \times 10^{-3} + j7.972 \times 10^{-3}$
299.458	$-4.560 \times 10^{-5} + j7.696 \times 10^{-5}$	⋯	$-1.035 \times 10^{-3} + j7.994 \times 10^{-3}$	$-1.036 \times 10^{-3} + j8.010 \times 10^{-3}$
300	$-4.498 \times 10^{-5} + j7.706 \times 10^{-5}$	⋯	$-1.015 \times 10^{-3} + j7.978 \times 10^{-3}$	$-1.016 \times 10^{-3} + j7.994 \times 10^{-3}$

Table 4.6 $-N^{-1}(A_k, \omega_i)$ look-up table for Q-HOSM ($r = 3$)

ω	A			
	0.0001	⋯	0.00998	0.01
30	$9.441 \times 10^{-6} + j0.155 \times 10^{-3}$	⋯	$8.515 \times 10^{-3} + j2.468 \times 10^{-3}$	$8.532 \times 10^{-3} + j2.471 \times 10^{-3}$
30.541	$1.375 \times 10^{-5} + j0.155 \times 10^{-3}$	⋯	$8.501 \times 10^{-3} + j2.419 \times 10^{-3}$	$8.517 \times 10^{-3} + j2.422 \times 10^{-3}$
⋮	⋮	⋮	⋮	⋮
298.917	$8.069 \times 10^{-5} + j1.345 \times 10^{-5}$	⋯	$7.852 \times 10^{-3} + j0.513 \times 10^{-3}$	$7.868 \times 10^{-3} + j0.514 \times 10^{-3}$
299.458	$8.082 \times 10^{-5} + j1.302 \times 10^{-5}$	⋯	$7.860 \times 10^{-3} + j0.435 \times 10^{-3}$	$7.876 \times 10^{-3} + j0.435 \times 10^{-3}$
300	$8.071 \times 10^{-5} + j1.295 \times 10^{-5}$	⋯	$7.858 \times 10^{-3} + j0.447 \times 10^{-3}$	$7.873 \times 10^{-3} + j0.448 \times 10^{-3}$

Table 4.7 $-N^{-1}(A_k, \omega_i)$ look-up table for Q-HOSM ($r = 4$)

ω	A			
	0.0001	...	0.00998	0.01
30	$5.268 \times 10^{-5} - j9.213 \times 10^{-5}$	...	$0.942 \times 10^{-3} - j8.083 \times 10^{-3}$	$0.943 \times 10^{-3} - j8.099 \times 10^{-3}$
30.541	$5.112 \times 10^{-5} - j9.181 \times 10^{-5}$	...	$0.918 \times 10^{-3} - j8.076 \times 10^{-3}$	$0.920 \times 10^{-3} - j8.092 \times 10^{-3}$
⋮	⋮	⋮	⋮	⋮
298.917	$5.813 \times 10^{-6} - j8.045 \times 10^{-5}$	...	$0.480 \times 10^{-3} - j8.007 \times 10^{-3}$	$0.481 \times 10^{-3} - j8.023 \times 10^{-3}$
299.458	$4.961 \times 10^{-6} - j8.090 \times 10^{-5}$	...	$0.310 \times 10^{-3} - j8.053 \times 10^{-3}$	$0.311 \times 10^{-3} - j8.069 \times 10^{-3}$
300	$4.906 \times 10^{-6} - j8.079 \times 10^{-5}$	...	$0.335 \times 10^{-3} - j8.042 \times 10^{-3}$	$0.335 \times 10^{-3} - j8.058 \times 10^{-3}$

HB Eq. (4.9) is rewritten as

$$W(j\omega_i) + N^{-1}(A_k, \omega_i) = 0, \tag{4.15}$$

where $N^{-1}(A_k, \omega_i)$ is presented in the table look-up format, see Table 4.1, where A_k and ω_i are the amplitude and the frequency of the predicted limit cycle that are to be identified. By equating real and imaginary parts of the both sides of Eq. (4.15) we obtain

$$\begin{aligned} F_1(\omega_i, A_k) &= \mathrm{Re}\{W(j\omega_i)\} + \mathrm{Re}\left\{N^{-1}(A_k, \omega_i)\right\} = 0, \\ F_2(\omega_i, A_k) &= \mathrm{Im}\{W(j\omega_i)\} + \mathrm{Im}\left\{N^{-1}(A_k, \omega_i)\right\} = 0. \end{aligned}$$

Next, two equations with two variables are solve numerically using the Newton–Raphson method [29].

$$\omega_{i+1} = \omega_i - \frac{1}{J_{i,k}}\left[F_1\frac{\partial F_2}{\partial A_k} - F_2\frac{\partial F_1}{\partial A_k}\right], \tag{4.16}$$

$$A_{k+1} = A_k + \frac{1}{J_{i,k}}\left[F_1\frac{\partial F_2}{\partial \omega_i} - F_2\frac{\partial F_1}{\partial \omega_i}\right], \tag{4.17}$$

where $J_{i,k} = \frac{\partial F_1}{\partial \omega_i}\frac{\partial F_2}{\partial A_k} - \frac{\partial F_1}{\partial A_k}\frac{\partial F_2}{\partial \omega_i} \neq 0$. Since analytical representation of F_1 and F_2 is not available, partial derivatives are obtained numerically as well,

$$\begin{aligned} \frac{\partial F_{1,2}}{\partial \omega} &= \frac{F_{1,2}(\omega_{i-1}, A_k) - F_{1,2}(\omega_i, A_k)}{\omega_{i-1} - \omega_i}, \\ \frac{\partial F_{1,2}}{\partial A} &= \frac{F_{1,2}(\omega_i, A_{k-1}) - F_{1,2}(\omega_i, A_k)}{A_{i-1} - A_i}. \end{aligned} \tag{4.18}$$

Given an initial solution guess $A_k = A_0$ and $\omega_i = \omega_0$, Eqs. (4.16) and (4.17) are solved iteratively. The iterations stop when the conditions

$$\left|\frac{\omega_{i+1} - \omega_i}{\omega_{i+1}}\right| < \epsilon_\omega; \left|\frac{A_{k+1} - A_k}{A_{k+1}}\right| < \epsilon_A,$$

re satisfied, where $\epsilon_\omega > 0$ and $\epsilon_A > 0$ are tolerance errors for frequency and amplitude, respectively, and the solution of Eq. (4.15) is approximately identified. If the error conditions are not satisfied in a reasonable amount of iterations, the iterative algorithm is to start from a new initial solutions guess, A_0 and ω_0.

4.3.3.3 Stability of Predicted Limit Cycles

Proposition 4.1 *If the predicted limit cycle of dynamically perturbed linearized system* (4.5) *with $W(j\omega)$ satisfying the assumptions A1 and A2 and controlled by FTCC is locally stable, then the following inequality holds,*

$$(\rho\Theta + \gamma\Gamma)|_{A_o,\omega_o} > |W(j\omega_o)|^2\ \Xi|_{A_o,\omega_o}, \tag{4.19}$$

where A_o, ω_o are the parameters of the limit cycle,

$$\begin{aligned}
\rho &= \mathrm{Re}\{W(j\omega)\}\mathrm{Re}\{N(A,\omega)\} + \mathrm{Im}\{W(j\omega)\}\mathrm{Im}\{N(A,\omega)\},\\
\gamma &= \mathrm{Re}\{W(j\omega)\}\mathrm{Im}\{N(A,\omega)\} - \mathrm{Im}\{W(j\omega)\}\mathrm{Re}\{N(A,\omega)\},\\
\Theta &= \frac{\partial(\mathrm{Re}\{N(A,\omega)\})}{\partial A}\frac{\partial(\mathrm{Im}\{W(j\omega)\})}{\partial\omega} - \frac{\partial(\mathrm{Im}\{N(A,\omega)\})}{\partial A}\frac{\partial(\mathrm{Re}\{W(j\omega)\})}{\partial\omega},\\
\Gamma &= \frac{\partial(\mathrm{Re}\{N(A,\omega)\})}{\partial A}\frac{\partial(\mathrm{Re}\{W(j\omega)\})}{\partial\omega} + \frac{\partial(\mathrm{Im}\{N(A,\omega)\})}{\partial A}\frac{\partial(\mathrm{Im}\{W(j\omega)\})}{\partial\omega},\\
\Xi &= \frac{\partial(\mathrm{Im}\{N(A,\omega)\})}{\partial A}\frac{\partial(\mathrm{Re}\{N(A,\omega)\})}{\partial\omega} - \frac{\partial(\mathrm{Re}\{N(A,\omega)\})}{\partial A}\frac{\partial(\mathrm{Im}\{N(A,\omega)\})}{\partial\omega},
\end{aligned}$$

$N(A,\omega)$ is the DF of FTCC and $W(j\omega)$ is the frequency characteristics of linearized system (4.5).

Proof Consider the Cartesian complex form of HB Eq. (4.9),

$$\begin{aligned}
H = &\overbrace{1 + \mathrm{Re}\{N(A,\omega)\}\mathrm{Re}\{W(j\omega)\} - \mathrm{Im}\{N(A,\omega)\}\mathrm{Im}\{W(j\omega)\}}^{U(A,\omega)} +\\
&j\underbrace{[\mathrm{Re}\{N(A,\omega)\}\mathrm{Im}\{W(j\omega)\} + \mathrm{Im}\{N(A,\omega)\}\mathrm{Re}\{W(j\omega)\}]}_{V(A,\omega)},
\end{aligned} \tag{4.20}$$

From Loeb criterion (see [21] pages 122–123), the analysis of Eq. (4.20) in the presence of the small amplitude disturbances gives the next necessary condition of stability for a limit cycle

$$\left.\frac{\partial U}{\partial A}\frac{\partial V}{\partial \omega}-\frac{\partial U}{\partial \omega}\frac{\partial V}{\partial A}\right|_{A_o,\omega_o}>0 \tag{4.21}$$

Then, the stability condition of Eq. (4.19) is obtained by replacing U and V in inequality (4.21).

Remark 4.1 The Eq. (4.19) gives necessary condition only, since the Loeb criterion, which Eq. (4.19) is based on, represents the necessary stability condition.

4.3.3.4 Numerical Algorithms for Computing Practical Stability Margins

In this section, the numerical algorithms for the identification of FTCC robustness metrics ($PSPM$ and $PSGM$) to unmodeled dynamics are developed and proposed.

I. $PSPM$ computation method

The phase shift $\theta \geq 0°$ is introduced into the linearized system (4.5) due to unmodeled dynamics. Then, the HB Eq. (4.9) becomes

$$e^{-j\theta}W(j\omega_i)=-N^{-1}(A_k,\omega_i). \tag{4.22}$$

From the Eq. (4.22),

$$|W(j\omega_i)|=\left|N^{-1}(A_k,\omega_i)\right|, \tag{4.23}$$

$$\arg\{W(j\omega_i)\}-\theta=\arg\left\{-N^{-1}(A_k,\omega_i)\right\}, \tag{4.24}$$

where θ is identified based on the given system TL, amplitude A_c and frequency ω_c. The $PSPM$ computational algorithm consists of the following steps:

Step 1. Assume that $A_k = A_c$, then a solution $\omega_i = \omega_{A_c}$ of Eq. (4.23) is to be obtained. If $\omega_{A_c} \geq \omega_c$, then the $PSPM$ is to be identified from Eq. (4.24) as

$$PSPM=\arg\{W(j\omega_{A_c})\}-\arg\left\{-N^{-1}(A_c,\omega_{A_c})\right\}. \tag{4.25}$$

If $\omega_{A_c} < \omega_c$, then proceed to Step 2.

Step 2. Assume that $\omega_i = \omega_c$, then a solution $A_k = A_{\omega_c}$ of Eq. (4.23) is to be obtained. Here, $A_{\omega_c} \leq A_c$ and the $PSPM$ is to be identified from Eq. (4.24) as

$$PSPM=\arg\{W(j\omega_c)\}-\arg\left\{-N^{-1}(A_{\omega_c},\omega_c)\right\}. \tag{4.26}$$

II. $PSGM$ computation method

The gain $K \neq 1$ is introduced into the linearized system (4.5) due to unmodeled dynamics. Then, the HB Eq. (4.9) becomes

$$KW(j\omega_i)=-N^{-1}(A_k,\omega_i). \tag{4.27}$$

From the Eq. (4.27),

$$K\,|W(j\omega_i)| = \left|N^{-1}(A_k,\omega_i)\right|, \tag{4.28}$$

$$\arg\{W(j\omega_i)\} = \arg\left\{-N^{-1}(A_k,\omega_i)\right\}, \tag{4.29}$$

where K is identified based on the given system TL, amplitude A_c and frequency ω_c. The $PSGM$ computational algorithm consists of the following steps:

Step 1. Assume that $A_k = A_c$, then a solution $\omega_i = \omega_{A_c}$ of Eq. (4.23) is to be obtained. If $\omega_{A_c} \geq \omega_c$, then the $PSGM$ is to be identified from Eq. (4.28) as

$$PSGM = \frac{\left|N^{-1}(A_c,\omega_{A_c})\right|}{\left|W(j\omega_{A_c})\right|}. \tag{4.30}$$

If $\omega_{A_c} < \omega_c$, then proceed to Step 2.

Step 2. Assume that $\omega_i = \omega_c$, then a solution $A_k = A_{\omega_c}$ of Eq. (4.23) is to be obtained. Here, $A_{\omega_c} \leq A_c$ and the $PSGM$ is to be identified from Eq. (4.28) as

$$PSGM = \frac{\left|N^{-1}(A_{\omega_c},\omega_c)\right|}{|W(j\omega_c)|}. \tag{4.31}$$

4.3.4 Achieving Prescribed Values of Practical Stability Margins by Cascading FTCC with Linear Compensator

The practical stability margins ($PSPM$ and $PSGM$) for system (4.1) controlled by FTCC may not satisfy the prescribed values due to the presence of unmodeled dynamics. For instance, the prescribed values can be (see page 33 of [6]): $PSPM = 30^\circ$ and $PSGM = 12\,\text{dB}$. Then, the prescribed values on practical stability margins can be achieved in couple of ways.

1. By cascading FTCC with linear compensator, when the transfer function of linearized system (4.5) is available.
2. By artificially increasing the order of FTCC.

Following the option 1, the cascade linear compensator to FTCC takes the form

$$W_c(s) = \frac{s+\frac{1}{\mu\tau}}{s+\frac{1}{\tau}}, \tag{4.32}$$

where μ is the attenuation parameter. If $0 < \mu < 1$, then the Eq. (4.32) is a phase-lag compensator. If $\mu > 1$, then the Eq. (4.32) is a phase-lead compensator.

The design procedure for a phase-lead dynamic compensator [30, 31], which is based on the classical Bode plots methodology [32], is as follows:

Step 1. Obtain the $PSPM$ of dynamically perturbed system (4.1), (4.4) controlled by FTCC: $PSPM_{un}$.

Step 2. Determine the maximum phase-lead angle of the compensator as

$$\phi_m = PSPM_c^\circ - PSPM_{un}^\circ + \langle 5, 12\rangle^\circ,$$

where $PSPM_c$ is the required $PSPM$ and $\langle 5, 12\rangle$ is an interval.

Step 3. Obtain the parameter μ, which satisfies

$$\sin\phi_m = \frac{\mu - 1}{\mu + 1}.$$

Step 4. From the Bode magnitude plot of the dynamically perturbed system (4.1), (4.4), identify the frequency ω_m corresponding to the magnitude

$$-\left[10\log\mu + 20\log|N(A_c, \omega_c)|\right].$$

Step 5. Find the pole and zero of the compensator $W_c(s)$ using

$$\text{Pole:}\frac{1}{\tau} = \omega_m\sqrt{\mu};\ \text{Zero:}\frac{1}{\mu\tau} = \frac{\text{Pole}}{\mu}.$$

Step 6. Compute the $PSPM_{new}$, after cascading the dynamic compensator with the FTCC for verification.

4.4 Simulation Examples

In this section, a couple of examples are presented validating the proposed theoretical developments on quantifying robustness of FTCC to unmodeled dynamics in terms of practical stability margins ($PSPM$ and $PSGM$).

4.4.1 Robustness Study of FTCC to Unmodeled Dynamics

It has been shown in Example 1 that the dynamically perturbed system (4.6), (4.8) with $T = 0.01$ controlled by FTCC (4.7) exhibits chattering. The robustness metrics for FTCC (4.7) to cascade unmodeled dynamics (4.8) are obtained, while analyzing the chattering, as follows.

(a) Computing parameters of chattering

DF is numerically obtained for FTCC (4.7) with $\alpha = 0.5$ and $\mu_1, \mu_2 = 1$ based on the proposed numerical algorithm in the Sect. 4.3.3.1. Note that the DF method is applied with the assumptions A1 and A2 are hold.

Then, the transfer function for the system dynamics (4.6) in sliding mode are

$$W(s) = \frac{\sigma(s)}{u(s)} = \frac{1}{s^2 + 3s + 8}. \tag{4.33}$$

Next, the transfer function for dynamically perturbed system (4.6), (4.8) is

$$W(s) = \left(\frac{1}{T^2s^2 + Ts + 1}\right) \cdot \left(\frac{1}{s^2 + 3s + 8}\right), \tag{4.34}$$

where $T = 0.01$.

The chattering parameters, amplitude A and frequency ω of main harmonic, are predicted by solving HB Eq. (4.9) via NR method, as shown in the Sect. 4.3.3.2, for the dynamically perturbed system (4.6), (4.8) controlled by FTCC (4.7). Table 4.8 shows the comparison of σ-limit cycle parameters that are obtained numerically using NR method with the simulations and the accuracy of the proposed method. Note that the predicted limit cycle using DF-HB method is stable in accordance with the Loeb's criterion in Proposition 4.1.

(b) Practical stability margins

The Tolerance Limits (TL), reasonable amplitude and frequency of chattering, for the system (4.35) controlled by FTCC (4.7) are defined as $A_c = 0.001$ and $\omega_c = 40\,\text{rad/s}$. Then, the proposed robustness metrics ($PSPM$ and $PSGM$) are identified for the FTC-controlled dynamically perturbed system (4.6), (4.8) by following the steps presented in the Sect. 4.3.3.4.

To compute $PSPM$: Step 1. The solution of Eq. (4.23) for $A_c = 0.001$ is obtained as: $\omega_{A_c} = 44.073\,\text{rad/s} > \omega_c$. Therefore,

$$PSPM = 2.709 - 2.369 = 0.3405\,\text{rad} = 19.51^\circ.$$

To compute $PSGM$: Step 1. The solution of Eq. (4.29) for $A_c = 0.001$ is obtained as: $\omega_{A_c} = 63.328 > \omega_c$. Therefore,

Table 4.8 Parameters of σ-limit cycle in FTCC (4.7) system

Parameter	Simulations	Numerical solution
A	5.21×10^{-4}	5.05×10^{-4}
ω (rad/s)	62.832	63.784

$$PSGM = \frac{5.639 \times 10^{-4}}{2.863 \times 10^{-4}} = 5.88\,\text{dB}.$$

The prescribed values of practical stability margins for the system (4.6) are defined as: $PSPM \geq 30^\circ$ and $PSGM \geq 5\,\text{dB}$. Note that the computed $PSPM$ has not satisfied the required value. The prescribed value on $PSPM$ is achieved by cascading a phase-lead compensator with the designed FTCC (4.7).

(c) Cascade linear compensator for FTCC

The cascaded phase-lead compensator is designed by following the steps presented in the Sect. 4.3.4 as follows.

Step 1. $PSPM_{un} = 19.51^\circ$; Step 2. $\phi_m = 30^\circ - 19.51^\circ + 9^\circ = 19.49^\circ$; Step 3. $\mu = 2.01$; Step 4. $|W(j\omega_m)| = -(3.013 + 65.02) = -68.033\,\text{dB} \implies \omega_m = 54\,\text{rad/s}$; Step 5. Pole = 76.39 and Zero = 38.17. Therefore, the phase-lead compensator to FTCC is obtained as

$$W_c(s) = \frac{76.39}{38.17} \times \frac{s + 38.17}{s + 76.39};$$

Step 6. Then, $PSPM_{new} = 30.54^\circ$ is obtained. Note that the $PSPM_{new}$ satisfies the required specification on $PSPM$, i.e., $PSPM \geq 30^\circ$.

4.4.2 Robustness Study of HOSM to Unmodeled Dynamics

A nonlinear SISO system is considered as

$$\begin{aligned} \dot{x}_1 &= x_2, \dot{x}_2 = x_3, \\ \dot{x}_3 &= -50x_1 - 180x_2 - 20x_3 + 35\sin(x_1) + 50u_1, \ \sigma = x_1, \end{aligned} \tag{4.35}$$

where u_1 is an unmodeled dynamics control input, and σ is a system output.

Then, input-output dynamics for system (4.35) are obtained as

$$\sigma^{(3)} = -50\sigma - 180\dot{\sigma} - 20\ddot{\sigma} + 35\sin(\sigma) + 50u_1. \tag{4.36}$$

The dynamics in Eq. (4.36) are rewritten as

$$\frac{1}{50} \times \left(\sigma^{(3)} + 20\ddot{\sigma} + 180\dot{\sigma} + 50\sigma\right) = u_1(t) + \zeta, \tag{4.37}$$

where $\zeta = 0.7\sin(\sigma)$, and $|\zeta| \leq 0.7$.

Note that the system (4.35) has relative degree $r_d = 3$. By taking u_1 from N-HOSM control in Eq. (4.12) with $r = 3$ and $\lambda = 1$, the system (4.35) is simulated

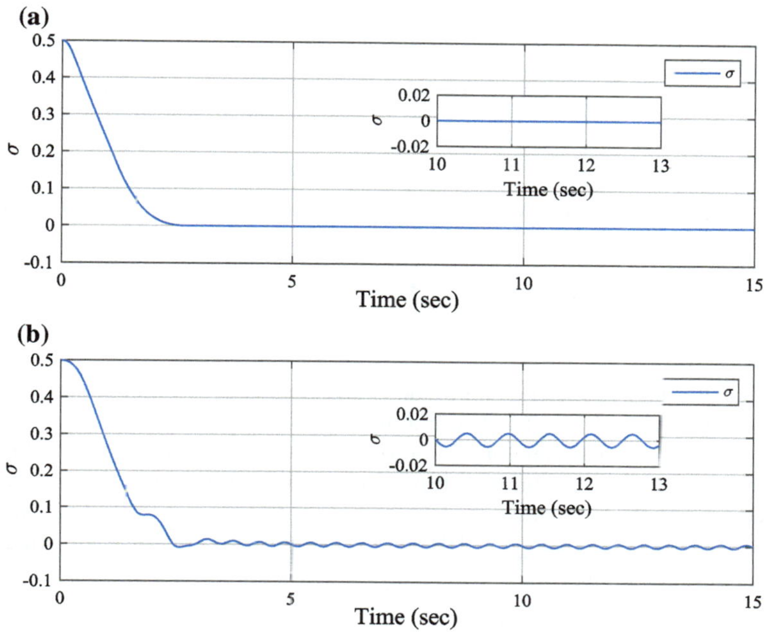

Fig. 4.2 σ in **a** dynamically unperturbed system (4.35) and **b** dynamically perturbed system (4.35), (4.8) controlled by N-HOSM controller

for $\sigma(0) = 0.5$. The N-HOSM control has efficiently driven σ to zero in finite-time in the presence of bounded disturbance ζ as shown in Fig. 4.2a.

By defining the same N-HOSM control to u, the sliding variable σ in the dynamically perturbed system (4.35), (4.8) with $T = 0.2$ exhibits chattering as shown in Fig. 4.2b.

The robustness metrics for N-HOSM control (4.12) to cascade unmodeled dynamics (4.8) are obtained, while analyzing the chattering, as follows.

(a) Computing Parameters of Chattering

DF is numerically obtained for N-HOSM (r=3) in Eq. (4.12) with $\lambda = 1$ based on the proposed numerical algorithm in the Sect. 4.3.3.1. Note that the DF method is applied by assuming $\zeta(t)$ in Eq. (4.37) has exactly been compensated by the designed N-HOSM control and the assumptions A1 and A2 are hold.

Then, the transfer function for the system dynamics (4.37) are written as

$$W(s) = \frac{\sigma(s)}{u(s)} = \frac{50}{s^3 + 20s^2 + 180s + 50}. \tag{4.38}$$

Table 4.9 Parameters of σ-limit cycle in N-HOSM controlled system

Parameter	Simulations	Numerical solution
A	5.81×10^{-3}	4.98×10^{-3}
ω (rad/s)	11.220	11.473

Next, the transfer function for dynamically perturbed system (4.35), (4.8) is

$$W(s) = \left(\frac{1}{T^2 s^2 + Ts + 1}\right) \cdot \left(\frac{50}{s^3 + 20s^2 + 180s + 50}\right), \tag{4.39}$$

where $T = 0.2$.

The chattering parameters, amplitude A and frequency ω of main harmonic, are predicted by solving HB Eq. (4.9) via NR method, as shown in the Sect. 4.3.3.2, for the dynamically perturbed system (4.35), (4.8) controlled by N-HOSM control. Table 4.9 shows the comparison of σ-limit cycle parameters that are obtained numerically using NR method with the simulations and the accuracy of the proposed method. Note that the predicted limit cycle using DF-HB method is stable in accordance with the Loeb's criterion in Proposition 4.1.

(b) Practical Stability Margins

The Tolerance Limits (TL), reasonable amplitude and frequency of chattering, for the system (4.35) controlled by N-HOSM control are defined as $A_c = 0.01$ and $\omega_c = 8\,\text{rad/s}$. Then, the proposed robustness metrics ($PSPM$ and $PSGM$) are identified for the N-HOSM controlled dynamically perturbed system (4.35), (4.8) by following the steps presented in the Sect. 4.3.3.4.

To compute $PSPM$: Step 1. The solution of Eq. (4.23) for $A_c = 0.01$ is obtained as: $\omega_{A_c} = 9.537\,\text{rad/s} > \omega_c$. Therefore,

$$PSPM = 1.078 - 0.6607 = 0.4168\,\text{rad} = 23.88°.$$

To compute $PSGM$: Step 1. The solution of Eq. (4.29) for $A_c = 0.01$ is obtained as: $\omega_{A_c} = 13.476 > \omega_c$. Therefore,

$$PSGM = \frac{7.852 \times 10^{-3}}{2.050 \times 10^{-3}} = 11.67\,\text{dB}.$$

The prescribed values of practical stability margins for the system (4.35) are defined as: $PSPM \geq 35°$ and $PSGM \geq 6\,\text{dB}$. Note that the computed $PSPM$ has not satisfied the required value. The prescribed value on $PSPM$ is achieved by cascading a phase-lead compensator with the designed N-HOSM controller.

(c) Cascade Linear Compensator for HOSM Controller

The cascaded phase-lead compensator is designed by following the steps presented in the Sect. 4.3.4 as follows.

Step 1. $PSPM_{un} = 23.88°$; Step 2. $\phi_m = 35° - 23.88° + 5° = 16.12°$; Step 3. $\mu = 1.77$; Step 4. $|W(j\omega_m)| = -(2.476 + 42.10) = -44.576\,\text{dB} \Longrightarrow \omega_m = 10.3\,\text{rad/s}$; Step 5. Pole = 13.70 and Zero = 7.745. Therefore, the phase-lead compensator to N-HOSM control is obtained as

$$W_c(s) = \frac{13.70}{7.745} \times \frac{s + 7.745}{s + 13.70};$$

Step 6. Then, $PSPM_{new} = 35.01°$ is obtained. Note that the $PSPM_{new}$ satisfies the required specification on $PSPM$, i.e., $PSPM \geq 35°$.

4.5 Case Study: Attitude HOSM Control of F-16 Aircraft

A nonlinear model of an F-16 aircraft [33] at Mach $= 0.7$ and height $h = 10,000$ ft is considered as:

$$\begin{aligned}
\dot{\theta} &= q\cos(\varphi) - r\sin(\varphi),\\
\dot{\varphi} &= p + q\sin(\varphi)\tan(\theta) + r\cos(\varphi)\tan(\theta),\\
\dot{\alpha} &= -\beta p + 0.0427\cos(\theta)\cos(\varphi) + 0.083589 + \tilde{Z}_\alpha\alpha + \tilde{Z}_q q + \tilde{Z}_\delta\delta_e,\\
\dot{\beta} &= -0.9973r + \alpha p + 0.0427\cos(\theta)\sin(\varphi) + \tilde{Y}_\beta\beta + \tilde{Y}_p p + \tilde{Y}_{\delta_r}\delta_r + \tilde{Y}_{\delta a}\delta_a,\\
\dot{p} &= -0.1345pq - 0.8225qr + \tilde{L}_\beta\beta + \tilde{L}_p p + \tilde{L}_r r - 50.933\delta_a + \tilde{L}_{\delta_r}\delta_r + \zeta_p(t),\\
\dot{q} &= 0.9586pr - 0.0833(r^2 - p^2) - 1.94166 + \tilde{M}_\alpha\alpha + \tilde{M}_q q + \tilde{M}_\delta\delta_e + \zeta_q(t),\\
\dot{r} &= -0.7256pq + 0.1345qr + \tilde{N}_\beta\beta + \tilde{N}_p p + \tilde{N}_r r + 4.125\delta_a + \tilde{N}_{\delta_r}\delta_r + \zeta_r(t),
\end{aligned} \tag{4.40}$$

where $\theta, \varphi, \alpha, \beta$ are pitch, roll, attack, and sideslip angles, respectively; p, q, r are roll, attack, and sideslip angular rates, respectively; δ_a, δ_e, and δ_r are aileron, elevator, and rudder deflections, respectively; $\alpha_{trim} = \theta_{trim} = 0.106803\,\text{rad}$, $\delta_{e_{trim}} = -0.0295\,\text{rad}$, $\varphi_{trim} = \beta_{trim} = p_{trim} = q_{trim} = r_{trim} = \delta_{a_{trim}} = \delta_{r_{trim}} = 0$; $\zeta_p(t)$, $\zeta_q(t)$, and $\zeta_r(t)$ represent disturbance terms due to external wind gusts, and are defined for simulations as: $\zeta_p(t) = 0.005\sin(t)$, $\zeta_q(t) = 0.005\cos(t)$, $\zeta_r(t) = 0.005(\sin(t) + \cos(t))$, and $|\zeta_p(t)| \leq 0.005$, $|\dot{\zeta}_p(t)| \leq 0.005$, $|\zeta_q(t)| \leq 0.005$, $|\dot{\zeta}_q(t)| \leq 0.005$, $|\zeta_r(t)| \leq 0.01$, $|\dot{\zeta}_r(t)| \leq 0.01$; $\tilde{Z}_\alpha = -1.15$, $\tilde{Z}_q = 0.9937$, $\tilde{Y}_\beta = -0.297$, $\tilde{Y}_p = 0.00085$, $\tilde{L}_\beta = -53.48$, $\tilde{L}_p = -4.324$, $\tilde{L}_r = -0.224$, $\tilde{L}_{\delta_r} = 10.177$, $\tilde{M}_\alpha = 3.724$, $\tilde{M}_q = -1.26$, $\tilde{M}_\delta = -19.5$, $\tilde{N}_\beta = 17.67$, $\tilde{N}_p = 0.234$, $\tilde{N}_r = -0.649$, $\tilde{N}_{\delta_r} = -6.155$; $\tilde{Z}_\delta, \tilde{Y}_{\delta_r}, \tilde{Y}_{\delta a}$ that have small control perturbations are neglected, and thus forming a square cascade structure of the aircraft model.

The unmodeled dynamics are given as

$$\eta_\rho \dot{\delta}_\rho = -(\delta_\rho - u_\rho) \quad \forall \rho = a, e, r, \tag{4.41}$$

where $\eta_\rho = [0.02, 0.02, 0.02]^T$.

The deflection and deflection rate limits of unmodeled dynamics are given as

$$|\delta_\rho| \le 0.37\,\text{rad}, \quad |\dot{\delta}_\rho| \le 1\,\text{rad/s} \quad \forall \rho = a, e, r. \tag{4.42}$$

The following filters are used to obtain the desired command profiles α_c, β_c, and φ_c by applying α_{ref}, β_{ref}, and φ_{ref}, respectively:

$$\frac{\alpha_c}{\alpha_{ref}} = \frac{4}{s^2 + 3s + 4}, \quad \frac{\beta_c}{\beta_{ref}} = \frac{16.98}{s^2 + 6.18s + 16.98}, \quad \frac{\varphi_c}{\varphi_{ref}} = \frac{2}{s^2 + 2.2s + 2}. \tag{4.43}$$

The problem formulation for this case study is as follows:

- To design single-loop trajectory tracking FTCC, specifically, Q-HOSM control that asymptotically drives attitude angles φ, α, β to angular commands $\varphi_c, \alpha_c, \beta_c$, respectively, in presence of bounded perturbations.
- To certify the aircraft attitude FTCC robustness to unmodeled dynamics using proposed robustness metrics ($PSPM$ and $PSGM$).

4.5.1 Design of Aircraft Attitude HOSM Control

Based on relative degree approach, the aircraft model (4.40) is expressed as

$$\begin{aligned}
\ddot{\varphi} &= -53.48\beta - 4.324p - 0.224r - 0.1345pq - 0.8225qr \\
&\quad -50.9333\delta_a + 10.177\delta_r + F_\varphi(x,t), \\
\ddot{\alpha} &= -1.15\dot{\alpha} + 3.7\alpha - 1.252q + 0.9526pr \\
&\quad -0.0828(r^2 - p^2) - 1.9294 - 19.3772\delta_e + F_\alpha(x,t), \\
\ddot{\beta} &= -0.297\dot{\beta} - 17.66775\beta - 0.237p + 0.6471r \\
&\quad +0.7235pq - 0.1348qr - 4.1572\delta_a + 6.147\delta_r + F_\beta(x,t),
\end{aligned} \tag{4.44}$$

where $x = [\varphi, \alpha, \beta, p, q, r]^T$, and the perturbation terms are defined

$$F_\varphi(x,t) = \zeta_p(t) + \frac{\mathrm{d}f_\varphi(x,t)}{\mathrm{d}t}, \; F_\alpha(x,t) = 0.9937\zeta_q(t) + \frac{\mathrm{d}f_\alpha(x,t)}{\mathrm{d}t},$$

$$
\begin{aligned}
F_\beta(x,t) &= 0.00085\zeta_p(t) - 0.9973\zeta_r(t) + \frac{\mathrm{d}f_\beta(x,t)}{\mathrm{d}t},\\
f_\varphi(x,t) &= q\sin(\varphi)\tan(\theta) + r\cos(\varphi)\tan(\theta),\\
f_\alpha(x,t) &= -\beta p + 0.0427\cos(\theta)\cos(\varphi) + 0.083589,\\
f_\beta(x,t) &= \alpha p + 0.0427\cos(\theta)\sin(\varphi).
\end{aligned}
$$

The trajectory tracking errors are defined as

$$e_\Lambda = \Lambda_c - \Lambda \quad \forall\Lambda = \varphi, \alpha, \beta. \tag{4.45}$$

Then, the sliding variables are defined as

$$\sigma_\Lambda = \dot{e}_\Lambda + c_\Lambda e_\Lambda \quad \forall\Lambda = \varphi, \alpha, \beta, \tag{4.46}$$

where $c_\varphi, c_\alpha, c_\beta > 0$ give asymptotic convergence of sliding variables to zero.

In this case study, Q-HOSM control (4.13) is used to control the attitude of F-16 aircraft (4.40). Since the Q-HOSM control is a discontinuous control, the HOSM control is designed in terms of angular deflection rates $\dot{\delta}_a, \dot{\delta}_e, \dot{\delta}_r$.

The system (4.44) is rewritten as

$$
\begin{aligned}
\sigma_\varphi^{(2)} &= \Psi_{11} + \Psi_{12} + 50.933\dot{\delta}_a - 10.177\dot{\delta}_r,\\
\sigma_\alpha^{(2)} &= \Psi_{21} + \Psi_{22} - 1.15\dot{\sigma}_\alpha + 3.7\sigma_\alpha + 19.37715\dot{\delta}_e,\\
\sigma_\beta^{(2)} &= \Psi_{31} + \Psi_{32} - 0.297\dot{\sigma}_\beta - 17.6677\sigma_\beta + 4.1572\dot{\delta}_a - 5.147\dot{\delta}_r,
\end{aligned} \tag{4.47}
$$

where $\Psi_{11}, \Psi_{21}, \Psi_{31}$ are known terms; $\Psi_{12}, \Psi_{22}, \Psi_{32}$ are unknown terms, but are assumed to be bounded in a reasonable flight domain $|\Psi_{12}| \le \lambda_a$, $|\Psi_{22}| \le \lambda_e$, and $|\Psi_{32}| \le \lambda_r$. Then, Eq. (4.47) is written in input-output decouple format as

$$
\begin{aligned}
\sigma_\varphi^{(2)} &= \Psi_{11} + \Psi_{12} + \nu_a,\\
\sigma_\alpha^{(2)} &= \Psi_{21} + \Psi_{22} - 1.15\dot{\sigma}_\alpha + 3.7\sigma_\alpha + \nu_e,\\
\sigma_\beta^{(2)} &= \Psi_{31} + \Psi_{32} - 0.297\dot{\sigma}_\beta - 17.6677\sigma_\beta + \nu_r,
\end{aligned} \tag{4.48}
$$

where

$$
\begin{bmatrix}\nu_a\\ \nu_e\\ \nu_r\end{bmatrix} = \begin{bmatrix}50.933 & 0 & -10.177\\ 0 & 19.3772 & 0\\ 4.1572 & 0 & -6.147\end{bmatrix}\begin{bmatrix}\dot{\delta}_a\\ \dot{\delta}_e\\ \dot{\delta}_r\end{bmatrix}. \tag{4.49}
$$

In this case study, the Q-HOSM (4.14) with $r = 2$ is taken and is defined as

$$\nu_a = -\Psi_{11} + \hat{\nu}_a, \quad \nu_e = -\Psi_{21} + \hat{\nu}_e, \quad \nu_r = -\Psi_{31} + \hat{\nu}_r,$$

where

$$\hat{\nu}_a = -\lambda_a \left(\dot{\sigma}_\varphi + |\sigma_\varphi|^{1/2} \operatorname{sign}[\sigma_\varphi]\right) / \left(|\dot{\sigma}_\varphi| + |\sigma_\varphi|^{1/2}\right),$$
$$\hat{\nu}_e = -\lambda_e \left(\dot{\sigma}_\alpha + |\sigma_\alpha|^{1/2} \operatorname{sign}[\sigma_\alpha]\right) / \left(|\dot{\sigma}_\alpha| + |\sigma_\alpha|^{1/2}\right),$$
$$\hat{\nu}_r = -\lambda_r \left(\dot{\sigma}_\beta + |\sigma_\beta|^{1/2} \operatorname{sign}[\sigma_\beta]\right) / \left(|\dot{\sigma}_\beta| + |\sigma_\beta|^{1/2}\right).$$

The unknown terms $\Psi_{12}, \Psi_{22}, \Psi_{32}$ are assumed to be exactly compensated by the HOSM controllers $\hat{\nu}_a, \hat{\nu}_e, \hat{\nu}_r$, and then Eq. (4.48) becomes

$$\sigma_\varphi^{(2)} = \hat{\nu}_a, \quad \sigma_\alpha^{(2)} = -1.15\dot{\sigma}_\alpha + 3.7\sigma_\alpha + \hat{\nu}_e,$$
$$\sigma_\beta^{(2)} = -0.297\dot{\sigma}_\beta - 17.6677\sigma_\beta + \hat{\nu}_r. \tag{4.50}$$

The single-loop trajectory tracking FTCC in terms of angular deflections is obtained as

$$\begin{bmatrix} \delta_a \\ \delta_e \\ \delta_r \end{bmatrix} = \begin{bmatrix} 50.933 & 0 & -10.177 \\ 0 & 19.3772 & 0 \\ 4.1572 & 0 & -6.147 \end{bmatrix}^{-1} \begin{bmatrix} \int_0^t \nu_a d\tau \\ \int_0^t \nu_e d\tau \\ \int_0^t \nu_r d\tau \end{bmatrix}. \tag{4.51}$$

The system (4.40) is simulated with the designed HOSM controller (4.51), while analyzing the robustness of the HOSM control to cascade unmodeled dynamics (4.41).

Note that in the two scenarios, dynamically unperturbed (4.40) and dynamically perturbed (4.40), (4.41) systems, the controller gains are taken as $\lambda_a = 7$, $\lambda_e = 16$, and $\lambda_r = 7$ to satisfy the angular rate limits in Eq. (4.42), $c_\varphi = c_\alpha = c_\beta = 3$, and simulated for initial conditions $\varphi(0) = 0.2$, $\alpha(0) = 0.1$, and $\beta(0) = 0.1$. In other words, the robustness of the HOSM controlled aircraft system (4.40), (4.51) to cascade unmodeled dynamics (4.41) is studied by having these parameters unchanged.

The high tracking accuracy given by the continuous single-loop aircraft attitude HOSM control (4.51) in the presence of bounded external perturbations is shown in Fig. 4.3a–c, and corresponding angular deflections that satisfying constraints (4.42) are shown in Fig. 4.3d.

The evolution of sliding variables σ_φ, σ_α, and σ_β in the dynamically unperturbed system (4.40) controlled by HOSM control (4.51) are shown in Fig. 4.4. The HOSM control (4.51) is successful in driving the sliding variables to zero in finite-time in the presence of bounded external perturbations.

The evolution of sliding variables σ_φ, σ_α, and σ_β in the dynamically perturbed system (4.40), (4.41) controlled by HOSM control (4.51) are shown in Fig. 4.5. The sliding variables are converged to acceptable limit cycles. The robustness of HOSM control (4.51) to unmodeled dynamics (4.41) is quantified using $PSPM$ and $PSGM$ in the following section.

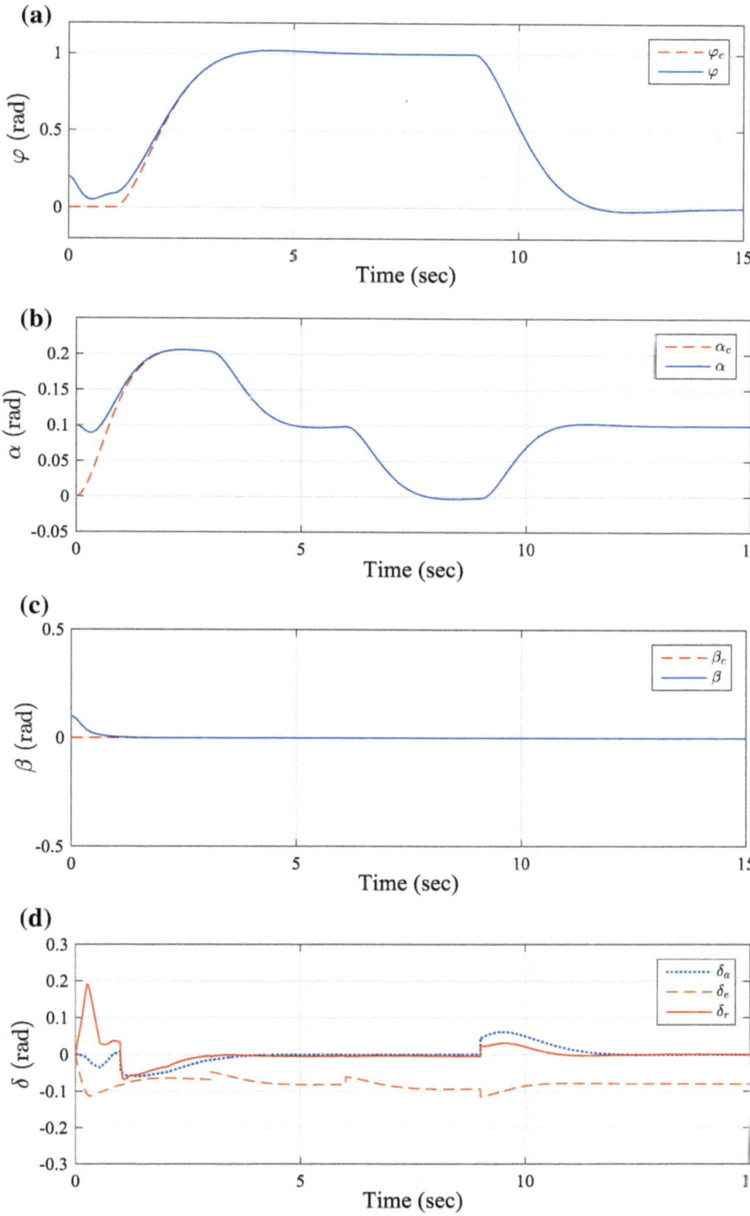

Fig. 4.3 **a–c** Trajectory tracking and **d** angular deflections of dynamically unperturbed aircraft (4.40)

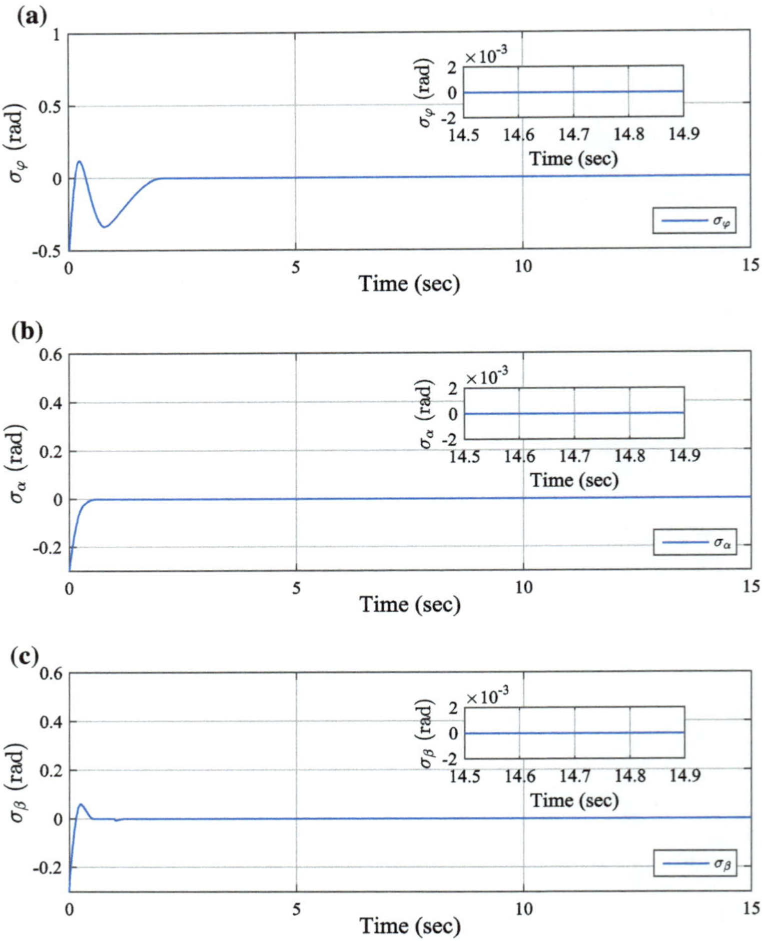

Fig. 4.4 Sliding variables in the dynamically unperturbed system (4.40)

4.5.2 Robustness Study of Aircraft Attitude FTCC to Cascade Unmodeled Dynamics

In this section, the robustness of designed HOSM controller (4.51) to cascade unmodeled dynamics (4.41) is analyzed using the tools developed in the Sect. 4.3.3.

(a) Computing Parameters of Chattering/Limit-Cycles

With the help of the proposed numerical algorithm for computing DFs in Sect. 4.3.3.1, the DFs for Q-HOSM controllers $\hat{v}_\rho$ ($\rho = a, e, r$) are obtained numerically, and the negative reciprocals of DFs are tabulated as Table 4.1.

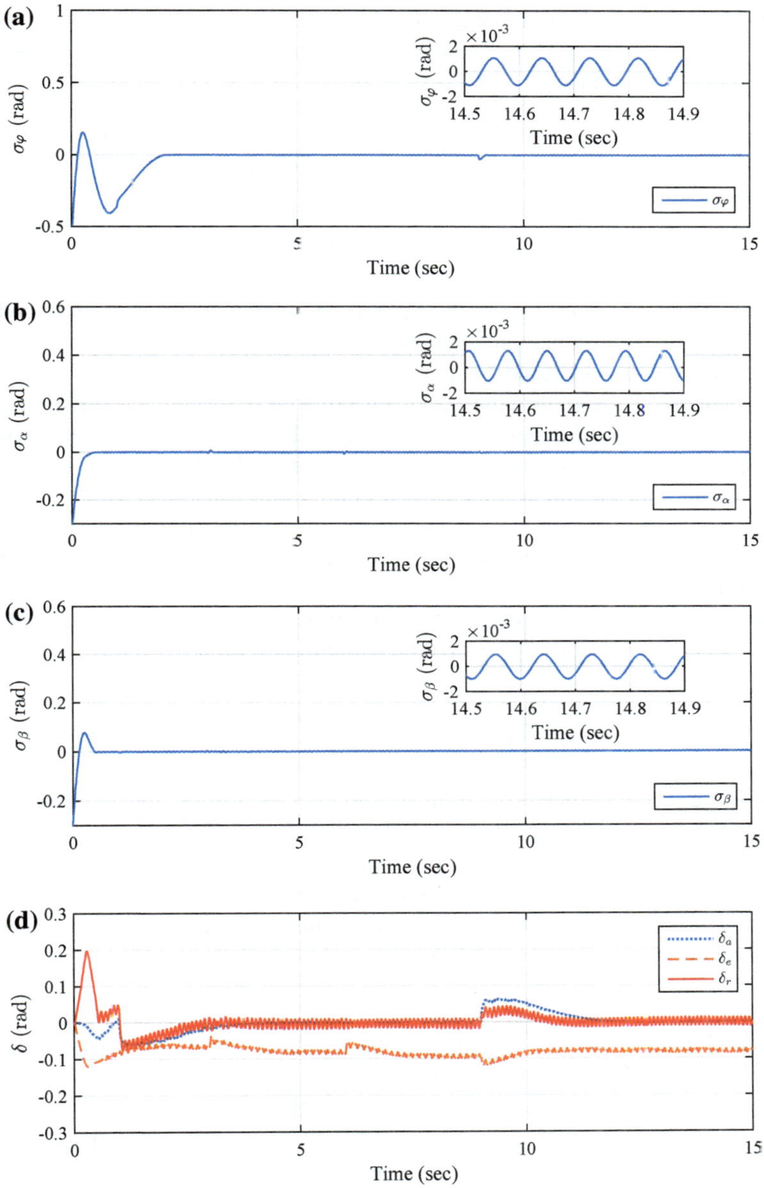

Fig. 4.5 **a–c** Sliding variables and **d** angular deflections in the dynamically perturbed system (4.40), (4.41)

Note that the perturbations are assumed to be compensated exactly by the HOSM control $\hat{\nu}_\rho$ $\forall \rho = a, e, r$. Then, the dynamically perturbed system (4.40), (4.41) in sliding mode is written as

$$W_\varphi(j\omega) = \frac{1}{0.02s+1}\cdot\frac{1}{s^2}\bigg|_{s=j\omega}, \quad W_\alpha(j\omega) = \frac{1}{0.02s+1}\cdot\frac{1}{s^2+1.15s-3.7}\bigg|_{s=j\omega},$$
$$W_\beta(j\omega) = \frac{1}{0.02s+1}\cdot\frac{1}{s^2+0.297s+17.66775}\bigg|_{s=j\omega}. \tag{4.52}$$

Using the proposed procedure in the Sect. 4.3.3.2, the parameter of predicted limit cycles for sliding variables are numerically obtained and are compared with the simulations as shown in Table 4.10.

Note that the TL are some what easy to introduce as the acceptable limit cycle parameters, amplitude and frequency, that may occur due to cascade unmodeled dynamics for attitude angle tracking errors. In this case study, the admissible Tolerance Limits for the attitude angle tracking errors are defined as TL: $A_{e_\Lambda} = 4.5 \times 10^{-5}$ and $\omega_{e_\Lambda} = 45\,\text{rad/s}$ $\forall \Lambda = \varphi, \alpha, \beta$. However, the proposed analysis estimates the parameters of predicted limit cycles for sliding variables σ_φ, σ_α and σ_β, but not for the attitude angle tracking errors.

Then, the frequency characteristics of the errors e_Λ $\forall \Lambda = \varphi, \alpha, \beta$ that correspond to inputs $\sigma_\Lambda = A_{\sigma_\Lambda}\sin(\omega_{\sigma_\Lambda} t)$ are written using Eq. (4.46) as

$$\frac{e_\Lambda(j\omega_{\sigma_\Lambda})}{\sigma_\Lambda(j\omega_{\sigma_\Lambda})} = \frac{1}{(j\omega_{\sigma_\Lambda}) + c_\Lambda} \quad \forall \Lambda = \varphi, \alpha, \beta. \tag{4.53}$$

Therefore, the parameters of limit cycles for the attitude angle tracking errors are obtained from Eq. (4.53) as $A_{e_\Lambda} = A_{\sigma_\Lambda}/(c_\Lambda^2 + \omega_{\sigma_\Lambda}^2)^{1/2}$ and $\omega_{e_\Lambda} = \omega_{\sigma_\Lambda}$, where A_{e_Λ} and ω_{e_Λ} are the amplitude and the frequency of limit cycle, respectively. Then, corresponding parameters are computed as in Table 4.11.

Next, TL for sliding variables (4.46) are computed from the TL for the attitude angle tracking errors using Eq. (4.53) as

Table 4.10 Parameters of limit cycles σ_φ, σ_α, and σ_β

Sliding variable	Simulations	Numerical solution
σ_φ	$A_{\sigma_\varphi} = 1.094 \times 10^{-3}$ $\omega_{\sigma_\varphi} = 69.81\,\text{rad/s}$	$A_{\sigma_\varphi} = 9.461 \times 10^{-4}$ $\omega_{\sigma_\varphi} = 69.03\,\text{rad/s}$
σ_α	$A_{\sigma_\alpha} = 1.169 \times 10^{-3}$ $\omega_{\sigma_\alpha} = 89.75\,\text{rad/s}$	$A_{\sigma_\alpha} = 1.108 \times 10^{-3}$ $\omega_{\sigma_\alpha} = 88.82\,\text{rad/s}$
σ_β	$A_{\sigma_\beta} = 9.735 \times 10^{-4}$ $\omega_{\sigma_\beta} = 69.81\,\text{rad/s}$	$A_{\sigma_\beta} = 9.303 \times 10^{-4}$ $\omega_{\sigma_\beta} = 69.55\,\text{rad/s}$

Table 4.11 Parameters of limit cycles e_φ, e_α, and e_β

Tracking error	Simulations	Numerical solution
e_φ	$A_{e_\varphi} = 1.565 \times 10^{-5}$ $\omega_{e_\varphi} = 69.81\,\text{rad/s}$	$A_{e_\varphi} = 1.369 \times 10^{-5}$ $\omega_{e_\varphi} = 69.03\,\text{rad/s}$
e_α	$A_{e_\alpha} = 1.302 \times 10^{-5}$ $\omega_{e_\alpha} = 89.75\,\text{rad/s}$	$A_{e_\alpha} = 1.246 \times 10^{-5}$ $\omega_{e_\alpha} = 88.82\,\text{rad/s}$
e_β	$A_{e_\beta} = 1.393 \times 10^{-5}$ $\omega_{e_\beta} = 69.81\,\text{rad/s}$	$A_{e_\beta} = 1.331 \times 10^{-5}$ $\omega_{e_\beta} = 69.55\,\text{rad/s}$

Table 4.12 Practical stability margins for σ_φ, σ_α, and σ_β

Sliding variable	$PSPM$	$PSGM$ (dB)
σ_φ	9.669°	24.01
σ_α	7.048°	18.66
σ_β	9.993°	24.44

Table 4.13 $PSPM$ of σ_φ, σ_α, and σ_β after cascading linear compensators with the designed HOSM controllers

Sliding variable	$PSPM_{new}$
σ_φ	33.07°
σ_α	31.75°
σ_β	32.92°

$$
\begin{aligned}
A_{\sigma_\Lambda} &= A_{\epsilon_\Lambda}(c_\Lambda^2 + \omega_{e_\Lambda}^2)^{1/2} = 2 \times 10^{-3}, \\
\omega_{\sigma_\Lambda} &= \omega_{\epsilon_\Lambda} = 45\,\text{rad/s}\ \forall \Lambda = \varphi, \alpha, \beta.
\end{aligned} \tag{4.54}
$$

(b) Practical Stability Margins

The robustness metrics are obtained for the aircraft (4.40) controlled by HOSM control to cascade unmodeled dynamics (4.41) using Sect. 4.3.3.4.

Table 4.12 presents the robustness metrics, $PSPM$ and $PSGM$, that are deduced using the TL for the sliding variables in Eq. (4.54).

Note that the identified $PSPM$s do not satisfy the prescribed value [6] (see page 33 of [6]): $PSPM_\Lambda \geq 30°\ \forall \Lambda = \varphi, \alpha, \beta$. Then, the required $PSPM$s are achieved (see Table 4.13) by cascading phase-lead compensators with the designed HOSM controllers.

(c) Cascade Compensator for HOSM Controlled F-16 Aircraft

The phase-lead compensators that cascade with the HOSM controllers are designed for the roll angle φ, angle of attack α, and sideslip angle β, respectively as follows:

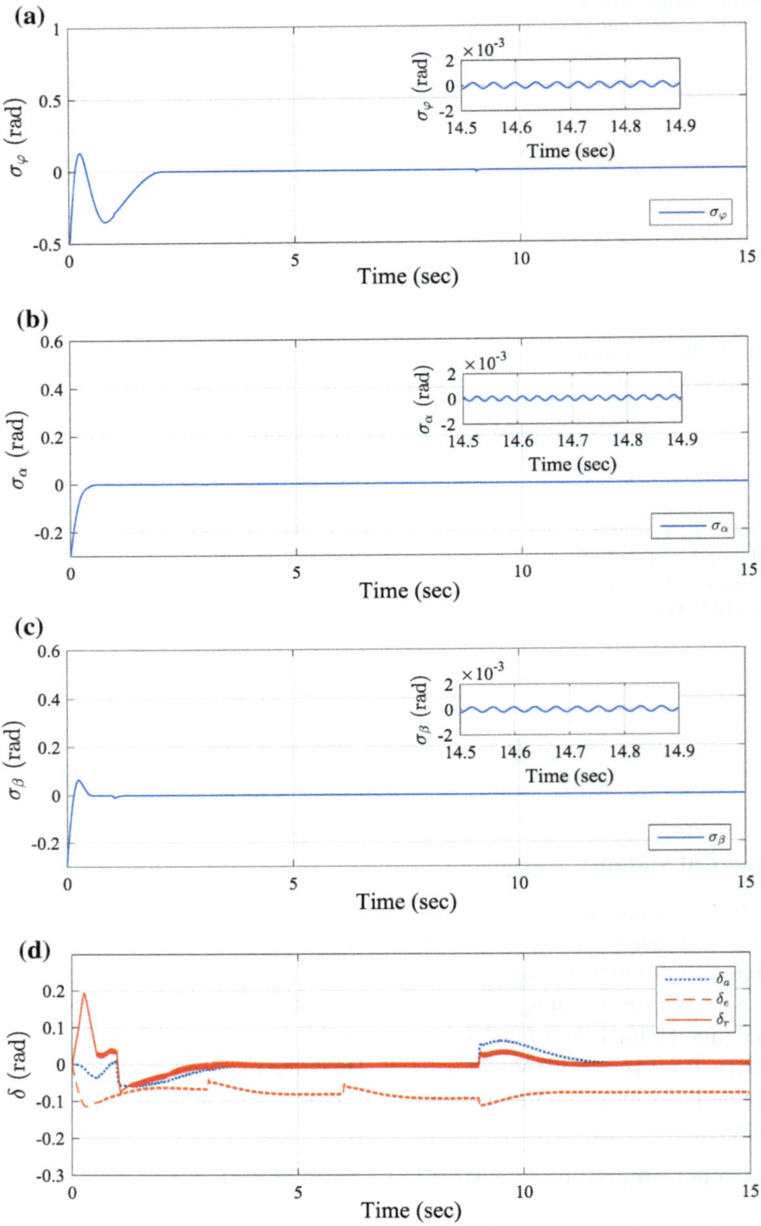

Fig. 4.6 **a–c** Sliding variables and **d** angular deflections in the dynamically perturbed aircraft system (4.40), (4.41) after cascading compensator with HOSM control

$$W_{dc_{\varphi}}(s) = \frac{97.94}{39.25} \times \frac{s + 39.25}{s + 97.94}, \quad W_{dc_{\alpha}}(s) = \frac{118.10}{42.70} \times \frac{s + 42.70}{s + 118.10},$$
$$W_{dc_{\beta}}(s) = \frac{95.76}{38.86} \times \frac{s + 38.86}{s - 95.76}.$$

The obtained $PSPM_{new}$s (see Table 4.13) are satisfied the prescribed value of $PSPM$s for the attitude angle control. Moreover, Fig. 4.6a–c shows that the chattering in the dynamically perturbed aircraft system (4.40), (4.41) is alleviated, while satisfying the prescribed values of $PSPM$ and $PSGM$, after cascading phase-lead compensator with corresponding HOSM control. The chattering of angular deflections is also alleviated as shown in Fig. 4.6d.

4.6 Conclusion

The gap in the certification of Finite Time Convergent Control (FTCC) for robustness to unmodeled dynamics was closed for the first time in this work. The robustness metrics, Practical Stability Phase Margin ($PSPM$) and Practical Stability Gain Margin ($PSGM$), for FTCC to cascade unmodeled dynamics were introduced, presiding the tool for FTCC certification. The tools/algorithms to identify these robustness metrics using Describing Function-Harmonic Balance (DF-HB) method were developed. A numerical algorithm to compute DFs for FTCC was proposed. A database of DFs for FTCC, specifically for nested, quasi-continuous Higher Order Sliding Mode (HOSM) controllers was developed. A numerical algorithm that solves HB equation using Newton Raphson method to obtain predicted chattering parameters was proposed. The computational algorithms to identify the FTCC robustness metrics ($PSPM$ and $PSGM$) were proposed. When the obtained values of $PSPM$ and $PSGM$ do not satisfy the prescribed values, a cascade linear compensator to FTCC was suggested to achieve the prescribed values. The proposed technique was applied to certify the FTC attitude controller of an F-16 aircraft for robustness to cascade unmodeled dynamics as a case study. Several other simulation examples were presented to validate the proposed method.

References

1. Utkin, V., Guldner, J., Shi, J.: Sliding Modes in Electromechanical Systems. Taylor and Francis, London (1999)
2. Shtessel, Y.B., Edwards, C., Fridman, L., Levant, A.: Sliding Mode Control and Observation. Springer, New York (2014)
3. Boiko, I.: Discontinuous Control Systems: Frequency-Domain Analysis and Design. Birkhuser, Boston (2009)
4. Oza, H.B., Orlov, Y.V., Spurgeon, S.K.: Continuous uniform finite time stabilization of planar controllable systems. SIAM J. Control Optim. **53**(3), 1154–1181 (2015)

5. Bhat, S.P., Bernstein, D.S.: Geometric homogeneity with applications to finite-time stability. Math. Control Signals Syst. **17**, 101–127 (2005)
6. Goddard.: Rules for the Design, Development, Verification, and Operation of Flight Systems: GSFC-STD-1000E (2009)
7. Boiko, I.M.: On relative degree, chattering and fractal nature of parasitic dynamics in sliding mode control. J. Franklin. Inst. **351**(4), 1939–1952 (2014)
8. Utkin, V.I.: Sliding modes in control and optimization. Springer, Berlin (1992)
9. Shtessel, Y.B., Lee, Y.-J.: New approach to chattering analysis in systems with sliding modes. In: Proceedings of the 35th IEEE Conference on Decision and Control, pp. 4014–4019, Dec (1996)
10. Lee, H., Utkin, V.: Chattering suppression methods in sliding mode control systems. Annu. Rev. Control **31**, 179–188 (2007)
11. Fridman, L.: An averaging approach to chattering. IEEE Trans. Autom. Control **46**(8), 1260–1265 (2001)
12. Fridman, L.: Singularly perturbed analysis of chattering in relay control systems. IEEE Trans. Autom. Control **47**(12), 2079–2084 (2002)
13. Boiko, I.: Analysis of sliding mode in frequency domain. Int. J. Control **78**(13), 969–981 (2005)
14. Boiko, I., Fridman, L.: Analysis of chattering in continuous sliding-mode controllers. IEEE Trans. Autom. Control **50**(9), 1442–1446 (2005)
15. Levant, A.: Chattering analysis. IEEE Trans. Autom. Control **55**(6), 1380–1389 (2010)
16. Levant, A., Fridman, L.: Accuracy of homogeneous sliding modes in the presence of fast actuators. IEEE Trans. Autom. Control **55**(3), 810–814 (2010)
17. Pisano, A., Usai, E.: Sliding mode control: a survey with applications in math. Math. Comput. Simul. **81**(5), 954–979 (2011)
18. Boiko, I., Fridman, L., Pisano, A., Usai, E.: Analysis of chattering in systems with second-order sliding-modes. IEEE Trans. Autom. Control **52**(11), 2085–2102 (2007)
19. Boiko, L., Fridman, L., Pisano, A., Usai, E.: Performance analysis of second-order sliding-mode control systems with fast actuators. IEEE Trans. Autom. Control **52**(6), 1053–1059 (2007)
20. Tsypkin, YaZ: Relay Control Systems. Cambridge University Press, England (1984)
21. Gelb, A., Vander Velde, W.E.: Multiple-input describing functions and nonlinear system design. Mc Graw-Hill, New York (1968)
22. Atherton, D.P.: Nonlinear Control Engineering-Describing Function Analysis and Designing. Van Nostrand Company Limited, UK (1975)
23. Levant, A.: Sliding order and sliding accuracy in sliding mode control. Int. J. Control **58**, 1247–1263 (1993)
24. Levant, A.: Robust exact differentiation via sliding mode technique. Automatica **34**(3), 379–384 (1998)
25. Boiko, I., Fridman, L., Castellanos, M.I.: Analysis of second-order sliding-mode algorithms in the frequency domain. IEEE Trans. Autom. Control **49**(6), 946–950 (2004)
26. Levant, A.: High-order sliding modes: differentiation and output-feedback control. Int. J. Control **76**(9–10), 924–941 (2003)
27. Levant, A.: Quasi-continuous high-order sliding-mode controllers. IEEE Trans. Autom. Control **50**(11), 1812–1816 (2005)
28. Schwartz, C., Gran, R.: Describing function analysis using matlab and simulink. IEEE Control Syst. Mag. **21**(4), 19–26 (2001)
29. Rosloniec, S.: Fundamental Numerical Methods for Electrical Engineering. Springer, Berlin (2008)
30. Rosales, A., Shtessel, Y., Fridman, L.: Analysis and design of systems driven by finite-time convergent controllers: practical stability approach. Int. J. Control (2017). doi:10.1080/00207179.2016.1255354
31. Rosales, A., Shtessel, Y., Fridman, L., Panathula, C.B.: Chattering analysis of HOSM controlled systems: frequency domain approach. IEEE Trans. Autom. Control (2017). doi:10.1109/TAC.2016.2619559
32. Dorf, R.C., Bishop, R.H.: Modern Control Systems, 12th edn. Prentice Hall, USA (2011)
33. Shtessel, Y., Buffington, S., Banda, S.: Multiple time scale flight control using reconfigurable sliding modes. J. Guid. Control Dyn. **22**(6), 873–883 (1999)

Chapter 5
On Inherent Gain Margins of Sliding-Mode Control Systems

Igor Boiko

5.1 Introduction

Stability margins, and specifically gain and phase margins, play a crucial role in linear systems design. Through the selection of margins the trade-off between stability and robustness of a linear system can be ensured. The situation is quite different in sliding mode (SM) control. If parasitic dynamics of a certain form are present in the loop of the sliding mode control system (and they are always present in real control systems) then a stable equilibrium point does not exist, and chattering occurs instead. Therefore, the same approach to stability analysis, which is used in linear control, cannot be applied to a sliding mode control system. However, similar to the approach used in linear systems analysis can also be used in SM systems if, instead of the actual dynamics, averaged dynamics are analyzed. Therefore, this problem is more complex than the one of the linear system analysis. As a result, it has become a common practice that most of research dealing with SM control systems is usually limited to the ideal sliding mode. And only in a relatively small number of publications, attempts are made to address the effect of chattering. Yet, very rare publications try to deal with the effects of the parasitic dynamics on the closed-loop performance. It was shown in [1] that parasitic dynamics, which inevitably exist in real control systems, result not only in generation of chattering but also in non-ideal closed-loop performance with respect to the propagation of external input signals and disturbances (averaged dynamics): non-ideal tracking and non-ideal disturbance rejection. The mechanism that explains the deterioration of the closed-loop performance is revealed through either the describing function (DF) analysis [2] or the locus of a perturbed relay system (LPRS) analysis [3]. This mechanism involves finding the so-called equivalent gain of the relay, which describes propagation of signals that are slower (have much lower spectrum) than chattering through the SM control systems. The use of the equivalent gain results in the averaged (on the period of chattering) dynamics of the SM system. The mathematical description of the obtained averaged dynamics is linear differential equations (subject to the plant being linear),

I. Boiko (✉)
The Petroleum Institute, Abu Dhabi, United Arab Emirates
e-mail: i.boiko@ieee.org

S. Li et al. (eds.), *Advances in Variable Structure Systems and Sliding Mode Control—Theory and Applications*, Studies in Systems, Decision and Control 115, DOI 10.1007/978-3-319-62896-7_5

which allows one to analyze such characteristics of this linear dynamics as gain and phase stability margins. Another approach to obtaining a model of averaged motions, through the singular perturbation method, provides results that are well correlated with the previous two approaches [4].

The same analysis can be done in second- and higher-order SM control systems [5–9]. In terms of the method of obtaining an averaged model of a second-order SM system, it is not much different from that of the first-order SM system. However, the technique of analysis may be significantly more complex: the use of the LPRS method becomes difficult (if possible at all), and the DF analysis becomes complex. It should be noted that some attempts of obtaining an averaged model of a second-order SM system were made in the past: the model of the sub-optimal algorithm was obtained in [10].

In this book chapter, analysis is done via the describing function method and, where possible, through the locus of a perturbed relay system method. First-order SM, the hysteresis relay control, the twisting algorithm and the sub-optimal algorithm are analyzed and compared in terms of the gain margins of the averaged dynamics that can describe the propagation of external signals through a SM control system. The chapter is illustrated by examples and concluded with observations derived from the presented analysis.

5.2 Gain Stability Margin of Averaged Dynamics of a Relay Feedback System

5.2.1 LPRS Analysis of a Relay Feedback System

We shall consider the following analysis of the relay feedback system, which illustrates the approach to treatment of chattering and averaged motions. Let the system be described by the following equations (Fig. 5.1):

$$\begin{aligned} \dot{\mathbf{x}} &= \mathbf{A}\mathbf{x} + \mathbf{B}u \\ y &= \mathbf{C}\mathbf{x} \end{aligned}, \tag{5.1}$$

$$u(t) = \begin{cases} c \text{ if } \sigma = f - y > b \quad \text{or} \quad \sigma > -b\,,\ u(t-) = c \\ -c \text{ if } \sigma = f - y < -b \quad \text{or} \quad \sigma < b\,,\ u(t-) = -c \end{cases}$$

where $\mathbf{x} \in \mathbb{R}^n$ is a state vector, y output, σ error signal, f input (set point), u control, c relay amplitude, $2b$ hysteresis value of the relay, $\mathbf{A} \in \mathbb{R}^{n\times n}$, $\mathbf{B} \in \mathbb{R}^{n\times 1}$, $\mathbf{C} \in \mathbb{R}^{1\times n}$ are matrices, $\mathbf{A}$ is nonsingular, the time $t-$ refers to time immediately preceding time t.

A method of analysis of self-excited oscillations and averaged (on the period of the self-excited oscillation) motions in relay feedback systems (Fig. 5.1) was proposed in [3]. The method is developed based on the assumption of the infinitely small

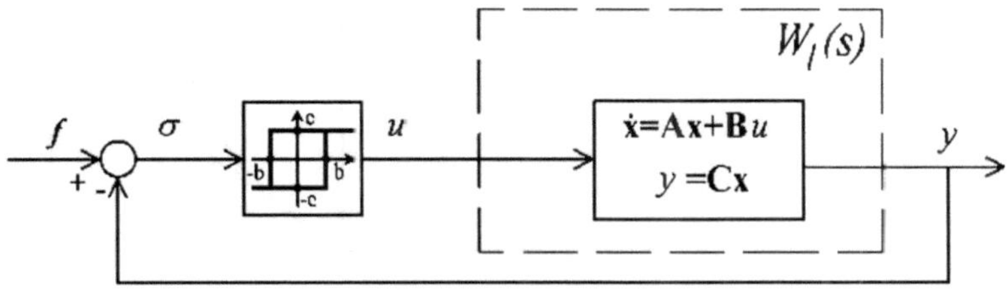

Fig. 5.1 Relay feedback system

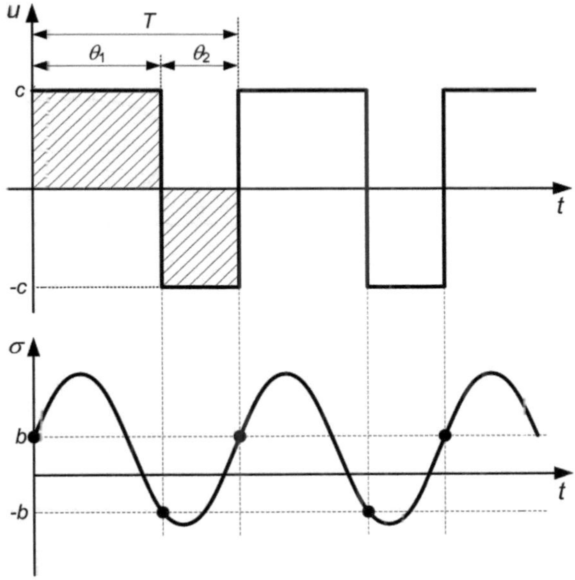

Fig. 5.2 Asymmetric oscillations in relay feedback system

asymmetry of the oscillations in the system, which is a result of the infinitely small value of the input signal f. If a constant input signal $f \neq 0$ is applied to the system, self-excited oscillations that feature unequally-spaced switchings (Fig. 5.2) will be generated in the system.

With asymmetric oscillations as shown in Fig. 5.2, the averaged on the period of these oscillations control $u(t)$ can be found as

$$u_0 = \frac{1}{\theta_1 + \theta_2} \int_0^{\theta_1+\theta_2} c(\theta_1 - \theta_2)\mathrm{d}t$$

A function showing the dependence of u_0 on the value of the constant input $f_0 \equiv \text{const}$ is named the *bias function*. A plot of a typical bias function is given in Fig. 5.3.

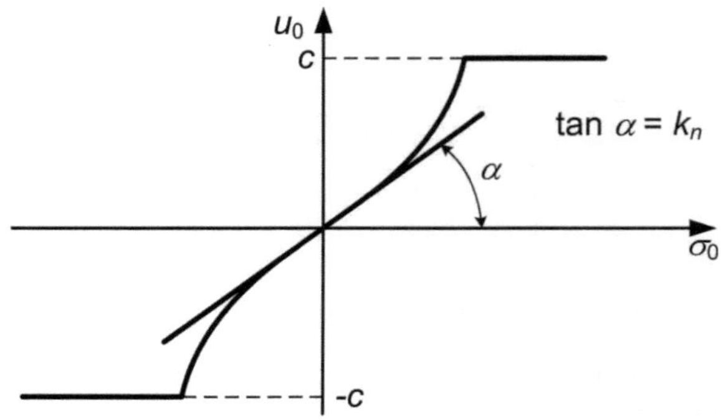

Fig. 5.3 Bias function

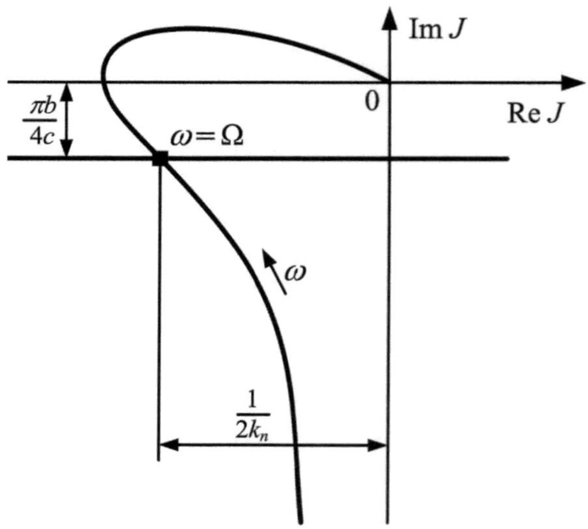

Fig. 5.4 Analysis through locus of a perturbed relay system

A complex function named LPRS was introduced in [3] as follows to carry out analysis of self-excited oscillations and of averaged dynamics:

$$\begin{aligned} J(\omega) &= -0.5\mathbf{C}[\mathbf{A}^{-1} + \tfrac{2\pi}{\omega}(\mathbf{I} - e^{\frac{2\pi}{\omega}\mathbf{A}})^{-1}e^{\frac{\pi}{\omega}\mathbf{A}}]\mathbf{B} \\ &+ j\tfrac{\pi}{4}\mathbf{C}(\mathbf{I} + e^{\frac{\pi}{\omega}\mathbf{A}})^{-1}(\mathbf{I} - e^{\frac{\pi}{\omega}\mathbf{A}})\mathbf{A}^{-1}\mathbf{B}\,. \end{aligned} \tag{5.2}$$

A typical plot of an LPRS is given in Fig. 5.4.

It was shown in [3] that exact value of the frequency of self-excited oscillations Ω can be found through LPRS from the following equation:

$$\operatorname{Im} J(\Omega) = -\frac{\pi b}{4c}. \tag{5.3}$$

or by finding the point of intersection of the LPRS and of the horizontal line drawn below the real axis at the distance of $\frac{\pi b}{4c}$ (Fig. 5.4).

Averaged dynamics can be modeled by the so-called equivalent gain of the non-linearity, with respect to the constant (or slowly varying) component in the error signal. The equivalent gain of the relay k_n that describes propagation of signals that are slow in comparison with the self-excited oscillations through the relay can be computed as:

$$k_{n(LPRS)} = -\frac{1}{2\mathrm{Re}J(\Omega)} = \lim_{f \to 0} \frac{u_0}{\sigma_0}, \tag{5.4}$$

where Ω is the frequency of the self-excited oscillations found per (5.3), and σ_0 is the average on the period of the self-excited oscillations value of the error:

$$\sigma_0 = \frac{1}{\theta_1 + \theta_2} \int_0^{\theta_1+\theta_2} \sigma(t)\mathrm{d}t.$$

The graphic interpretation of Eq. (5.4) is that the equivalent gain is inversely proportional to the distance between the point Ω of the LPRS and the real axis (Fig. 5.4).

As a result, propagation of the harmonic input signal $f(t)$ of frequency ω_{in} (subject to $\omega_{in} \ll \Omega$) through the relay feedback system, which provides the averaged on the period of the oscillations motions, can be described by the following complex frequency response function (transfer function having s replaced with $j\omega_{in}$):

$$W(j\omega_{in}) = \frac{k_n W_l(j\omega_{in})}{1 + k_n W_l(j\omega_{in})}, \tag{5.5}$$

where $W_l(s) = \mathbf{C}(\mathbf{I}s - \mathbf{A})^{-1}\mathbf{B}$.

Analysis of averaged motions, therefore, includes analysis of self-excited oscillations (frequency Ω). And these two aspects make overall analysis of a SM system complete.

5.2.2 Gain Stability Margins

Let us analyze the gain stability margin of the relay system. We start with the system having ideal relay ($b = 0$). According to the definition of the gain margin γ_m, the following equation must hold:

$$\gamma_m k_{n(LPRS/DF)} W_l(j\omega_\pi) = -1 \tag{5.6}$$

Using (5.4) and considering the equality $\mathrm{Im}\, W_l(j\omega_\pi) = 0$, we transform it into:

$$\frac{\gamma_m}{2}\frac{\mathrm{Re}W_l(j\omega_\pi)}{\mathrm{Re}J(\Omega)} = 1 \tag{5.7}$$

Considering the series formula of the LPRS [11]:

$$\begin{aligned} J(\omega) = &\sum_{k=1}^{\infty}(-1)^{k+1}\mathrm{Re}\ W_l(k\omega) \\ &+j\sum_{k=1}^{\infty}\mathrm{Im}\ W_l[(2k-1)\omega]/(2k-1) \end{aligned}, \tag{5.8}$$

we can now rewrite formula (5.7) as

$$\frac{\gamma_m}{2}\frac{\mathrm{Re}W_l(j\omega_\pi)}{\mathrm{Re}W_l(j\Omega)+\sum_{k=2}^{\infty}(-1)^{k+1}\mathrm{Re}\ W_l(k\Omega)} = 1. \tag{5.9}$$

Given the closeness of the two frequencies: $\omega_\pi \approx \Omega$, (5.9) is rewritten as:

$$\gamma_m \approx 2\frac{\mathrm{Re}W_l(j\Omega)+\sum_{k=2}^{\infty}(-1)^{k+1}\mathrm{Re}W_l(k\Omega)}{\mathrm{Re}W_l(j\Omega)}. \tag{5.10}$$

For systems having relative degree three (if relative degree is less than three, oscillations cannot be excited), which are considered as an example, all terms in the infinite series of (5.10) are negative: $\mathrm{Re}W_l(k\Omega) < 0$, with diminishing magnitudes. However, they are multiplied by either 1 or -1, with the first term of the series multiplied by -1. As a result, the whole sum is a positive quantity:

$$\sum_{k=2}^{\infty}(-1)^{k+1}\mathrm{Re}W_l(k\Omega) > 0.$$

Therefore, given that $\mathrm{Re}W_l(j\Omega) < 0$, the gain margin is:

$$\gamma_m < 2. \tag{5.11}$$

Approximate DF analysis can be produced from the development presented above by simple discounting of the higher terms in the series of (5.8). In this case, LPRS can be simply replaced with the DF that is defined as:

$$N(a) = \frac{2}{\pi a}\int_0^\pi f(a\sin\psi)\sin\psi \mathrm{d}\psi + j\frac{2}{\pi a}\int_0^\pi f(a\sin\psi)\cos\psi \mathrm{d}\psi.$$

Application of this formula to the relay nonlinearity leads to the following describing function formula:

$$N(a) = \frac{4c}{\pi a}. \tag{5.12}$$

The equivalent gain of the nonlinearity, with respect to the constant (or slowly varying) component in the error signal, computed within the DF method, is defined by the following equation:

$$k_{n(DF)}(a) = \left.\frac{\partial u_0}{\partial \sigma_0}\right|_{\sigma_0=0} = \frac{2}{\pi}\left.\frac{\partial}{\partial \sigma_0}\right|_{\sigma_0=0} \int_0^{\pi} f(\sigma_0 + a\sin\psi)d\psi$$

For the ideal relay, the equivalent gain is computed as:

$$k_{n(DF)} = \frac{2c}{\pi a} \tag{5.13}$$

Considering together the harmonic balance condition

$$W_l(j\Omega)N(a) = -1, \tag{5.14}$$

the equality $\omega_\pi = \Omega$, and Eqs. (5.6), (5.12), (5.13), we come to the following conclusion:

$$\gamma_m = 2. \tag{5.15}$$

This remarkable property of the relay feedback system, which is approximate due to the DF analysis, was noted in [12]. The relay feedback system has some adaptive properties: parameters of the plant may change, but the gain stability margin would always stay equal to 2.

5.3 Gain Stability Margins in Systems with Second-Order SM Algorithms

Let us analyze stability margins of two second-order SM (SOSM) algorithms, in particular. We shall carry out analysis of the twisting and the sub-optimal, and compare gain margins of these algorithms with that of the relay control.

5.3.1 Gain Stability Margin of Averaged Dynamics of System with Sub-Optimal Algorithm

The version of the sub-optimal algorithm called generalized sub-optimal algorithm was proposed in [13, 14], which is given as follows:

$$u = c\,\mathrm{sign}\,(\sigma - \beta\sigma_{Mi})\,,$$

where c is control amplitude, $\sigma = f - y$ an error signal, β positive constant parameter of the algorithm, σ_{Mi} is the latest "singular point" of σ, i.e. the value of σ at the most resent time t_i such that $\dot{\sigma}\,(t_i) = 0$.

DF of the generalized sub-optimal algorithm was found in [15]:

$$N(a) = \frac{4c}{\pi a}\left(\sqrt{1-\beta^2} + j\beta\right). \tag{5.16}$$

And the input-output properties of the algorithm were studied in [16]. Specifically, it was found that the equivalent gain can be computed through the LPRS as:

$$k_{n(LPRS)} = \frac{1+\beta}{-2\mathrm{Re}J(\Omega) + 2\beta R(\Omega)}, \tag{5.17}$$

where $R(\Omega)$ is computed as an infinite series containing even harmonics of the frequency response $W_l(jk\Omega)$ [16]. DF analysis of the input-output properties of the sub-optimal algorithms can be obtained from the LPRS analysis by setting R to zero and replacing the LPRS with the transfer function (frequency response), which leads to:

$$k_{n(DF)} = \frac{2c}{\pi a}(1+\beta). \tag{5.18}$$

Gain stability margin, as in the case of the relay controller, is given by Eq. (5.6), which, with account of the DF-based formulas (5.16) and (5.18), will result in:

$$\begin{aligned}\gamma_m &= -\frac{1}{W_l(j\omega_\pi)}\frac{\pi a}{2c}\frac{1}{1+\beta} = \\ &= -\frac{1}{W_l(j\omega_\pi)}\frac{\pi a}{4c}\frac{2}{1+\beta}\frac{\sqrt{1-\beta^2}+j\beta}{\sqrt{1-\beta^2}+j\beta} = \\ &= -\frac{1}{W_l(j\omega_\pi)}\frac{2\left(\sqrt{1-\beta^2}+j\beta\right)}{(1+\beta)N(a)}\end{aligned} \tag{5.19}$$

Considering the harmonic balance Eq. (5.14), which is valid for the sub-optimal algorithm too, we transform (5.19) into:

$$\gamma_m = \frac{W_l(j\Omega)}{W_l(j\omega_\pi)}\frac{2\left(\sqrt{1-\beta^2}+j\beta\right)}{1+\beta} \tag{5.20}$$

In (5.20), unlike in the case of the relay controller, the two frequencies are different: $\Omega > \omega_\pi$. Also, it should be noted that gain margin γ_m is a real but not a complex quantity. Therefore, complex quantities in (5.20) can be replaced with their absolute values:

$$\gamma_m = \frac{|W_l(j\Omega)|}{W_l(j\omega_\pi)}\frac{2}{1+\beta} \tag{5.21}$$

Given the fact that, due to the decreasing character of the magnitude characteristic of the plant, the following inequality normally holds: $|W_l(j\Omega)| < |W_l(j\omega_\pi)|$. As a result, *gain margin of the generalized suboptimal algorithm is always smaller than that of the relay control (smaller than 2).*

5.3.2 Gain Stability Margin of Averaged Dynamics of System with Twisting Algorithm

The twisting algorithm is given as follows [5]:

$$u = c_1 \text{sign}(\sigma) + c_2 \text{sign}(\dot{\sigma}),$$

where $c_1 > 0$ and $c_2 > 0$ are control amplitudes, $\sigma = f - y$ an error signal.
DF of the twisting algorithm was found in [17]:

$$N(a) = \frac{4}{\pi a}(c_1 + jc_2). \tag{5.22}$$

Analyze now the input-output properties of a system with the twisting algorithm by finding first the equivalent gain. We shall consider the twisting algorithm as a parallel connection of two ideal relays. Each of these relays can be brought (separately) to an asymmetric oscillation if a certain constant bias signal is applied to each relay input. Propagation of this bias signal through a particular relay is described as an equivalent gain of this relay. Therefore, the methodology of a conventional relay system can be used in this case too. This allows us to write the formula for the equivalent gain of the twisting algorithm as:

$$k_{n(DF)} = \frac{2c_1}{\pi a} + j\omega_{in}\frac{2c_2}{\pi a_2}, \tag{5.23}$$

where a_2 is the amplitude of the oscillation of $\dot{\sigma}$; ω_{in} comes from the s at the input of the second relay (differentiation of σ). It should be noted that the frequency of the input signal must be used when replacing s with $j\omega$ because k_n describes the propagation of the forced component of the motion. Taking into account that $a_2 = \Omega a$ (see [5]), we rewrite (5.23) as follows:

$$k_{n(DF)} = \frac{2}{\pi a}\left(c_1 + j\frac{\omega_{in}}{\Omega}c_2\right). \tag{5.24}$$

Considering the gain margin Eq. (5.6) together with the harmonic balance Eq. (5.14), and the expressions for the DF (5.22) and the equivalent gain (5.24), we obtain a formula for the gain margin as follows:

$$\gamma_m = \frac{W_l(j\Omega)}{W_l(j\omega_\pi)} \frac{2\,(c_1 + jc_2)}{c_1 + j\frac{\omega_{in}}{\Omega}c_2} \tag{5.25}$$

Chattering is a results of the presence of parasitic dynamics. Lower effect of parasitic dynamics would correspond to a higher frequency of chattering. In fact, no parasitic dynamics would correspond to infinite frequency of chattering. Therefore, one could assume that parasitic dynamics has low contribution to overall dynamics of the system, or equivalently, that the frequency of chattering is high. At least one can assume that the frequency of chattering is much higher than the spectrum of possible input signals. The following assumption regarding the relationship between the frequency of the input signal and the frequency of chattering $\omega_{in} \ll \Omega$ results in the following approximate formula:

$$\gamma_m = 2\frac{W_l(j\Omega)}{W_l(j\omega_\pi)}\left(1 + j\frac{c_2}{c_1}\right), \tag{5.26}$$

which is produced from (5.25) by setting ω_{in}/Ω to zero. Considering that gain margin γ_m is a real but not a complex quantity, we take absolute values of the complex multipliers in (5.26):

$$\gamma_m = 2\frac{|W_l(j\Omega)|}{W_l(j\omega_\pi)}\sqrt{1 + \frac{c_2^2}{c_1^2}} \tag{5.27}$$

Usually the following holds$|W_l(j\Omega)| < |W_l(j\omega_\pi)|$, which may lead to both possibilities: of the gain margin being larger ($\gamma_m > 2$) and smaller ($\gamma_m < 2$) than that of the relay control.

5.4 Inherent Gain Margin and Closed-Loop Performance of SM Control Systems

In this section, we aim to relate the derived formulas of the gain stability margins for the analyzed three controllers and the closed-loop performance with respect to tracking of set points and rejection of external disturbances. In doing so, we proceed from the following assumptions and statements:

- Parasitic dynamics always exist in real SM control systems. Therefore both first-order and higher-order SM control systems exhibit chattering.
- Amplitude of oscillations due to chattering in the control signal, which is equal (first harmonic) to $a_u = \frac{4c}{\pi}$ for the first-order SM control and sub-optimal algorithm, and $a_u = \frac{4\sqrt{c_1^2+c_2^2}}{\pi}$ for the twisting algorithm, can be set to a necessary value and, therefore, the amplitude of chattering of the system output can also be adjusted by the same means: $a_y = a_u\,|W_l(j\Omega)|$. As a result, the amplitude of chattering a_y cannot be used as a comprehensive metric allowing one to compare performance of different SM control algorithms.

- On the other hand, if the amplitude of chattering of control a_u is set the same for all algorithms being compared then the amplitude of chattering of the output a_y characterizes performance of an algorithm: the smaller a_y, subject to a certain fixed value of a_u, the better.
- Yet, amplitude of chattering does not describe input-output properties of an algorithm. Low-frequency loop gain, which is related with the gain stability margin, can serve as a characteristic of closed-loop performance: a steady-state error as a reaction to a unity step of the control signal is given by $\sigma_0 = 1/(1 + k_n W_l(0))$.

In linear control systems, the meaning of gain and phase stability margins is two-fold. Firstly, it is related to the *robustness* of a system: the larger these gains the higher changes of the plant dynamics (due to operating conditions, aging, etc.) are allowed without loss of stability of the system. And secondly, gain and phase margins are related with the quality of *transient processes*: the smaller the margins the more oscillatory the transient motion would be. Therefore, there are some reasonable lower limits for these margins even if the plant dynamics are not subject to any variations.

In SM control systems, as shown above, the DF $N(a)$ and equivalent gain k_n are related. This relationship can be characterized by the gain margin, which is given by (5.15), (5.20), (5.26) for the relay control, the sub-optimal algorithm and the twisting algorithm, respectively. Formulas (5.15), (5.20), (5.26) show that the gain margin of SM control systems does not reflect robustness the way linear systems do: if the gain of the plant is increased, the value of γ_m would stay unchanged. However, gain margin defines the closed-loop performance, because its value would reflect the disturbance-attenuation properties.

Let us write the transfer function relating the averaged motions in the error signal and the system input:

$$W_e(j\omega_{in}) = \frac{1}{1 + k_n W_l(j\omega_{in})}, \tag{5.28}$$

where the equivalent gain is given by formulas (5.13), (5.18), or (5.24) for the relay control, the sub-optimal algorithm and the twisting algorithm, respectively. At zero and low frequencies of the input signal ($\omega_{in} \ll \Omega$) and given that the plant is *type-0 servo* the following holds:

$$W_e(j\omega_{in}) \approx W_e(0) = \frac{1}{1 + k_n W_l(0)}. \tag{5.29}$$

It is obvious that higher values of the equivalent gain entail smaller errors or better tracking quality. However, equivalent gains are not suitable for comparison of different SM control algorithms in terms of their closed-loop performance. Yet, gain margins are well suitable for that purpose, because instead of showing the error in a particular case, it shows how much the errors would differ if the plant is controlled by different SM algorithms. Because the gain stability margin cannot be changed by increasing or decreasing the plant gain, we shall call it *inherent gain margin*. This term refers to the fact that the gain depends on the control algorithms rather than on the plant or process. The following summary of the results can be given as follows:

- The lower the value of the *inherent gain margin* the higher the loop gain is $k_n W_l(0)$, and the lower the steady-state error σ_0 is.
- Since stability with respect to external signals is maintained automatically in a SM system, a smaller value of the *inherent gain margin* is beneficial, because it does not endanger stability but results in a better closed-loop performance with respect to external signals. However, the quality of transients, that may be affected by the gain margin, must be considered too.
- Other means of enhancement of closed-loop performance, such as introduction of phase lag and lead-lag filters, can be used too, so that advantages of using a particular SM control algorithm must be weighed for every particular application against the use of compensating filters; their use in combination can be considered too.

5.5 Examples

We shall consider the following three examples of applications of the considered controllers to a plant having principal dynamics of relative degree two $W_p(s) = \frac{1}{s^2+s+1}$ and parasitic dynamics (actuator) given by the following first-order-plus–time-delay transfer function: $W_a(s) = \frac{e^{-0.1s}}{0.02s+1}$. Principal and parasitic dynamics together results in the following dynamics of the linear part of the system:

$$W_l(s) = \frac{e^{-0.1s}}{(0.02s+1)\left(s^2+s+1\right)}, \tag{5.30}$$

To design the three controllers that would provide results suitable for comparison, we need to satisfy two requirements:

- The amplitude of the oscillations of the control signal $a_u = a\,|N(a)|$ must be the same for all three controllers;
- The sub-optimal and the twisting controllers must provide the same phase shift $\phi_N = \arg N(a) = -\pi + \frac{\mathrm{Im}N(a)}{\mathrm{Re}N(a)}$ of the frequency of the oscillations (this cannot be required from the relay control by definition).

Therefore, we can write for the amplitudes of the control signal:

$$\frac{4c}{\pi} = \frac{4}{\pi}\sqrt{c_1^2 + c_2^2} \tag{5.31}$$

and for the phase angles:

$$\frac{\beta}{\sqrt{1-\beta^2}} = \frac{c_2}{c_1}. \tag{5.32}$$

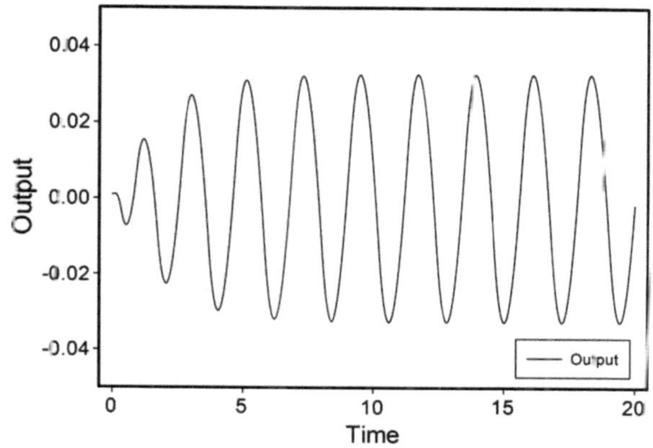

Fig. 5.5 Symmetric oscillations in relay feedback system

Considering (5.31) and (5.32) as equations for c_1 and c_2, one can find equivalent parameters for the twisting controller from known c and β of the sub-optimal controller:

$$c_1 = c\sqrt{1-\beta^2}, \tag{5.33}$$

$$c_2 = c\beta. \tag{5.34}$$

Simulations for $c = 0.2$, $\beta = 0.2$ of the relay and sub-optimal controllers, and $c_1 = 0.196$ and $c_2 = 0.040$ computed per (5.33), (5.34) for the twisting controller, showing symmetric oscillations (chattering), are presented in Figs. 5.5, 5.6 and 5.7, respectively. The observed values of the oscillation parameters are: for the relay controller $a = 0.0327$, $\Omega = 2.848$; for the sub-optimal controller $a = 0.0191$, $\Omega = 3.696$; and for the twisting controller $a = 0.0188$, $\Omega = 3.696$ rad/s. One can see that the parameters of the oscillations for the sub-optimal and twisting controllers are nearly identical, which is an expected outcome of the design through (5.33), (5.34). Even the transients are similar in the sub-optimal and twisting algorithms, that can be explained by the same shape of the describing functions [16, 17].

What concerns the inherent gain margins, if the parameters of the twisting algorithm are selected per (5.33) and (5.34) then substitution into (5.27) yields for the twisting algorithm:

$$\gamma_{mTwist} = 2\frac{|W_l(j\Omega)|}{W_l(j\omega_\pi)}\frac{1}{\sqrt{1-\beta^2}}, \tag{5.35}$$

which is always larger than the inherent gain margin for the sub-optimal algorithm (5.21).

Bias functions and equivalent gains of the systems with the considered three SM algorithms are presented in [1, 3, 16, 17].

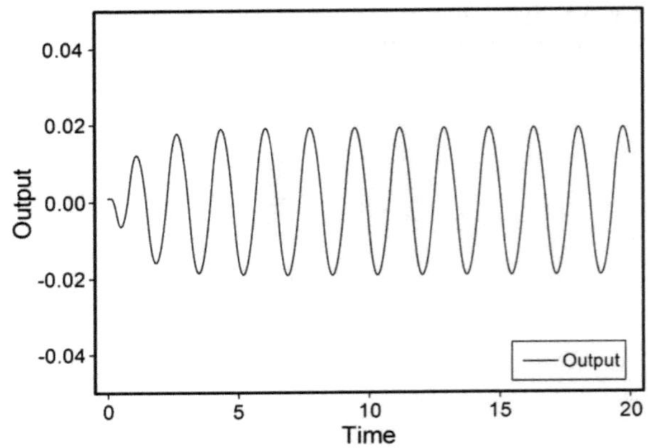

Fig. 5.6 Chattering in the system with sub-optimal controller chattering in the system with sub-optimal controller

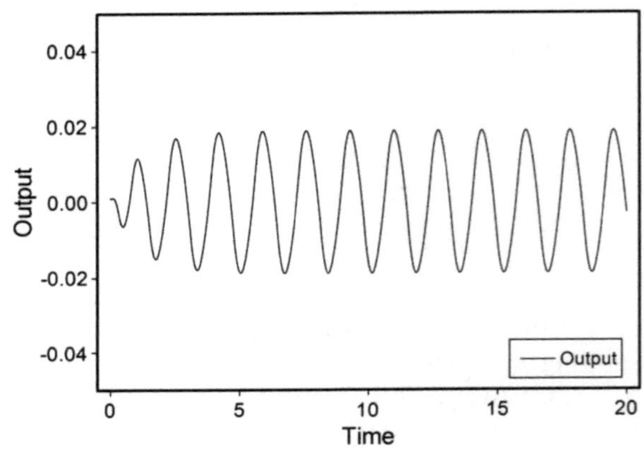

Fig. 5.7 Chattering in the system with twisting controller

5.6 Conclusion

The notion of *inherent gain margin*, applicable to sliding mode control algorithm performance, is introduced. Analysis of three different SM control algorithms: the relay control, the sub-optimal algorithm and the twisting algorithm, with respect to their closed-loop performance, is presented. It is shown that these algorithms have different values of inherent gain margin. Particularly, the sub-optimal algorithm has the lowest value of all. It is shown that having a low value of inherent gain margin is beneficial, as leading to higher accuracy of tracking and higher degree of disturbance rejection. However, other effects, such as the quality of transients, and other possible methods of increasing the open-loop gain, such as phase-lag and lead–lag filters, must be considered too when conclusion is made about closed-loop performance of different control algorithms. This would make the comparison more comprehensive, yet the comparison process more complex too. This is not done in the present chapter, and can be considered as a possible direction for future research.

References

1. Boiko, I.: Analysis of closed-loop performance and frequency-domain design of compensating filters for sliding mode control systems. IEEE Trans. Autom. Control. **52**(10), 1882–1891 (2007)
2. Atherton, D.: Nonlinear Control Engineering - Describing Function Analysis and Design, Workingham, Berks. Van Nostrand Company Limited, UK (1975)
3. Boiko, I.: Oscillations and transfer properties of relay servo systems - the locus of a perturbed relay system approach. Automatica **41**(4), 677–683 (2005)
4. Fridman, L.: An averaging approach to chattering. IEEE Trans. Autom. Control. **46**(8), 1260–1265 (2001)
5. Levant, A.: Sliding order and sliding accuracy in sliding mode control. Int. J. Control. **58**(6), 1247–1263 (1993)
6. Levant, A.: Universal SISO sliding-mode controllers with finite-time convergence. IEEE Trans. Autom. Control. **46**(9), 1447–1451 (2001)
7. Bartolini, G., Ferrara, A., Pisano, A., Usai, E.: On the convergence properties of a 2-sliding control algorithm for nonlinear uncertain systems. Int. J. Control. **74**, 718–731 (2001)
8. Man, Z., Paplinski, A.P., Wu, H.R.: A robust MIMO terminal sliding mode control for rigid robotic manipulators. IEEE Trans. Autom. Control. **39**(12), 2464–2468 (1994)
9. Shtessel, Y.B., Shkolnikov, I.A., Brown, M.J.: An asymptotic second-order smooth sliding mode control. Asian J. Control. **5**(4), 498–504 (2003)
10. Boiko, I., Fridman, L., Pisano, A., Usai, E.: On the transfer properties of the generalized sub-optimal second-order sliding mode control algorithm. IEEE Trans. Autom. Control. **54**(2), 399–403 (2009)
11. Boiko, I.: Input-output analysis of limit cycling relay feedback control systems. In: Proceedings of 1999 American Control Conference. San Diego, USA, 542–546 (1999)
12. Hsu, J.C., Meyer, A.U.: Modern Control Principles and Applications. McGraw Hill, New York (1968)
13. Bartolini, G., Pisano, A., Punta, E., Usai, E.: A survey of applications of second order sliding mode control to mechanical systems. Int. J. Control. **76**(9/10), 875–892 (2003)
14. Bartolini, G., Ferrara, A., Usai, E.: Output tracking control of uncertain nonlinear second-order systems. Automatica **33**(12), 2203–2212 (1997)
15. Boiko, I., Fridman, L., Pisano, A., Usai, E.: Analysis of chattering in systems with second-order sliding mode. IEEE Trans. Autom. Control. **52**(11), 2085–2102 (2007)
16. Boiko, I., Fridman, L., Pisano, A., Usai, E.: Analysis of input-output performance of second-order sliding mode control algorithms. IEEE Trans. Autom. Control. **54**(2), 399–403 (2009)
17. Boiko, I., Fridman, L., Castellanos, M.: Analysis of second order sliding mode algorithms in the frequency domain. IEEE Trans. Autom. Control. **49**(6), 946–950 (2004)

Chapter 6
Adaptive Sliding Mode Control Based on the Extended Equivalent Control Concept for Disturbances with Unknown Bounds

Tiago Roux Oliveira, José Paulo V.S. Cunha and Liu Hsu

6.1 Introduction

Many authors have been dedicating their work to remove the obstacle of considering bounded disturbances with known bounds in the sliding mode control design [2–5, 14, 17, 18, 22, 23]. This research topic is of practical significance because the control magnitude is often related with the chattering intensity in actual applications. Overestimated disturbances may lead to nonminimum chattering [22].

Generally, adaptive sliding mode controllers can be classified as follows:

1. Schemes based on monotonically increasing gains [14, 17, 23];
2. Strategies with increasing and decreasing gains [2, 5, 18]; and
3. Approaches based on the equivalent control [3, 4, 22].

The benefit of the latter approach is that once the sliding motion begins, an accurate estimate of the disturbance can be obtained. If the disturbance decreases, the equivalent control does automatically the same. It means that if the control gain is updated according to the equivalent control, a less conservative controller in terms of control effort and improved performance can be obtained. Although the equivalent control cannot be measured precisely in practice due to model mismatches, a good approximation is the average control signal [21].

In this chapter, a novel adaptive sliding mode control strategy based on the *extended equivalent control* [8, 9] is developed [15]. The adaptation rule combines the qualities of monotonically increasing gains and the equivalent control. During the reaching phase, a positive feedback loop increases the control gain (modulation signal) until the disturbances are dominated. After that, the conditions for the sliding

T.R. Oliveira (✉) · J.P.V.S. Cunha
State University of Rio de Janeiro, Rio de Janeiro, Brazil
e-mail: tiagoroux@uerj.br

J.P.V.S. Cunha
e-mail: jpaulo@ieee.org

L. Hsu
COPPE/Federal University of Rio de Janeiro, Rio de Janeiro, Brazil
e-mail: liu@coep.ufrj.br

S. Li et al. (eds.), *Advances in Variable Structure Systems and Sliding Mode Control—Theory and Applications*, Studies in Systems, Decision and Control 115, DOI 10.1007/978-3-319-62896-7_6

mode phase can be satisfied. Thus, the disturbances bounds can be satisfactorily estimated by the equivalent control which updates the proposed modulation function. The modulation gain is adapted in such a way that it is small as possible to mitigate chattering effects, but large enough to ensure the existence of the sliding mode in the presence of bounded exogenous disturbances.

The new adaptation law seems simpler than earlier methods found in the literature. The restriction of assuming only smooth disturbances in previous works [3, 4, 22] can be overcome. In addition, the new adaptive scheme does not require knowledge of the minimum and maximum allowed values of the adaptive gain [22], and do not require any information about the bound on the disturbances [3]. When compared with the adaptive methodologies applied to [2, 5, 18], no mechanism is necessary to detect the sliding mode and adapt the control gain in order to make it increase or decrease. Additionally, perfect stabilization and ideal sliding mode can be guaranteed rather than only the convergence to a neighborhood of the origin and intermittent loss of the sliding motion.

6.2 Discontinuous Differential Equations

Since we shall deal with discontinuous systems, it is necessary to define precisely the meaning of a solution for such systems. Here, Filippov's definition is assumed [6]. Note that the control signal u is not necessarily a function of t in the usual sense when sliding modes take place. In order to avoid clutter, we will denote by $u(t)$ the locally integrable functions which are equivalent to u, in the sense of *equivalent control* [21], along any given Filippov solution $z(t)$ of the closed-loop system. It should be stressed that $z(t)$ is, *by definition*, absolutely continuous. This definition is motivated by the adequate representation of the behavior of physical systems when the actual switching mechanism tends to an ideal switching device which corresponds to the given discontinuous differential equation [21]. Then, along any such solution, u can be replaced by $u(t)$ in the right-hand side of the governing differential equations. Although the equivalent control $u(t) = u_{\mathrm{eq}}(t)$ is not directly available, for affine systems filtering u with any causal linear time-invariant filter with impulse response $g(t)$ gives $g(t) * u = g(t) * u(t) = g(t) * u_{\mathrm{eq}}(t)$. Here, the symbol "$*$" stands for the convolution operator defined as:

$$g(t) * u(t) := \int_0^t g(t-s)u(s)\mathrm{d}s \,. \tag{6.1}$$

6.2.1 *Extended Equivalent Control*

The *extended equivalent control* [8, 9] is defined as an equivalent control which applies for a complete system motion, i.e., on and outside the sliding mode surface $\sigma(x(t), t) = 0$, $\sigma : \mathbb{R}^n \times \mathbb{R}_+ \to \mathbb{R}^m$. Consider the class of nonlinear systems described by

$$\dot{x} = f(x, t) + B(x, t)u\,, \tag{6.2}$$

where $f : \mathbb{R}^n \times \mathbb{R}_+ \to \mathbb{R}^n$ and $B : \mathbb{R}^n \times \mathbb{R}_+ \to \mathbb{R}^{n\times m}$ are smooth. Let $x(t)$ be a solution of this system, for $t \in [0, T)$. Then, the extended equivalent control is a locally integrable function, defined almost everywhere in the interval $[0, T)$, and is given by

$$u_{\text{eq}}(t) = -[GB(x(t), t)]^{-1}\left[Gf(x(t), t) - \frac{\mathrm{d}}{\mathrm{d}t}\sigma(x(t), t) + \frac{\partial}{\partial t}\sigma(x(t), t)\right], \tag{6.3}$$

where $G = \frac{\partial}{\partial x}\sigma(x(t), t)$. The above expression is well defined since the solution $x(t)$ is absolutely continuous by definition and, thus, has derivatives almost everywhere.

6.2.2 *Average Control*

The ideal equivalent control signal is not available for controller realization. Since the extended equivalent control $u_{\text{eq}}(t)$ in systems affine in control (such as (6.2)) coincides with the slow component of the actual sliding mode control u, the *average control* denoted by $u_{\text{av}}(t)$ can be estimated by applying the real control signal to a low-pass filter with unit DC gain and time constant $\tau > 0$ small enough as compared with the slow component, but yet large enough to attenuate the high-frequency switching in u [21]. Thus, the low-pass filter

$$\tau\dot{u}_{\text{av}}(t) = -u_{\text{av}}(t) + u\,, \tag{6.4}$$

gives the average control which is an estimate of $u_{\text{eq}}(t)$ satisfying [22]

$$|u_{\text{av}}(t) - u_{\text{eq}}(t)| \leq \mathcal{O}(\tau) \to 0\,, \tag{6.5}$$

as $\tau \to +0$, provided that $u_{\text{eq}}(t)$ is a continuous signal.

6.3 Motivating Example

In order to illustrate the proposed approach, consider the simplest scalar integrator example inspired by [22]:

$$\dot{x} = u + d(t) \tag{6.6}$$

subject to an input disturbance $d(t)$, which is piece-wise continuous and uniformly bounded by an unknown constant $\bar{d} > 0$ such that $|d(t)| \leq \bar{d}$, $\forall t > 0$. Differently from [22], no known constant upper bound is imposed on $|\dot{d}(t)|$.

Our objective is to apply the first-order sliding mode controller

$$u = -\rho(t)\text{sign}[x]\,, \tag{6.7}$$

into the system (6.6) and design the modulation function $\rho(t) > 0$ in an adaptive fashion so that the state x can be driven to zero.

Basically, the modulation function ρ must satisfy the differential equation:

$$\dot{\rho}(t) = -\gamma_f \rho(t) + c_f |u_{\text{eq}}(t)|\,, \quad \rho(0) > 0\,, \tag{6.8}$$

with constants satisfying the inequalities

$$c_f > \gamma_f > 0\,. \tag{6.9}$$

The signal $u_{\text{eq}}(t)$ is the *extended equivalent control* [9], which is assumed available to simplify this presentation.

The idea is very intuitive: outside of the sliding motion $|u_{\text{eq}}(t)| = \rho(t)$, almost everywhere, whereas during the sliding motion $|u_{\text{eq}}(t)| = |d(t)|$. Thus, during the reaching phase, the differential equation (6.8) can be rewritten as

$$\dot{\rho}(t) = \left(c_f - \gamma_f\right) \rho(t)\,. \tag{6.10}$$

Since $c_f - \gamma_f > 0$, ρ increases exponentially in order to dominate the disturbance $(\rho(t) \geq \bar{d})$ in some finite time, such that the sliding surface $x \equiv 0$ is reached.

Once the sliding mode is achieved, the adaptation is given by the input-to-state stable (ISS, see Appendix) filter (6.8) with input being the estimate of $|d(t)|$ given by the equivalent control, that is $|u_{\text{eq}}(t)| = |d(t)|$. The modulation signal is the solution of (6.8) given by

$$\rho(t) = e^{-\gamma_f t}\rho(0) + c_f e^{-\gamma_f t} * |u_{\text{eq}}(t)|\,. \tag{6.11}$$

If

$$c_f e^{-\gamma_f t} * |d(t)| > |d(t)|\,, \quad \forall t \geq t_0\,, \tag{6.12}$$

for some $t_0 \geq 0$, then it can be concluded from (6.11) and $\rho(0) > 0$ that $\rho(t) > |d(t)|$, which is a sufficient condition for the existence of sliding modes. Inequality (6.12) means that the output signal of a first order low-pass filter with transfer function $c_f/(s + \gamma_f)$ dominates its input signal $|d(t)|$.

In the proposed adaptation law, if the amplitude of the disturbance $d(t)$ decreases, then $\rho(t)$ will eventually decrease due to the forgetting factor $-\gamma_f \rho$ in the adaptation law (6.8). This is an advantage over earlier adaptive schemes [14, 17, 23] based on non decreasing gains.

6.4 Problem Formulation

6.4.1 Nonlinear Plant

Consider a nonlinear system in the *normal form* [11]:

$$\dot{\eta} = f_0(\eta, \xi, t) , \tag{6.13}$$
$$\dot{\xi}_1 = \xi_2 , \tag{6.14}$$
$$\vdots$$
$$\dot{\xi}_{r-1} = \xi_r , \tag{6.15}$$
$$\dot{\xi}_r = f(x, t) + g(x, t)\,[u + d(t)] , \tag{6.16}$$

where $\eta \in \mathbb{R}^{n-r}$, $\xi = [\xi_1, \ldots, \xi_r]^T \in \mathbb{R}^r$ and $x = [\eta^T, \xi^T]^T \in \mathbb{R}^n$ is the state, $u \in \mathbb{R}$ is the control signal, $f(x, t) \in \mathbb{R}$ is a state dependent nonlinear term, $d(t) \in \mathbb{R}$ is an exogenous input disturbance, $g(x, t) > \underline{g}$ ($\forall x \in \mathbb{R}^n$, $\forall t \in \mathbb{R}_+$) is the input gain, and $\underline{g} > 0$ is a constant lower bound. In order to satisfy the minimum-phase condition, the nonlinear subsystem (6.13) with partial state η and input signal ξ is assumed to be ISS (see Appendix).

6.4.2 Allowable Disturbance Signals

The fundamental assumption for the unknown input disturbance $d(t)$ is stated below:

(A1) The input disturbance $d(t)$ is assumed to be unknown, locally integrable and norm bounded by $|d(t)| \leq \bar{d}, \forall t$, where $\bar{d} \geq 0$ is an *unknown* scalar. Moreover, there exist a finite time $t_0 \geq 0$ and *known* constants $c_f > \gamma_f > 0$ such that

$$|d(t)| \leq c_f e^{-\gamma_f t} * |d(t)| , \quad \forall t \geq t_0 . \tag{6.17}$$

Note that (A1) relaxes the knowledge of upper bounds for disturbances usually assumed in standard sliding mode control designs. In addition, it allows for some classes of piece-wise continuous disturbance signals. Inequality (6.17) does not impose extra conditions on the disturbance time derivatives either, as the boundedness of $\dot{d}(t)$ assumed in [4, 22].

It seems not easy to fully characterize the classes of disturbance signals, their waveforms and relationship with the constants c_f and γ_f, such that assumption (A1) can be verified. Some examples of disturbances are presented in the simulation tests of Sect. 6.8. In order to illustrate a class of disturbance signals that verify inequality (6.17), assume that the time derivative $\frac{\mathrm{d}|d(t)|}{\mathrm{d}t}$ exists almost everywhere. If at a given instant $t_1 \geq t_0$ the modulation signal verifies $\rho(t_1) > |d(t_1)|$ and the sliding mode exists, then $|u_{\text{eq}}(t_1)| = |d(t_1)|$. A sufficient condition to maintain $\rho(t_1+\varepsilon) > |d(t_1+\varepsilon)|$ $(\varepsilon \to 0+)$ is $\dot{\rho}(t) \geq \frac{\mathrm{d}|d(t)|}{\mathrm{d}t}$ when $t = t_1$, thus, from (6.8), one has

$$\dot{\rho}(t) = -\gamma_f \rho(t) + c_f |u_{\mathrm{eq}}(t)| \geq \frac{\mathrm{d}|d(t)|}{\mathrm{d}t} \,. \tag{6.18}$$

Since $\rho(t_1) \geq |d(t_1)| = |u_{\mathrm{eq}}(t_1)|$, the previous inequality is implied by

$$-\gamma_f |d(t)| + c_f |d(t)| \geq \frac{\mathrm{d}|d(t)|}{\mathrm{d}t} \,, \tag{6.19}$$

when $t = t_1$. Therefore, the following inequality can be obtained:

$$\frac{1}{|d(t)|} \frac{\mathrm{d}|d(t)|}{\mathrm{d}t} = \frac{\mathrm{d}\left[\ln|d(t)|\right]}{\mathrm{d}t} \leq c_f - \gamma_f \quad (>0) \,. \tag{6.20}$$

From this inequality, we can conclude that some classes of non smooth and even unbounded exponentially growing disturbances satisfying

$$|d(t)| < d_0 \, e^{(c_f - \gamma_f)t} \,, \quad \forall t \geq t_1 \,, \tag{6.21}$$

$d_0 > 0$, could be considered, provided that the coefficients in (6.9) are properly chosen. Piecewise-continuous disturbances $d(t)$ with continuous $|d(t)|$ and $\frac{\mathrm{d}|d(t)|}{\mathrm{d}t}$ defined almost everywhere may belong to this class, e.g., square wave and sawtooth.

6.5 Adaptive Sliding Mode Control

First, define the relative degree one sliding variable:

$$\sigma(t) = S\,\xi(t) \,, \qquad S = [s_0 \,, s_1 \,, \ldots \,, s_{r-1}] \,, \tag{6.22}$$

where S is chosen such that the polynomial $s_{r-1}\lambda^{r-1} + \cdots + s_1\lambda + s_0$ is Hurwitz. Without loss of generality, we will fix $s_{r-1} = 1$ to obtain a monic polynomial.

By computing $\dot{\sigma}$, one has

$$\dot{\sigma} = f_\sigma(x,t) + g(x,t)\,[u + d(t)] \,, \tag{6.23}$$

with $f_\sigma(x,t) = \sum_{i=0}^{r-2} s_i \xi_{i+2} + f(x,t)$.

Let us consider the control law:

$$u = u_{\mathrm{c}} + u_{\mathrm{s}} \,, \tag{6.24}$$

$$u_{\mathrm{c}} = -\frac{f_\sigma(x,t)}{g(x,t)} \,, \tag{6.25}$$

$$u_{\mathrm{s}} = -\rho(t)\mathrm{sign}[\sigma] \,. \tag{6.26}$$

If the modulation function is designed such that $\rho(t) > |d(t)| + \delta_d$ ($\forall t \geq t_1 \geq 0$, $\delta_d > 0$), so that it can be concluded that $\sigma\dot{\sigma} \leq -\delta_\sigma|\sigma| < 0$ for some constant $\delta_\sigma > 0$, the sliding surface $\sigma = 0$ will be reached in finite time. Since the polynomial $\lambda^{r-1} + \cdots + s_1\lambda + s_0$ is Hurwitz, then the reduced order equation which governs the closed-loop system during the sliding motion is exponentially stable and, consequently, the state ξ will converge exponentially to the origin. Due to the ISS condition imposed on the internal dynamics (6.13), we can also conclude that $\eta \to 0$ as $\xi \to 0$ and guarantee global stabilization.

On the other hand, if no upper bound for the input disturbance is known as assumed in (A1), we apply the following adaptive modulation function described by

$$\dot{\rho}(t) = -\gamma_f\rho(t) + c_f\left(|u_{\text{seq}}(t)| + \delta\right), \quad \rho(0) \geq 0, \tag{6.27}$$

where $u_{\text{seq}}(t)$ is the extended equivalent control of u_s, and $\delta > 0$ is a constant which guarantees a desired minimum control level for start-up.

As discussed before, the condition $c_f > \gamma_f > 0$ is used in (6.27) to guarantee the control loop is unstable and exponentially growing before the sliding mode takes place, so that it can be indeed attained. The convolution inequality in (A1) is a sufficient condition for the sliding mode existence, that is, it guarantees the modulation function (6.27) to become an upper bound for $|d(t)|$ after the sliding mode takes place.

Figure 6.1 describes the proposed adaptive sliding mode controller. Since the extended equivalent control signal $u_{\text{seq}}(t)$ is unavailable, it is approximated by the average control signal $u_{\text{av}}(t)$ given by the low-pass filter

$$\tau\dot{u}_{\text{av}}(t) = -u_{\text{av}}(t) + u_s, \tag{6.28}$$

for the purpose of implementation of the adaptive law (6.27).

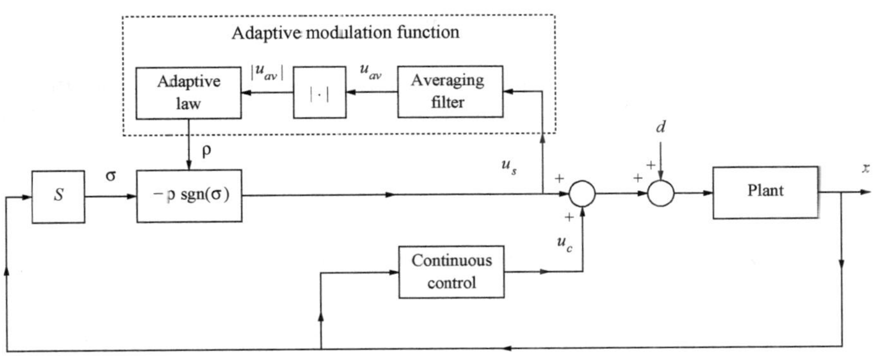

Fig. 6.1 Block diagram of the proposed adaptive sliding mode controller

6.6 Stability Analysis

The main stability results of the proposed adaptive sliding mode controller are summarized in the following theorem assuming the availability of the extended equivalent control signal $u_{\text{seq}}(t)$ needed in the adaptive modulation function (6.27).

Theorem 6.1 *Consider the nonlinear system (6.13)–(6.16) presented in Sect. 6.4.1, where the unknown matched disturbance $d(t)$ satisfies Assumption (A1). The state-feedback based sliding mode control law u is given by (6.24)–(6.26), with adaptive modulation function ρ (6.27) and sliding variable σ in (6.22). Under these conditions, the sliding surface $\sigma = 0$ is reached in finite time and the closed-loop system is uniformly globally exponentially stable in the sense that the state $x = [\eta^T, \xi^T]^T$ converges exponentially to the origin and all remaining signals are uniformly bounded.*

Proof The term u_c given by (6.25) in the control law (6.24) cancels out the harmful nonlinearities in the system (6.16) and eliminates any possibility of finite-time escape of the state trajectories. Thus, the system signals will be *regular* and can grow at most exponentially [20]. This fact leads us to the second step of the proof.

The modulation signal $\rho(t)$ verifies the differential equation (6.27). Assuming no initial sliding mode, $u_{\text{seq}}(t) = u_s(t)$. Thus, $|u_{\text{seq}}(t)| = \rho(t)$ and the solution of (6.27) is given by

$$\rho(t) = e^{(c_f - \gamma_f)t}\left[\rho(0) + c_f\delta\right] - c_f\delta\,, \tag{6.29}$$

grows exponentially because $c_f - \gamma_f > 0$. Since assumption (A1) holds, there exists a finite-time instant $T_1 > 0$ such that $\rho(t) \geq |d(t)|$ is satisfied $\forall t > T_1$. Then, from [10, Lemma 1], the ideal sliding mode $\sigma(t) \equiv 0$ is achieved in finite time.

The signal σ is a relative degree one output for (6.13)–(6.16). Therefore, it is possible to apply an invertible transformation and rewrite it into a particular relative degree one *regular form* [21] such that all the state variables of the transformed nonlinear system are exponentially ISS with respect to σ. Thus, $x = [\eta^T, \xi^T]^T$ tends exponentially to zero since the sliding mode on the manifold $\sigma = 0$ is reached and the polynomial $\lambda^{r-1} + \cdots + s_1\lambda + s_0$ is Hurwitz.

Finally, from the ISS nature of the adaptive law (6.27), it can be concluded that the modulation signal $\rho(t)$ is also uniformly bounded since in finite time $u_{\text{seq}}(t) = d(t)$, and $d(t)$ is uniformly bounded. □

6.7 Extensions of the Proposed Adaptive Law

In this section, some extensions of the proposed adaptive law are discussed.

6.7.1 Global Differentiation and Output-Feedback

The generalization of the tools introduced here to the output-feedback framework are still possible if we apply the global higher-order sliding mode (HOSM) differentiator with dynamic gains recently developed in [16]. As usual, the idea is to define the output signal $y = \xi_1$ of relative degree r and replace its upper derivatives $\xi_2 = \dot{y}\,,\ \xi_3 = \ddot{y}\,, \ldots\,,\ \xi_r = y^{(r-1)}$ by their estimates $\hat{\xi}_2\,, \hat{\xi}_3\,, \ldots\,, \hat{\xi}_r$ obtained with the exact differentiator [12]. The stability analysis is straightforward since the finite-time convergence of the differentiator with dynamic gains is demonstrated and the separation principle is fulfilled independently [13].

6.7.2 Non Input-to-State-Stable Zero Dynamics

Consider $r = 1$ and $\xi = \xi_1$ in (6.13)–(6.16), and assume the internal dynamics $\dot{\eta} = f_0(\eta, \xi, t)$ is linear and time invariant such that it could be written as $\dot{\eta} = A_0\eta + B_0\xi$. Then, the proposed approach can be extended to systems with non ISS zero dynamics (A_0 non-Hurwitz) [1] by redefining the sliding variable $\sigma = \xi - K\eta$, as in [7]. Assuming the pair $(A_0\,, B_0)$ is controllable, when the state variables are restricted to the manifold $\sigma = 0$, the reduced-order model becomes $\dot{\eta} = (A_0 + B_0 K)\eta$, which is exponentially stable for an appropriate feedback matrix $K^T \in \mathbb{R}^{n-1}$. Once $\eta \to 0$, the convergence of ξ to zero would also be proved. The only change needed in the control design is the modification of the term $f_\sigma(x, t)$ which appears in (6.23) and (6.24) to

$$f_\sigma(x, t) = f(x, t) - K\,(A_0\eta + B_0\xi)\ . \tag{6.30}$$

6.7.3 Adaptive Twisting Algorithm

The first and simplest second order sliding mode (SOSM) algorithm is the so-called *"twisting algorithm (TA)"* [14]. For (6.13)–(6.16) with $r = 2$, our adaptive version of the TA would be given by:

$$u = -a(t)\mathrm{sign}[\xi_2] - b(t)\mathrm{sign}[\xi_1]\,, \quad b(t) \geq 2a(t)\,, \tag{6.31}$$

where $a(t) = \rho(t)$ in (6.27) is defined analogously to satisfy $\rho(t) \geq |d(t)|$. The proposed adaptive twisting algorithm would ensure the finite-time exact convergence of both ξ_1 and ξ_2, *i.e.*, there exists $T > 0$ such that $\xi_1(t) = \xi_2(t) = 0$, $\forall t > T$.

6.8 Simulation Results

We seize the opportunity to consider the integrator example (6.6) already studied in Sect. 6.3 to focus the features of the proposed adaptation scheme and present some numerical results.

In the following simulations, the controller parameters applied to (6.24)–(6.28) and (6.22) were: $f_\sigma(x,t) = 0, g(x,t) = 1, S = 1, \delta = 0.2, c_f = 1.7, \gamma_f = 0.8\,\text{rad/s}$ and $\tau = 0.01\,\text{s}$. Notice that $c_f > \gamma_f > 0$, such that inequality (6.9) is verified.

Distinct profiles for the disturbance signal $d(t)$ were evaluated, including smooth and non smooth disturbances.

6.8.1 Smooth Disturbances

Figure 6.2 presents simulation results obtained for a sinusoidal disturbance $d(t) = 5\sin(4t)$. It can be seen that after an initial exponential growth of the modulation signal $\rho(t)$, it becomes larger than the absolute value of the sinusoidal disturbance ($|d(t)|$), then the sliding mode takes place when $\sigma \equiv 0$. Thereafter, the amplitude of the switching control signal u is maintained larger than the amplitude of the disturbance.

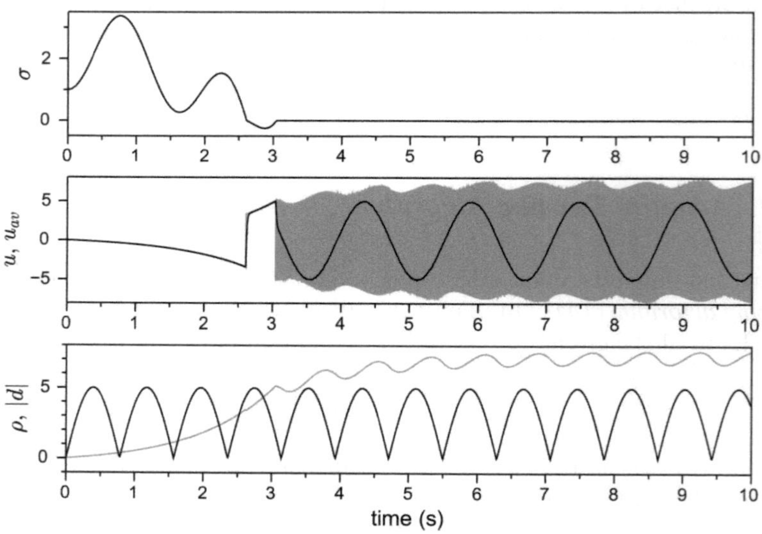

Fig. 6.2 Performance of the proposed adaptation scheme under sinusoidal disturbance: sliding variable σ, control signal u (*gray*), average control u_{av} (*black*), modulation signal ρ (*gray*) and absolute value of the disturbance $|d|$ (*black*)

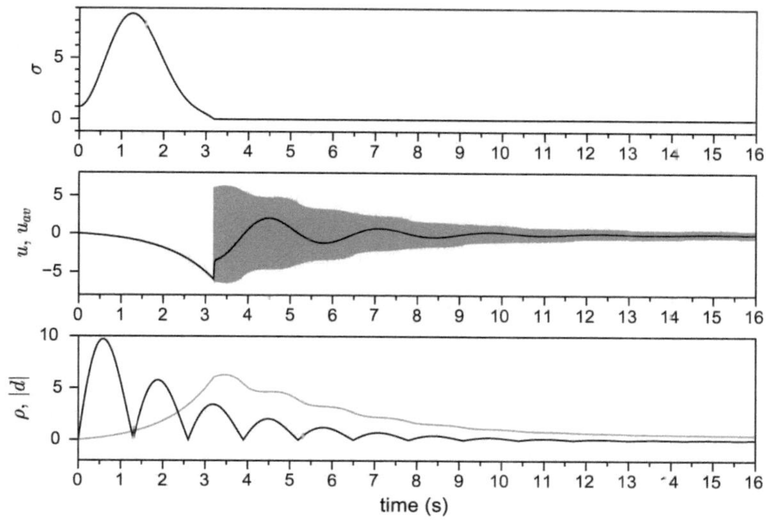

Fig. 6.3 Performance of the proposed adaptation scheme under an exponentially decaying sinusoidal disturbance: sliding variable σ, control signal u *(gray)*, average control u_{av} *(black)*, modulation signal ρ *(gray)* and absolute value of the disturbance $|d|$ *(black)*

Figure 6.3 presents simulation results obtained for an exponentially decaying sinusoidal disturbance. The sliding mode takes place ($\sigma \equiv 0$) after the modulation signal $\rho(t)$ becomes larger than the absolute value of the disturbance ($|d(t)|$). It can be seen that the proposed adaptation scheme reduces the modulation signal amplitude as the disturbance vanishes exponentially.

6.8.2 Non Smooth Disturbances

Figures 6.4, 6.5 and 6.6 present simulation results obtained for discontinuous periodic disturbances of rectangular shape, peak value 2, always positive, frequency 2 Hz, and different duty cycles.

In Fig. 6.4 the duty cycle is 50%, such that the disturbance verifies assumption (A1). It can be seen that after an initial exponential growth of the modulation signal $\rho(t)$, it becomes larger than the disturbance $d(t)$, which allows the sliding mode in the manifold $\sigma = 0$. Therefore, the amplitude of the switching control signal u is maintained larger than the peak of the pulsed disturbance.

In Fig. 6.5 the duty cycle is 80%, such that its mean value is greater than in the previous case. Thus, the amplitude of the modulation signal $\rho(t)$ becomes much larger than the disturbance $d(t)$.

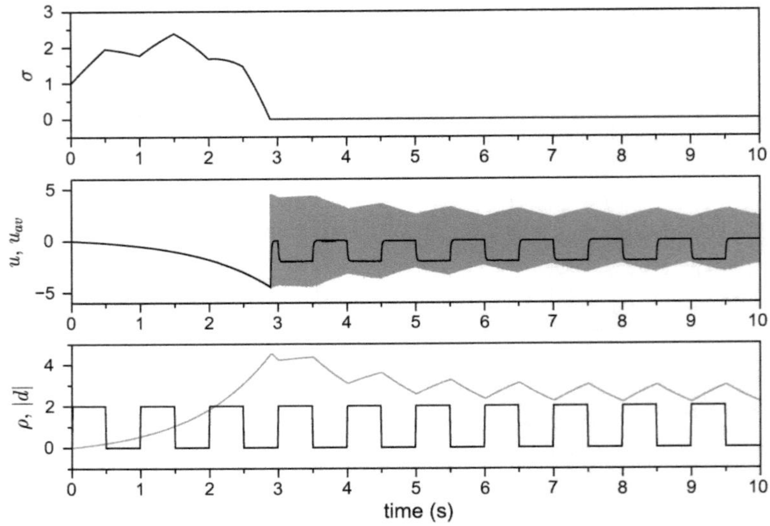

Fig. 6.4 Performance of the proposed adaptation scheme under rectangular pulsed disturbances with 50% duty cycle: sliding variable σ, control signal u (*gray*), average control u_{av} (*black*), modulation signal ρ (*gray*) and disturbance d (*black*)

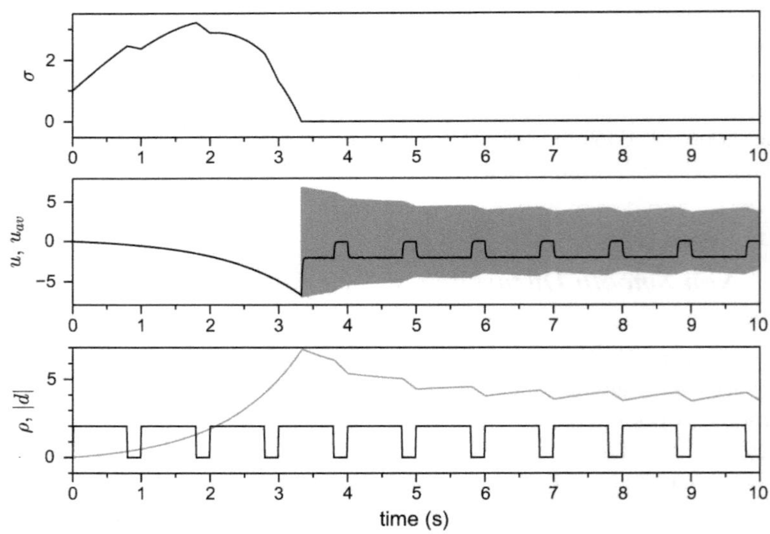

Fig. 6.5 Performance of the proposed adaptation scheme under rectangular pulsed disturbances with 80% duty cycle: sliding variable σ, control signal u (*gray*), average control u_{av} (*black*), modulation signal ρ (*gray*) and disturbance d (*black*)

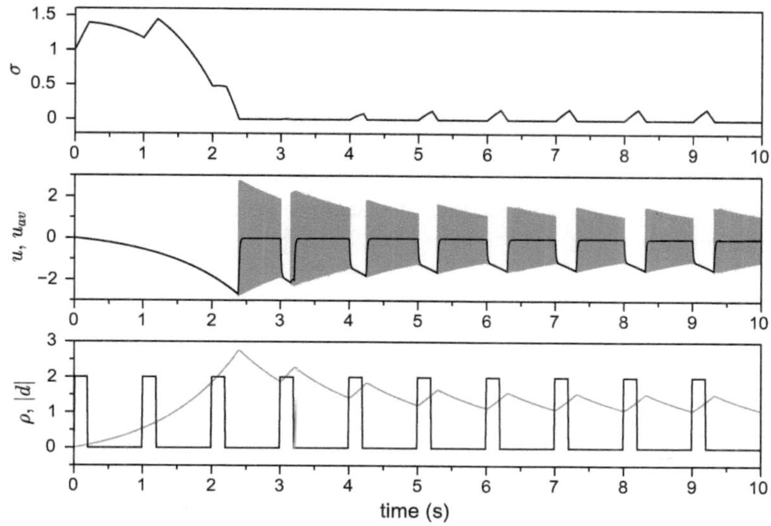

Fig. 6.6 Performance of the proposed adaptation scheme under rectangular pulsed disturbances with 20% duty cycle: sliding variable σ, control signal u (*gray*), average control u_{av} (*black*), modulation signal ρ (*gray*) and disturbance d (*black*)

In Fig. 6.6 the duty cycle is 20%. In this case, the disturbance does not verify assumption (A1), which impairs the effectiveness of the adaptation scheme and the sliding mode cannot be guaranteed. This causes the repetitive pulses in the sliding variable σ.

6.9 Conclusion

A new adaptive sliding mode controller to circumvent (non) smooth input exogenous disturbances with unknown constant bounds has been addressed. The cases of state dependent disturbances and uncertain parameters are left for future research.

The proposed controller is based on the extended equivalent control concept. The modulation function (control gain) is designed to dominate the disturbances. It decreases with the disturbance and, consequently, the precision of the stabilization is improved, while preserving in theory the sliding mode occurrence. In practice, if the gains decrease, the sensitivity of the overall closed-loop system is also reduced.

The main steps of the proof for uniform and global exponential stability as well as perfect disturbance rejection were carried out assuming the availability of the equivalent control. The effects of the averaging filter needed to estimate the equivalent control signal is a current research topic.

Although only the stabilization under full-state feedback and no parametric uncertainties are being considered to keep the focus on disturbances with unknown bounds, the developed method can be extended to more general uncertain systems where trajectory tracking via output-feedback is also pursued. We will leave these generalizations for future works. The generalization of the proposed tools for fixed-time algorithms [19] or HOSM based controllers and their combination with other non-overestimated adaptive strategies also seem to be interesting topics for future investigation.

Acknowledgements The authors thank the Brazilian funding agencies CAPES, CNPq and FAPERJ for the financial support.

Appendix

To define the concept of input-to-state stability/stable (ISS) [11, Sect. 4.9], consider the system

$$\dot{x} = f(x, u, t)\,, \tag{6.32}$$

where $f : \mathbb{R}^n \times \mathbb{R}^m \times \mathbb{R}_+ \to \mathbb{R}^n$ is piecewise continuous in t and locally Lipschitz in x and u. The input signal $u(t)$ is piecewise continuous and uniformly bounded. Assume that this system has a globally uniformly asymptotically stable equilibrium point at $x = 0$ when $u(t) \equiv 0$.

Definition 6.1 The system (6.32) is said to be input-to-state stable (ISS) if there exist a class $\mathscr{K}\mathscr{L}$ function β and a class $\mathscr{K}$ function γ such that for any initial state $x(t_0)$ and any bounded input signal $u(t)$, the solution $x(t)$ exists $\forall t \geq t_0 \geq 0$ and satisfies

$$\|x(t)\| \leq \beta\left(\|x(t_0)\|, t - t_0\right) + \gamma\left(\sup_{t_0 \leq \tau \leq t} \|u(\tau)\|\right). \tag{6.33}$$

References

1. Baev, S., Shtessel, Y., Shkolnikov, I.: Nonminimum-phase output tracking in causal systems using higher-order sliding modes. Int. J. Robust Nonlinear Control. **18**(4–5), 454–467 (2008)
2. Bartolini, G., Levant, A., Plestan, F., Taleb, M., Punta, E.: Adaptation of sliding modes. IMA J. Math. Control. Inf. **30**(3), 285–300 (2013)
3. Bartoszewicz, A.: A new technique to compensate for disturbance in sliding mode control systems. In: Proccedings of 24th Conference IEEE Indutrial Electronics Society, pp. 1708–1711 (1998)
4. Edwards, C., Shtessel, Y.B.: Adaptive continuous higher order sliding mode control. Automatica **65**, 183–190 (2016)
5. Estrada, A., Plestan, F., Allouche, B.: An adaptive version of a second order sliding mode output feedback controller. In: Proccedings of European Control Conference, pp. 3228–3233 (2013)

6. Filippov, A.F.: Differential Equations with Discontinuous Righthand Sides. Kluwer Academic Publishers, Dordrecht (1988)
7. Gonzalez, T., Moreno, J.A., Fridman, L.: Variable gain super-twisting sliding mode control. IEEE Trans. Autom. Control. **57**(8), 2100–2105 (2012)
8. Hsu, L., Costa, R.R.: Adaptive control with sliding modes: theory and applications. In: Minicourse Proceedings XI Congresso Brasileiro de Automática, pp. 39–60 (1996)
9. Hsu, L., Cunha, J.P.V.S., Costa, R.R., Lizarralde, F.: Multivariable output-feedback sliding mode control. In: Yu, X., Xu, J.X. (eds.) Variable Structure Systems: Towards the 21st Century, pp. 283–313. Springer, Berlin (2002)
10. Hsu, L., Lizarralde, F., Araújo, A.D.: New results on output-feedback variable structure model-reference adaptive control: design and stability analysis. IEEE Trans. Autom. Control. **42**(3), 386–393 (1997)
11. Khalil, H.K.: Nonlinear Systems, 3rd edn. Prentice-Hall, Upper Saddle River (2002)
12. Levant, A.: Higher-order sliding modes, differentiation and output-feedback control. Int. J. Control. **76**(9), 924–941 (2003)
13. Levant, A., Livne, M.: Exact differentiation of signals with unbounded higher derivatives. IEEE Trans. Autom. Control. **57**(4), 1076–1080 (2012)
14. Moreno, J.A., Negrete, D.Y., Torres-González, V., Fridman, L.: Adaptive continuous twisting algorithm. Int. J. Control. **89**(9), 1798–1806 (2016)
15. Oliveira, T.R., Cunha, J.P.V.S., Hsu, L.: Adaptive sliding mode control for disturbances with unknown bounds. In: Proccedings of 14th International Workshop on Variable Structure Systems, pp. 59–64 (2016)
16. Oliveira, T.R., Estrada, A., Fridman, L.M.: Global exact differentiator based on higher-order sliding modes and dynamic gains for globally stable output-feedback control. In: Proccedings of IEEE Conference on Decision and Control, pp. 4109–4114 (2015)
17. Oliveira, T.R., Leite, A.C., Peixoto, A.J., Hsu, L.: Overcoming limitations of uncalibrated robotics visual servoing by means of sliding mode control and switching monitoring scheme. Asian J. Control. **16**(3), 752–764 (2014)
18. Plestan, F., Shtessel, Y., Brégeault, V., Poznyak, A.: New methodologies for adaptive sliding mode control. Int. J. Control. **83**(9), 1907–1919 (2010)
19. Polyakov, A., Efimov, D., Perruquetti, W.: Finite-time and fixed-time stabilization: implicit Lyapunov function approach. Automatica **51**, 332–340 (2015)
20. Sastry, S., Bodson, M.: Adaptive Control: Stability, Convergence and Robustness. Prentice-Hall, Englewood Cliffs (1989)
21. Utkin, V.I.: Sliding Modes in Control and Optimization. Springer, Berlin (1992)
22. Utkin, V.I., Poznyak, A.S.: Adaptive sliding mode control with application to super-twist algorithm: equivalent control method. Automatica **49**(1), 39–47 (2013)
23. Yan, L., Hsu, L., Xiuxia, S.: A variable structure MRAC with expected transient and steady-state performance. Automatica **42**(5), 805–813 (2006)

Chapter 7
Indirect Adaptive Sliding-Mode Control Using the Certainty-Equivalence Principle

Alexander Barth, Markus Reichhartinger, Kai Wulff, Johann Reger, Stefan Koch and Martin Horn

7.1 Introduction

For many controller design problems a precise mathematical model of the plant to be controlled is not available. This may be due to a lack of insight into the plant dynamics, too complex relations between too many interacting components or simply uncertain parameters or unknown disturbances. Against this background, in adaptive control, it is distinguished between *parametric* or *structured uncertainty* and *inherent* or *unstructured uncertainty*, see e.g. [7, 24]. Indeed, the former describes the effect of uncertain system parameters, the latter accounts for unmodeled dynamics or external disturbances.

Sliding-mode controller design, typically, does not distinguish these natures of uncertainty. In fact, all uncertainties are regarded as unstructured. Robustness is ensured as long as certain conditions with respect to the uncertainties are fulfilled.

A. Barth · K. Wulff · J. Reger (✉)
Computer Science and Automation Department, Technische Universität Ilmenau, 98684 Ilmenau, Germany
e-mail: johann.reger@tu-ilmenau.de

A. Barth
e-mail: alexander.barth@tu-ilmenau.de

K. Wulff
e-mail: kai.wulff@tu-ilmenau.de

M. Reichhartinger · S. Koch · M. Horn
Faculty of Electrical and Information Engineering, Graz University of Technology, 8010 Graz, Austria
e-mail: markus reichhartinger@tugraz.at

S. Koch
e-mail: stefan.koch@tugraz.at

M. Horn
e-mail: martin.horn@tugraz.at

S. Li et al. (eds.), *Advances in Variable Structure Systems and Sliding Mode Control—Theory and Applications*, Studies in Systems, Decision and Control 115, DOI 10.1007/978-3-319-62896-7_7

The common design paradigm requires worst case bounds on the uncertainty for an appropriate controller tuning. Consequently, this approach often results in unnecessary high controller gains and may lead to undesired pronounced chattering [5]. Adaptive and variable-gain controllers based on sliding-mode concepts tackle this issue by adapting the gains of the control law depending on the actual perturbations, e.g. [8, 20–22]. The main idea of the prominent adaptive-gain algorithms is to increase the sliding-mode controller gains whenever the sliding variable leaves a certain domain around the sliding surface, see e.g. [8, 20, 21]. This approach generally results in smaller controller gains compared to constant-gain algorithms. In any case, the class of uncertainty, the algorithms can cope with, cannot be extended by these adaptive-gain methods. Consider exemplarily the simple dynamical system

$$\frac{\mathrm{d}x}{\mathrm{d}t} = \Delta(t, x) - k(t)\,\mathrm{sign}x, \tag{7.1}$$

with the uncertainty $\Delta(t, x)$. The time-varying parameter k is governed by any of the adaptation laws discussed e.g. in [17]. Hence, as long as the absolute value of the uncertainty does not exceed a possibly unknown upper bound, system (7.1) exhibits a sliding-mode at $x = 0$. However, even in the case of $\Delta(t, x) = \delta x$, with a constant δ, the required boundedness is violated and global stability of $x = 0$ cannot be guaranteed. The main idea of our approach is to exploit the available information about the nature of the uncertainty as much as possible by distinguishing between its structured and unstructured parts. The principle idea presented in this work is to design a sliding-mode controller coping with the unstructured uncertainty whereas the structured part of the uncertainty is compensated by an adaptive part. While this concept may work with any stabilizing sliding-mode algorithm we choose the *super-twisting algorithm (STA)* as nominal controller in a *certainty-equivalence principle* design of the adaptive controller. The super-twisting algorithm is particularly suitable for such combination as there have been a number of Lyapunov functions proposed in the recent past, that are inevitably needed for the design procedure of the adaptive part pursued in this chapter. As a result the class of admissible uncertainties covered by the proposed control concept is considerably enlarged with respect to the considered sliding-mode controls, including adaptive-gain super-twisting algorithms. Initial results of this approach have shown that the gains of the sliding-mode controller may be reduced significantly while maintaining its robustness [1–3].

For the remainder of this section we discuss classes of uncertainties that can be handled by popular sliding mode concepts. An introductory example illustrates how unstructured uncertainties affect the robustness properties of the controller and may lead to unstable motions in the case of a super-twisting controller applied to a simple DC-motor control problem. Section 7.2 introduces a formal problem description and casts the robustness problem into the sliding-mode control concept. In Sect. 7.3 we present the design procedure of the certainty-equivalence based super-twisting algorithm (CESTA). A central role in the design of this controller plays the Lyapunov function for the nominally stabilizing part of the controller. In Sect. 7.4 we thus consider various Lyapunov functions recently proposed for sliding-mode controllers.

We derive the respective different adaptation laws and discuss their impact on the resulting control law. Finally in Sect. 7.6 we present a case study to demonstrate the efficiency of the proposed control concept applied to an experimental test rig and compare its performance with conventional super-twisting algorithms.

7.1.1 Classes of Uncertainties

Consider the scalar control system given by

$$\dot{x} = u + \Delta \tag{7.2}$$

with $x,\ u,\ \Delta \in \mathbb{R}$. Therein Δ is an external, Lebesgue-measurable time-dependent disturbance. The objective is to design the control u such that the state x vanishes within finite time despite the disturbance. Note that even in the nominal case, i.e. $\Delta = 0$, the specification of finite time convergence of x requires a control law $u = u(x)$ that is non-Lipschitz continuous at the origin $x = 0$. Popular options are the algorithms proposed in [4]. However, in the general case robustness against $\Delta \neq 0$ may be achieved by a sliding-mode controller for specific classes of disturbances. Bounded disturbances, i.e.

$$\sup_t |\Delta(t, x)| \leq \delta_b, \tag{7.3}$$

where δ_b is a known positive constant, without any additional restriction may be handled by a conventional sliding-mode controller

$$u = -k_C \,\mathrm{sign}[x] \qquad \text{with} \qquad k_0 > \delta_b, \tag{7.4}$$

see [23]. A system (7.2) subject to disturbances bounded by

$$|\Delta(t, x)| \leq \delta_r \sqrt{|x|}, \tag{7.5}$$

may be controlled by the super-twisting algorithm

$$\begin{aligned} u &= -\, k_1 \sqrt{|x|} \mathrm{sign}[x] \ + v \\ \frac{\mathrm{d}v}{\mathrm{d}t} &= -\, k_2 \,\mathrm{sign}[x] \end{aligned} \tag{7.6}$$

where the controller gains k_1 and k_2 are all positive and satisfy $k_1 > \delta_r$. Applying control law (7.6) to system (7.2) yields the closed loop dynamics

$$\begin{aligned}\frac{\mathrm{d}x}{\mathrm{d}t} &= -\,k_1\sqrt{|x|}\operatorname{sign}[x]\ +v+\Delta,\\ \frac{\mathrm{d}v}{\mathrm{d}t} &= -\,k_2\operatorname{sign}[x].\end{aligned} \tag{7.7}$$

This control algorithm also is capable of rejecting Lipschitz-continuous disturbances characterized by

$$\sup_t\left|\frac{\mathrm{d}\Delta(t)}{\mathrm{d}t}\right| \le \delta_1 \qquad \text{with} \qquad \delta_1 > 0. \tag{7.8}$$

This requires a choice of the gains k_1 and k_2 as discussed in [10, 12, 13]. In contrast to the discontinuous control signal (7.4) the super-twisting algorithm provides an absolutely continuous control signal which may be beneficial in many real world applications. The adaptive gains version of the super-twisting algorithm presented in [6] also copes with an unknown upper bound δ_1. Note that the disturbances satisfying inequality (7.8) are assumed to be independent of the state variable x. If this assumption does not hold, the control signal u appears in the time derivative of Δ, i.e.

$$\frac{\mathrm{d}\Delta(t,x)}{\mathrm{d}t} = \frac{\partial\Delta}{\partial x}\frac{dx}{dt} + \frac{\partial\Delta}{\partial t} = \frac{\partial\Delta}{\partial x}\left(u+\Delta(t,x)\right) + \frac{\partial\Delta}{\partial t}. \tag{7.9}$$

Proving stability then relies on an appropriate choice of the controller gains which depend on some upper bound on the uncertainty. However, its derivative (7.9) itself contains k_1, k_2. Typically, the state dependency of Δ results only in local stability [11]. This issue is addressed in more detail in the following introductory example.

7.1.2 Introductory Example (Control of a DC Motor)

In this example a standard DC motor control setup is discussed. It consists of an inner current loop controller and an outer angular velocity feedback loop. The overall control task is to make the angular velocity track a given reference. The design of the current controller depends on the known electrical characteristics of the motor. Hence, the current controller including the selection of the gains often is designed independent of the underlying application of the motor. In the example the inner loop controller is realized by a super-twisting algorithm. The outer loop controller is a standard PI controller where the tuning is left to the operating engineer.

A lumped parameter model which describes the dynamics of the DC motor is derived from the equivalent circuit model shown in Fig. 7.1. Applying Kirchhoff's voltage law yields the first-order differential equation

$$u(t) - R\,i(t) - L_\mathrm{m}\frac{\mathrm{d}i(t)}{\mathrm{d}t} - k_\mathrm{m}\,\omega(t) = 0 \tag{7.10}$$

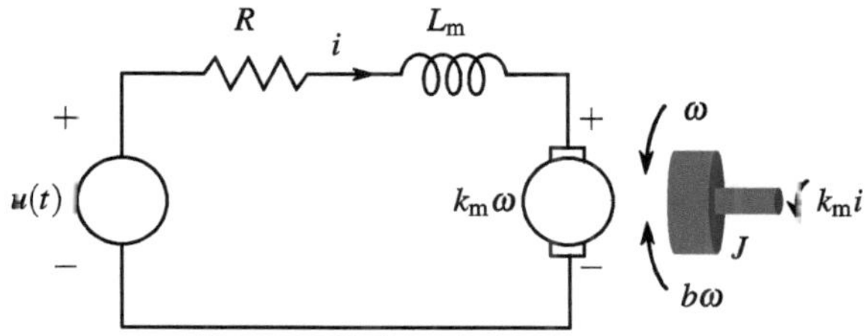

Fig. 7.1 Schematic diagram of the DC motor consisting of an ideal resistor R, an inductance L_m and the control voltage u

where the positive parameters R, L_m, k_m are the winding resistance, the inductance and the motor constant of the DC-drive, respectively. Making use of Newton's second law of motion yields the differential equation

$$J\frac{d\omega(t)}{dt} = -b\,\omega(t) + k_m\,i(t) \tag{7.11}$$

that describes the dynamics of the mechanical subsystem. The parameter b is a positive friction coefficient which represents viscous friction, J is the moment of inertia of the rotating parts, i.e. the armature and the external load.

The design requirement of the sliding-mode based current control loop is to track a reference input i_d which is generated by the outer control loop. Therefore, the sliding variable is introduced as the difference of the measured and the reference current, i.e.

$$s := i_d - i. \tag{7.12}$$

Taking the derivative of the sliding variable with respect to time and using (7.10) yields

$$\frac{ds}{dt} = \frac{di_d}{dt} + \frac{R}{L_m}i + \frac{k_m}{L_m}\omega - \frac{1}{L_m}u. \tag{7.13}$$

Assuming that the motor inductance is known, a control law of the form

$$u = -L_m(u_s + u_c) \tag{7.14}$$

shall be proposed. The control law is composed of two parts, where u_s denotes the sliding-mode based control and u_c accounts for the known dynamics. The control u_s is obtained from the super-twisting algorithm

$$\begin{aligned} u_s &= -\,k_1\sqrt{|s|}\text{sign}[s] \; + v, \\ \frac{dv}{dt} &= -\,k_2\,\text{sign}[s]. \end{aligned} \tag{7.15}$$

Substituting control law (7.14) in view of (7.15) into (7.13) and selecting

$$u_c = -\frac{R}{L_\mathrm{m}} i \tag{7.16}$$

leads to the closed loop dynamics

$$\begin{aligned} \frac{\mathrm{d}s}{\mathrm{d}t} &= -k_1\sqrt{|s|}\,\mathrm{sign}[s] + v + \underbrace{\frac{\mathrm{d}i_d}{\mathrm{d}t} + \frac{k_\mathrm{m}}{L_\mathrm{m}}\omega}_{\Delta}, \\ \frac{\mathrm{d}v}{\mathrm{d}t} &= -k_2\,\mathrm{sign}[s], \end{aligned} \tag{7.17}$$

which matches the super-twisting structure presented in (7.7).

The super-twisting algorithm achieves finite time stabilization of the sliding variable s and keeps $s \equiv 0$ for all subsequent times if the perturbation Δ satisfies any condition discussed in the previous paragraph. In the problem at hand, the perturbation involves the term $\frac{\mathrm{d}i_d}{\mathrm{d}t}$ which is introduced by the outer control loop. Hence the properties of the perturbation regarding growth conditions and boundedness will obviously depend on the design of this loop. Therefore, providing an estimate of the bounds in advance represents a challenging task. As mentioned above, the motor angular velocity tracking error

$$e := \omega_d(t) - \omega(t) \tag{7.18}$$

is regulated by a PI controller. In this regard, the reference current signal is

$$i_d(t) = k_p\, e(t) + k_i \int_0^t e(\tau)\,\mathrm{d}\tau. \tag{7.19}$$

In order to investigate the performance of the cascaded control structure, three different simulations are carried out. The motor is supposed to track a smooth reference signal where $|\dddot{\omega}_d| \leq \delta_1 = 2\,\mathrm{rad/s}^3$. The initial conditions are chosen as $i(0) = -0.5\,\mathrm{A}$ and $\omega(0) = -2\,\mathrm{rad/s}$. The parameters of the super-twisting algorithm are chosen as

$$k_1 = 1.5\sqrt{N}, \qquad k_2 = 1.1N \tag{7.20}$$

and $N = 20$, see e.g. [10]. These parameters remain unchanged throughout the experiments. For the first experiment the parameters of the outer control loop are selected as $k_{p,1} = 0.45$ and $k_{i,1} = 0.9$.

The simulation results are shown in Fig. 7.2. The plot on the top shows the angular velocity and the desired trajectory, the lower plot depicts the current tracking error, i.e. the sliding variable s. The left column provides the results obtained by the first experiment. It is observed that the super-twisting algorithm is capable to maintain the sliding variable at zero, i.e. track the reference current in the presence of the

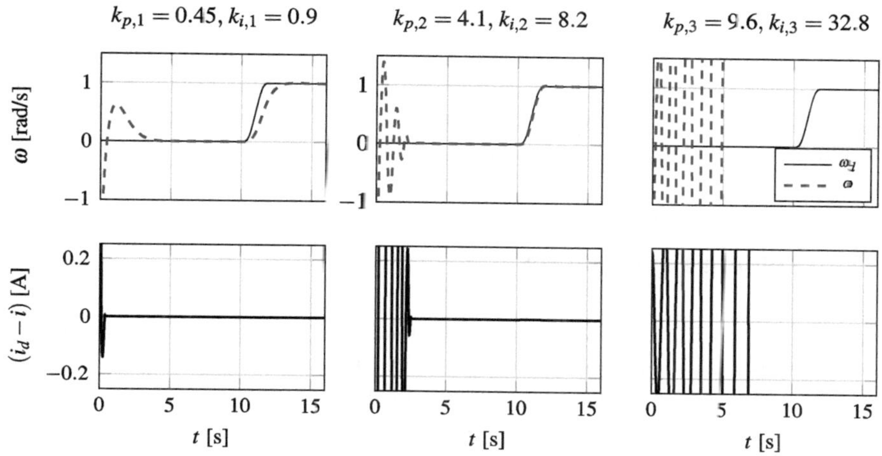

Fig. 7.2 Results obtained with a cascaded control structure with various settings for the outer PI control loop

perturbation. However, depending on the specific application, the velocity tracking error for the ramp reference signal might be undesirably large. Therefore, one might seek for a more "aggressive" tuning of the PI control loop. Selecting $k_{p,2} = 4.1$ and $k_{i,2} = 8.2$ yields the results provided in the second column. With this tuning, the performance regarding the velocity tracking error is significantly improved. However, the initial transient shows very little damping. Increasing the controller gains further to $k_{p,3} = 9.6$ and $k_{i,3} = 32.8$ causes the closed-loop system to get unstable, see third column in Fig. 7.2.

The reason for this unstable motion is found in the structure of the perturbation Δ of the closed-loop dynamics (7.17). Taking a closer look at the perturbation and substitute (7.11) into (7.19). With (7.18) and differentiation we obtain

$$\frac{\mathrm{d}i_d}{\mathrm{d}t} = k_p \left(\dot{\omega}_d + \frac{b}{J}\omega - \frac{k_\mathrm{m}}{J} i \right) + k_i e. \tag{7.21}$$

Thus, the perturbation acting on the closed-loop dynamics (7.17) is

$$\Delta = \left(\frac{k_p b}{J} + \frac{k_\mathrm{m}}{L_\mathrm{m}} \right) \omega - \frac{k_p k_\mathrm{m}}{J} i + k_i e + k_p \dot{\omega}_d . \tag{7.22}$$

Note that the perturbation in fact depends on the state of the system and in particular grows linearly with $s = i_d - i$. Therefore, neither of the requirements (7.5), (7.8) of the super-twisting algorithm are met without further assumptions. Of course, so as to insure (7.5) we might consider some worst-case scenario for the system states and select the sliding-mode controller parameters accordingly. However, this foils one major advantage that this method enjoys over classical sliding mode control.

Furthermore, the perturbation Δ in (7.22) also depends on the controller gains k_p, k_i of the outer loop. In a decentralized design scenario where the inner loop is designed without any knowledge about the outer loop it seems impossible to guarantee the requirement (7.5) or (7.8) in all situations and hence, unstable behavior as demonstrated in our simulation example cannot be ruled out.

7.2 Problem Formulation

The major aim of the approach presented in this contribution is to extend the class of uncertainties that can be compensated in the closed-loop control (7.7) by distinguishing between a structured and an unstructured part. Before we present our approach in Sect. 7.3 we shall give a formal problem definition in the following.

We consider nonlinear systems of the form

$$\dot{x} = f(x) + g(x)\,\big(\underline{\Delta}(t,x) + u\big)\,, \qquad x(0) = x_0 \in \mathbb{R}^n \tag{7.23}$$

where $u \in \mathbb{R}$ denotes a scalar input and $x \in \mathbb{R}^n$ the state vector. The vector fields $g, f : \mathbb{R}^n \to \mathbb{R}^n$ are supposed to be known. Furthermore, the system is affected by a matched disturbance $\underline{\Delta}(t,x) \in \mathbb{R}$ that is composed of two parts:

$$\underline{\Delta}(t,x) = \Delta_{\mathrm{s}}(t,x) + \Delta_{\mathrm{u}}(t)\,. \tag{7.24}$$

$\Delta_{\mathrm{s}}(t,x)$ describes the structured uncertainty and $\Delta_{\mathrm{u}}(t)$ denotes an unstructured uncertainty. The latter shall be norm bounded,[1] i.e.

$$\|\ \Delta_{\mathrm{u}}(t)\ \| < \delta \quad \text{for all} \quad t > 0\,, \tag{7.25}$$

and the structured uncertainty can be decomposed into the form

$$\Delta_{\mathrm{s}}(t,x) = \Theta^{\mathrm{T}}\phi(t,x) \tag{7.26}$$

where $\Theta \in \mathbb{R}^p$ is an unknown but constant parameter vector and $\phi : \mathbb{R} \times \mathbb{R}^n \to \mathbb{R}^p$ a known base function.

The design goal for the control law is to force the system (7.23) onto some desired manifold $s(t,x) = 0$ for all possible uncertainties $\underline{\Delta}(t,x)$. Note, that the overall uncertainty $\underline{\Delta}(t,x)$ considered in (7.24) is *not* norm bounded.

For the control design we shall adopt a sliding-mode control scheme in combination with a certainty-equivalence approach. For the sliding variable we choose $s : \mathbb{R} \times \mathbb{R}^n \to \mathbb{R}$ such that $|s(t,x)|$ describes the distance to the desired manifold $s(t,x) = 0$. The function $s(t,x)$ should be designed such that s is differentiable with

[1] This requirement may be relaxed according to the sliding-mode algorithm chosen for the stabilizing part in Sect. 7.3.

respect to t and x and imposes desirable, stable dynamic behavior of the system states x for $s \equiv 0$. Moreover we assume relative-degree one of the sliding variable with respect to the input u, i.e.

$$\frac{\partial s}{\partial x} g(x) \neq 0. \tag{7.27}$$

Considering the evolution of the sliding variable s,

$$\begin{aligned} \dot{s}(t,x) &= \frac{\partial s}{\partial t} + \frac{\partial s}{\partial x}\big(f(x) + g(x)\left(\Delta_{\mathrm{s}}(t,x) + \Delta_{\mathrm{u}}(t) + u\right)\big) \\ &= \frac{\partial s}{\partial t} + \frac{\partial s}{\partial x} f(x) + \frac{\partial s}{\partial x} g(x)\Delta_{\mathrm{s}}(t,x) + \frac{\partial s}{\partial x} g(x)\Delta_{\mathrm{u}}(t) + \frac{\partial s}{\partial x} g(x) u, \end{aligned} \tag{7.28}$$

and using the description (7.26) for the structured uncertainty, we obtain

$$\dot{s}(t,x) = \frac{\partial s}{\partial t} + \frac{\partial s}{\partial x} f(x) + \Theta^{\mathrm{T}}\phi(t,x)\frac{\partial s}{\partial x} g(x) + \frac{\partial s}{\partial x} g(x)\Delta_{\mathrm{u}}(t) + \frac{\partial s}{\partial x} g(x) u. \tag{7.29}$$

In terms of the abbreviations

$$\begin{aligned} w(t,x,u) &:= \frac{\partial s}{\partial t} + \frac{\partial s}{\partial x} f(x) + \frac{\partial s}{\partial x} g(x) u\,, \\ \alpha(t,x) &:= \phi(t,x)\frac{\partial s}{\partial x} g(x)\,, \\ \beta(t,x) &:= \frac{\partial s}{\partial x} g(x)\Delta_{\mathrm{u}}(t) \end{aligned}$$

for system (7.29) we obtain

$$\dot{s}(t,x) = \Theta^{\mathrm{T}}\alpha(t,x) + \beta(t,x) + w(t,x,u)\,, \tag{7.30}$$

where w acts as new control input and s may be seen as some virtual output.

The control objective is now to reach the manifold $s(t,x) \equiv 0$ and to render the internal dynamics of (7.30) with respect to the output s asymptotically stable for all uncertainties of the class (7.25)–(7.26). For this task we shall consider w as the control signal to be designed. The original control is then obtained by

$$u = \left(\frac{\partial s}{\partial x} g(x)\right)^{-1}\left(-\frac{\partial s}{\partial t} - \frac{\partial s}{\partial x} f(x) + w\right)$$

which reveals that s needs to satisfy the common relative-degree-one condition given in Eq. (7.27).

Comparing this to the standard sliding-mode control problem (7.2) we see that

$$\dot{s}(t,x) = \Delta(t,x) + w\,, \tag{7.31}$$

where the (modified) uncertainty

$$\Delta(t,x) = \Theta^{\mathrm{T}}\alpha(t,x) + \beta(t,x). \tag{7.32}$$

Note that this uncertainty $\Delta(t,x)$ is neither norm bounded nor bounded by some function of the sliding variable s, which is a common requirement for sliding-mode control as shown in Sect. 7.1.1. We will show in the following that our control approach only requires an upper bound on the unstructured uncertainty $\beta(t,x)$. Note further, that the linearity in the parameter Θ is also visible in the dynamics (7.30). This will help us in the design of a suitable adaptation law.

7.3 Combining Certainty-Equivalence and Super-Twisting SMC

This section presents the main idea of combining certainty-equivalence adaptation and sliding-mode control. We give a brief description of the steps in the design procedure.

Consider the new control

$$w = w_{\mathrm{CE}} + w_{\mathrm{SMC}} \tag{7.33}$$

where w_{SMC} is generated by some sliding-mode control law and w_{CE} denotes the certainty-equivalence adaptive control given by

$$w_{\mathrm{CE}} = -\,\hat{\Theta}^{\mathrm{T}}\alpha(t,x)\,, \tag{7.34}$$

where $\hat{\Theta} \in \mathbb{R}^p$ denotes an estimate of the parameter vector Θ.

The sliding-mode control part will cope with the unstructured uncertainty, the adaptive control part is to compensate the structured uncertainty. We may use any sliding-mode control method for w_{SMC} that is able to deal with relative-degree-one systems. For example, consider again the super-twisting algorithm [10]

$$\begin{aligned} w_{\mathrm{SMC}} &= -\,k_1\,|s|^{\frac{1}{2}}\,\mathrm{sign}[s] + v \\ \dot{v} &= -\,k_2\,\mathrm{sign}[s] \end{aligned} \tag{7.35}$$

with constant parameters $k_1,\ k_2 > 0$. Note, that the application of the super-twisting algorithm requires that

$$|\beta(t,x)| < \bar{\beta}\sqrt{|s(t,x)|}, \quad \text{for all } t > 0\,, \tag{7.36}$$

where $\bar{\beta}$ is some positive constant. To shorten formulas and improve readability, in the sequel we use the notation

$$\lceil a \rfloor^b := |a|^b \operatorname{sign}[a]\,. \tag{7.37}$$

Substituting (7.34) and (7.35) into (7.33) we obtain for the complete control law

$$\begin{aligned} w &= -\,k_1 \lceil s \rfloor^{\frac{1}{2}} + v - \hat{\Theta}^{\mathrm{T}}\alpha(t,x) \\ \dot{v} &= -\,k_2 \operatorname{sign}[s]\,. \end{aligned} \tag{7.38}$$

Inserting control law (7.38) into the system (7.30) we obtain the closed-loop dynamics

$$\begin{aligned} \dot{s} &= -\,k_1 \lceil s \rfloor^{\frac{1}{2}} + v + \tilde{\Theta}^{\mathrm{T}}\alpha(t,x) + \beta(t,x) \\ \dot{v} &= -\,k_2 \operatorname{sign}[s] \end{aligned} \tag{7.39}$$

where $\tilde{\Theta} = \Theta - \hat{\Theta}$ denotes the parameter estimation error.

In order to obtain the adaptation law for the parameter estimate $\hat{\Theta}$, we will resort to the *certainty-equivalence principle*, see e.g. [9, 15]. The main idea of this approach can be summarized by the following steps:

1. Localize the structured uncertainty and separate it according to (7.24) by choosing an appropriate base function $\phi(t,x)$.
2. Design a nominal controller assuming perfect knowledge of the parameter vector Θ in the structured uncertainty. This controller has to stabilize the system for all unstructured uncertainties considered.
3. Choose a nominal Lyapunov function for the nominal closed-loop system.
4. Extend the nominal Lyapunov function by a quadratic term of the estimation error $\tilde{\Theta}$.
5. Calculate the time-derivative of the extended Lyapunov function and separate all terms containing the parameter estimation error $\tilde{\Theta}$. Choose the adaptation law for $\hat{\Theta}$ to cancel all unknown terms of the structured uncertainty, rendering the convergence to the sliding manifold independent of the parameter estimation error.

The idea is to combine this design procedure with sliding-mode controllers and improve their robustness features. The focus lies mainly on disturbances that do not have an upper bound or are bounded by a function of the sliding variable.

7.4 Impact of the Choice of Lyapunov Function on the Adaptive Part

The design procedure for the certainty-equivalence approach reveals the key role of the nominal Lyapunov function chosen in step 3. In our case the stabilizing controller is given by the sliding-mode control w_{SMC}. While any design method may be used, the super-twisting algorithm considered above is particularly convenient for this purpose because various Lyapunov functions have been proposed for this method in the recent past, [12, 13, 16, 18]. In the following we design adaptation laws using different Lyapunov functions that have been suggested for sliding-mode algorithms and discuss their properties.

7.4.1 Weak Lyapunov Function

The first Lyapunov function that we consider has been proposed in [16] for the STA. The authors consider the following, Lipschitz continuous, Lyapunov function

$$V_{\mathrm{nom},1} = k_2|s| + \frac{1}{2}v^2 \,. \tag{7.40}$$

We shall use this approach for the closed-loop system with certainty-equivalence super-twisting algorithm (CESTA) in (7.39).

Taking the derivative and substituting (7.39) yields

$$\begin{aligned}
\dot{V}_{\mathrm{nom},1} &= k_2 \,\mathrm{sign}[s] \left(-k_1 \lceil s \rfloor^{\frac{1}{2}} + v + \tilde{\Theta}^{\mathrm{T}}\alpha(t,x) + \beta(t,x)\right) - k_2 \,\mathrm{sign}[s]\, v & (7.41)\\
&= k_2 \,\mathrm{sign}[s] \left(-k_1 \lceil s \rfloor^{\frac{1}{2}} + \tilde{\Theta}^{\mathrm{T}}\alpha(t,x) + \beta(t,x)\right). & (7.42)
\end{aligned}$$

We now assume that the parameters of the plant are perfectly known. Then $\tilde{\Theta}$ is equal to zero and the structured uncertainty disappears from the right-hand side. With (7.36) we may rewrite (7.42) as

$$\dot{V}_{\mathrm{nom},1} \leq k_2 \left(-k_1 + \bar{\beta}\right) |s|^{\frac{1}{2}}. \tag{7.43}$$

If the controller parameter k_1 is chosen larger than the upper bound $\bar{\beta}$ and $k_2 > 0$ then the time-derivative of the Lyapunov function will always be negative semi-definite in the variables s and v.

In a next step let us extend the Lyapunov function by a quadratic form in terms of the parameter estimation error, i.e.

$$V_{\mathrm{ex},1} = k_2|s| + \frac{1}{2}v^2 + \frac{1}{2\gamma}\tilde{\Theta}^{\mathrm{T}}\tilde{\Theta}. \tag{7.44}$$

The variable γ is a positive constant that can be used to tune the dynamic behavior of the adaptation law. Differentiating equation (7.44) with respect to time yields

$$\dot{V}_{\text{ex},1} = k_2 \operatorname{sign}[s] \left(-k_1 \lceil s \rfloor^{\frac{1}{2}} + \tilde{\Theta}^{\mathrm{T}} \alpha(t,x) + \beta(t,x)\right) + \frac{1}{\gamma} \tilde{\Theta}^{\mathrm{T}} \dot{\tilde{\Theta}} \tag{7.45}$$

$$= k_2 \operatorname{sign}[s] \left(-k_1 \lceil s \rfloor^{\frac{1}{2}} + \tilde{\Theta}^{\mathrm{T}} \alpha(t,x) + \beta(t,x)\right) - \frac{1}{\gamma} \tilde{\Theta}^{\mathrm{T}} \dot{\hat{\Theta}}. \tag{7.46}$$

Note that $\dot{\tilde{\Theta}} = -\dot{\hat{\Theta}}$ due to the assumption of a constant parameter Θ. Rendering the derivative of the Lyapunov function independent of the parameter estimation error we may show convergence to the desired sliding-manifold, whatever the structured uncertainty is. We may achieve this by choosing

$$\dot{\hat{\Theta}} = \gamma\, k_2 \operatorname{sign}[s] \alpha(t,x). \tag{7.47}$$

This adaptation law eliminates the influence of the parameter estimation error, i.e.

$$\dot{V}_{\text{ex},1} = -k_1 k_2 |s|^{\frac{1}{2}} + k_2 \operatorname{sign}[s] \beta(t,x) \leq k_2 \left(-k_1 + \bar{\beta}\right) |s|^{\frac{1}{2}} \tag{7.48}$$

in which neither the estimation error $\tilde{\Theta}$ nor the parameter Θ is appearing. As in the nominal case, given by (7.43), the right-hand side of (7.48) can be rendered non-positive by choosing $k_1 > \bar{\beta}$. A closer look at (7.48) reveals that it neither contains $\tilde{\Theta}$ nor v. Thus, convergence can only be guaranteed for the sliding variable s. The other states, namely v and $\tilde{\Theta}$ may not vanish. Therefore the convergence of the parameter estimation $\hat{\Theta}$ to the actual value is in general not given.

Moreover note that already the right-hand side of (7.43) does not depend on the controller variable v. This makes clear that also the first Lyapunov function is not suited for showing asymptotic convergence of both closed-loop state variables s and v. This issue is addressed in the following sections.

7.4.2 Non-differentiable Strict Lyapunov Function

In this section we consider the Lyapunov function candidate first proposed in [13]:

$$V_{\text{nom},2} = z^{\mathrm{T}} P\, z \tag{7.49}$$

with $P = P^{\mathrm{T}} > 0$ and $z \in \mathbb{R}^2$ constructed from the state variables s and v of the closed-loop system (7.39):

$$\begin{aligned} z_1 &= \lceil s \rfloor^{\frac{1}{2}} \\ z_2 &= v\,. \end{aligned} \tag{7.50}$$

It turns out that $V_{\text{nom},2}$ is not continuously differentiable along the solutions of (7.39). Nevertheless it can be shown that $V_{\text{nom},2}$ constitutes a strict Lyapunov function for (7.39) using a theorem of Zubov when following the reasoning of [14, Proof of Theorem 1]. It can be shown that $V_{\text{nom},2}$ is absolutely continuous in t along the solutions of (7.39).[2] It remains to be shown that the derivative $\dot{V}_{\text{nom},2}$ is negative definite almost everywhere with respect to t.

At points t with $z_1(t) \neq 0$ the derivative is given by:

$$\dot{V}_{\text{nom},2} = z^{\mathrm{T}} P\, \dot{z} + \dot{z}^{\mathrm{T}} P\, z\,. \tag{7.51}$$

where

$$\dot{z} = \begin{pmatrix} \dfrac{\partial z_1}{\partial s}\dot{s} \\ \dfrac{\partial z_2}{\partial v}\dot{v} \end{pmatrix} = \begin{pmatrix} \dfrac{\dot{s}}{2\sqrt{|s|}} \\ -k_2 \text{sign}[s] \end{pmatrix} = \frac{1}{|z_1|}\begin{pmatrix} \frac{1}{2}\dot{s} \\ -k_2\, z_1 \end{pmatrix}. \tag{7.52}$$

Using inequality (7.36), we may rewrite $\beta(t, x)$ as

$$\beta(t, x) = \tilde{\beta}(t, x, s)\lceil s \rfloor^{\frac{1}{2}} \tag{7.53}$$

with the function $\tilde{\beta}(t, x, s)$ satisfying the inequality

$$|\tilde{\beta}(t, x, s)| \leq \bar{\beta}\,. \tag{7.54}$$

In order to show that (7.49) is a strict Lyapunov function for the closed-loop system (7.39) in the nominal case, i.e. $\tilde{\Theta} \equiv 0$, we have to proof that

$$\dot{V}_{\text{nom},2} = \frac{1}{|z_1|} z^{\mathrm{T}} \left(A^{\mathrm{T}}(t, x, s) P + P\, A(t, x, s)\right) z < 0 \tag{7.55}$$

for all $z,\ z_1 \neq 0$ with the time-varying matrix $A(t, x, s)$ defined as

$$A(t, x, s) := \begin{pmatrix} -\frac{1}{2}(k_1 - \tilde{\beta}(t, x, s)) & \frac{1}{2} \\ -k_2 & 0 \end{pmatrix}. \tag{7.56}$$

The derivative $\dot{V}_{\text{nom},2}$ is negative if P defines a quadratic Lyapunov function for $A(t, x, s)$ with arbitrary $\tilde{\beta}(t, x, s) \in [-\bar{\beta}, \bar{\beta}]$. Note that $A(t, x, s)$ is within the convex hull of

$$A_1 = \begin{pmatrix} -\frac{1}{2}(k_1 + \bar{\beta}) & \frac{1}{2} \\ -k_2 & 0 \end{pmatrix} \quad \text{and} \quad A_2 = \begin{pmatrix} -\frac{1}{2}(k_1 - \bar{\beta}) & \frac{1}{2} \\ -k_2 & 0 \end{pmatrix}.$$

[2] The solutions of (7.39) either cross points of $s = 0$ or remain on $s = 0$, $v = 0$. In each case $V_{\text{nom},2}$ is absolutely continuous and monotonically decreasing in t.

Thus $\dot{V}_{\text{nom},2}$ in (7.55) is negative if there exists a common quadratic Lyapunov function for A_1 and A_2. The following theorem comes handy for this problem [19].

Theorem 7.1 *Given* $A_1, A_2 \in \mathbb{R}^{n\times n}$, *Hurwitz with* $\operatorname{rank}(A_1 - A_2) = 1$. *There exists a common quadratic Lyapunov function for the LTI Systems with* A_1 *and* A_2 *if and only if the matrix product* $A_1 A_2$ *has no eigenvalue on the negative real axis.*

For A_1, A_2 being Hurwitz we require $k_1, k_2 > 0$. Applying the theorem above we obtain for the matrix product:

$$A_1 A_2 = \begin{pmatrix} \frac{1}{4}(k_1^2 - \bar{\beta}^2) - \frac{k_2}{2} & -\frac{1}{4}(k_1 + \bar{\beta}) \\ \frac{k_2}{2}(k_1 - \bar{\beta}) & -\frac{k_2}{2} \end{pmatrix}.$$

It can be readily verified that $A_1 A_2$ has complex conjugate eigenvalues if $k_1^2 > \bar{\beta}^2$. For the case of two real eigenvalues the product $A_1 A_2$ will always have at least one negative eigenvalue for $k_2 > 0$. Thus for stability of the nominal system we require for the controller parameters: $k_1 > \bar{\beta}, k_2 > 0$.

We obtain the same conditions on the controller parameters as for the first nominal Lyapunov function $V_{\text{nom},1}$ in (7.43). Note however, that the derivative of $V_{\text{nom},2}$ is *negative definite* almost everywhere. Thus $V_{\text{nom},2}$ is a strict Lyapunov function for (7.39) and guarantees convergence of both states s and v, [14].

In view of this, we may now repeat the design procedure for the certainty-equivalence adaptive part. Again, the nominal Lyapunov function (7.49) is extended by a quadratic term:

$$V_{\text{ex},2} = z^{\mathrm{T}} P z + \frac{1}{2\gamma} \tilde{\Theta}^{\mathrm{T}} \tilde{\Theta} \tag{7.57}$$

with a constant parameter $\gamma > 0$. At points t with $z_1(t) \neq 0$ the derivative is given by:

$$\dot{V}_{\text{ex},2} \leq -\frac{1}{|z_1|} z^{\mathrm{T}} Q_0 z + \frac{1}{|z_1|} z^{\mathrm{T}} P\, \Lambda(t,x) \tilde{\Theta} + \frac{1}{\gamma} \dot{\tilde{\Theta}}^{\mathrm{T}} \tilde{\Theta} \tag{7.58}$$

with the matrix $\Lambda(t, x)$ defined as

$$\Lambda(t, x) := \begin{pmatrix} \alpha(t, x) \\ 0 \end{pmatrix} \in \mathbb{R}^{2\times p}. \tag{7.59}$$

In order to eliminate the influence of the structured uncertainty contained in $\Lambda(t, x)$ from (7.58), we choose

$$\dot{\hat{\Theta}} = \frac{\gamma}{|z_1|} \Lambda(t, x)^{\mathrm{T}} P z \tag{7.60}$$

for the adaptation law. This yields $\dot{V}_{\text{ex},2} = \dot{V}_{\text{nom},2}$ and thus makes the convergence of s and v independent of the parameter estimation error. But again convergence of the parameter estimation to the actual values cannot be ensured as $\dot{V}_{\text{ex},2}$ does not contain the estimation error $\tilde{\Theta}$.

Note that this adaptation law contains a singularity for z_1 at zero. In [3] the authors present two possible solutions for this issue: Choosing P as diagonal matrix will cancel the singularity due to the structure of $\Lambda(t, x)$. However, this reproduces exactly the first nominal Lyapunov function (7.40) exhibiting only a negative semi-definite time-derivative. The second approach imposes the following additional condition on the structured uncertainty:

$$\alpha_i(t, x) \leq |s|^{\rho_i} \tag{7.61}$$

where $\rho_i > \frac{1}{2}$ for all elements $\alpha_i(t, x)$ in $\alpha(t, x)$. If (7.61) holds, then $\alpha(t, x)$ goes faster to zero than $|z_1| = \sqrt{|s|}$. Consequently, the adaptation law (7.60) remains bounded.

7.4.3 Differentiable Strict Lyapunov Function

So far the presented adaptation laws show properties that may be troublesome in some cases: In the first approach the adaptation law is devoid of the controller state v, which is caused by the negative semi-definiteness of the right-hand side of (7.43). Thus, convergence of both, the sliding variable s and the controller state v, cannot be shown. The second Lyapunov function (7.49) resolves this problem. However, the adaptation law (7.60) which is imposed by the specific choice of quadratic Lyapunov function may exhibit a singularity. In order to avoid this, an additional condition on the structured uncertainty is required. In the following we advocate a third approach that is able to tackle the afore-mentioned issues.

Stability Proof for the Nominal System

We use the Lyapunov function

$$V_{\text{nom},3} = \frac{2}{3} k_1 |s|^{\frac{3}{2}} - s\, v + \frac{2}{3\, k_1^2} |v|^3 \tag{7.62}$$

for proving stability of the nominal closed loop system. This Lyapunov function was first presented in [18] together with a detailed proof of its positive definiteness.

Let us briefly recapitulate the stability analysis of the nominal system, i.e. $\tilde{\Theta} \equiv 0$. Without loss of generality we further assume $\beta \equiv 0$ for brevity of the expressions. As first step we calculate the derivative of (7.62) and obtain

$$\dot{V}_{\text{nom},3} = -\left(k_1^2 - k_2\right) |s| + 2 k_1 \lceil s \rfloor^{\frac{1}{2}} v - |v|^2 - \frac{2\, k_2}{k_1^2} \lceil s\, v \rfloor^0 |v|^2 \tag{7.63}$$

with $k_1, k_2 > 0$ controller parameters regarding (7.35). Using

$$\eta_1 := k_1^2 - k_2, \qquad \eta_2 := 2k_1, \\ \eta_3 := 1 - \frac{2k_2}{k_1^2}, \qquad \eta_4 := 1 + \frac{2k_2}{k_1^2} \tag{7.64}$$

and the state vector z as in (7.50), we may express (7.63) as

$$\dot{V}_{\mathrm{nom},3} \leq \begin{cases} -\eta_1 z_1^2 + \eta_2 |z_1||z_2| - \eta_3 |z_2|^2, & z_1 z_2 \geq 0 \\ -\eta_1 z_1^2 - \eta_2 |z_1||z_2| - \eta_4 |z_2|^2, & z_1 z_2 < 0 \end{cases} \tag{7.65}$$

which can be also written as

$$\dot{V}_{\mathrm{nom},3} \leq \begin{cases} -z^{\mathrm{T}} W_1 z, & z_1 z_2 \geq 0 \\ -z^{\mathrm{T}} W_2 z, & z_1 z_2 < 0 \end{cases} \tag{7.66}$$

where

$$W_1 =: \begin{pmatrix} \eta_1 & -\frac{\eta_2}{2} \\ -\frac{\eta_2}{2} & \eta_3 \end{pmatrix} \quad \text{and} \quad W_2 := \begin{pmatrix} \eta_1 & \frac{\eta_2}{2} \\ \frac{\eta_2}{2} & \eta_4 \end{pmatrix}. \tag{7.67}$$

We are left to show that the matrices W_1 and W_2 are positive definite and then conclude that $\dot{V}_{\mathrm{nom}}$ is negative definite and finally the nominal system asymptotically stable. As shown in [18], if the parameters satisfy the inequality

$$k_1^2 > 2\,k_2 \tag{7.68}$$

then both matrices W_1 and W_2 are positive definite. Note, that $\dot{V}_{\mathrm{nom},3}$ does not contain any singular point as we obtained in (7.55). The result will be discussed in the following section.

Adaptive Extension

The intention of using the third Lyapunov function (7.62) is to avoid the drawbacks in the adaptation law that stymied the first two design approaches. Again we rely on the certainty-equivalence principle as our design guideline. As before, we extend the nominal Lyapunov function by a quadratic term, i.e.

$$V_{\mathrm{ex},3} = V_{\mathrm{nom},3} + \frac{1}{2\,\gamma} \tilde{\Theta}^{\mathrm{T}}\, \tilde{\Theta}\,. \tag{7.69}$$

The derivative reads

$$\dot{V}_{\mathrm{ex},3} = \dot{V}_{\mathrm{nom},3} + \tilde{\Theta}^{\mathrm{T}} \left(\frac{1}{\gamma}\, \dot{\tilde{\Theta}} + \alpha(t,x) \left(k_1 \lceil z_1 \rfloor^{\frac{1}{2}} - z_2 \right) \right). \tag{7.70}$$

By choosing the following adaptation law

$$\dot{\hat{\Theta}} = \gamma\, \alpha(t, x) \left(k_1 \lceil z_1 \rfloor^{\frac{1}{2}} - z_2 \right) \tag{7.71}$$

we are able to remove the influence of $\tilde{\Theta}$ on (7.70). Consequently, the convergence of the system states s, v to zero is independent of the parameter estimation error $\tilde{\Theta}$.

Choosing $V_{\text{nom},3}$ for the nominal case yields two advantage over the previous approaches. The derivative of the nominal Lyapunov function is strictly negative definite, thus, convergence of both system states is guaranteed. Furthermore, the adaptation law exhibits no singularity. As before, the derivative of the extended Lyapunov function (7.70) does not contain the estimation error $\tilde{\Theta}$ and thus convergence of the parameter estimation $\hat{\Theta}$ to the actual value is not guaranteed.

7.4.4 Discussion

In the previous section we presented three different approaches to design an adaptation law to the same nominal STA. Choosing different Lyapunov function approaches for the nominal controller results in three different adaptation laws for the certainty equivalence part: In the first case we consider a Lyapunov function with only negative semi-definite derivative (7.43) along the nominal closed-loop systems' solutions. As a result convergence is only guaranteed for the controller state s but not for v. Furthermore, the adaptation law (7.47) for the certainty-equivalence approach does not contain the controller state v.

This problem is addressed by the quadratic Lyapunov function in the second approach. For the nominal system this strict Lyapunov function renders the origin asymptotically stable. However the adaptation law obtained (7.60) has a singularity at $z_1 = s = 0$. This requires an additional bound on the structured uncertainty in order to keep the adaptation law bounded. On the other hand, the quadratic Lyapunov function offers the possibility to combine the certainty-equivalence adaptation with the adaptive-gain super-twisting controller as nominal part, see e.g. [5]. A detailed discussion of this combination is found in [3].

Avoiding the singularity is the main motivation for using the third Lyapunov function (7.62). In conjunction with the closed-loop system, $V_{\text{nom},3}$ is differentiable everywhere with negative definite time-derivative. The adaptation law (7.71) generated by this Lyapunov function has no singularity and contains all states of the nominal closed-loop system, which guarantees convergence of the sliding variable s and the controller state v to the origin even for the extended controller.

All three approaches have in common that they allow us to combine sliding-mode and certainty-equivalence adaptive control. In contrast to existing adaptive approaches for sliding-mode control, our design makes explicit use of the structure of the uncertainty. Instead of varying the gains of the sliding-mode part directly, we propose to adapt the parameters of the uncertainty. This procedure has two main advantages: Firstly, we can compensate uncertainties of the form (7.26) that are neither required to be bounded by a constant value nor bounded with respect to the sliding variable. Secondly, the sliding-mode gains remain constant until we are able to dominate the disturbance. This may avoid effects that come with high gains in sliding-mode control.

The latter two approaches use the parametrization z in (7.50) for the Lyapunov function. Existing Lyapunov approaches [14, 18, 22] for the conventional super-twisting algorithm use a slightly different parametrization, namely

$$z_2 = v - \beta(t) \tag{7.72}$$

where the unstructured disturbance $\beta(t)$ is only required to be Lipschitz continuous in time (as opposed to (7.36)), see e.g. [22] for a detailed discussion of this requirement. In [14, 18], the Lyapunov functions (7.49) and (7.62) have been studied to show the robustness of the super-twisting controller against such Lipschitz time varying disturbances. In our approaches presented above, we are not able to use this parametrization (7.72), as z_2 will appear in the adaptation laws (7.60) and (7.71). With $\beta(t)$ unknown such adaptation cannot be applied. Therefore our requirement (7.36) on the unstructured uncertainty $\beta(t, x)$ differs from the conventional condition.

The following sections illustrates the application of our proposed approach. First we revisit the simulation example from Sect. 7.1.2 followed by a case study using a laboratory test-bench. We will discuss the differences between the conventional super-twisting algorithm and our proposed approach.

7.5 Introductory Example - Revisited

In order to illustrate the control concept proposed we resume the DC motor example of Sect. 7.1.2. Recall the explicit formulation of the disturbance (7.22) acting on the closed-loop system:

$$\Delta = \left(\frac{k_p b}{J} + \frac{k_\mathrm{m}}{L_\mathrm{m}} \right) \omega - \frac{k_p k_\mathrm{m}}{J} i + k_i e + k_p \dot{\omega}_d \,. \tag{7.73}$$

In context of our problem formulation this disturbance acts on the system (7.31), that is

$$\dot{s}(t, x) = \Delta(t, x) + w \,,$$

where w is the control signal and

$$\Delta(t, x) = \Theta^\mathrm{T} \alpha(t, x) + \beta(t, x).$$

We identify the unstructured part of the disturbance as $\beta(t) := k_p \dot{\omega}_d$ leaving the remainder for the structured disturbance. For the certainty-equivalence part we choose the following parametrisation

$$\Theta^{\mathrm{T}} \alpha(t,x) = \begin{bmatrix} \theta_1 \ \theta_2 \ \theta_3 \end{bmatrix} \begin{bmatrix} \omega \\ i \\ e \end{bmatrix} . \tag{7.74}$$

The complete control law (7.33)

$$w = w_{\mathrm{CE}} + w_{\mathrm{SMC}}$$

with super-twisting control part (7.35) and certainty-equivalence part (7.34) using the adaptation law (7.47) reads:

$$\begin{aligned} w_{\mathrm{SMC}} &= -k_1 \, |s|^{\frac{1}{2}} \operatorname{sign}[s] + v, \\ \dot{v} &= -k_2 \operatorname{sign}[s], \\ w_{\mathrm{CE}} &= -\hat{\Theta}^{\mathrm{T}} \alpha(t,x) \,, \\ \dot{\hat{\Theta}} &= \gamma \, k_2 \operatorname{sign}[s] \alpha(t,x). \end{aligned}$$

The controller parameters k_1, k_2 are selected as in the introductory example of Sect. 7.1.2. For the adaptation rate we choose $\gamma = 50$.

Figure 7.3 shows the simulation result with PI controller gains $k_{p,3} = 9.6$ and $k_{i,3} = 32.8$. The certainty-equivalence adaptation combined with the super-twisting sliding-mode controller is capable of driving the sliding variable to the origin despite that no bound for the overall perturbation is known. The third plot in Fig. 7.3 illustrates the evolution of $\hat{\Theta}$. Note that the estimation errors $\tilde{\Theta}_i$ may not converge to zero. For the overall control performance we note that the initial transients as well as the tracking behavior is superior to the results of the simple super-twisting algorithms, c.f. Fig. 7.2.

Note that applying solely the constant gains super-twisting algorithm, only local stability results have been achieved. This becomes evident by inspecting the uncertainty $\Delta(t,x)$ which has to have a bounded time-derivative, i.e.

$$\sup_t \left| \frac{\mathrm{d}\Delta(t,x)}{\mathrm{d}t} \right| \le \delta_1, \quad \forall x. \tag{7.75}$$

Hence the motion of the system has to be restricted to some area of the state space which also depends on the controller parameters of the super-twisting algorithm.

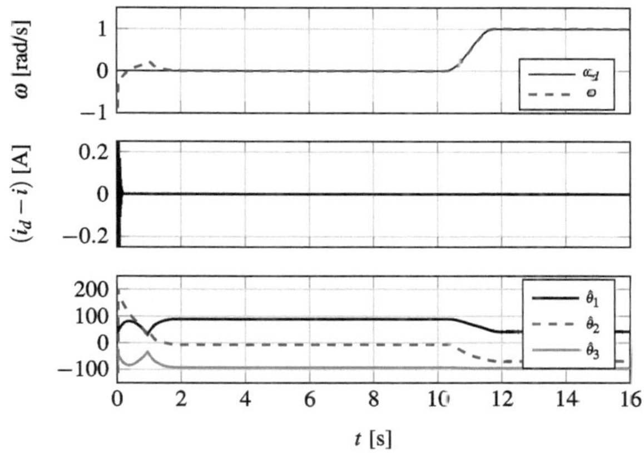

Fig. 7.3 Results obtained for the cascaded DC motor control problem with the certainty equivalence based super-twisting algorithm

7.6 Case Study (Speed Control with Unbalanced Load)

Using a laboratory test-bench we illustrate some advantages of our proposed controller design in comparison with existing sliding-mode approaches. The test-bench employed is shown in Fig. 7.4. It consists mainly of a DC motor to which an unbalanced mass is attached. The unbalanced mass is a metal cuboid with center of mass not in line with the rotary axis, similar to a washing machine whose loaded drum rotates at a certain angular velocity.

The system may be represented by the following differential equation

$$J\,\dot{\omega} + d\,\sin[\varphi] + \Delta(\varphi, \omega) = \tau \tag{7.76}$$

where ω is the angular velocity, φ the angle of the imbalance and τ the torque generated by the motor. The system parameters are the moment of inertia J and the displacement d of the imbalance's center of mass from the rotation axis. It is assumed that the parameter d is unknown at the time of the controller design. Moreover there exists an uncertainty $\Delta(\varphi, \omega)$ that cannot be modeled accurately due to friction effects.

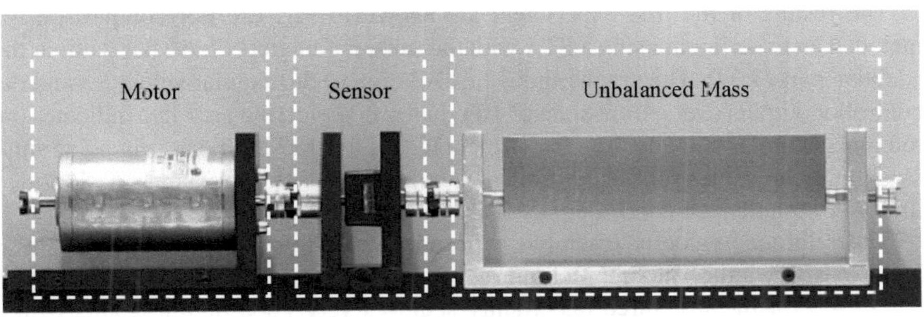

Fig. 7.4 Image of the test-bench

7.6.1 Controller Design

The control task is to track a specified angular velocity ω_d. Therefore we define the sliding variable as the difference between the actual velocity and its reference, i.e.

$$s := \omega - \omega_d \,. \tag{7.77}$$

We identify the structured and unstructured uncertainty by analyzing the system (7.76). The unbalanced load can be modeled as structured uncertainty since it can be rewritten in the form (7.30) with

$$\alpha(t, x) = \sin[\varphi] \tag{7.78}$$

where $\Theta \in \mathbb{R}$ and φ are directly available from measurements. The remaining parts in (7.76) are modeled as unstructured uncertainty that the sliding-mode part of the controller has to take care of.

For the first experiment we implemented a CESTA according to the methodology presented in Sect. 7.3. The adaptation law is generated using the first version (7.47) of the three approaches presented. We chose $\gamma = 2$ for the adaptation rate. The controller parameters of the sliding-mode part were selected as

$$k_1 = 0.12, \quad k_2 = 1.2 \,. \tag{7.79}$$

The controller parameters are a result of experimental tuning. Note that the design rule (7.20) in [10] considers a class of disturbances different from (7.36) and thus is not directly applicable for the CESTA.

7.6.2 Experimental Results

The controllers have been implemented on a real-time platform running at a sampling frequency of 5 kHz. An underlying current controller was used in order to command torques directly.

The results of the first experiment are shown in Fig. 7.5. For comparison we included an experiment using a STA with the same parameters (7.79) but without the adaptive part (7.34). The graphs show the evolution of the angular velocities and the controller outputs over a time-span of 10 s. In the first plot, the grey line indicates the reference angular velocity. The experimental data for the CESTA is printed as solid black line, the values for the controller without adaptation approach as dash-dotted line. Additionally, the controller states are shown in the third plot.

The reference velocity resembles a rapid acceleration and is kept constant at 360 deg/s afterwards. The first plot in Fig. 7.5 shows that the CESTA with adaptation law (7.47) for the structured uncertainty achieves very good output results with a

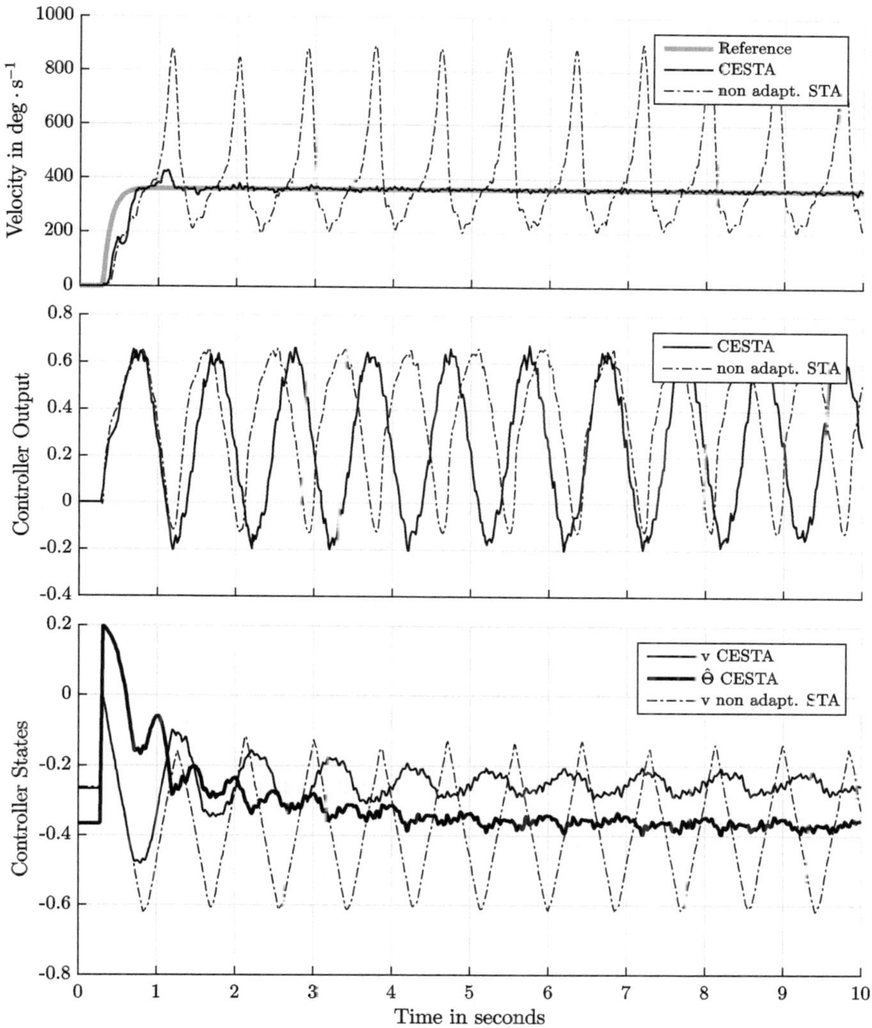

Fig. 7.5 Tracking performance comparison between conventional STA and CESTA

slight overshoot at around 1 s and shows very little deviation from the reference value afterwards. Without the adaptive part the STA is not able to solve the given task. The velocity oscillates around the reference value with peaks up to 900 $^{\text{deg}}/_{\text{s}}$.

The second plot in Fig. 7.5 displays the control signals of the CESTA as solid line and the non-adaptive STA as dash-dotted line. The graphic underscores that the difference in the overall control signal amplitude between both controllers is very small. However, using the information about the structured uncertainty, the CESTA generates a control signal that fits almost perfectly to the uncertainty and therefore keeps the velocity at the desired level.

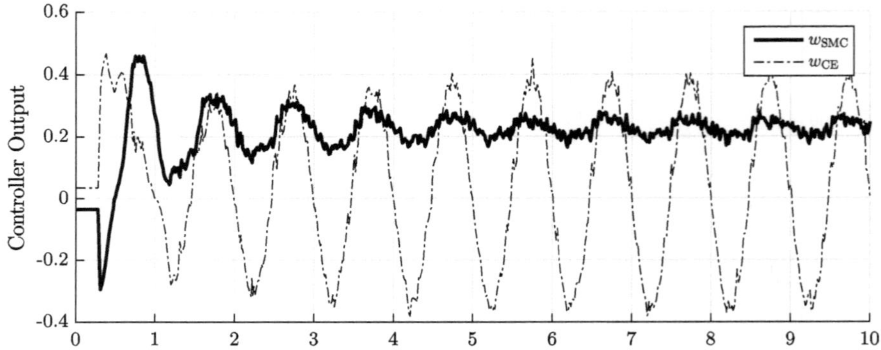

Fig. 7.6 Different components of the CESTA control signal

For a better illustration of this effect, we printed the two components w_{SMC} and w_{CE} according to (7.33) of the CESTA separately in Fig. 7.6. Comparing the w_{SMC} in Fig. 7.6 with the control signal of the non-adaptive STA in Fig. 7.5 leads to an interesting observation: A much lower amplitude is needed to track the desired reference speed since the structured uncertainty is mostly compensated by the certainty-equivalence part w_{CE} of the controller. This is caused by the small error of the parameter estimation $\tilde{\Theta}$. The actual value of $\approx$0.4 for Θ was estimated in an additional experiment where we hold the unbalanced load in a steady state at an angle of $\pi/2$. Comparing this value to the estimated value $\hat{\Theta}$ in the third plot in Fig. 7.5 indicates a small estimation error. Therefore the sliding-mode part of the CESTA only compensates the unstructured part of the disturbance which has significantly lower amplitude.

The parameters of the STA (7.79) in the first experiment where tuned for good performance in conjunction with the adaptive part. In order to challenge our advocated approach we tried to find the best set of parameters for the conventional STA without adaptation in a second experiment.

Considering the parameter design rule (7.20) we obtain an estimate of N from the controller output in Fig. 7.5. The CESTA generates a control signal from about -0.2 to 0.6 for a good tracking of the reference speed. For the desired speed of 360 $^{deg}/_{s}$ the disturbance from (7.78) is estimated by $N \approx 2.2$. The design rule (7.20) results in the parameter combination

$$k_1 = 2.2, \quad k_2 = 2.42\,. \tag{7.80}$$

However, we were not able to verify this controller experimentally due to strong chattering effects.

The best output-performance we obtained for the parameter combination

$$k_1 = 0.4, \quad k_2 = 2.4\,. \tag{7.81}$$

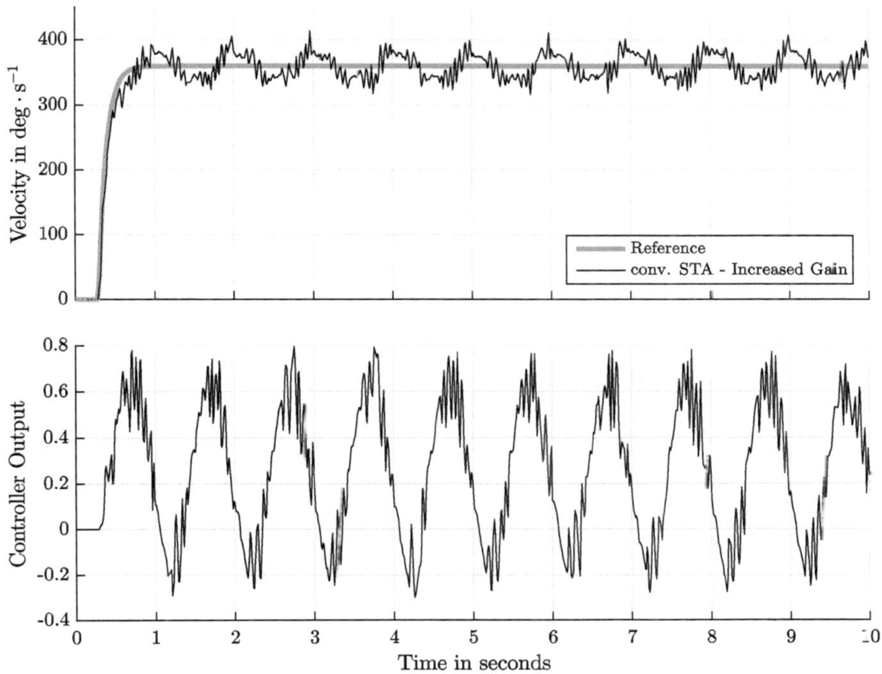

Fig. 7.7 Tracking of the STA with increased gains

A further increase led to chattering which generates a loud noise in the test-bench indicating strong physical strain. This audible effects can also be seen in the time response of the second experiment shown in Fig. 7.7 including the velocity and the control signal of the conventional STA with higher controller gains compared to the first implementation.

The tracking error is significantly reduced compared to the conventional STA in the first experiment and reaches almost the same level as the CESTA. As mentioned before, this does not come for free with the conventional STA. Taking a closer look at the second plot of Fig. 7.7 indicates that the control signal contains a larger portion of high frequency parts compared to the controllers in Fig. 7.5. This chattering phenomenon is caused by the higher controller gains and the non-Lipschitz continuous terms in the control law (7.35).

Note that this experiment also emulates the behavior of adaptive- and variable-gain sliding-mode control approaches, [6, 20]. The adaptive-gain super-twisting control (AGSTA) presented in [20] increases the gains k_1, k_2 of the sliding-mode control law (7.7) until the disturbance is dominated and decreases them subsequently if possible. Another approach is the variable-gain STA (VGSTA), see [6], where the gains of the sliding-mode part of the controller are a function of the sliding variable or time itself and thus allows to adjust the gains accordingly. However, both approaches lead to high gains similar to the ones given in (7.81) since they are required for dominating the overall disturbance. Due to the periodicity of the disturbance, the gains cannot back off. Essentially, this will lead to similar chattering effects as shown in Fig. 7.7.

From the results it may be concluded that the CESTA approach is able to improve the control performance significantly. If the structured uncertainty can be expressed linear in an unknown parameter and a known function then an adaptive extension is given to the nominal STA. The variable-structure part of this combined controller only has to take care of the unstructured uncertainty and allows to reduce the controller gains which also reduces the occurrence of undesired effects like chattering.

7.7 Concluding Remarks

In this chapter we have presented a method to improve the robustness of sliding-mode control with an adaptive extension. Contrary to existing adaptive methods, we do not adapt the gains of the sliding-mode controller itself. Instead we extend the controller by an adaptive part that fits to the structured uncertainty. We have shown the influence of the Lyapunov function for the nominal controller on the adaptation law that we generate with our method. This should be understood as motivation to keep searching for other Lyapunov functions for higher-order sliding-mode control since this is the key ingredient to our method. The improved robustness properties of our approach are demonstrated via a simple but effective case study. Here we showed that our method is able to achieve better reference tracking with even lower controller gains compared to the conventional approach. In addition, the smaller controller gains of the sliding-mode part of the controller helps to reduce chattering effects in the closed-loop system.

Acknowledgements The authors gratefully acknowledge the advice of Jaime Moreno on the construction of Lyapunov functions for the super-twisting control and in particular for pointing out a subtle, but fatal error in an earlier proof of the negative definiteness of $\dot{V}_{\mathrm{nom},2}$. Furthermore, the authors kindly acknowledge support by the German Academic Exchange Service (DAAD) with financial means of The German Federal Ministry for Education and Research (BMBF) through the project "Lyapunov-Function based Adaptive Sliding-Mode Control" and through funding from the European Union Horizon 2020 research and innovation programme under the Marie Skłodowska-Curie grant agreement No. 734832.

References

1. Barth, A., Reichhartinger, M., Reger, J., Horn, M., Wulff, K.: Lyapunov-design for a super-twisting sliding-mode controller using the certainty-equivalence principle. In: Proceedings of the 1st IFAC MICNON, pp. 860–865 (2015)
2. Barth, A., Reichhartinger, M., Reger, J., Horn, M., Wulff, K.: Certainty-equivalence based super-twisting control using continuous adaptation laws. In: Proceedings of 2016 14th International Workshop on Variable Structure Systems, pp. 92–97 (2016)

3. Barth, A., Reichhartinger, M., Wulff, K., Horn, M., Reger, J.: Certainty-equivalence adaptation combined with super-twisting sliding-mode control. Int. J. Control **89**(9), 1767–1776 (2016)
4. Bhat, S., Bernstein, D.: Finite-time stability of continuous autonomous systems. SIAM J. Control Optim. **38**(3), 751–766 (2000)
5. Brégeault, V., Plestan, F., Shtessel, Y., Poznyak, A.: Adaptive sliding mode control for an electropneumatic actuator. In: Proceedings of the 11th International Workshop on Variable Structure Systems, pp. 260–265 (2010)
6. González, T., Moreno, J., Fridman, L.: Variable gain super-twisting sliding mode control. IEEE Trans. Autom. Control **57**(8), 2100–2105 (2012)
7. Ioannou, P., Sun, J.: Robust Adpative Control. Prentice Hall, Upper Saddle River (1996)
8. Kochalummoottil, J., Shtessel, Y., Moreno, J., Fridman, L.: Adaptive twist sliding mode control: a Lyapunov design. In: Proceedings of the 50th IEEE Conference on Control and Decision, pp. 7623–7628 (2011)
9. Krstić, M., Kokotović, P., Kanellakopoulos, I.: Nonlinear and Adaptive Control Design. Wiley, New York (1995)
10. Levant, A.: Sliding order and sliding accuracy in sliding mode control. Int. J. Control **58**(6), 1247–1263 (1993)
11. Levant, A.: Higher-order sliding modes, differentiation and output-feedback control. Int. J. Control **76**(9–10), 924–941 (2003)
12. Moreno, J.: A linear framework for the robust stability analysis of a generalized super-twisting algorithm. In: Proceedings of the 6th International Conference on Electrical Engineering, Computing Science and Automatic Control (CCE), pp. 1–6 (2009)
13. Moreno, J., Osorio, M.: A Lyapunov approach to second-order sliding mode controllers and observers. In: Proceedings of the 47th IEEE Conference on Control and Decision, pp. 2856–2861 (2008)
14. Moreno, J., Osorio, M.: Strict Lyapunov functions for the super-twisting algorithm. IEEE Trans. Autom. Control **57**(4), 1035–1040 (2012)
15. Narendra, K., Valavani, L.: Stable adaptive controller design - direct control. IEEE Trans. Autom. Control **23**(4), 570–583 (1978)
16. Orlov, Y.: Finite time stability and robust control synthesis of uncertain switched systems. SIAM J. Control Optim. **43**(4), 1253–1271 (2005)
17. Plestan, F., Shtessel, Y., Bregeault, V., Poznyak, A.: New methodologies for adaptive sliding mode control. Int. J. Control **83**(9), 1907–1919 (2010)
18. Sánchez, T., Moreno, J.: A constructive Lyapunov function design method for a class of homogeneous systems. In: Proceedings of the 53rd IEEE Conference on Control and Decision, pp. 5500–5505 (2014)
19. Shorten, R., Mason, O., Cairbre, F., Curran, P.: A unifying framework for the SISO circle criterion and other quadratic stability criteria. Int. J. Control **77**, 1–8 (2004)
20. Shtessel, Y., Moreno, J., Plestan, F., Fridman, L., Poznyak, A.: Super-twisting adaptive sliding mode control: a Lyapunov design. In: Proceedings of the 49th IEEE Conference on Control and Decision, pp. 5109–5113 (2010)
21. Shtessel, Y., Plestan, F., Taleb, M.: Lyapunov design of adaptive super-twisting controller applied to a pneumatic actuator. In: Proceedings of the 18th IFAC World Congress, pp. 3051–3056 (2011)
22. Shtessel, Y., Taleb, M., Plestan, F.: A novel adaptive-gain super-twisting sliding mode controller: methodology and application. Automatica **48**(5), 759–769 (2012)
23. Utkin, V.: Sliding Modes in Control and Optimization. Springer Science & Business Media, Berlin (2013)
24. Zhou, K., Doyle, J., Glover, K.: Robust and Optimal Control. Prentice Hall, Upper Saddle River (1996)

Part II
The Usage of VSS/SMC Techniques for Solutions of Different Control Problems

Chapter 8
Variable Structure Observers for Nonlinear Interconnected Systems

Mokhtar Mohamed, Xing-Gang Yan, Sarah K. Spurgeon and Zehui Mao

8.1 Introduction

The advancement of modern technologies has produced many complex systems. An important class of such systems, which is frequently called a system of systems or large-scale system, can be expressed by sets of lower-order ordinary differential equations which are linked through interconnections. Such models are typically called large-scale interconnected systems (see, e.g. [4, 22, 44, 46]). In order to achieve the desired performance of the closed-loop system in the presence of uncertainties, robust control methods are needed. In recent decades, much of the literature has focused on designing advanced robust controllers for such systems using H_∞ control [5], backstepping techniques [17], robust adaptive control [28], and sliding mode control [8, 30].

Increasing requirements for system performance have resulted in increasing complexity within system modelling and it becomes of interest to consider nonlinear large-scale interconnected systems. Such models are then used for controller design. In order to obtain required levels of performance from the controllers, it is desirable to have knowledge of all the system states for use by the control scheme. This state information may be difficult or expensive to obtain and it becomes of interest to

M. Mohamed · X.-G. Yan (✉)
School of Engineering and Digital Arts, University of Kent, Canterbury CT2 7NT, UK
e-mail: x.yan@kent.ac.uk

M. Mohamed
e-mail: mskm3@kent.ac.uk

S.K. Spurgeon
Department of Electronic and Electrical Engineering, University College London, London WC1E 7JE, UK
e-mail: s.spurgeon@ucl.ac.uk

Z. Mao
College of Automation Engineering, Nanjing University of Aeronautics and Astronautics, Nanjing 211106, People's Republic of China
e-mail: zehuimao@163.com

S. Li et al. (eds.), *Advances in Variable Structure Systems and Sliding Mode Control—Theory and Applications*, Studies in Systems, Decision and Control 115, DOI 10.1007/978-3-319-62896-7_8

design an observer to estimate all the system states using only the subset of information available from the measured and known inputs and outputs of the system.

8.1.1 Interconnected Systems

Large-scale interconnected systems have been studied since the 1960s (see [2] and references therein) due to their relevance in a number of practical application areas and the availability of pertinent theoretical results. Large-scale interconnected systems widely exist in the real world, for example, power networks, ecological systems, transportation networks, biological systems and information technology networks [21, 22]. A large-scale system is composed of several subsystems with interconnections, whereby the dynamics interact [13]. The application of centralised control [48] to prescribe stability of an interconnected system particularly when the system is spread over a wide geographical area may require additional costs for implementation and careful consideration of the required information sharing between subsystems. This motivates considering the design of decentralized control strategies whereby each subsystem has a local control based only upon locally available information.

Early work focussed on linear systems [3, 27]. However, due to the uncertainties and disturbances present in large-scale interconnected systems, study of the stability of such systems is a very challenging task [45]. Subsequent results used decentralised control frameworks for nonlinear large-scale interconnected systems. The study of such decentralised controllers has stimulated a wide literature (e.g. [26, 50]) and recently [19, 52]. In much of this work, however, it is assumed that all the system state variables are available for use by the controller [4, 22, 39, 53]. However, this assumption can limit practical application as usually only a subset of state variables may be available/measureable [49]. It becomes of interest to establish observers to estimate the system states and then use the estimated states to replace the true system states in order to implement state feedback decentralised controllers. It should also be noted that such observer design has been applied for fault detection and isolation [25, 43, 49]. This further motivates the study of observer design for nonlinear large-scale interconnected systems.

8.1.2 Observer Design

The concept of an observer was first introduced by Luenberger (1964) where the difference between the output measurements from the actual plant and the output measurements of a corresponding dynamical model were used to develop an injection signal to force the resulting output error to zero [20]. In the 1970s the problem of designing observers for estimating system states for large-scale interconnected systems was addressed in [3]. Subsequently, many methods have been developed to design observers for large-scale interconnected linear systems [31, 33]. The methodology of Luenberger was extended to nonlinear systems [51] where uncertainties are rarely considered. However, many approaches such as the sliding mode approach in [47], the adaptive technique in [40] and an error linearisation approach in [41] have been successfully used in observer design. Note, results concerning observer design for nonlinear interconnected systems are very few when compared with the corresponding results available on controller design for interconnected systems.

Sliding mode techniques have been used to design observers for nonlinear interconnected power systems in [24]. In [18] state estimation and sliding mode control for a special class of stochastic dynamic systems, so-called semi Markovian jump systems, is presented. The authors designed a state observer to generate an estimate of the unmeasured state components, and then a sliding mode control law is synthesized based on these estimated states. The authors in [38] discussed the regulation problem of a permanent magnet synchronous motor servo system based on adaptive fuzzy sliding mode control. They used an adaptive method to estimate the upper bound of the approximation error between the equivalent control law and the fuzzy controller. An adaptive observer is designed for a class of interconnected systems in [40] in which it is required that the isolated nominal subsystems are linear. Observer schemes for interconnected systems are proposed in [16, 25, 29, 43] where the obtained results are unavoidably conservative as it is required that the designed observer can be used for certain fault detection and isolation problems. For example, it is required that the uncertainty can be decoupled from any faults in [43] and the considered system is not an interconnected system. Robust observer design is considered in [23] for a class of linear large scale dynamical systems where it is required that the interconnections satisfy quadratic constraints. In [35], a decentralized control scheme which uses estimated states obtained from a decentralised observer is proposed. This design framework is based on linear matrix inequalities and is thus only applicable to linear systems. A robust observer for nonlinear interconnected systems based on a constrained Lyapunov equation has been developed in [42]. A proportional integral observer is utilized for nonlinear interconnected systems for disturbance attenuation in [12] and interconnected nonlinear dynamical systems are considered in [6] where the authors combine the advantages of input-to-state dynamical stability and use reduced order observers to obtain quantitative information of the state estimation error. This work does not, however, consider uncertainties. It should be noted that in all the existing work relating to observer design for large scale interconnected systems, it is required that either the isolated subsystems are linear or the interconnections are linear. Moreover, most of the designed observers are used for special purposes such as fault detection or stabilization and thus some specific limitations are unavoidably required on the interconnected systems as a result of the problem class considered.

8.1.3 Contribution

In this chapter, a class of nonlinear interconnected systems with disturbances is considered where both the nominal isolated subsystems and interconnections are nonlinear. It is not required that either the nominal isolated subsystems or the interconnections are linearisable. A robust variable structure observer is established based on a simplified system structure by using a Lyapunov analysis methodology. The structure of the internal dynamics, the structure of uncertainties and the bounds on uncertainties are fully used in the observer design to reduce conservatism. These bounds are allowed to have a general nonlinear form. The observer states converge to the system states asymptotically. A numerical simulation example is given to demonstrate the application of the proposed approach. A case study of a coupled inverted pendulum system shows the practicality of the designed observer.

8.1.4 Notation

For a square matrix A, $A > 0$ denotes a symmetric positive definite matrix, and $\lambda_{\min}(A)(\lambda_{\max}(A))$ denotes the minimum (maximum) eigenvalues of A. The symbol I_n represents the nth order unit matrix and $\mathbb{R}^+$ represent the set of nonnegative real numbers. The set of $n \times m$ real matrices will be denoted by $\mathbb{R}^{n \times m}$. The Lipschitz constant of the function f will be written as ℓ_f. For any mapping $h(x) : \mathbb{R}^n \to \mathbb{R}$ and vector field $f(x) : \mathbb{R}^n \to \mathbb{R}^n$, $L_f h(x)$ denotes the Lie derivative of $h(x)$ along the vector field $f(x)$ and L_f^r denotes the r-th order Lie derivative of $h(x)$ along the vector field $f(x)$. Finally, $\|\cdot\|$ denotes the Euclidean norm or its induced norm.

8.2 Preliminaries

Consider the single-input-single-output nonlinear system

$$\dot{x}(t) = f(x) + g(x)u \tag{8.1}$$

$$y(t) = h(x) \tag{8.2}$$

where $x \in \Omega \subset \mathbb{R}^n$ (Ω is a neighborhood of the origin), $y \in \mathbb{R}$ and $u \in U \subset \mathbb{R}$ (U is an admissible control set) are the state, output and input respectively, $f(x), g(x) \in \mathbb{R}^n$ are smooth vector fields defined in the domain Ω, and $h(x) \in \mathbb{R}$ is a smooth function in the domain Ω.

First, some key elements of the geometric approach in [15] are recalled. These will be used in the later analysis.

Definition 8.1 ([15]) System (8.1) and (8.2) is said to have uniform relative degree r in the domain Ω if for any $x \in \Omega$,

(i) $L_g L_f^k h(x) = 0, \quad \text{for} \quad k = 1, 2, \ldots, r-2$
(ii) $L_g L_f^{r-1} h(x) \neq 0$

Now consider system (8.1) and (8.2). It is assumed that system (8.1) and (8.2) has uniform relative degree r in the domain Ω. Construct a mapping $\phi : x \to z$ as follows:

$$\phi(\cdot): \begin{cases} z_1 = h(x) \\ z_2 = L_f h(x) \\ \quad \vdots \\ z_r = L_f^{r-1} h(x) \\ z_{r+1} = \phi_{r+1} \\ \quad \vdots \\ z_n = \phi_n(x) \end{cases} \tag{8.3}$$

where $\phi(\cdot) = \text{col}(\phi_1(x), \phi_2(x), \ldots, \phi_n(x))$, $\phi_1(x) = h(x)$, $\phi_2(x) = L_f h(x), \ldots,$ $\phi_r(x) = L_f^{r-1} h(x)$ and the functions $\phi_{r+1}(x), \ \ldots, \ \phi_n(x)$ need to be selected such that

$$L_g \phi_i(x) = 0, \qquad i = r+1, r+2, \ldots, n$$

and the Jacobian matrix

$$J_\phi := \frac{\partial \phi(x)}{\partial x}$$

is nonsingular in the domain Ω. Then the mapping $\phi : x \to z$ forms a diffeomorphism in the domain Ω (see Proposition 4.1.3 in [15]). For the sake of simplicity, let

$$\begin{aligned} \zeta &= \left[\zeta_1 \ \zeta_2 \ \ldots \ \zeta_r\right]^T := \left[z_1 \ z_2 \ \ldots \ z_r\right]^T \\ \eta &= \left[\zeta_{r+1} \ \zeta_{r+2} \ \cdots \ \zeta_n\right]^T := \left[z_{r+1} \ z_{r+2} \ \cdots \ z_n\right]^T \end{aligned}$$

Then, from [15], it follows that in the new coordinates z, system (8.1) and (8.2) can be described by

$$\begin{aligned} \dot{\zeta}_1 &= \zeta_2 \\ \dot{\zeta}_2 &= \zeta_3 \\ &\vdots \\ \dot{\zeta}_{r-1} &= \zeta_r \\ \dot{\zeta}_r &= a(\zeta, \eta) + b(\zeta, \eta) u \\ \dot{\eta} &= q(\zeta, \eta) \end{aligned} \tag{8.4}$$

where

$$a(\zeta,\eta) = L_f^r h(\phi^{-1}(\zeta,\eta))$$
$$b(\zeta,\eta) = L_g L_f^{r-1} h(\phi^{-1}(\zeta,\eta))$$

and

$$q(\zeta,\eta) = \begin{bmatrix} q_{r+1}(\zeta,\eta) \\ q_{r+2}(\zeta,\eta) \\ \vdots \\ q_n(\zeta,\eta) \end{bmatrix} = \begin{bmatrix} L_f\phi_{r+1}(\phi^{-1}(\zeta,\eta)) \\ L_f\phi_{r+2}(\phi^{-1}(\zeta,\eta)) \\ \vdots \\ L_f\phi_n(\phi^{-1}(\zeta,\eta)) \end{bmatrix}$$

It should be noted that the coordinate transformation (8.3) will be available if $\phi_i(x)$ are available for $i = r+1, \ldots, n$, and in this case, the system (8.4) can be obtained directly.

8.3 Problem Formulation

Consider the set of nonlinear interconnected systems

$$\dot{x}_i(t) = f_i(x_i) + g_i(x_i)u_i + \Delta f_i(x_i) + \sum_{\substack{j=1 \\ j\neq i}}^{N} D_{ij}(x_j) \tag{8.5}$$

$$y_i(t) = h_i(x_i), \quad i = 1, 2, \ldots, N \tag{8.6}$$

where $x_i \in \Omega_i \subset \mathbb{R}^{n_i}$ (Ω_i is a neighbourhood of the origin), $y_i \in \mathbb{R}$ and $u_i \in U_i \subset \mathbb{R}$ (U_i is an admissible control set) are the state, output and input of the i-th subsystem respectively, $f_i(x_i) \in \mathbb{R}^{n_i}$ and $g_i(x_i) \in \mathbb{R}^{n_i}$ are smooth vector fields defined in the domain Ω_i, and $h_i(x_i) \in \mathbb{R}$ are smooth in the domain Ω_i for $i = 1, 2, \ldots, N$. The term $\Delta f_i(x_i)$ includes all the uncertainties experienced by the i-th subsystem. The term $\sum_{\substack{j=1 \\ j\neq i}}^{N} D_{ij}(x_j)$ is the nonlinear interconnection of the i-th subsystem.

Definition 8.2 The systems

$$\dot{x}_i(t) = f_i(x_i) + g_i(x_i)u_i + \Delta f_i(x_i) \tag{8.7}$$
$$y_i(t) = h_i(x_i), \qquad i = 1, 2, \ldots, N \tag{8.8}$$

are called the isolated subsystems of the systems (8.5) and (8.6), and the systems

$$\dot{x}_i(t) = f_i(x_i) + g_i(x_i)u_i \tag{8.9}$$
$$y_i(t) = h_i(x_i), \qquad i = 1, 2, \ldots, N \tag{8.10}$$

are called the nominal isolated subsystems of the systems (8.5) and (8.6).

In this chapter, under the assumption that the isolated subsystems (8.9) and (8.10) have uniform relative degree r_i in the considered domain Ω_i, the interconnected systems (8.5) and (8.6) are to be analysed. The objective is to explore the system structure based on a geometric transformation to design a robust asymptotic observer for the interconnected system (8.5) and (8.6).

It should be noted that the following results can be extended to the case where the isolated subsystems are multi-input and multi-output using the corresponding framework in Sect. 8.2 for the multi-input and multi-output case provided in [15].

8.4 System Analysis and Assumptions

In this section, some assumptions are imposed on the system (8.5) and (8.6) to facilitate the observer design.

Assumption 8.1 The nominal isolated subsystem (8.9) and (8.10) has uniform relative degree r_i in the domain $x_i \in \Omega_i$ for $i = 1, 2, \ldots, N$.

Under Assumption 8.1, it follows from Sect. 8.2 that there exists a coordinate transformation

$$T_i : x_i \to \mathrm{col}(\zeta_i, \eta_i) \tag{8.11}$$

where

$$\zeta_i = \begin{bmatrix} \zeta_{i1} \\ \zeta_{i2} \\ \vdots \\ \zeta_{ir_i} \end{bmatrix} = \begin{bmatrix} h_i(x_i) \\ L_f h_i(x_i) \\ \vdots \\ L_f^{r_i-1} h_i(x_i) \end{bmatrix} \in \mathbb{R}^{r_i} \tag{8.12}$$

and $\eta_i \in \mathbb{R}^{n_i - r_i}$ is defined by

$$\eta_i = \begin{bmatrix} \eta_{i1} \\ \eta_{i2} \\ \vdots \\ \eta_{n_i - r_i} \end{bmatrix} = \begin{bmatrix} \phi_{i(r_i+1)}(x_i) \\ \phi_{i(r_i+2)}(x_i) \\ \vdots \\ \phi_{in_i}(x_i) \end{bmatrix} \in \mathbb{R}^{n_i - r_i} \tag{8.13}$$

for $i = 1, 2, \ldots, N$. The functions $\phi_{i(r_i+1)}(x_i)$, $\phi_{i(r_i+2)}(x_i), \ldots, \phi_{in_i}(x_i)$ can be obtained by solving the following partial differential equations:

$$L_{g_i} \phi_i(x_i) = 0, \quad x_i \in \Omega_i, \quad i = 1, 2, \ldots, N. \tag{8.14}$$

From Sect. 8.2, it follows that in the new coordinate system (ζ_i, η_i), the nominal isolated subsystem (8.9) and (8.10) can be described by

$$\dot{\zeta}_i = A_i \zeta_i + \beta_i(\zeta_i, \eta_i, u_i) \tag{8.15}$$

$$\dot{\eta}_i = q_i(\zeta_i, \eta_i) \tag{8.16}$$

$$y_i = C_i \zeta_i \tag{8.17}$$

where

$$A_i = \begin{bmatrix} 0 & 1 & 0 & \dots & 0 \\ 0 & 0 & 1 & \dots & 0 \\ \vdots & \vdots & \vdots & \vdots & \vdots \\ 0 & 0 & 0 & \dots & 1 \\ 0 & 0 & 0 & \dots & 0 \end{bmatrix} \in \mathbb{R}^{r_i \times r_i}, \qquad C_i = \begin{bmatrix} 1 & 0 & \dots & 0 \end{bmatrix} \in \mathbb{R}^{1 \times r_i} \tag{8.18}$$

$$\beta_i(\zeta_i, \eta_i, u_i) = \begin{bmatrix} 0 \\ \vdots \\ 0 \\ L_{f_i}^{r_i} h_i(T_i^{-1}(\zeta_i, \eta_i)) + L_{g_i} L_{f_i}^{r_i - 1} h_i(T_i^{-1}(\zeta_i, \eta_i)) u_i \end{bmatrix} \tag{8.19}$$

It is clear to see that the pair (A_i, C_i) is observable. Thus, there exists a matrix L_i such that $A_i - L_i C_i$ is Hurwitz stable. This implies that, for any positive definite matrix $Q_i \in \mathbb{R}^{r_i \times r_i}$, the Lyapunov equation

$$(A_i - L_i C_i)^T P_i + P_i (A_i - L_i C_i) = -Q_i \tag{8.20}$$

has a unique positive-definite solution $P_i \in \mathbb{R}^{r_i \times r_i}$ for $i = 1, 2, \dots, N$.

Assumption 8.2 The uncertainty $\Delta f_i(x_i)$ in (8.5) satisfies

$$\frac{\partial T_i}{\partial x_i} \Delta f_i(x_i) = \begin{bmatrix} E_i \Delta \Psi_i(x_i) \\ 0 \end{bmatrix}. \tag{8.21}$$

where $T_i(\cdot)$ is defined in (8.11), $E_i \in \mathbb{R}^{r_i \times r_i}$ is a constant matrix satisfying

$$E_i^T P_i = H_i C_i \tag{8.22}$$

for some matrix H_i, with P_i satisfying (8.20), and $\|\Delta \Psi_i(x_i)\| \le \kappa_i(x_i)$, where $\kappa_i(x_i)$ is continuous and Lipschitz about x_i in the domain Ω_i for $i = 1, 2, \dots, N$.

Remark 8.1 Solving the Lyapunov equation (8.20) in the presence of the constraint (8.22) constitutes the well known constrained Lyapunov problem [10]. Although there is no general solution available for this problem, associated discussion and an algorithm can be found in [9].

Remark 8.2 Assumption 8.2 is a limitation on the uncertainty $\Delta f_i(x_i)$, which is used to guarantee the existence of asymptotic observers. Denote the nonlinear uncertain term $\Delta\Psi_i(x_i)$ in (8.21) in the new coordinate frame (ζ_i, η_i) by $\Delta\Phi_i(\zeta_i, \eta_i)$ i.e.

$$\Delta\Phi_i(\zeta_i, \eta_i) = [\Delta\Psi_i(x_i)]_{x_i = T_i^{-1}(\zeta_i, \eta_i)} \tag{8.23}$$

From Assumption 8.2, there exists a function $\rho_i(\zeta_i, \eta_i)$ such that

$$\|\Delta\Phi_i(\zeta_i, \eta_i)\| \le \rho_i(\zeta_i, \eta_i) \tag{8.24}$$

and $\rho_i(\zeta_i, \eta_i)$ satisfies the Lipschitz condition in $T_i(\Omega_i)$. Thus for any (ζ_i, η_i) and $(\hat{\zeta}_i, \hat{\eta}_i) \in T_i(\Omega_i)$,

$$\|\rho_i(\zeta_i, \eta_i) - \rho_i(\hat{\zeta}_i, \hat{\eta}_i)\| \le l_i^a \|\zeta_i - \hat{\zeta}_i\| + l_i^b \|\eta_i - \hat{\eta}_i\| \tag{8.25}$$

where both l_i^a and l_i^b are nonnegative constants.

Consider the interconnections $D_{ij}(x_j)$ in system (8.5). Partition $\frac{\partial T_i}{\partial x_i} D_{ij}(x_j)$ as follows

$$\frac{\partial T_i}{\partial x_i} D_{ij}(x_j)\big|_{x_j = T_j^{-1}(\zeta_j, \eta_j)} = \begin{bmatrix} \Gamma_{ij}^a(\zeta_j, \eta_j) \\ \Gamma_{ij}^b(\zeta_j, \eta_j) \end{bmatrix} \tag{8.26}$$

where $\Gamma_{ij}^a(\zeta_j, \eta_j) \in \mathbb{R}^{r_i}$, $\Gamma_{ij}^b(\zeta_j, \eta_j) \in \mathbb{R}^{n_i - r_i}$ for $i = 1, 2, \ldots, N$ and $i \ne j$.

Assumption 8.3 The nonlinear terms $\Gamma_{ij}^a(\zeta_j, \eta_j) \in \mathbb{R}^{r_i}$ and $\Gamma_{ij}^b(\zeta_j, \eta_j) \in \mathbb{R}^{n_i - r_i}$ in (8.26) satisfy the Lipschitz condition in $T_i(\Omega_i)$.

Assumption 8.3 implies that there exist nonnegative constants α_{ij}^a, α_{ij}^b, μ_{ij}^a and μ_{ij}^b such that

$$\|\Gamma_{ij}^a(\zeta_i, \eta_i) - \Gamma_{ij}^a(\hat{\zeta}_i, \hat{\eta}_i)\| \le \alpha_{ij}^a \|\zeta_j - \hat{\zeta}_j\| + \alpha_{ij}^b \, \| \eta_j - \hat{\eta}_j\| \tag{8.27}$$

$$\|\Gamma_{ij}^b(\zeta_i, \eta_i) - \Gamma_{ij}^b(\hat{\zeta}_i, \hat{\eta}_i)\| \le \mu_{ij}^a \|\zeta_j - \hat{\zeta}_j\| + \mu_{ij}^b \, \| \eta_j - \hat{\eta}_j\| \tag{8.28}$$

for $i = 1, 2, \ldots, N$ and $i \ne j$

Remark 8.3 It should be noted that Assumptions 8.2 and 8.3 have limitations to the coordinate transformation T_i in (8.11). However, in local case, the corresponding limitation is trivial and the Lipschitz conditions in Assumptions 8.2 and 8.3 can be satisfied if all the associated functions are smooth.

From (8.15)–(8.17) and the analysis above, it follows that under Assumption 8.2, in the new coordinate system (ζ_i, η_i), the system (8.5) and (8.6) can be described by

$$\dot{\zeta}_i = A_i \zeta_i + \beta_i(\zeta_i, \eta_i, u_i) + E_i \Delta\Psi_i(\zeta_i, \eta_i) + \sum_{\substack{j=1 \\ j \ne i}}^{N} \Gamma_{ij}^a(\zeta_j, \eta_j) \tag{8.29}$$

$$\dot{\eta}_i = q_i(\zeta_i, \eta_i) + \sum_{\substack{j=1 \\ j \neq i}}^{N} \Gamma_{ij}^b(\zeta_j, \eta_j) \tag{8.30}$$

$$y_i = C_i \zeta_i \tag{8.31}$$

where A_i and C_i are given in (8.18), $\beta_i(\cdot)$ is defined in (8.19) and $\Gamma_{ij}^a(\cdot)$ and $\Gamma_{ij}^b(\cdot)$ are defined in (8.26).

Remark 8.4 Since $\beta_i(\cdot)$ is continuous in the domain $T_i(\Omega_i)$, it is straightforward to see that there exists a subset in the domain $T_i(\Omega_i)$ such that the function $\beta_i(\cdot)$ is Lipschitz in the subset

$$\| \beta_i(\zeta_i, \eta_i, u_i) - \beta_i(\hat{\zeta}_i, \hat{\eta}_i, u_i) \| \leq v_i^a(u_i) \| \zeta_i - \hat{\zeta}_i \| + v_i^b(u_i) \| \eta_i - \hat{\eta}_i \| \tag{8.32}$$

where $v_i^a(u_i)$ and $v_i^b(u_i)$ are known nonnegative functions of u_i for $i = 1, 2, \ldots, N$.

Assumption 8.4 The function $q_i(\zeta_i, \eta_i)$ in Eq. (8.30) has the following decomposition

$$q_i(\zeta_i, \eta_i) = M_i \eta_i + \theta_i(\zeta_i, \eta_i) \tag{8.33}$$

where $M_i \in \mathbb{R}^{(n_i - r_i) \times (n_i - r_i)}$ is a Hurwitz matrix and $\theta_i(\zeta_i, \eta_i)$ is Lipschitz in the domain $T_i(\Omega_i)$.

Under Assumption 8.4, there exist nonnegative constants τ_i^a and τ_i^b such that.

$$\| \theta_i(\zeta_i, \eta_i) - \theta_i(\hat{\zeta}_i, \hat{\eta}_i) \| \leq \tau_i^a \| \zeta_i - \hat{\zeta}_i \| + \tau_i^b \| \eta_i - \hat{\eta}_i \| \tag{8.34}$$

for $i = 1, 2, \ldots, N$. Further, from the fact that M_i is Hurwitz stable for $\Lambda_i > 0$, the following Lyapunov equation has a unique solution $\Pi_i > 0$

$$M_i^T \Pi_i + \Pi_i M_i = -\Lambda_i, \qquad i = 1, 2, \ldots, N. \tag{8.35}$$

8.5 Nonlinear Observer Synthesis

In this section, an observer is designed for the transformed systems (8.29)–(8.31) which provides asymptotic estimation of the states of the interconnected systems (8.29)–(8.31). For the system (8.29)–(8.31), construct dynamical systems

$$\begin{aligned} \dot{\hat{\zeta}}_i = {} & A_i \hat{\zeta}_i + L_i(y_i - C_i \hat{\zeta}_i) + \beta_i(\hat{\zeta}_i, \hat{\eta}_i, u_i) + K_i(y, \hat{\zeta}_i, \hat{\eta}_i) \\ & + \sum_{\substack{j=1 \\ j \neq i}}^{N} \Gamma_{ij}^a(\hat{\zeta}_j, \hat{\eta}_j) \end{aligned} \tag{8.36}$$

$$\dot{\hat{\eta}}_i = M_i\hat{\eta}_i + \theta_i(\hat{\zeta}_i, \hat{\eta}_i) + \sum_{\substack{j=1\\ j\neq i}}^{N} \Gamma_{ij}^b(\hat{\zeta}_j, \hat{\eta}_j) \tag{8.37}$$

where the term $K_i(y_i, \hat{\zeta}_i, \hat{\eta}_i)$ is defined by

$$K_i(y_i, \hat{\zeta}_i, \hat{\eta}_i) = \begin{cases} \frac{P_i^{-1}C_i^T(y_i - C_i\hat{\zeta}_i)}{\|y_i - C_i\hat{\zeta}_i\|} \parallel H_i \parallel \rho_i(\hat{\zeta}_i, \hat{\eta}_i), & y_i - C_i\hat{\zeta}_i \neq 0 \\ 0, & y_i - C_i\hat{\zeta}_i = 0 \end{cases} \tag{8.38}$$

where P_i and H_i satisfy (8.20) and (8.22) respectively.

The following results are ready to be presented.

Theorem 8.1 *Suppose Assumptions 8.1– 8.4 hold. Then, the dynamical system (8.36) and (8.37) is a robust asymptotic observer of system (8.29), (8.30) and (8.31), if the function matrix* $\mathbb{W}^T(\cdot) + \mathbb{W}(\cdot)$ *is positive definite in the domain* $T(\Omega) \times U := T(\Omega_1) \times U_1 \times T(\Omega_2) \times U_2 \times \ldots \times T(\Omega_N) \times U_N$, *where the matrix* $\mathbb{W}(\cdot) = \left[w_{ij}(\cdot)\right]_{2N\times 2N}$, *and its entries* $w_{ij}(\cdot)$ *are defined by*

$$w_{ij} = \begin{cases} \lambda_{\min}(Q_i) - 2\lambda_{\max}(P_i)v_i^a(\cdot) - 2l_i^a\|C_i\|\|H_i\|, \; i = j, \; 1 \le i \le N \\ -2\lambda_{\max}(P_i)\alpha_{ij}^a, \; i \neq j, \; 1 \le i \le N, 1 \le j \le N \\ \lambda_{\min}(\Lambda_{i-N}) - 2\lambda_{\max}(\Pi_{i-N})\tau_{i-N}^b, \; i = j, \; N+1 \le i \le 2N \\ -2\lambda_{\max}(\Pi_{(i-N)})\mu_{(i-N)(j-N)}^b, \\ \qquad\qquad i \neq j, \; N+1 \le i \le 2N, N+1 \le j \le 2N \\ -2[\lambda_{\max}(P_i)v_i^b(\cdot) + l_i^b\|C_i\|\|H_i\| + \lambda_{\max}(\Pi_i)\tau_i^a], \\ \qquad\qquad j - i = N, \; 1 \le i \le N, N+1 \le j \le 2N \\ -2\lambda_{\max}(P_i)\alpha_{i(j-N)}^b, \qquad j - i \neq N, \; 1 \le i \le N, N+1 \le j \le 2N \\ 0, \qquad i - j = N, \; N+1 \le i \le 2N, 1 \le j \le N \\ -2\lambda_{\max}(\Pi_{i-N})\mu_{(i-N)j}^a, \; i - j \neq N, \; N+1 \le i \le 2N, 1 \le j \le N \end{cases}$$

Proof Let $e_{\zeta_i} = \zeta_i - \hat{\zeta}_i$ and $e_{\eta_i} = \eta_i - \hat{\eta}_i$ for $i = 1, 2, \ldots, N$. Compare systems (8.29), (8.30) and (8.36), (8.37). It follows that the error dynamical systems are described by

$$\begin{aligned} \dot{e}_{\zeta_i} &= (A_i - L_iC_i)e_{\zeta_i} + \beta_i(\zeta_i, \eta_i, u_i) - \beta_i(\hat{\zeta}_i, \hat{\eta}_i, u_i) + E_i\Delta\Psi_i(\zeta_i, \eta_i) \\ &\quad - K_i(y_i, \hat{\zeta}_i, \hat{\eta}_i) + \sum_{\substack{j=1\\ j\neq i}}^{N} \Gamma_{ij}^a(\zeta_j, \eta_j) - \sum_{\substack{j=1\\ j\neq i}}^{N} \Gamma_{ij}^a(\hat{\zeta}_j, \hat{\eta}_j) \end{aligned} \tag{8.39}$$

$$\dot{e}_{\eta_i} = M_ie_{\eta_i} + \theta_i(\zeta_i, \eta_i) - \theta_i(\hat{\zeta}_i, \hat{\eta}_i) + \sum_{\substack{j=1\\ j\neq i}}^{N} \Gamma_{ij}^b(\zeta_j, \eta_j)$$

$$-\sum_{\substack{j=1\\j\neq i}}^{N}\Gamma_{ij}^{b}(\hat{\zeta}_j,\hat{\eta}_j) \tag{8.40}$$

Now, for the system (8.39) and (8.40) consider the following candidate Lyapunov function

$$V=\sum_{i=1}^{N}e_{\zeta_i}^{T}P_ie_{\zeta_i}+\sum_{i=1}^{N}e_{\eta_i}^{T}\Pi_ie_{\eta_i} \tag{8.41}$$

Then, the time derivative of the candidate Lyapunov function can be described by

$$\dot{V}=\sum_{i=1}^{N}[(\dot{e}_{\zeta_i}^{T}P_ie_{\zeta_i}+e_{\zeta_i}^{T}P_i\dot{e}_{\zeta_i})+(\dot{e}_{\eta_i}^{T}\Pi_ie_{\eta_i}+e_{\eta_i}^{T}\Pi_i\dot{e}_{\eta_i})] \tag{8.42}$$

Substituting both $\dot{e}_{\zeta_i}$ in (8.39) and $\dot{e}_{\eta_i}$ in (8.40) into Eq. (8.42), it follows by direct computation that the time derivative of the function V in (8.41) can be described by

$$\begin{aligned}\dot{V}=\sum_{i=1}^{N}\Big\{&e_{\zeta_i}^{T}[(A_i-L_iC_i)^TP_i+P_i(A_i-L_iC_i)]e_{\zeta_i}+2e_{\zeta_i}^{T}P_i[\beta_i(\zeta_i,\eta_i,u_i)\\&-\beta_i(\hat{\xi}_i,\hat{\eta}_i,u_i)]+2[e_{\zeta_i}^{T}P_iE_i\Delta\Psi_i(\zeta_i,\eta_i)-e_{\zeta_i}^{T}P_iK_i(y_i,\hat{\xi}_i,\hat{\eta}_i)]+2e_{\zeta_i}^{T}P_i\\&\times\sum_{\substack{j=1\\j\neq i}}^{N}[\Gamma_{ij}^{a}(\zeta_j,\eta_j)-\Gamma_{ij}^{a}(\hat{\zeta}_j,\hat{\eta}_j)]+e_{\eta_i}^{T}(M_i^T\Pi_i+\Pi_iM_i)e_{\eta_i}+2e_{\eta_i}^{T}\Pi_i\\&\times[\theta_i(\zeta_i,\eta_i)-\theta_i(\hat{\xi}_i,\hat{\eta}_i)]+2e_{\eta_i}^{T}\Pi_i\sum_{\substack{j=1\\j\neq i}}^{N}[\Gamma_{ij}^{b}(\zeta_j,\eta_j)-\Gamma_{ij}^{b}(\hat{\zeta}_j,\hat{\eta}_j)]\Big\}\end{aligned} \tag{8.43}$$

From (8.22) and (8.38), it follows that:
(i) If $y_i-C_i\hat{\xi}_i=0$, then from (8.22) and $e_{\zeta_i}^{T}C_i^{T}=(y_i-C_i\hat{\xi}_i)^T$

$$e_{\zeta_i}^{T}P_iE_i\Delta\Phi_i(\zeta_i,\eta_i)-e_{\zeta_i}^{T}P_iK_i(y_i,\hat{\xi}_i,\hat{\eta}_i)=e_{\zeta_i}^{T}C_i^{T}H_i^{T}\Delta\Phi_i(\zeta_i,\eta_i) \tag{8.44}$$

$$=(H_i(y_i-C_i\hat{\xi}_i))^T\Delta\Phi_i(\zeta_i,\eta_i)=0 \tag{8.45}$$

(ii) If $y_i-C_i\hat{\xi}_i\neq 0$, then from (8.22), (8.24), (8.25) and (8.38)

$$\begin{aligned}&e_{\zeta_i}^{T}P_iE_i\Delta\Phi_i(\zeta_i,\eta_i)-e_{\zeta_i}^{T}P_iK_i(y_i,\hat{\xi}_i,\hat{\eta}_i)\\&=e_{\zeta_i}^{T}C_i^{T}H_i^{T}\Delta\Phi_i(\zeta_i,\eta_i)-e_{\zeta_i}^{T}P_i\frac{P_i^{-1}C_i^{T}(y_i-C_i\hat{\xi}_i)}{\|y_i-C_i\hat{\xi}_i\|}\parallel H_i\parallel\rho_i(\hat{\xi}_i,\hat{\eta}_i)\end{aligned}$$

$$
\begin{aligned}
&= (C_i e_{\zeta_i})^T H_i^T \Delta\Phi_i(\zeta_i, \eta_i) - \frac{e_{\zeta_i}^T C_i^T C_i e_{\zeta_i}}{\|C_i e_{\zeta_i}\|} \|H_i\| \rho_i(\hat{\zeta}_i, \hat{\eta}_i) \\
&\le \|C_i e_{\zeta_i}\| \|H_i\| \{\rho_i(\zeta_i, \eta_i) - \rho_i(\hat{\zeta}_i, \hat{\eta}_i)\} \\
&\le \|C_i e_{\zeta_i}\| \|H_i\| \{l_i^a \|\zeta_i - \hat{\zeta}_i\| + l_i^b \|\eta_i - \hat{\eta}_i\|\}
\end{aligned}
$$

Then, from (i) and (ii) above, it follows that

$$
\begin{aligned}
e_{\zeta_i}^T P_i E_i \Delta\Phi_i(\zeta_i, \eta_i) - e_{\zeta_i}^T P_i K_i(y_i, \hat{\zeta}_i, \hat{\eta}_i)& \\
\le \|C_i e_{\zeta_i}\| \|H_i\| \big(l_i^a \|e_{\zeta_i}\| + l_i^b \|e_{\eta_i}\|\big)& \qquad (8.46)
\end{aligned}
$$

Substituting (8.27), (8.28), (8.32), (8.34), and (8.46) into (8.43) yields

$$
\begin{aligned}
\dot{V} \le \sum_{i=1}^{N} \Big\{ &- e_{\zeta_i}^T Q_i e_{\zeta_i} + 2\|e_{\zeta_i}\| \|P_i\| \big[v_i^a \|e_{\zeta_i}\| + v_i^b \|e_{\eta_i}\|\big] + 2\|e_{\zeta_i}\| \|C_i\| \|H_i\| \\
&\times \big[l_i^a \|e_{\zeta_i}\| + l_i^b \|e_{\eta_i}\|\big] + 2\|e_{\zeta_i}\| \|P_i\| \sum_{\substack{j=1\\ j\neq i}}^{N} \big[\alpha_{ij}^a \|e_{\zeta_j}\| + \alpha_{ij}^b \|e_{\eta_j}\|\big] - e_{\eta_i}^T \Lambda_i e_{\eta_i} \\
&+ 2e_{\eta_i}^T \|\Pi_i\| \big[\tau_i^a \|e_{\zeta_i}\| + \tau_i^b \|e_{\eta_i}\|\big] + 2e_{\eta_i}^T \|\Pi_i\| \sum_{\substack{j=1\\ j\neq i}}^{N} \big[\mu_{ij}^a \|e_{\zeta_j}\| + \mu_{ij}^b \|e_{\eta_j}\|\big] \Big\} \\
\le \sum_{i=1}^{N} \Big\{ &- e_{\zeta_i}^T Q_i e_{\zeta_i} + 2v_i^a \|e_{\zeta_i}\|^2 \|P_i\| + 2v_i^b \|e_{\zeta_i}\| \|e_{\eta_i}\| \|P_i\| + 2l_i^a \|e_{\zeta_i}\|^2 \|C_i\| \\
&\times \|H_i\| + 2l_i^b \|e_{\zeta_i}\| \|e_{\eta_i}\| \|C_i\| \|H_i\| + \sum_{\substack{j=1\\ j\neq i}}^{N} \big[2\alpha_{ij}^a \|e_{\zeta_i}\| \|e_{\zeta_j}\| \|P_i\| + 2\alpha_{ij}^b \|e_{\zeta_i}\| \\
&\times \|e_{\eta_j}\| \|P_i\|\big] - e_{\eta_i}^T \Lambda_i e_{\eta_i} + 2\tau_i^a \|\Pi_i\| \|e_{\zeta_i}\| \|e_{\eta_i}\| + 2\tau_i^b \|\Pi_i\| \|e_{\eta_i}\|^2 \\
&+ \sum_{\substack{j=1\\ j\neq i}}^{N} \big[2\mu_{ij}^a \|\Pi_i\| \|e_{\zeta_j}\| \|e_{\eta_i}\| + 2\mu_{ij}^b \|\Pi_i\| \|e_{\eta_i}\| \|e_{\eta_j}\|\big] \Big\} \\
\le \sum_{i=1}^{N} \Big\{ &- \big[\lambda_{\min}(Q_i) - 2\lambda_{\max}(P_i) v_i^a - 2l_i^a \|C_i\| \|H_i\|\big] \|e_{\zeta_i}\|^2 \\
&+ \Big[\sum_{\substack{j=1\\ j\neq i}}^{N} 2\lambda_{\max}(P_i) \alpha_{ij}^a\Big] \|e_{\zeta_i}\| \|e_{\zeta_j}\| + \big[2\lambda_{\max}(P_i) v_i^b + 2l_i^b \|C_i\| \|H_i\| \\
&+ 2\lambda_{\max}(\Pi_i) \tau_i^a\big] \|e_{\zeta_i}\| \|e_{\eta_i}\| + 2\sum_{\substack{j=1\\ j\neq i}}^{N} \lambda_{\max}(P_i) \alpha_{ij}^b \|e_{\zeta_i}\| \|e_{\eta_j}\|
\end{aligned}
$$

$$+\sum_{\substack{j=1\\j\neq i}}^{N} 2\lambda_{\max}(\Pi_i)\mu_{ij}^{a}\|e_{\zeta_j}\|\|e_{\eta_i}\| - [\lambda_{\min}(\Lambda_i) - 2\lambda_{\max}(\Pi_i)\tau_i^{b}]\|e_{\eta_i}\|^2$$

$$+\Big[\sum_{\substack{j=1\\j\neq i}}^{N} 2\lambda_{\max}(\Pi_i)\mu_{ij}^{b}]\|e_{\eta_i}\|\|e_{\eta_j}\|\Big\}$$

$$\leq -\sum_{i=1}^{N}\Big\{\big[\lambda_{\min}(Q_i) - 2\lambda_{\max}(P_i)\nu_i^{a} - 2l_i^{a}\|C_i\|\|H_i\|\big]\|e_{\zeta_i}\|^2$$

$$-\Big[\sum_{\substack{j=1\\j\neq i}}^{N} 2\lambda_{\max}(P_i)\alpha_{ij}^{a}]\|e_{\zeta_i}\|\|e_{\zeta_j}\| - [2\lambda_{\max}(P_i)\nu_i^{b} + 2l_i^{b}\|C_i\|\|H_i\|$$

$$+2\lambda_{\max}(\Pi_i)\tau_i^{a}]\|e_{\zeta_i}\|\|e_{\eta_i}\|]\| - 2\sum_{\substack{j=1\\j\neq i}}^{N}\lambda_{\max}(P_i)\alpha_{ij}^{b}\|e_{\zeta_i}\|\|e_{\eta_j}\|]$$

$$-\sum_{\substack{j=1\\j\neq i}}^{N} 2\lambda_{\max}(\Pi_i)\mu_{ij}^{a}\|e_{\zeta_j}\|\|e_{\eta_i}\| + [\lambda_{\min}(\Lambda_i) - 2\lambda_{\max}(\Pi_i)\tau_i^{b}]\|e_{\eta_i}\|^2$$

$$-[\sum_{\substack{j=1\\j\neq i}}^{N} 2\lambda_{\max}(\Pi_i)\mu_{ij}^{b}]\|e_{\eta_i}\|\|e_{\eta_j}\|\Big\}$$

Then, from the definition of the matrix $\mathbb{W}(\cdot)$ and the inequality above, it follows that

$$\dot{V} \leq -\frac{1}{2}X^T[\mathbb{W}^T(\cdot) + \mathbb{W}(\cdot)]X$$

where $X = [\|e_{\zeta_1}\|, \|e_{\zeta_2}\|, \ldots, \|e_{\zeta_N}\|, \|e_{\eta_1}\|, \|e_{\eta_2}\|, \ldots, \|e_{\eta_N}\|]^T$. Since $\mathbb{W}^T(\cdot) + \mathbb{W}(\cdot)$ is positive definite in the domain $T(\Omega) \times U$, it is clear that $\dot{V}|_{(8.39)-(8.40)}$ is negative definite. Therefore, the error system (8.39) and (8.40) is asymptotically stable, that is,

$$\lim_{t\to\infty}\|\zeta_i(t) - \hat{\zeta}_i(t)\| = 0 \quad \text{and} \quad \lim_{t\to\infty}\|\eta_i(t) - \hat{\eta}_i(t)\| = 0 \tag{8.47}$$

Hence, the conclusion follows.

Remark 8.5 It is clear that the structure of system (8.36) and (8.37) is variable due to the term in (8.38). Theorem 8.1 shows that system (8.36) and (8.37) is an asymptotic observer of the interconnected system (8.29)–(8.31). Therefore, system (8.36) and (8.37) is called a variable structure observer throughout this chapter.

Now, consider the interconnected system (8.5) and (8.6). Assume that $\frac{\partial T_i^{-1}(\zeta_i,\eta_i)}{\partial(\zeta_i,\eta_i)}$ is bounded in $T_i(\Omega_i)$ for $i = 1, 2, \ldots, N$. There exists a positive constant γ_i such that

$$\left\| \frac{\partial T_i^{-1}(\zeta_i, \eta_i)}{\partial(\zeta_i, \eta_i)} \right\| \le \gamma_i, \qquad (\zeta_i, \eta_i) \in T_i(\Omega_i), \; i = 1, 2, \ldots, N$$

Define $\hat{x}_i = T_i^{-1}(\hat{\zeta}_i, \hat{\eta}_i), \; i = 1, 2, \ldots, N$. Then,

$$\|x_i - \hat{x}_i\| = \|T_i^{-1}(\zeta_i, \eta_i) - T_i^{-1}(\hat{\zeta}_i, \hat{\eta}_i)\| \le \gamma_i(\|\zeta_i - \hat{\zeta}_i\| + \|\eta_i - \hat{\eta}_i\|) \quad (8.48)$$

From (8.47) and (8.48), it follows that

$$\lim_{t \to \infty} \|x_i(t) - \hat{x}_i(t)\| = 0$$

This implies that $\hat{x}_i$ is an asymptotic estimate of x_i for $i = 1, 2, \ldots, N$. Therefore,

$$\hat{x}_i = T_i^{-1}(\hat{\zeta}_i, \hat{\eta}_i)$$

provides an asymptotic estimate of the states x_i of the system (8.5) and (8.6), where $\hat{\zeta}_i$ and $\hat{\eta}_i$ are given by (8.36) and (8.37) for $i = 1, 2, \ldots, N$.

Remark 8.6 From the analysis above, it is clear to see that, in this chapter, it is not required that either the nominal isolated subsystems or the interconnections are linearisable. The uncertainties are bounded by nonlinear functions and are fully used in the observer design in order to reject the effects of the uncertainties, and thus robustness is enhanced. Although the designed observer is a local asymptotic observer, the developed results can be extended to the global case if the associated conditions hold globally.

8.6 Simulation Examples

8.6.1 A Numerical Example

Consider the nonlinear interconnected systems:

$$\dot{x}_1 = \underbrace{\begin{bmatrix} x_{12} \\ -0.1 \sin x_{12} \\ -3x_{11}^2 - 3.25x_{13} - 2x_{12} \end{bmatrix}}_{f_1(x_1)} + \underbrace{\begin{bmatrix} 0 \\ 1 \\ 0 \end{bmatrix}}_{g_1(x_1)} u_1 + \underbrace{\begin{bmatrix} \Delta\sigma_1 \\ 0.5\Delta\sigma_1 \\ -2\Delta\sigma_1 \end{bmatrix}}_{\Delta f_1(x_1)}$$
$$+ \underbrace{\begin{bmatrix} 0.2(x_{21}^2 + x_{22}) \\ 0 \\ 0.1 \sin x_{21} \end{bmatrix}}_{D_{12}(x_2)} \quad (8.49)$$

$$y_1 = \underbrace{x_{11}}_{h_1(x_1)} \tag{8.50}$$

$$\dot{x}_2 = \underbrace{\begin{bmatrix} -x_{21} \\ -x_{21}^2 - 3x_{22} + \cos(x_{21}^2 + x_{22}) - 1 \\ -2x_{23} + 0.2x_{21}^2 \end{bmatrix}}_{f_2(x_2)} + \underbrace{\begin{bmatrix} 1 \\ -2x_{21} \\ 0 \end{bmatrix}}_{g_2(x_2)} u_2 + \underbrace{\begin{bmatrix} -\Delta\sigma_2 \\ 2x_{21}\Delta\sigma_2 \\ 0 \end{bmatrix}}_{\Delta f_2(x_2)}$$

$$+ \underbrace{\begin{bmatrix} 0 \\ 0.1\sin(x_{13} + 2x_{11}) \\ 0 \end{bmatrix}}_{D_{21}(x_1)} \tag{8.51}$$

$$y_2 = \underbrace{x_{21}}_{h_2(x_2)} \tag{8.52}$$

where $x_1 = \text{col}(x_{11}, x_{12}, x_{13})$ and $x_2 = \text{col}(x_{21}, x_{22}, x_{23})$, $h_1(x_1)$ and $h_2(x_2)$, and $u_1(t)$ and $u_2(t)$ are the system state, output and input respectively, $D_{12}(\cdot)$ and $D_{21}(\cdot)$ are interconnected terms and $\Delta f_1(x_1)$ and $\Delta f_2(x_2)$ are the uncertainties experienced by the system which satisfy

$$||\Delta f_1(x_1)|| = 0.1|x_{13} + 2x_{11}|\sin^2 t \tag{8.53}$$

$$||\Delta f_2(x_2)|| = 0.1x_{21}^2|\cos t| \tag{8.54}$$

The domain considered is

$$\Omega = \Big\{(x_{11},\ x_{12},\ x_{13},\ x_{21},\ x_{22},\ x_{23}),\ \Big|\ |x_{11}| < 3,$$
$$|x_{21}| \le 1.3,\ x_{12},\ x_{13},\ x_{22},\ x_{23} \in \mathbb{R}\Big\} \tag{8.55}$$

By direct computation, it follows that the first subsystem has uniform relative degree 2, and the second subsystem has uniform relative degree 1. The corresponding transformations are obtained as follows:

$$T_1 : \begin{cases} \zeta_{11} = x_{11} \\ \zeta_{12} = x_{12} \\ \eta_1 = x_{13} + 2x_{11} \end{cases}, \qquad T_2 : \begin{cases} \zeta_2 = x_{21} \\ \eta_{21} = x_{21}^2 + x_{22} \\ \eta_{22} = x_{23} \end{cases}$$

In the new coordinates, the system (8.49)–(8.52) can be described by:

$$\dot{\zeta}_1 = \underbrace{\begin{bmatrix} 0 & 1 \\ 0 & 0 \end{bmatrix}}_{A_1} \begin{bmatrix} \zeta_{11} \\ \zeta_{12} \end{bmatrix} + \underbrace{\begin{bmatrix} 0 \\ -0.1\sin\zeta_{11} + u_1 \end{bmatrix}}_{\beta_1} + \underbrace{\begin{bmatrix} \Delta\sigma_1(\zeta_1, \eta_1) \\ 0.5\Delta\sigma_1(\zeta_1, \eta_1) \end{bmatrix}}_{E_1\Delta\Psi(\zeta_1, \eta_1)}$$

$$+\underbrace{\begin{bmatrix} 0.2\eta_{21} \\ 0 \end{bmatrix}}_{\Gamma_{12}^a} \tag{8.56}$$

$$\dot{\eta}_1 = \underbrace{-3.25\eta_1 + 0.25\zeta_{11}^2}_{q_1(\zeta_1,\eta_1)} + \underbrace{0.4\eta_{21} + 0.1\sin\zeta_2}_{\Gamma_{12}^b} \tag{8.57}$$

$$y_1 = \begin{bmatrix} 1 & 0 \end{bmatrix} \begin{bmatrix} \zeta_{11} \\ \zeta_{12} \end{bmatrix} \tag{8.58}$$

$$\dot{\zeta}_2 = \underbrace{-}_{A_2}\zeta_2 + \underbrace{u_2}_{\beta_2} - \underbrace{\Delta\sigma_2(\zeta_2, \eta_2)}_{E_2\Delta\Psi(\zeta_2,\eta_2)} \tag{8.59}$$

$$\dot{\eta}_2 = \underbrace{\begin{bmatrix} -3 & 0 \\ 0 & -2 \end{bmatrix} \begin{bmatrix} \eta_{21} \\ \eta_{22} \end{bmatrix} + \begin{bmatrix} \cos\eta_{21} - 1 \\ 0.2\zeta_2^2 \end{bmatrix}}_{q_2(\zeta_2,\eta_2)} + \underbrace{\begin{bmatrix} 0.1\sin\eta_1 \\ 0 \end{bmatrix}}_{\Gamma_{21}^b} \tag{8.60}$$

$$y_2 = \zeta_2 \tag{8.61}$$

where $\zeta_1 = (\zeta_{11}, \zeta_{12})^T$, $\eta_1 \in \mathbb{R}$, $\zeta_2 \in \mathbb{R}$, and $\eta_2 = (\eta_{21}, \eta_{22})^T$.
From (8.53) and (8.54)

$$\|\Delta\Psi_1(\zeta_1, \eta_1)\| \leq ||\Delta\sigma_1(\zeta_1, \eta_1)|| \leq \underbrace{0.1|\eta_1|^2 \cos t}_{\rho_1(\cdot)}$$

$$\|\Delta\Psi_2(\zeta_2, \eta_2)\| \leq ||\Delta\sigma_2(\zeta_2, \eta_2)|| \leq \underbrace{0.1\zeta_2|\sin t|^2}_{\rho_2(\cdot)}$$

Then, for the first subsystem, choose $L_1 = \begin{bmatrix} 3 & 2 \end{bmatrix}^T$ and $Q = I$. It follows that the Lyapunov Eq. (8.20) has a unique solution:

$$P_1 = \begin{bmatrix} 0.5 & -0.5 \\ -0.5 & 1 \end{bmatrix}$$

and the solution to Eq. (8.22) is $H_1 = 0.25$. As $M_1 = -3.25$, let $\Lambda_1 = 3.25$. Thus the solution of equation (8.35) is $\Pi_1 = 0.5$. Now, for the second subsystem, choose $L_2 = 0$ and $Q_2 = 2$. It follows that the Lyapunov equation (8.20) has a unique solution $P_2 = 1$ and the solution to Eq.(8.22) is $H_2 = -1$. As

$$M_2 = \begin{bmatrix} -3 & 0 \\ 0 & -2 \end{bmatrix}, \quad \text{let} \quad \Lambda_2 = \begin{bmatrix} 1 & 0 \\ 0 & 1 \end{bmatrix}$$

Then

$$\Pi_2 = \begin{bmatrix} 0.1667 & 0 \\ 0 & 0.25 \end{bmatrix}$$

By direct computation, it follows that the matrix $\mathbb{W}^T(\cdot) + \mathbb{W}(\cdot)$ is positive definite in the domain Ω defined in (8.55). Thus, all the conditions of Theorem 8.1 are satisfied. This implies that the dynamical system

$$\dot{\hat{\zeta}}_1 = \begin{bmatrix} 0 & 1 \\ 0 & 0 \end{bmatrix} \begin{bmatrix} \hat{\zeta}_{11} \\ \hat{\zeta}_{12} \end{bmatrix} + \begin{bmatrix} 3 \\ 2 \end{bmatrix} (y_1 - C_1 \hat{\zeta}_1) + \begin{bmatrix} 0 \\ u_1 \end{bmatrix} + K_1(\cdot) + \begin{bmatrix} 0.2\hat{\eta}_{21} \\ 0 \end{bmatrix} \quad (8.62)$$

$$\dot{\hat{\eta}}_1 = -3.25\hat{\eta}_1 + 0.25\hat{\zeta}_{11}^2 + 0.4\hat{\eta}_{21} + 0.1 \sin \hat{\zeta}_2 \quad (8.63)$$

$$\dot{\hat{\zeta}}_2 = -\hat{\zeta}_2 + u_2 + K_2(\cdot) \quad (8.64)$$

$$\dot{\hat{\eta}}_2 = \begin{bmatrix} -3 & 0 \\ 0 & -2 \end{bmatrix} \begin{bmatrix} \hat{\eta}_{21} \\ \hat{\eta}_{22} \end{bmatrix} + \begin{bmatrix} \cos \hat{\eta}_{21} - 1 \\ 0.2\hat{\zeta}_2^2 \end{bmatrix} + \begin{bmatrix} 0.1 \sin \hat{\eta}_1 \\ 0 \end{bmatrix} \quad (8.65)$$

is a robust observer of the system (8.56)–(8.61) where $\hat{\zeta}_1 = \text{col}(\hat{\zeta}_{11}, \hat{\zeta}_{12})$, $\hat{\eta}_2 = \text{col}(\hat{\eta}_{21}, \hat{\eta}_{22})$, and $K_1(\cdot)$ and $K_2(\cdot)$ defined in (8.38) are as follows

$$K_1(y_1, \hat{\zeta}_1, \hat{\eta}_1) = \begin{cases} \begin{bmatrix} 0.1 \\ 0.05 \end{bmatrix} \frac{(\zeta_{11} - \hat{\zeta}_{11})}{\|\zeta_{11} - \hat{\zeta}_{11}\|} |\eta_1| \sin^2 t, & \zeta_{11} - \hat{\zeta}_{11} \neq 0 \\ 0, & \zeta_{11} - \hat{\zeta}_{11} = 0 \end{cases}$$

$$K_2(y_2, \hat{\zeta}_2, \hat{\eta}_2) = \begin{cases} 0.1 \frac{(\zeta_2 - \hat{\zeta}_2)}{\|\zeta_2 - \hat{\zeta}_2\|} \zeta_2^2 |\cos t|, & \zeta_2 - \hat{\zeta}_2 \neq 0 \\ 0, & \zeta_2 - \hat{\zeta}_2 = 0 \end{cases}$$

Therefore,

$$\begin{aligned} \hat{x}_{11} &= \hat{\zeta}_{11} \\ \hat{x}_{12} &= \hat{\zeta}_{12} \\ \hat{x}_{13} &= \hat{\eta}_1 - 2\hat{\zeta}_{11} \end{aligned} \qquad \text{and} \qquad \begin{aligned} \hat{x}_{21} &= \hat{\zeta}_2 \\ \hat{x}_{22} &= \hat{\eta}_{21} - \hat{\zeta}_2^2 \\ \hat{x}_{23} &= \hat{\eta}_{22} \end{aligned}$$

with $\hat{\zeta}_1 = \text{col}(\hat{\zeta}_{11}, \hat{\zeta}_{12})$, $\hat{\eta}_1$, $\hat{\zeta}_2$ and $\hat{\eta}_2 = \text{col}(\hat{\eta}_{21}, \hat{\eta}_{22})$ given by system (8.62)–(8.65), provide an asymptotic estimate for x_1 and x_2 of system (8.49)–(8.52).

For simulation purposes, the controllers are chosen as:

$$u_1 = -\zeta_{11} - 2\zeta_{12} \quad \text{and} \quad u_2 = -\zeta_{21} \quad (8.66)$$

The simulation results in Figs. 8.1 and 8.2 show that the designed observer estimates the states of the interconnected system in (8.49)–(8.52), $x_1 = \text{col}(x_{11}, x_{12}, x_{13})$ and $x_2 = \text{col}\ (x_{21}, x_{22}, x_{23})$ respectively, very well. Figure 8.3 shows the error between the actual states and the estimated states in both subsystems.

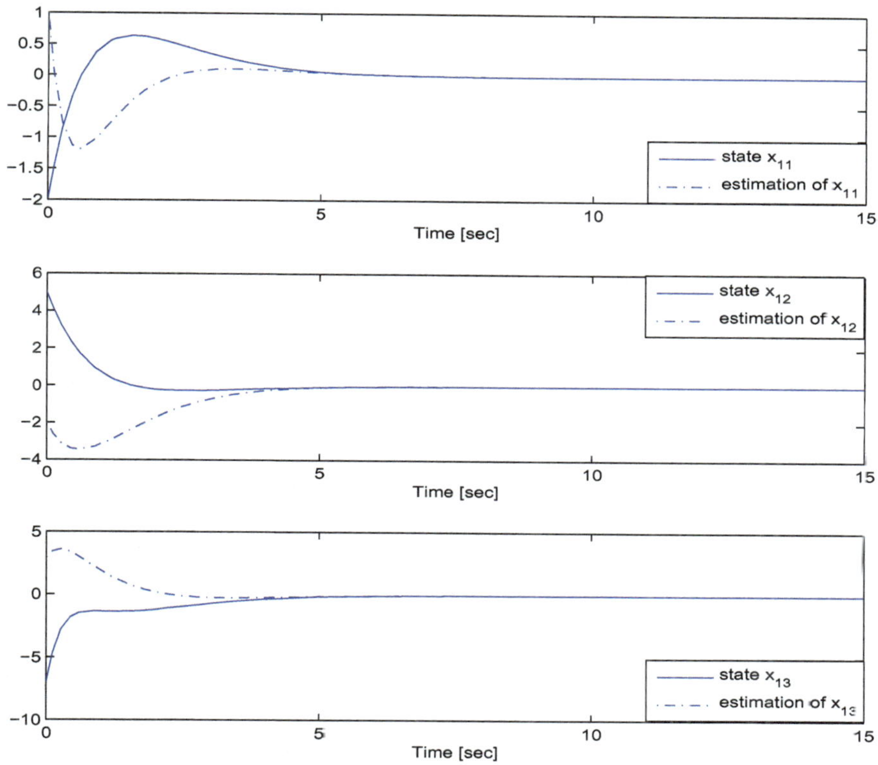

Fig. 8.1 The time response of the states of the first subsystem, $x_1 = \text{col}\,(x_{11}, x_{12}, x_{13})$, and their estimates $\hat{x}_1 = \text{col}\,(\hat{x}_{11}, \hat{x}_{12}, \hat{x}_{13})$

8.6.2 Case Study: Observer Design for Coupled Inverted Pendula

Consider the system given in Fig. 8.4 formed by two identical inverted pendula which are connected by a spring and subject to distinct inputs u_1 and u_2 (see, e.g. [11]). A salient feature of the system is that the point of attachment of the spring can change along the full length (l) of the pendulum. The input to each pendulum is the torque u_i applied at the pivot point. The pay-loads are assumed to be both known and equal to m. Let $x_i = \text{col}(x_{i1}, x_{i2}) = \text{col}(\theta_i, \theta_i - \omega_i)$ for $i = 1, 2$ where $\omega_i := \dot{\theta}_i$ is the corresponding angle velocity. From [11], the dynamic equations of the system can be described as

$$\dot{x}_1 = \underbrace{\begin{bmatrix} 1 & -1 \\ 1 - \frac{g}{l} & -1 \end{bmatrix}}_{A_1} \begin{bmatrix} x_{11} \\ x_{12} \end{bmatrix} + \underbrace{\begin{bmatrix} 0 \\ \frac{-1}{ml^2} \end{bmatrix} u_1 + \begin{bmatrix} 0 \\ \frac{ka^2}{ml^2} x_{11} \end{bmatrix}}_{\beta_1} + \underbrace{\begin{bmatrix} 0 \\ 1.8033\ \Delta\sigma_1(\zeta_1, \eta_1) \end{bmatrix}}_{E_1 \Delta\Psi(x_1)}$$

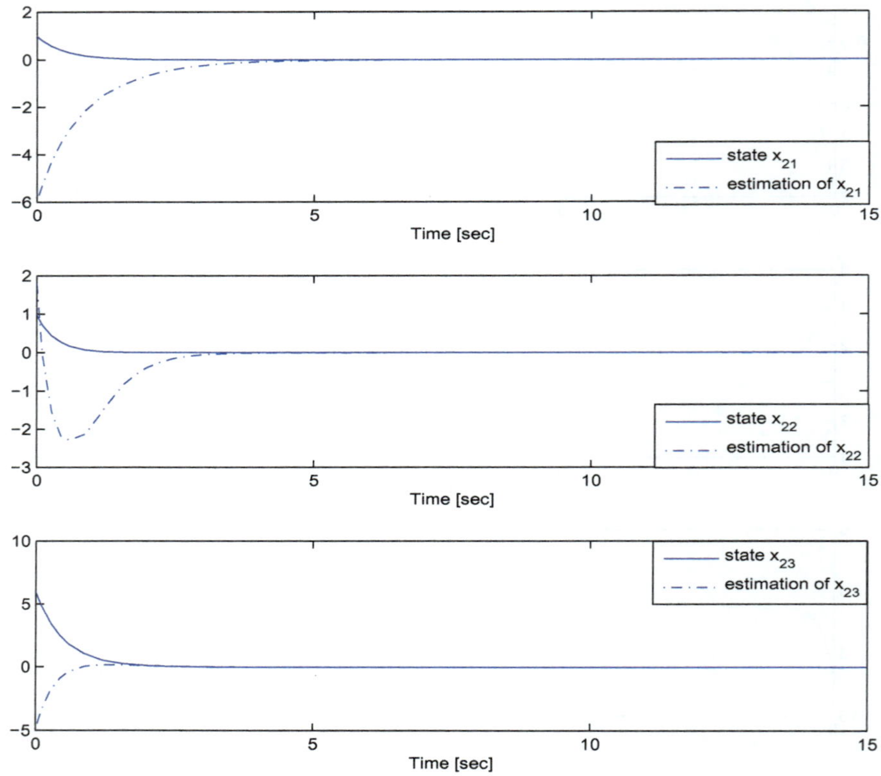

Fig. 8.2 The time response of the states of the second subsystem, $x_2 = \text{col}\,(x_{21}, x_{22}, x_{23})$, and their estimates $\hat{x}_2 = \text{col}\,(\hat{x}_{21}, \hat{x}_{22}, \hat{x}_{23})$

$$+\underbrace{\begin{bmatrix} 0 & 0 \\ \frac{-ka^2}{ml^2} & 0 \end{bmatrix}}_{\Gamma_{12}^a} x_2 \tag{8.67}$$

$$y_1 = \begin{bmatrix} 1 & 0 \end{bmatrix} \begin{bmatrix} x_{11} \\ x_{12} \end{bmatrix} \tag{8.68}$$

$$\dot{x}_2 = \underbrace{\begin{bmatrix} 1 & -1 \\ 1 - \frac{g}{l} & -1 \end{bmatrix}}_{A_2} \begin{bmatrix} x_{21} \\ x_{22} \end{bmatrix} + \underbrace{\begin{bmatrix} 0 \\ \frac{-1}{ml^2} \end{bmatrix} u_1 + \begin{bmatrix} 0 \\ -\frac{ka^2}{ml^2} x_{21} \end{bmatrix}}_{\beta_2} + \underbrace{\begin{bmatrix} 0 \\ 1.8033\ \Delta\sigma_2(\zeta_2, \eta_2) \end{bmatrix}}_{E_2 \Delta\Psi(x_2)}$$

$$+\underbrace{\begin{bmatrix} 0 & 0 \\ \frac{ka^2}{ml^2} & 0 \end{bmatrix}}_{\Gamma_{21}^a} x_1 \tag{8.69}$$

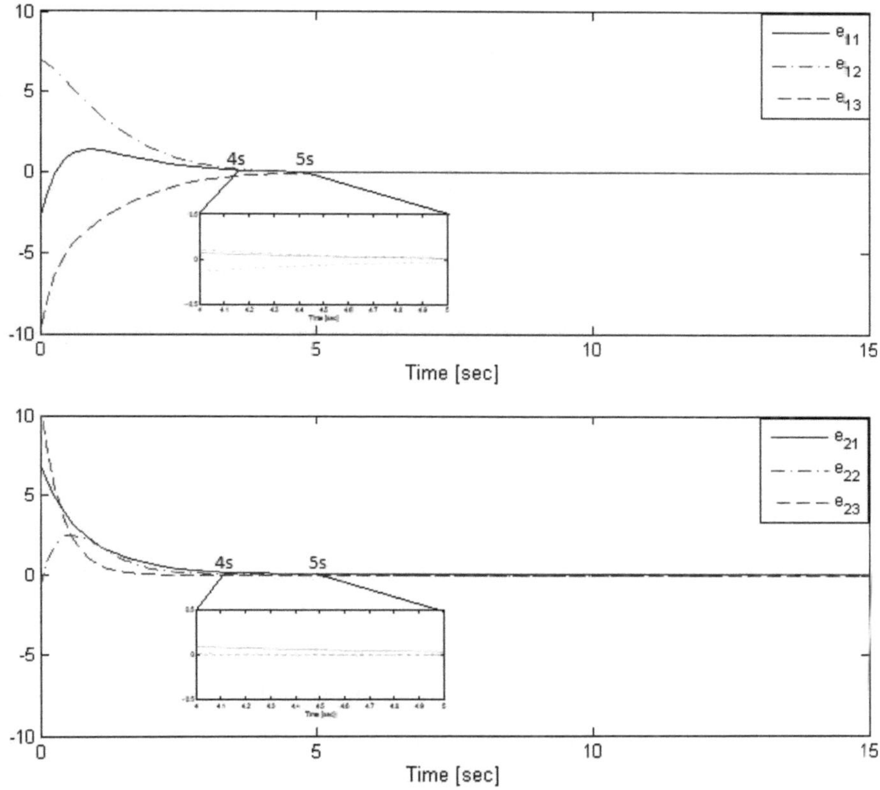

Fig. 8.3 The time response of the estimates errors

$$y_2 = \begin{bmatrix} 1 \ 0 \end{bmatrix} \begin{bmatrix} x_{21} \\ x_{22} \end{bmatrix} \tag{8.70}$$

where k and g are the spring and gravity constants, and a is a parameter explained in Fig. 8.4. It is assumed the only measurable state is

$$y_i = \begin{bmatrix} 1 & 0 \end{bmatrix} x_i, \qquad i = 1, 2 \tag{8.71}$$

and the parameters are chosen as

$$\frac{g}{l} = 9.8, \qquad \frac{1}{ml^2} = 0.5, \qquad \frac{ka^2}{ml^2} = 2.268 \tag{8.72}$$

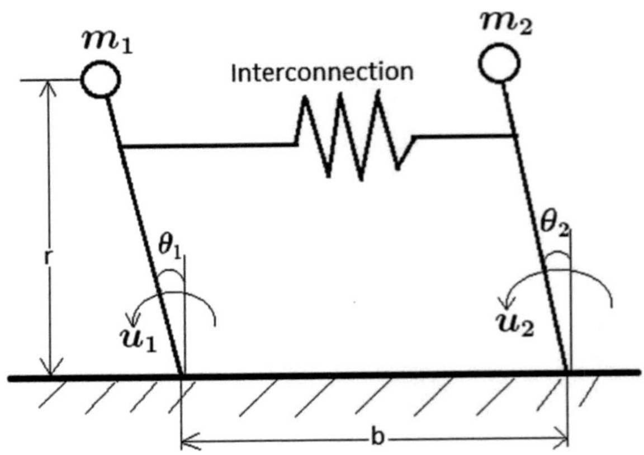

Fig. 8.4 Coupled inverted pendula

$$\|\Delta\Psi_1(x_1)\| \leq ||\Delta\sigma_1(x_1)|| \leq \underbrace{0.5\cos(\text{t})}_{\rho_1(\cdot)}$$

$$\|\Delta\Psi_2(x_2)\| \leq ||\Delta\sigma_2(x_2)|| \leq \underbrace{0.1\sin(\text{t})}_{\rho_2(\cdot)}$$

Then, for both subsystems, choose $L_i = [11 \quad -8.8]^T$ and $Q_i = I$ where $i = 1, 2$. It follows that the Lyapunov equation (8.20) has unique solutions:

$$P_i = \begin{bmatrix} 0.0545 & -0.0455 \\ -0.0455 & 0.5 \end{bmatrix}, \qquad i = 1, 2 \tag{8.73}$$

and the solutions to equation (8.22) are $H_i = [0 \quad 1]^T$ for $i = 1, 2$.

By direct computation, it follows that the matrix $\mathbb{W}^T(\cdot) + \mathbb{W}(\cdot)$ is symmetric positive definite. Thus, all the conditions of Theorem 8.1 are satisfied. This implies that the dynamical system

$$\begin{aligned}\dot{\hat{x}}_1 = &\begin{bmatrix} 1 & -1 \\ -8.81 & -1 \end{bmatrix}\begin{bmatrix} \hat{x}_{11} \\ \hat{x}_{12} \end{bmatrix} + \begin{bmatrix} 11 \\ -8.8 \end{bmatrix}(y_1 - C_1\hat{x}_1) + \begin{bmatrix} 0 \\ -0.5 \end{bmatrix} u_1 \\ &+ \begin{bmatrix} 0 \\ 2.268\hat{x}_{11} \end{bmatrix} + K_1(\cdot) + \begin{bmatrix} 0 & 0 \\ -2.268 & 0 \end{bmatrix}\begin{bmatrix} \hat{x}_{21} \\ \hat{x}_{22} \end{bmatrix}\end{aligned} \tag{8.74}$$

$$\begin{aligned}\dot{\hat{x}}_2 = &\begin{bmatrix} 1 & -1 \\ -8.81 & -1 \end{bmatrix}\begin{bmatrix} \hat{x}_{21} \\ \hat{x}_{22} \end{bmatrix} + \begin{bmatrix} 11 \\ -8.8 \end{bmatrix}(y_2 - C_2\hat{x}_2) + \begin{bmatrix} 0 \\ -0.5 \end{bmatrix} u_2 \\ &+ \begin{bmatrix} 0 \\ -2.268\hat{x}_{21} \end{bmatrix} + K_2(\cdot) + \begin{bmatrix} 0 & 0 \\ 2.268 & 0 \end{bmatrix}\begin{bmatrix} \hat{x}_{11} \\ \hat{x}_{12} \end{bmatrix}\end{aligned} \tag{8.75}$$

is a robust observer of the system (8.67)–(8.70) where $\hat{x}_1 = \text{col}(\hat{x}_{11}, \hat{x}_{12})$, $\hat{x}_2 = \text{col}(\hat{x}_{21}, \hat{x}_{22})$, and the $K_1(\cdot)$ and $K_2(\cdot)$ defined in (8.38) are

$$K_1(y_1, \hat{x}_1) = \begin{cases} \begin{bmatrix} 9.918 \\ 0.9016 \end{bmatrix} \frac{(x_{11} - \hat{x}_{11})}{\|x_{11} - \hat{x}_{11}\|} \cos t), & x_{11} - \hat{x}_{11} \neq 0 \\ 0, & x_{11} - \hat{x}_{11} = 0 \end{cases}$$

$$K_2(y_2, \hat{x}_2) = \begin{cases} \begin{bmatrix} 1.9836 \\ 0.1803 \end{bmatrix} \frac{(x_{21} - \hat{x}_{21})}{\|x_{21} - \hat{x}_{21}\|} \sin t), & x_{21} - \hat{x}_{21} \neq 0 \\ 0, & x_{21} - \hat{x}_{21} = 0 \end{cases}$$

For simulation purposes, the controllers are chosen as:

$$u_i = -61.6x_{i1} + 22x_{i2}, \qquad i = 1, 2 \tag{8.76}$$

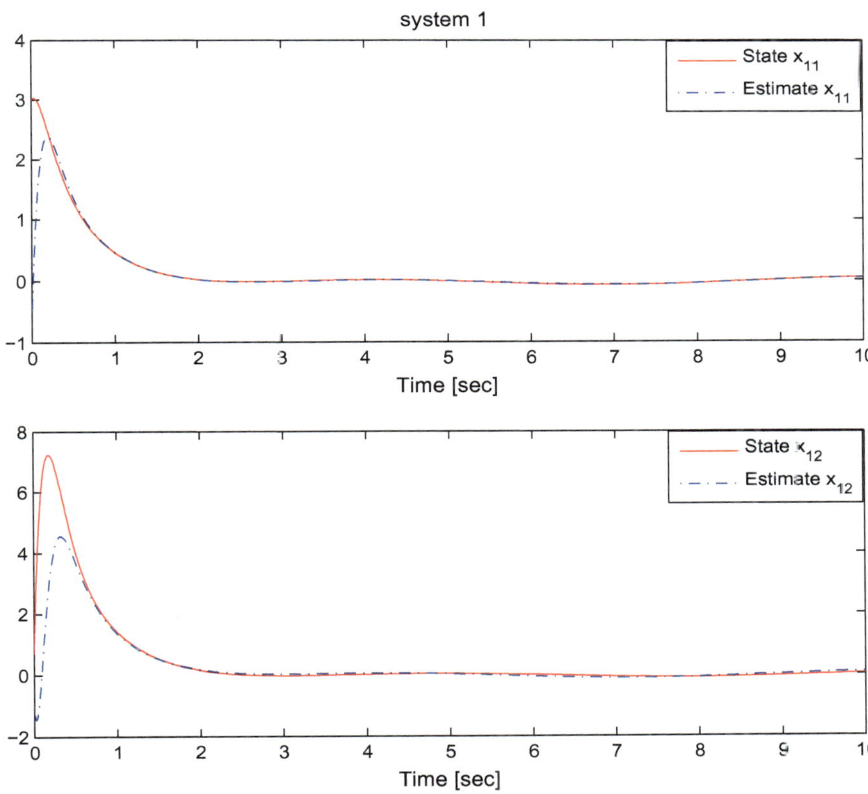

Fig. 8.5 The time response of the states of the first subsystem, $x_1 = \text{col}\ (x_{11}, x_{12})$, and their estimates $\hat{x}_1 = \text{col}\ (\hat{x}_{11}, \hat{x}_{12})$

24. Modarres, A.A., Momeni, H., Yazdizadeh, A.: A new decentralized sliding mode observer design for interconnected power systems with unknown and time-varying delay. In: Proceeding of the 4th Conference on Thermal Power Plants, pp. 1–6 (2012)
25. Reppa, V., Polycarpou, M.M., Panayiotou, C.G.: Decentralized isolation of multiple sensor faults in large-scale interconnected nonlinear systems. IEEE Trans. Autom. Control **60**(6), 1582–1596 (2014)
26. Saksena, V., Cruz, J.: An approach to decentralized control of large scale systems using aggregation methods. IEEE Trans. Autom. Control **30**(9), 912–914 (1985)
27. Sandell, N., Varaiya, P., Athans, M., Safonov, M.: Survey of decentralized control methods for large scale systems. IEEE Trans. Autom. Control **23**(2), 108–128 (1978)
28. Sastry, S., Bodson, M.: Adaptive Control: Stability. Convergence and robustness. Prentic-Hall, New Jersey (1994)
29. Sharma, R., Aldeen, M.: Fault and disturbance reconstruction in non-linear systems using a network of interconnected sliding mode observers. IET Control Theory Appl. **5**(6), 751–763 (2011)
30. Shtessel, Y., Edwards, C., Fridman, L., Levant, A.: Sliding Mode Control and Observation. Springer, New York (2014)
31. Siljak, D., Vukcevic, M.: Decentralization, stabilization, and estimation of large-scale linear systems. IEEE Trans. Autom. Control **21**(3), 363–366 (1976)
32. Spurgeon, S.K.: Sliding mode observers: a survey. Int. J. Sys. Sci. **39**(8), 751–764 (2008)
33. Sundareshan, M.K. Decentralized observation in large-scale systems. In: Proceedings in Decision and Control including the 15th Symposium on Adaptive Processes, 1976 IEEE Conference, pp. 659–664 (1976)
34. Sundareshan, M.K., Elbanna, R.M.: Design of decentralized observation schemes for large-scale interconnected systems: Some new results. Autom. **26**(4), 789–796 (1990)
35. Swarnakar, A., Marquez, H.J., Chen, T.: A new scheme on robust observer based control design for nonlinear interconnected systems with application to an industrial utility boiler. In: Proceedings of the American Control Conference, IEEE, New York City, USA, pp. 5601-5606 (2007)
36. Vadim, I.U.: Survey paper variable structure systems with sliding modes. IEEE Trans. Autom. Control **22**(2), 212–222 (1977)
37. Utkin, V., Guldner, J., Shi, J.: Sliding Mode Control in Electro-Mechanical Systems. CRC press, USA (2009)
38. Wang, Y., Fei, J.: Adaptive fuzzy sliding mode control for PMSM position regulation system. Int. J. Innov. Comput. Inf. Control. **11**(3), 881–891 (2015)
39. Wu, H.: Decentralized adaptive robust state feedback for uncertain large-scale interconnected systems with time delays. J. Optim. Theory Appl. **126**(2), 439–462 (2005)
40. Wu, H.: Adaptive robust state observers for a class of uncertain nonlinear dynamical systems with delayed state perturbations. IEEE Trans. Autom. Control **54**(6), 1407–1412 (2009)
41. Xia, X.H., Gao, W.B.: Nonlinear observer design by observer error linearization. SIAM J. Control Optim. **27**(1), 199–216 (1989)
42. Yan, X.G., Lam, J., Xie, L.: Robust observer design for non-linear interconnected systems using structural characteristics. Int. J. Control **76**(7), 741–746 (2003)
43. Yan, X.G., Edwards, C.: Robust decentralized actuator fault detection and estimation for large-scale systems using a sliding mode observer. Int. J. Control **81**(4), 591–606 (2008)
44. Yan, X.G., Lam, J., Dai, G.: Decentralized robust control for nonlinear large-scale systems with similarity. Comput. Elect. Eng. **25**(3), 169–179 (1999)
45. Yan, X.G., Spurgeon, S.K., Edwards, C.: Decentralized output feedback sliding mode control of nonlinear large-scale systems with uncertainties. J. Optim. Theory Appl. **119**(3), 597–614 (2003)
46. Yan, X.G., Spurgeon, S.K., Edwards, C.: Decentralised stabilisation for nonlinear time delay interconnected systems using static output feedback. Autom. **49**(2), 633–641 (2013)
47. Yan, X.G., Spurgeon, S.K., Edwards, C.: State and parameter estimation for nonlinear delay systems using sliding mode techniques. IEEE Trans. Autom. Control **58**(4), 1023–1029 (2013)

48. Yan, X.G., Spurgeon, S.K., Edwards, C.: Memoryless static output feedback sliding mode control for nonlinear systems with delayed disturbances. IEEE Trans. Autom. Control **59**(7), 1906–1912 (2014)
49. Yan, X.G., Spurgeon, S.K., Edwards, C.: Variable Structure Control of Complex Systems Analysis and Design. Springer, Berlin (2017)
50. Yang, G.H., Zhang, S.Y.: Decentralized control of a class of large-scale systems with symmetrically interconnected subsystems. IEEE Trans. Autom. Control **41**(5), 710–713 (1996)
51. Zeitz, M.: The extended Luenberger observer for nonlinear systems. Syst. Control Lett. **9**(2), 149–156 (1987)
52. Zhang, C., Li, H.: A generalized robust decentralized control methodology for a class of interconnected nonlinear systems subject to uncertainties and disturbances. Nonlinear Anal. Hybrid Syst. **22**, 55–71 (2016)
53. Zhao, G., Zhao, C., Cheng, J.: Decoupled Terminal Sliding-Mode Control for a Class of Under-Actuated Mechanical Systems With Hybrid Sliding Surfaces. Int. J. Innov. Comput Inf. Control. **10**(6), 2011–2023 (2014)

Chapter 9
A Unified Lyapunov Function for Finite Time Stabilization of Continuous and Variable Structure Systems with Resets

Harshal B. Oza, Yury V. Orlov and Sarah K. Spurgeon

9.1 Introduction

Uniform finite time stabilization of a perturbed double integrator system is considered in the presence of a unilateral constraint on the position variable. Unilaterally constrained systems give rise to hard non-linearities due to jumps or *resets* in the velocity and are quite challenging from the point of view of stabilization. Such systems occur in various feedback control disciplines [1, 2]. The practical motivation for proposing a synthesis framework for such systems has been demonstrated by applications to biped robots [3, 4], where the generalized velocities of the robot inherently undergo a reset when the swing leg collides with the ground. Studying systems with resets also finds natural application in the area of hybrid systems [5]. Developing a clear understanding of Lyapunov based stability and robustness properties of a closed-loop system when resets with (or without) a finite accumulation point is a challenging and interesting area of study [6].

It is well-known, as described in [7, Sect. 3], that real-life implementation of discontinuous control for synthesizing joint torques for robot manipulators causes high frequency oscillations. This work is motivated by the use of continuous second order sliding mode control for biped robots and extends previous work on discontinuous

H.B. Oza (✉)
School of Engineering and Applied Science, Ahmedabad University,
Gujarat, India
e-mail: harshal.oza@ahduni.edu.in

Y.V. Orlov
CICESE, Ensenada, Mexico
e-mail: yorlov@cicese.mx

S.K. Spurgeon
Department of Electronic and Electrical Engineering,
University College London, London, UK
e-mail: s.spurgeon@ucl.ac.uk

S. Li et al. (eds.), *Advances in Variable Structure Systems and Sliding Mode Control—Theory and Applications*, Studies in Systems, Decision and Control 115, DOI 10.1007/978-3-319-62896-7_9

stabilization published in [8] to the case of continuous homogeneous controllers. A recent survey in [9] details the chronology in this area of control engineering. The reference [10] is also relevant where the proofs of continuous finite time stability are achieved via a similar route to that proposed in this work for systems without impacts.

The main contribution of this work is that it unifies the Lyapunov stability proofs for both continuous and discontinuous cases while making use of a continuous homogeneous controller. Similar to the work in [8], it is assumed that the restitution or reset map relating to the velocities just before and after the time of impact is fully known. No such assumption is made on the time of impact. The main difference between the work in [8] and the results presented here is that the later generalizes the results of [8] in that both discontinuous and continuous controllers are admissible by the proposed Lyapunov analysis and synthesis.

A continuous homogeneous controller is used for finite time stabilization of the closed-loop system in the presence of resets. The Zhuravlev–Ivanov non-smooth transformation [11, 12] is used to first transform the system into a variable-structure system without jumps. Within engineering applications, this transformation is very useful in the analysis of vibro-impact systems [11, 13]. The transformed system is a time varying variable structure homogeneous system with negative homogeneity degree [14] where the solutions are well-defined in the sense of Filippov's definition [15], an attribute absent in the case of the original jump system (see [16] for the solution concept of systems with jumps and friction). It is important to note that the use of finite time stability of discontinuous systems [14] is the most natural method for proving uniform finite time stability due to the variable structure nature of the transformed system despite the continuous controller. This is because all the existing frameworks for continuous homogeneous systems [17, 18] require continuity of the vector field, a condition unavoidably violated at the time of jumps.

The remainder of the chapter is organized as follows. Section 9.2 provides all the mathematical preliminaries necessary for the development of the subsequent theoretical concepts. Section 9.3 states the mathematical problem formulation including the definition of the non-smooth transformation. Section 9.3.2 further motivates the developments by showing mathematically why existing approaches are not suitable for application to the problem under consideration. Section 9.4 identifies a parameterized set of Lyaunov functions to prove uniform finite time stability. Section 9.5 concludes the chapter.

9.2 Mathematical Preliminaries

This chapter will establish uniform finite time stabilisation of variable structure and continuous non-smooth systems in the presence of unilateral constraints that give rise to resets. It is important to first clearly describe what is meant by each of the mathematical terms that will be used.

Consider the discontinuous dynamical system

$$\dot{x} = \phi(x, t), \tag{9.1}$$

where $x = (x_1, x_2, \ldots, x_n)^T$ is the state vector, $t \in \mathbb{R}$ is the time variable and the function $\phi(x, t)$ is piece-wise continuous. The function $\phi : \mathbb{R}^{n+1} \to \mathbb{R}^n$ is piece-wise continuous iff $\mathbb{R}^{n+1}$ is partitioned into a finite number of domains $G_j \in \mathbb{R}^{n+1}, j = 1, ..., N$, with disjoint interiors and boundaries ∂G_j of measure zero such that ϕ is continuous within each of these domains and for all $j = 1, ..., N$ it has a finite limit $\phi^j(x, t)$ as the argument $(x^*, t^*) \in G_j$ approaches a boundary point $(x, t) \in \partial G_j$.

For example, the function $\phi(x, t) = \text{sign}(x)$ for $x \in \mathbb{R}$ defines two domains $G_1 = \{x : x \geq 0\}$, $G_2 = \{x : x \leq 0\}$ in $\mathbb{R}$ with the common boundary of zero measure $\partial G_j = \{x : x = 0\}, j = 1, 2$. Of course, $\phi(x)$ is a piece-wise continuous function since the limits $\lim_{x_1^* \to x^+} \partial G_1 = 1$ and $\lim_{x_2^* \to x^-} \partial G_2 = -1$ are finite $\forall x_1^* \in G_1, x_2^* \in G_2, x \in \partial G_j, j = 1, 2$.

Definition 9.1 (*Solutions in the sense of Filippov* [15]) Given the differential equation (9.1), let the smallest convex closed set $\Phi(x, t)$ be introduced for each point $(x, t) \in \mathbb{R}^n \times \mathbb{R}$ such that $\Phi(x, t)$ contains all the limit points of $\phi(x^*, t)$ as $x^* \to x$, $t =$ constant, and $(x^*, t) \in \mathbb{R}^{n+1} \backslash (\cup_{j=1}^{N} \partial G_j)$. An absolutely continuous function $x(\cdot)$ defined on the interval $\mathbb{I}$ is said to be a solution of (9.1) if it satisfies the differential inclusion

$$\dot{x} \in \Phi(x, t) \tag{9.2}$$

almost everywhere on the interval $\mathbb{I}$.

Solutions to differential equations will be understood as defined in Definition 9.1 throughout the chapter. For the example of $\phi(x, t) = \text{sign}(x)$ where the state $x \in \mathbb{R}$, the closed convex set Φ is obtained as $\Phi(x, t) = [-1, 1]$. The solution to the differential equation (9.1) exists for an arbitrary initial condition $x(t_0)$ on an interval $\mathbb{I} = [t_0, t_1)$ and is non-unique in general [15, Theorem 10].

This work studies robustness of variable structure and non-smooth systems in order for these controllers to be applied to unilaterally constrained systems. It is important to define the class of perturbations that will be studied. Let the perturbed version of (9.1) be given as follows:

$$\dot{x} = \phi(x, t) + \psi(x, t) \tag{9.3}$$

where, $\psi(x, t)$ is a piece-wise continuous function with components $\psi_1, \psi_2, \ldots, \psi_n$. When studying variable structure systems, the components $\psi_i, i = 1, 2, \ldots, n$ are assumed to be locally uniformly bounded by upperbounds

$$\text{ess sup}\, |\psi_i(x, t)| \leq M_i, \quad i = 1, 2, \ldots, n, \tag{9.4}$$

for almost all $(x, t) \in B_\delta \times \mathbb{R}$, where $B_\delta \subset \mathbb{R}$ is a ball of radius $\delta > 0$, and some positive constants $M_i \geq 0$ fixed a priori. When studying a continuous non-smooth vector field $\phi(x, t)$ along with piece-wise continuous perturbations $\psi(x, t)$, the components $\psi_i, i = 1, 2, \ldots, n$ are assumed to possess local uniform upperbounds

$$\text{ess sup}\,|\psi_i(x,t)| \leq M_i\bar{\alpha}(\|x\|)), \quad i = 1, 2, \ldots, n, \tag{9.5}$$

where $\bar{\alpha}(\|x(x,t)\|)$ is a continuous positive definite function such that

$$\lim_{\|x\|\to 0} \bar{\alpha}(\|x\|)) \to 0. \tag{9.6}$$

Definition 9.2 (*Differential inclusions for uncertain discontinuous systems* [14]) An absolutely continuous function $x(\cdot)$, defined on an interval $\mathbb{I}$, is said to be a solution of the uncertain differential equation (9.3) with the rectangular uncertainty constraints (9.4) (sectorial constraints (9.5)) iff it is a solution of (9.3) on the interval $\mathbb{I}$ in the sense of Definition 9.1 for some piece-wise continuous function ψ subject to (9.4) (respectively, (9.5)).

An uncertain system (9.3) can be represented as a differential inclusion of the form

$$\dot{x} \in \Phi(x,t) + \Psi(x,t), \tag{9.7}$$

where $\Phi(x,t)$ is the same as defined in Definition 9.1, Ψ is the cartesian product of the intervals $\Psi_i = [-M_i,\ M_i]$ for the uncertainty constraints (9.4), Ψ is cartesian product of the intervals $\Psi = [-M_i\bar{\alpha}(x,t),\ M_i\bar{\alpha}(x,t)]$ for the uncertainty constraints (9.5) with $(x,t) \in \partial G_j, j = 1, 2, \ldots, n$ representing the discontinuity (or limit) points of $\phi(x^*,t)$ as $x^* \to x$ and the set

$$\Phi(x,t) + \Psi(x,t) = \{\phi + \psi : \phi \in \Phi(x,t), \psi \in \Psi(x,t)\}. \tag{9.8}$$

Continuing with an example (see [19]) of a vector field to highlight where the above Definition 9.2 is applicable, a planar non-smooth system $\dot{x_1} = x_2$, $\dot{x}_2 = -|x_1|^{\alpha_1}\,\text{sign}(x_1) - |x_2|^{\alpha_2}\,\text{sign}(x_2)$ where α_1, α_2 are positive scalars belonging to the interval $(0, 1)$ has the right hand side $\phi_1 = x_2$, $\phi_2 = -|x_1|^{\alpha_1}\,\text{sign}(x_1) - |x_2|^{\alpha_2}\,\text{sign}(x_2)$. Consider the uncertainty $\psi_1 = 0$, $\psi_2(x,t) = M_2|x_2|^{\alpha_2}\,\text{sign}(x_1)$ $\sin(t)$, $M_2 > 0$. It can be verified that ψ_2 belongs to the class of uncertainties with upperbound (9.4) when $\alpha_2 = 0$ and to the class of uncertainties with upperbound (9.5) when $\alpha_2 \in (0, 1)$. In the latter case, the inclusion (9.7) can be obtained by considering Ψ as the cartesian product $(0, -M_2|x_2|^{\alpha_2}) \times (0, M_2|x_2|^{\alpha_2})$ on the set of all discontinuity points $\{(x,t) : x_1 = 0\}$.

The focus of this work from the viewpoint of stability analysis is on uniform finite time stability with respect to initial time t_0 as well as uncertainty $\psi(x,t)$. It is important to highlight what is meant by uniformity. This is a well studied area for systems with Lipschitz dynamics and many references are available [20–22] with regard to uniformity with respect to initial time. It can be seen from the above references, however, that emphasis is seldom given to uniformity with respect to uncertainty. Definitions [14, Definitons 2.3, 2.4, 2.5] of (uniform) stability, (uniform) asymptotic stability and (uniform) finite time stability of the inclusion (9.2) for the discontinuous vector field can be seen as the counterparts of the definitions available in the references [20–23] for similar stability concepts in the case of continuous vector field. Hence, definitions [14, Definitons 2.3, 2.4, 2.5] are not included here as they only focus on uniformity of stability with respect to initial conditions. The following definitions are inherited from [14, Definitons 2.6, 2.7, 2.8] and accommodate uniformity of stability with respect to the uncertainty ψ. It should be noted that the word 'equiuniform' appearing in [14] is utilised in the following definitions to refer to uniformity of various stability concepts with respect to the initial conditions as well as to the uncertainty ψ.

Suppose that $x = 0$ is an equilibrium point of the uncertain system (9.3), (9.4) (or similarly (9.3), (9.5)), i.e., $x = 0$ is a solution of (9.3) for some function ψ_0, admissible in the sense of either (9.4) or (9.5), and let $x(\cdot, t_0, x^0)$ denote a solution $x(\cdot)$ of (9.3) for some admissible function ψ under the initial conditions $x(t_0) = x^0$.

Definition 9.3 (*Equiuniform stability* [14]) The equilibrium point $x = 0$ of the uncertain system (9.3), (9.4) (or similarly (9.3), (9.5)) is equiuniformly stable iff for each $t_0 \in \mathbb{R}$, $\varepsilon > 0$ there exists $\delta = \delta(\varepsilon)$, dependent on ε and independent of t_0 and ψ, such that each solution $x(\cdot, t_0, x^0)$ of (9.3), (9.4) (or similarly (9.3), (9.5)) with the initial data $x^0 \in B_\delta$ exists on the semi-infinite time interval $[t_0, \infty)$ and satisfies the inequality

$$\|(x(t, t_0, x^0))\| \leq \varepsilon \quad \forall t \in [t_0, \infty). \tag{9.9}$$

Definition 9.4 (*Equiuniform asymptotic stability* [14]) The equilibrium point $x = 0$ of the uncertain system (9.3), (9.4) (or similarly (9.3), (9.5)) is said to be equiuniformly asymptotically stable if it is equiuniformly stable and the convergence

$$\lim_{t \to \infty} \|(x(t, t_0, x^0))\| \to 0 \tag{9.10}$$

holds for all the solutions $x(\cdot, t_0, x^0)$ of the uncertain system (9.3), (9.4) (or similarly (9.3), (9.5)) initialized within some B_δ (uniformly in the initial data t_0 and x^0). If this convergence remains in force for each $\delta > 0$, the equilibrium point is said to be globally equiuniformly asymptotically stable.

Definition 9.5 (*Equiuniform finite time stability* [14]) The equilibrium point $x = 0$ of the uncertain system (9.3), (9.4) (or similarly (9.3), (9.5)) is said to be globally equiuniformly finite time stable if, in addition to the global equiuniform asymptotical stability, the limiting relation

$$x(t, t_0, x^0) = 0 \tag{9.11}$$

holds for all the solutions $x(\cdot, t_0, x^0)$ and for all $t \geq t_0 + T(t_0, x^0)$, where the settling time function

$$T(t_0, x^0) = \sup_{x(\cdot, t_0, x^0)} \inf\{T \geq 0 : x(t, t_0, x^0) = 0 \quad \text{for all } t \geq t_0 + T\} \tag{9.12}$$

is such that

$$T(B_\delta) = \sup_{x^0 \in B_\delta, t_0 \in \mathbb{R}} T(t_0, x^0) < \infty \quad \text{for all } \delta > 0, \tag{9.13}$$

where $\delta = \delta(\varepsilon)$ is independent of t_0 and ψ.

The infimum in (9.12) is to detect the first instant $t = T$ such that $x(t, t_0, x^0) = 0$, $\forall t \geq t_0 + T$ and the supprenum is for taking the worst case trajectory that takes the longest time to arrive at the origin.

The chapter will also use the concept of geometric homogeneity. The following definitions are inherited from [14, Definitions 2.9, 2.10].

Definition 9.6 (*Homogeneity of differential inclusions and equations* [14]) The differential inclusion (9.2) (the differential equation (9.1), the uncertain systems (9.3), (9.4) or the uncertain systems (9.3), (9.5)) is called locally homogeneous of degree $q \in \mathbb{R}$ with respect to dilation $(r_1, r_2, \ldots, r_n)$, where $r_i > 0$, $i = 1, 2, \ldots, n$, if there exist a constant $c_0 > 0$, called a lower estimate of the homogeneity parameter, and a ball $B_\delta \subset \mathbb{R}^n$, called a homogeneity ball, such that any solution $x(\cdot)$ of (9.2) (respectively, that of the differential equation (9.1), the uncertain systems (9.3), (9.4) or the uncertain systems (9.3), (9.5)) evolving within the ball B_δ, generates a parameterised set of solutions $x^c(\cdot)$ with components

$$x_i^c(t) = c^{r_i} x_i(c^q t) \tag{9.14}$$

and any parameter $c \geq c_0$.

Definition 9.7 (*Homogeneous piece-wise continuous functions* [14]) A piece-wise continuous function $\phi : \mathbb{R}^{n+1} \to \mathbb{R}^n$ is called locally homogeneous of degree $q \in \mathbb{R}$ with respect to dilation $(r_1, r_2, \ldots, r_n)$, where $r_i > 0$, $i = 1, 2, \ldots, n$, if there exists a constant $c_0 > 0$ and a ball $B_\delta \subset \mathbb{R}^n$ such that

$$\phi_i(c^{r_1} x_1, c^{r_2} x_2, \ldots, c^{r_n} x_n, c^{-q} t) = c^{q+r_i} \phi_i(x_1, x_2, \ldots, x_n, t) \tag{9.15}$$

for all $c \geq c_0$ and almost all $(x, t) \in B_\delta \times \mathbb{R}$.

The global homogeneity concept for the inclusions (9.2) and the piece-wise continuous functions ϕ can be formally introduced using Definitions 9.6 and 9.7 by taking $\delta = \infty$. It should be noted that geometric homogeneity defined in Definition 9.7 is a stronger concept than that appearing in [24] and the references therein since the former additionally covers differential inclusions arising from discontinuous right hand sides. Definition 9.7 is inspired from [14, 25] which present homogeneity approaches for systems with discontinuous right hand sides.

Definition 9.8 (*Quasi-homogeneity principle* [26, Theorem 4.2]) Let the following conditions be satisfied:

1. a piece-wise continuous function ϕ is locally homogeneous of degree $q \in \mathbb{R}$ with respect to dilation $(r_1, r_2, \dots r_n)$,
2. components ψ_i, $i = 1, 2, \dots, n$ of a piece-wise continuous function ψ are locally uniformly bounded by constants $M_i \geq 0$,
3. $M_i = 0$ whenever $q + r_i > 0$.
4. the uncertain system (9.3), (9.4) is globally equiuniformly asymptotically stable around the origin.

Then, the uncertain system (9.3), (9.4) is globally equiuniformly finite time stable around the origin.

Several definitions relating to Lyapunov functions are also presented. The following three definitions relating to the *weak*, *strong* and *strict* Lyapunov functions can be found in various texts and papers [21], [22, Chap. 4], [14, 23, 27].

Definition 9.9 (*Weak Lyapunov function and uniform stability* [22, Theorem 4.8]) Consider the non-autonomous system (9.1) with $\phi(x, t)$ locally Lipschitz in x and piece-wise continuous in t. Let $x = 0$ be an equilibrium point for (9.1) and $D \subset \mathbb{R}^n$ be a domain containing $x = 0$. Let $V : [0, \infty) \times D \to \mathbb{R}$ be a continuously differentiable function such that

$$
\begin{aligned}
W_1(x) \leq &V(x,t) \leq W_2(x) \\
\frac{\partial V}{\partial t} + &\frac{\partial V}{\partial x}\phi(x,t) \leq 0
\end{aligned}
\tag{9.16}
$$

$\forall t \geq 0$ and $\forall x \in D$, where $W_1(x)$ and $W_2(x)$ are continuous positive definite functions on D. Then, $x = 0$ is uniformly stable.

Extension of the Definition 9.9 to Lipschitz-continuous (or locally Lipschitz) nonsmooth function $V(x, t)$ and to piece-wise continuous $\phi(x, t)$ can be found in [26, Theorem 3.1, Lemmas 3.1, 3.2].

Definition 9.10 (*Strong Lyapunov function and uniform asymptotic stability* [22, Theorem 4.9]) Consider the non-autonomous system (9.1) with $\phi(x, t)$ locally Lipschitz in x and piece-wise continuous in t. Let $x = 0$ be an equilibrium point for (9.1) and $D \subset \mathbb{R}^n$ be a domain containing $x = 0$. Let $V : [0, \infty) \times D \to \mathbb{R}$ be a continuously differentiable function such that

$$
\begin{aligned}
W_1(x) \leq &V(x,t) \leq W_2(x) \\
\frac{\partial V}{\partial t} + &\frac{\partial V}{\partial x}\phi(x,t) \leq -W_3(x)
\end{aligned}
\tag{9.17}
$$

$\forall t \geq 0$ and $\forall x \in D$, where $W_1(x)$, $W_2(x)$ and $W_3(x)$ are continuous positive definite functions on D. Then, $x = 0$ is uniformly asymptotically stable. If the conditions

$D = \mathbb{R}^n$, $W_1(0) = 0$ and $\lim_{\|x\|\to\infty} W_1(x) \to \infty$ hold true, then $x = 0$ is globally uniformly asymptotically stable.

Definition 9.11 (*Strict Lyapunov functions and uniform finite time stability* [23, Theorem 4.1]) Consider the non-autonomous system (9.1) with $\phi(x, t)$ continuous in x and t. Let $x = 0$ be an equilibrium point for (9.1) and $D \subset \mathbb{R}^n$ be a domain containing $x = 0$. Let $V : [0, \infty) \times D \to \mathbb{R}$ be a continuously differentiable function such that

$$\begin{aligned} W_1(x) \leq & V(x, t) \leq W_2(x) \\ \frac{\partial V}{\partial t} + & \frac{\partial V}{\partial x}\phi(x, t) \leq -k(t)(V(x, t))^{\alpha} \end{aligned} \tag{9.18}$$

$\forall t \geq 0$ and $\forall x \in D$, where $W_1(x)$ and $W_2(x)$ are continuous positive definite functions on D, $k : [0, \infty] \to \mathbb{R}^+$ such that $k(t) > 0 \,\forall\, t \geq 0$ and $\alpha \in (0, 1)$. Then, $x = 0$ is uniformly finite time stable. If the conditions $D = \mathbb{R}^n$, $W_1(0) = 0$ and $\lim_{\|x\|\to\infty} W_1(x) \to \infty$, $\lim_{\|x\|\to\infty} W_2(x) \to \infty$ hold true, then $x = 0$ is globally uniformly finite stable.

The reader is referred to [27] where finite time stability of autonomous systems was studied using strict Lyapunov functions. It can be seen from the definitions presented thus far that finite time stability is only meaningful when asymptotic stability is proven first. This chapter also uses the term [14] *semi-global Lyapunov functions* frequently. The following definition of semi-global stabilisation can be found in [22, Sect. 12.1]:

Definition 9.12 (*Semi-global stabilisation* [22, Sect. 12.1]) If feedback control does not achieve global stabilisation, but can be designed such that any given compact set $B_r = \{x : \|x\| < r\}$, with $0 < r < \infty$ chosen arbitrarily large, can be included in the region of attraction, the resulting stabilisation is said to be semi-global stabilisation.

9.3 Problem Statement

Consider the following system:

$$\dot{x}_1 = x_2 \tag{9.19}$$

$$\dot{x}_2 = u(x_1, x_2) + \omega(x_1, x_2, t) \tag{9.20}$$

$$x_1 \geq 0 \tag{9.21}$$

$$x_2(t_k^+) = -\bar{e}\, x_2(t_k^-) \quad \text{if} \quad x_2(t_k^-) < 0, x_1(t_k) = 0, \tag{9.22}$$

where x_1, x_2 are the position and the velocity respectively, u is the control input, $\omega(x_1, x_2, t)$ is a piece-wise continuous disturbance [14, Sect. 2], [15], t_k is the time instant of the kth jump where the velocity undergoes a reset or jump, $\bar{e}$ represents the loss of energy and $x_2(t_k^+)$ and $x_2(t_k^-)$ represent right and left limits respectively of x_2 at the jump time t_k. The equalities (9.19) and (9.20) represent the continuous dynamics without jumps in the velocity. The inequality (9.21) represents the unilateral constraint on the position x_1 which evolves in a domain with a boundary [1]. It is assumed that the jump event occurs instantaneously within an infinitesimally small time and hence mathematically can be represented by *Newton's restitution rule* [1, 28] given by (9.22) where it is assumed that $\bar{e} \in (0, 1)$.

The piece-wise continuous disturbance $\omega(x_1, x_2, t)$ is assumed to admit a uniform upper bound

$$\operatorname*{ess\,sup}_{t\geq 0} |\omega(x,t)| \leq M|x_2|^{\alpha} \tag{9.23}$$

on its magnitude such that $0 < M < \mu_1 < \mu_2 - M$. The finite time controller [18, 29]

$$u(x_1, x_2) = -\mu_2|x_1|^{\frac{\alpha}{2-\alpha}} \operatorname{sign}(x_1) - \mu_1|x_2|^{\alpha} \operatorname{sign}(x_2) \tag{9.24}$$

is used with $\alpha \in [0, 1)$ and $\mu_2 > \mu_1 > 0$. The aim is to establish uniform finite-time stability of the closed-loop system (9.19), (9.24). This work employs Zhuravlev–Ivanov's method of non-smooth transformation [11, 12], [1, Sect. 1.4.2] to transform the impact system (9.19) into a jump-free system. Although this method has been explained for a similar problem in [8], details are included here for completeness. Let the non-smooth coordinate transformation be defined as follows:

$$\begin{aligned} x_1 &= |s|, \quad x_2 = Rv \operatorname{sign}(s), \\ R &= 1 - k \operatorname{sign}(sv), \quad k = \frac{1-\bar{e}}{1+\bar{e}}. \end{aligned} \tag{9.25}$$

The variable structure transformed system

$$\begin{aligned} \dot{s} &= Rv \\ \dot{v} &= R^{-1} \operatorname{sign}(s) u(|s|, Rv \operatorname{sign}(s)) \\ &\quad + R^{-1} \operatorname{sign}(s)\omega(|s|, Rv \operatorname{sign}(s), t) \end{aligned} \tag{9.26}$$

is then obtained by employing (9.25) and using the dynamics (9.19), (9.20). The controller (9.24) can be represented in the transformed coordinates as follows:

$$\begin{aligned} u(|s|, Rv \operatorname{sign}(s)) = &-\mu_1|Rv \operatorname{sign}(s)|^{\alpha} \operatorname{sign}(Rv \operatorname{sign}(s)) \\ &- \mu_2\big||s|\big|^{\frac{\alpha}{2-\alpha}} \operatorname{sign}(|s|) \end{aligned} \tag{9.27}$$

Substituting (9.27) into (9.26), the closed-loop system in the coordinate frame (s, v) can be obtained as follows:

$$\begin{aligned}\dot{s} &= Rv \\ \dot{v} &= -\mu_1 R^{\alpha-1}|v|^\alpha \operatorname{sign}(v) - \mu_2 R^{-1}|s|^{\frac{\alpha}{2-\alpha}} \operatorname{sign}(s) \\ &\quad + R^{-1} \operatorname{sign}(s)\omega(|s|, Rv \operatorname{sign}(s), t)\end{aligned} \tag{9.28}$$

Furthermore, the upper-bound (9.23) can be revised in the new coordinates as follows:

$$\operatorname*{ess\,sup}_{t\geq 0} |\omega(s, v, t)| \leq M|Rv \operatorname{sign}(s)|^\alpha = MR^\alpha |v|^\alpha \tag{9.29}$$

Unlike the existing literature on hybrid systems [5], this work does not regularize the Zeno motion temporally or dynamically. In the context of the bouncing ball analogy, the temporal regularization as described in [5, Sect. 4] and subsequently utilized in [30, Sect. 5.C] means that the impact takes a small, but not infinitesimally small, time $\varepsilon > 0$ and the dynamic regularization means that the impact is elastic but is more like that with a highly stiff spring. No such regularization is employed here and ideal Zeno modes due to non-elastic impacts are allowed giving rise to instantaneous jumps in zero time[1]. Furthermore, existing Lyapunov approaches on the study of 'uniformly small ordinary time' [31] that lead to finite time stabilization results and computation of finite settling time inherently differ in that the jumps in a corresponding Lyapunov function [31, Theorem 3.3, Example 3.4] always need to be analyzed at the reset time instant. More importantly, the decrease in successive jumps also have to belong to class $\mathscr{K}_\infty$ (see [31, Theorem 3.3]). This work does not need such assumptions while proving robust finite time stability without analyzing the jumps in the Lyapunov function due to the jump-free system produced by the non-smooth transformation.

9.3.1 *Equivalence of Jump-Free and Unilateral Systems*

Consider the non-smooth change of coordinates (9.25). The original dynamical system is given by (9.19)–(9.22), (9.24). The transformed system is given by (9.28). It is of interest to ensure the equivalency of the systems.

Zhuravlev [11] first proposed the non-smooth transformation (9.25) to avoid jumps in the solutions. The concept was later utilised in [12] to study periodic solutions and stability of planar vibro-impact systems. All these theorems assumed differentiability of the planar vector field $f(x)$ where $\ddot{x} = f(x)$. However, differentiability was required for the proof of asymptotic stability and not by the definition of the non-smooth transformation itself (see [12, Theorem 1]).

[1] The measure of time is in fact better represented by the Dirac measure [1] and the Dirac distribution.

The main idea of the transformation (9.25) is to use the mirror of the original trajectory $x_1(t)$ with the mirror placed on the constraint $x_1 = 0$ [1, Chap. 1]. The following analysis is included here from [1, Chap. 1] and from [12, Sect. 1] for completeness. Consider the transformation $x = |s| = s \text{ sign}(s)$ for the position and $\dot{x} = Rv \text{ sign}(s)$ for the velocity. Let the dynamical equations of the planar system $\ddot{x} = f(t, x, \dot{s}) + \sigma_{\dot{x}}(t_k)\delta_{t_k}$ be analysed in transformed coordinates, where $\sigma_{\dot{x}}(t_k)$ represents a jump in the velocity at time t_k. The definition $x = |s| = s \text{ sign}(s)$ dictates that $\dot{s} = \frac{\mathrm{d}}{\mathrm{d}t}\{s(t) \text{ sign}(s(t))\} + \sigma_x(t_j)\delta_{t_j}$, where t_j specifies an instant when the sign of $s(t)$ undergoes a change, $\sigma_{(\cdot)}$ represents a jump in the quantity, $\delta_{(\cdot)}$ denotes the Dirac delta measure and $\{\dot{F}\}$ represents the derivative of any function $F(t)$ calculated ignoring the points of discontinuity and which is not defined at the points of discontinuity. From the fact that $\sigma_x(t_j^+) = s(t_j^+) \text{ sign}(s(t_j^+)) - s(t_j^-) \text{ sign}(s(t_j^-))$ and since $s(t)$ is continuous (due to continuity of $x(t)$), it follows that

$$\dot{x} = \frac{\mathrm{d}}{\mathrm{d}t}\{s(t) \text{ sign}(s(t))\} \text{ sign}(s) \triangleq Rv \text{ sign}(s), \tag{9.30}$$

where the definition of the new coordinate $\dot{x} \triangleq Rv \text{ sign}(s)$ is used. Since, there is no jump in x, $\sigma_x = \sigma_s = 0$. Furthermore, the equality $\frac{\mathrm{d}}{\mathrm{d}t}\{s \text{ sign}(s)\} = \dot{s} \text{ sign}(s)$ holds true ignoring the points of discontinuity (i.e. $s = 0$). Hence, (9.30) produces the first equation in (9.25) by canceling sign(s) on both sides, so that $\dot{s} = Rv$. As for the velocity equation, it can be obtained that

$$\ddot{x} = f(t, x, \dot{x}) + \sigma_{\dot{x}}(t_k)\delta_{t_k} = \frac{\mathrm{d}}{\mathrm{d}t}\{Rv \text{ sign}(s)\} + \sigma_{Rv \text{ sign}(s)}(t_j)\delta_{t_j}, \tag{9.31}$$

where t_j denotes an instant where $\dot{x} = Rv \text{ sign}(s)$ may be discontinuous. Inspection of (9.25) shows that this can occur when either $s(t)$ or $v(t)$ crosses zero. In the case when $v(t_j) = 0$, it follows that $\sigma_{\dot{x}}(t_j) = \sigma_{Rv \text{ sign}(s)}(t_j) = 0$. The following can be obtained from the last term of (9.31):

$$\ddot{x} = R\dot{v} \text{ sign}(s) \tag{9.32}$$

In other words, the jump $\sigma_{\dot{x}}(t_k)$ has to equal zero, since $\sigma_{Rv \text{ sign}(s)}(t_j)$ is equal to zero due to $v(j) = 0$, for the equality (9.31) to hold true. Substituting $\ddot{x} = f$ from (9.31) into (9.32) and simplifying, $\dot{v} = R^{-1} \text{ sign}(s) f$ is obtained as given in the second equation in (9.25). However, if the trajectory intersects the v-axis with $v \neq 0$, then the change of sign is due to s and the jump can be computed as $\sigma_{\dot{x}}(t_j) = 2\, v(t_j) \text{ sign}(s(t_j^+)) = \sigma_{\dot{x}}(t_k)$, where t_k and t_j coincide. The quantity $2\, v(t_j) \text{ sign}(s(t_j^+))$ is obtained as follows. The jump occurs only when the variable s changes sign [12, Sect. 1]. When s changes sign from positive to negative with $v < 0$ on the $s = 0$ axis (i.e. from the fourth quadrant of the (s, v) plane to the third quadrant) at time $t = t_j$, the following can be obtained:

$$
\begin{aligned}
v(t_j^+) &= \lim_{t\to t_j^+} v(t_j) = \lim_{t\to t_j^-} v(t_j) = v(t_j^-)\\
\lim_{t\to t_j^+} \operatorname{sign}(s(t_j)) &= -\lim_{t\to t_j^-} \operatorname{sign}(s(t_j)) = -1\\
R(t_j^+) &= \lim_{t\to t_j^+} (1 - k \operatorname{sign}(s(t_j^+)) \operatorname{sign}(v(t_j^+))) = 1 - k\\
R(t_j^-) &= \lim_{t\to t_j^-} (1 - k \operatorname{sign}(s(t_j^-)) \operatorname{sign}(v(t_j^-))) = 1 + k\\
\sigma_{R v \operatorname{sign}(s)}(t_j) &= R(t_j^+)\, v(t_j^+) \operatorname{sign}(s(t_j^+)) - R(t_j^-)\, v(t_j^-) \operatorname{sign}(s(t_j^-))\\
\Rightarrow \sigma_{R v \operatorname{sign}(s)}(t_j) &= -2\, v(t_j^+) = 2\, v(t_j^+) \operatorname{sign}(s(t_j^+))
\end{aligned}
\tag{9.33}
$$

The same analysis can be carried out when s changes sign from negative to positive with $v > 0$ and $\sigma_{R v \operatorname{sign}(s)}(t_j) = 2\, v(t_j^+) \operatorname{sign}(s(t_j^+))$ is obtained.

The coincidence of t_k and t_j above is due to the fact that there is no jump in s or x for all times and the fact that the contact occurs on the constraint $x = 0$ when s changes sign with non-zero v. In this case $\sigma_{\dot{x}}(t_k) = \sigma_{R v \operatorname{sign}(s)}(t_j) = 2\, v(t_j) \operatorname{sign}(s(t_j^+))$ holds true and the same second equation of (9.25) is arrived due to cancellation of the equal jump terms on both sides of (9.31). Substituting $\ddot{x}$ from (9.31) into (9.32) and simplifying again produces $\dot{v}$ as given in the second equation in (9.25).

Hence, starting from the transformed system in (9.25), it follows that no impact occurs if the (s, v) trajectory crosses the s-axis (i.e. $\{(s, v) : v = 0\}$). If it crosses the v-axis, then this occurs when $s = 0$ (i.e. $x = 0$, the constraint is attained), and an impact occurs with magnitude $2\, v(t_j) \operatorname{sign}(s(t_j^+))$.

Before beginning with the proof of finite time stability, a summary of the stabilization characteristics is given. The following four facts are summarized:

1. The point $(0, 0) = (x_1, x_2)$ is the equilibrium of the dynamical system (9.19)–(9.22), (9.24). This is because $\dot{x}_1 = 0$, $\dot{x}_2 = 0$ is obtained when, for example, $\omega(t) = 0$ and $(x_1(0), x_2(0)) = (0, 0)$. Similarly, the point $(0, 0) = (s, v)$ is the equilibrium of the dynamical system (9.28). This is because

$$\dot{s} = R\, v = 0, \quad \dot{x}_2 = R^{-1} \operatorname{sign}(s)\, u(|s|, R\, v \operatorname{sign}(s)) = 0$$

   is obtained, for example, when $\omega(t) = 0$ and $(s(0), v(0)) = (0, 0)$. Hence, the trivial solution $(0, 0)$ is a unique solution and is the equilibrium point for both dynamical systems.
2. The non-smooth transformation (9.25) is such that the coordinates (s, v) cannot be retrieved from the (x_1, x_2) as it is a singular transformation. However, the point $(0, 0)$ is an exception in that it is uniquely transformed from (x_1, x_2) system to the (s, v) representation and vice versa. This claim can be verified as follows: $v = 0 \Rightarrow \dot{x} = R\, v \operatorname{sign}(s) = 0$. Also, the expression $\dot{x} = 0 \Rightarrow v = (R \operatorname{sign}(s))^{-1}\, \dot{x} = 0$ holds true. Furthermore, expressions $x = 0 \Rightarrow s = \pm x = 0$ and $s = 0 \Rightarrow x = |s| = 0$ hold true.
3. The boundedness of (s, v) guarantees boundedness of (x_1, x_2). This can be shown as follows. Within the compact set $D_{\tilde{R}} = \{(s, v) \in \mathbb{R}^2 : V(s, v) \leq \tilde{R}\}$, the

following is obtained by utilizing the inequalities $|s| \leq \frac{\tilde{R}}{\mu_2}, \quad |v| \leq \sqrt{2\tilde{R}}$:

$$|x_1| = |s| \leq \frac{\tilde{R}}{\mu_2}, \quad |x_2| = |Rv\,\mathrm{sign}(s)| = |\sqrt{R^2v^2}| \leq R\sqrt{2\tilde{R}} \tag{9.34}$$

4. There exists, as shown in the following sections, a semi-global Lyapunov function proving equiuniform finite time stability of (9.28), thereby proving equiuniform finite time stability of (9.19)–(9.22), (9.24) due to points 1, 2 and 3.

9.3.2 Consideration of Existing Lyapunov Functions

This section analyzes a class of existing continuous finite time controllers with regard to their potential to solve the problem statement presented in Sect. 9.3. It turns out, as shown in the following, that underlying Lyapunov proofs do not hold. Recall that the control law in the original coordinates is given in [17] as follows:

$$\begin{aligned} u(x_1, x_2) = & -\mathrm{sign}(x_2)|x_2|^{\alpha} \\ & -\mathrm{sign}(\phi_\alpha(x_1, x_2))|\phi_\alpha(x_1, x_2)|^{\frac{\alpha}{2-\alpha}} \end{aligned} \tag{9.35}$$

where, $\phi_\alpha(x_1, x_2) = x_1 + \frac{1}{2-\alpha}\,\mathrm{sign}(x_2)|x_2|^{2-\alpha}$. The closed-loop system is then given by (9.19), (9.35). The following transformed closed-loop system can be obtained by applying the non-smooth coordinate transformation (9.25):

$$\begin{aligned} \dot{s} &= Rv \\ \dot{v} &= R^{-1}\,\mathrm{sign}(s)\,(u(|s|, Rv\,\mathrm{sign}(s))) \\ &= -R^{\alpha-1}\,\mathrm{sign}(v)|v|^{\alpha} - R^{-1}\,\mathrm{sign}(s)\,\mathrm{sign}(\phi_\alpha)|\phi_\alpha|^{\frac{\alpha}{2-\alpha}}, \end{aligned} \tag{9.36}$$

where $\phi_\alpha(s, v) = |s| + \frac{R^{2-\alpha}}{2-\alpha}\,\mathrm{sign}(sv)|v|^{2-\alpha}$. Let the Lyapunov function proposed in [17] be defined in the transformed coordinates as $V(s, v) = \frac{2-\alpha}{3-\alpha}|\phi_\alpha|^{\frac{3-\alpha}{2-\alpha}} + r_2 v\phi_\alpha + \frac{r_1}{3-\alpha}|v|^{3-\alpha}$ where $r_1 > 1, r_2 < 1$ are arbitrary scalars. The temporal derivative of $V(s, v)$ along the trajectories of the transformed closed-loop system (9.36) can be obtained as follows:

$$\begin{aligned} \dot{V}(s, v) = & -R^{\alpha-1}r_1v^2 - R^{1-\alpha}|v|^{1-\alpha}|\phi_\alpha|^{\frac{1+\alpha}{2-\alpha}} \\ & -R^{-1}r_2|\phi_\alpha|^{\frac{2}{2-\alpha}}\,\mathrm{sign}(s) \\ & -R^{\alpha-1}r_2\phi_\alpha\,\mathrm{sign}(v)|v|^{\alpha} \\ & -(R^{1-\alpha}r_2 + R^{-1}r_1\,\mathrm{sign}(s))\,\mathrm{sign}(v\phi_\alpha)|v|^{2-\alpha}|\phi_\alpha|^{\frac{\alpha}{2-\alpha}} \end{aligned} \tag{9.37}$$

Although the homogeneity properties

$$V(k^{2-\alpha}s, kv) = k^{3-\alpha}V(s, v), \dot{V}(k^{2-\alpha}s, kv) = k^2\dot{V}(s, v) \tag{9.38}$$

hold true for the transformed system, it is not mathematically correct to restrict the analysis on the closed curve $(s, v) : \max_{s,v \neq (0,0)} (|\phi_\alpha|^{\frac{1}{2-\alpha}}, |v|) = 1$ encircling the origin of the closed-loop system (9.36) as was done in [17]. This is because it is always possible to have initial conditions either starting from or intersecting with the semi-axis $(s, v) : v = 0, s < 0$ of the transformed system (9.36) thereby causing (9.37) to take positive values due to an additional 'sign(s)' in the third term on the right hand side of (9.37). This is clearly in contrast to the negative definiteness of the derivative of the Lyapunov function obtained in [17] which enabled analysis to be performed only on the closed curve encircling the origin when combined with the homogeneity property (9.38).

Thus, motivated by the fact that the Lyapunov framework presented in [17] is applicable neither to the original jump-system (9.19), (9.35) due to the violation of the continuity requirements at the time of a jump nor to the transformed system (9.36) due to the non-negative definiteness of the derivative of the Lyapunov function, a proof of finite time stability of the closed-loop system (9.19), (9.24) is now presented.

9.4 Uniform Finite Time Stability

This section achieves the aim as stated in Sect. 9.3. The following Lemma is first presented from [8, Lemma 1]:

Lemma 9.1 *Assume $\bar{e} \in (0, 1)$, then the following is true:*

$$\mathrm{sign}(sv)\ \mathrm{sign}(R - R^{-1}) = -1 \tag{9.39}$$

It is of interest to note that the discontinuity and in turn Filippov's inclusion [15] in (9.28) is caused by the fact that R switches between two positive values on sets $\{(s, v) : s = 0\}$, $\{(s, v) : v = 0\}$ of Lebesgue measure zero. Let the two values of R be defined as follows:

$$R = \begin{cases} R_1 = \frac{2}{1+\bar{e}}, & \text{if } \mathrm{sign}(sv) = -1; \\ R_2 = \frac{2\bar{e}}{1+\bar{e}}, & \text{if } \mathrm{sign}(sv) = 1. \end{cases} \tag{9.40}$$

Then, it is trivial to note that given $\bar{e} \in (0, 1)$, the following is true from the computations in Lemma 9.1:

$$\begin{aligned} &R_1 > R_2 > 0\,,\ R_1^{-1} < R_2^{-1},\ |R_1 - R_1^{-1}| < |R_2 - R_2^{-1}| \\ &|R_1 - R_1^{-1}| = \frac{3 + \bar{e}}{2}|k|\,,\ |R_2 - R_2^{-1}| = \frac{3\bar{e} + 1}{2\bar{e}}|k| \end{aligned} \tag{9.41}$$

It is now possible to state the main results of this chapter.

Theorem 9.1 *Given* $M = 0$, $\alpha \in [0, 1)$, *the impact system* (9.19), (9.20), (9.21), (9.22), (9.24) *and its transformed version* (9.28) *are globally finite time stable.*

Proof Lyapunov stability analysis can be performed in the transformed coordinates since both represent the same system. Let a Lyapunov function candidate be given as follows:

$$V(s, v) = \mu_2 \frac{2-\alpha}{2} |s|^{\frac{2}{2-\alpha}} + \frac{1}{2} v^2 \tag{9.42}$$

Note that the function $V(s, v)$ is a globally radially unbounded $\mathscr{C}^1$ smooth positive definite function. By computing the temporal derivative of this function along the system trajectories of (9.28) with $M = 0$, it is obtained that,

$$\begin{aligned} \dot{V} \leq\ & \mu_2 |v||s|^{\frac{\alpha}{2-\alpha}} |R - R^{-1}| \operatorname{sign}(sv) \operatorname{sign}(R - R^{-1}) \\ & - \mu_1 R^{\alpha-1} |v|^{\alpha+1} \end{aligned} \tag{9.43}$$

From Lemma 9.1, (9.43) can be simplified as follows:

$$\dot{V} \leq -\mu_2 |v||s|^{\frac{\alpha}{2-\alpha}} |R - R^{-1}| - \mu_1 R^{\alpha-1} |v|^{\alpha+1} \tag{9.44}$$

It can be verified that $R^{\alpha-1} > 0$ for either sign of $\operatorname{sign}(s\,v)$ since $\bar{e} \in (0, 1)$. Furthermore, the trajectories of the closed-loop system (9.28) in the (s, v) plane never generate a sliding mode on $v = 0$ since $v\dot{v} \not< 0$. Since the equilibrium point $s = v = 0$ is the only trajectory of (9.28) on the invariance manifold $v = 0$ where $\dot{V}(s, v) = 0$ and since (9.44) holds true for almost all t, the differential inclusion (9.28) is globally equiuniformly asymptotically stable by applying the invariance principle [32]. Moreover, the system described in (9.28) is globally homogeneous of negative degree $q = -1$ with respect to the dilation $r = \left(\frac{2-\alpha}{1-\alpha}, \frac{1}{1-\alpha}\right)$ and is globally equiuniformly finite time stable according to [14, Theorem 3.1]. The proof is complete by noting that the proof of Theorem 9.1 for the case $\alpha = 0$ coincides with that appearing in [8, Theorem 1]. □

The closed-loop system (9.28) is a globally homogeneous system if $\omega(x, t) = 0\, \forall t \geq 0$. Consider next the case when M takes a nonzero value. The control law (9.27) can reject any disturbance ω with a uniform upper bound (9.29). Global asymptotic stability was established in [18]. However, to establish finite time stability, *uniform* asymptotic stability is required [14]. The next theorem achieves this objective for the transformed system (9.28) (and equivalently for the jump system (9.19)–(9.22), (9.24)).

Theorem 9.2 *Given* $\alpha \in \{\{0\} \cup (\frac{1}{2}, 1)\}$, *the closed-loop impact system* (9.19)–(9.22), (9.24) *and its transformed version* (9.28) *are globally equiuniformly finite time stable, regardless of whichever disturbance* ω, *satisfying condition* (9.29) *(or equivalently* (9.23)*) with* $M < \mu_1 < \mu_2$, *affects the system.*

Proof The proof is given in several steps. Consider the case $\alpha \in (\frac{1}{2}, 1)$.

1. Global Asymptotic Stability

Under the conditions of the theorem, the time derivative of the Lyapunov function (9.42), computed along the trajectories of (9.28) is estimated as follows:

$$\dot{V} \leq -\mu_2 |v||s|^{\frac{\alpha}{2-\alpha}} |R - R^{-1}| - (\mu_1 - M)R^{\alpha-1}|v|^{\alpha+1} \tag{9.45}$$

The first term on the right hand side of (9.45) follows from Lemma 9.1. Since $M < \mu_1$ by a condition of the Theorem, the global asymptotic stability of (9.28) is then established by applying the invariance principle [32] as discussed in Theorem 9.1.

2. Semiglobal Strong Lyapunov Functions.

The goal of this step is to show the existence of a parameterized family of local Lyapunov functions $V_{\tilde{R}}(s, v)$, with an a-priori given $\tilde{R} > 0$, such that each $V_{\tilde{R}}(s, v)$ is well-posed on the corresponding compact set

$$D_{\tilde{R}} = \{(s, v) \in \mathbb{R}^2 : V(s, v) \leq \tilde{R}\}. \tag{9.46}$$

In other words, $V_{\tilde{R}}(s, v)$ is to be positive definite on $D_{\tilde{R}}$ and its derivative, computed along the trajectories of the uncertain system (9.28) with initial conditions within $D_{\tilde{R}}$, is to be negative definite in the sense that,

$$\dot{V}_{\tilde{R}}(s, v) \leq -W_{\tilde{R}}(s, v) \tag{9.47}$$

for all $(s, v) \in D_{\tilde{R}}$ and for some $W_{\tilde{R}}(s, v)$, positive definite on $D_{\tilde{R}}$. A parameterized family of Lyapunov functions $V_{\tilde{R}}(s, v)$, $\tilde{R} > 0$ with the properties defined above are constructed by combining the Lyapunov function V of (9.42), where the time derivative along the system motion is only negative semi-definite, with the indefinite function

$$U(s, v) = U_1(s, v) + U_2(s, v) = s\,v|v|^{\alpha} + \kappa_1 \alpha s^3 v, \tag{9.48}$$

as follows:

$$V_{\tilde{R}}(s, v) = V(s, v) + \kappa_{\tilde{R}} U(s, v), \tag{9.49}$$

where the weight parameters $\kappa_{\tilde{R}}$, κ_1 are chosen small enough, namely,

$$\begin{aligned} \kappa_{\tilde{R}} &< \min \left\{ \begin{matrix} \frac{1}{\left(\sqrt{2\tilde{R}}\right)^{2\alpha} + \kappa_1 \alpha}, & \frac{(2-\alpha)\mu_2}{\rho^{(1-\alpha)}\left(1+\kappa_1 \alpha \rho^{2(2-\alpha)}\right)}, \\ \frac{\mu_2 |R_1 - R_1^{-1}|}{\Theta}, & \frac{\mu_1 - M}{R_1 \sqrt{2\tilde{R}}} \end{matrix} \right\} \\ \kappa_1 &< \frac{\mu_2 R_1^{-1}(1+\alpha)}{3\alpha(\mu_1 + M) R_1^{\alpha-1} \rho^{\frac{4-3\alpha}{2}} \sqrt{2\tilde{R}}} \end{aligned} \tag{9.50}$$

$$\begin{aligned}\rho &= \frac{2\tilde{R}}{(2-\alpha)\mu_2} \\ \Theta &= (\mu_1 + M)(1+\alpha)R_1^{\alpha-1}(2\tilde{R})^{\frac{2\alpha-1}{2}}\rho^{1-\alpha} + \\ &3\kappa_1\alpha R_1\rho^{\frac{4-3\alpha}{2}}\sqrt{2\tilde{R}}\end{aligned} \tag{9.51}$$

and R_1 is defined in (9.40). Noting that, due to (9.45), all possible solutions of the uncertain system (9.28), initialized at $t_0 \in \mathbb{R}$ within the compact set (9.46), are a-priori estimated by

$$\sup_{t\in[t_0,\infty)} V(s,v) \leq \tilde{R}, \tag{9.52}$$

the following inequalities hold true:

$$|s|^{\frac{2}{2-\alpha}} \leq \rho, \quad |v| \leq \sqrt{2\tilde{R}}. \tag{9.53}$$

The Lyapunov function (9.49) is positive definite on the compact set (9.46), for all $(s,v) \in D_{\tilde{R}}$ and $\kappa_{\tilde{R}} > 0$ satisfying (9.50) as shown below:

$$\begin{aligned}V_{\tilde{R}}(s,v) &= \mu_2\frac{2-\alpha}{2}|s|^{\frac{2}{2-\alpha}} + \frac{1}{2}v^2 + \kappa_{\tilde{R}}s\,v|v|^{\alpha} + \kappa_{\tilde{R}}\kappa_1\alpha s^3 v \\ &\geq \mu_2\frac{2-\alpha}{2}|s|^{\frac{2}{2-\alpha}} + \frac{1}{2}v^2 - \frac{1}{2}\kappa_{\tilde{R}}s^2 - \frac{1}{2}\kappa_{\tilde{R}}|v|^{2\alpha}v^2 \\ &-\frac{1}{2}\kappa_{\tilde{R}}\kappa_1\alpha s^6 - \frac{1}{2}\kappa_{\tilde{R}}\kappa_1\alpha v^2 \geq L_{\tilde{R}}V(s,v)\end{aligned} \tag{9.54}$$

where,

$$L_{\tilde{R}} < \min\left\{\begin{array}{l}\mu_2\frac{2-\alpha}{2} - \frac{1}{2}\kappa_{\tilde{R}}\,\rho^{(1-\alpha)}\left(1+\kappa_1\alpha\rho^{2(2-\alpha)}\right), \\ 1 - \kappa_{\tilde{R}}(\sqrt{2\tilde{R}}^{2\alpha} + \kappa_1\alpha)\end{array}\right\},$$

the trivial inequality $2ab > -(a^2+b^2)$, $\forall a, b \in \mathbb{R}$, the equalities

$$s^6 = |s|^{\frac{2(1-\alpha)}{2-\alpha}}|s|^{\frac{2}{2-\alpha}}|s|^4, \quad |v|^{2(\alpha+1)} = v^2|v|^{2\alpha} \tag{9.55}$$

and the bound (9.53) are utilized. It should be noted that $L_{\tilde{R}} > 0$ due to (9.50) and hence positive definiteness of $V_{\tilde{R}}$ is ensured from (9.54) on $D_{\tilde{R}}$. Similarly, it can be shown that the following inequality holds:

$$V_{\tilde{R}}(s,v) \leq M_{\tilde{R}}V(s,v) \tag{9.56}$$

where,

$$M_{\tilde{R}} > \max \left\{ \begin{array}{l} \mu_2 \frac{2-\alpha}{2} + \frac{1}{2}\kappa_{\tilde{R}}\, \rho^{(1-\alpha)} \left(1 + \kappa_1 \alpha \rho^{2(2-\alpha)}\right), \\ 1 + \kappa_{\tilde{R}} (\sqrt{2\tilde{R}}^{2\alpha} + \kappa_1 \alpha) \end{array} \right\}. \tag{9.57}$$

The time derivative of the indefinite function $U(s, v)$ along the trajectories of the uncertain system (9.28) is obtained as follows:

$$\begin{aligned} \dot{U}_1(s, v) \leq{} & R|v|^{\alpha+2} + (\mu_1 + M)(1+\alpha)R^{\alpha-1}|s||v|^{2\alpha} \\ & - \mu_2(1+\alpha)R^{-1}|s|^{\frac{2}{2-\alpha}}|v|^{\alpha} \end{aligned} \tag{9.58}$$

Similarly,

$$\begin{aligned} \dot{U}_2(s, v) \leq{} & 3\kappa_1 \alpha R s^2 v^2 + \kappa_1 \alpha (\mu_1 + M) R^{\alpha-1}|s|^3|v|^{\alpha} \\ & - \kappa_1 \alpha R^{-1} \mu_2 |s|^{\frac{6-2\alpha}{2-\alpha}} \end{aligned} \tag{9.59}$$

It should be noted that the inequality

$$\begin{aligned} |s| &= |s|^{\frac{2(1-\alpha)}{2-\alpha}}|s|^{\frac{\alpha}{2-\alpha}} \leq \rho^{1-\alpha}|s|^{\frac{\alpha}{2-\alpha}}, \\ |v|^{2\alpha} &= |v||v|^{2\alpha-1} \leq |v|\sqrt{2\tilde{R}}^{2\alpha-1} \end{aligned} \tag{9.60}$$

holds true for the case $\alpha \in (\frac{1}{2}, 1)$ of the Theorem. The inequalities of (9.58) and (9.59) are re-written by utilizing (9.53) and (9.60) as follows:

$$\begin{aligned} \dot{U}_1(s, v) \leq{} & (\mu_1 + M)(1+\alpha)R^{\alpha-1}|s|^{\frac{\alpha}{2-\alpha}}|v|\sqrt{2\tilde{R}}^{2\alpha-1}\rho^{1-\alpha} \\ & + R|v|^{\alpha+1}\sqrt{2\tilde{R}} - \mu_2(1+\alpha)R^{-1}|s|^{\frac{2}{2-\alpha}}|v|^{\alpha} \end{aligned} \tag{9.61}$$

$$\begin{aligned} \dot{U}_2(s, v) \leq{} & 3\kappa_1 \alpha R |s|^{\frac{\alpha}{2-\alpha}}|v|\,\rho^{\frac{4-3\alpha}{2}}\sqrt{2\tilde{R}} - \kappa_1 \alpha \mu_2 R^{-1}|s|^{\frac{6-2\alpha}{2-\alpha}} \\ & + \kappa_1 \alpha (\mu_1 + M) R^{\alpha-1}|s|^3|v|^{\alpha} \end{aligned} \tag{9.62}$$

where the corresponding upper bound on $|v|$ and $|s|$ from (9.53) are utilized. The parameter R in (9.45) and (9.61) is a state function and keeps switching between the two values as shown in Lemma 9.1. This corresponds to the fact that the rate of decay of the Lyapunov function (9.49) switches depending on R. Hence, by combining (9.45) and (9.61) and by considering the slowest decay by utilizing (9.41), the time derivative of (9.49) can be obtained as follows:

$$\begin{aligned} \dot{V}_{\tilde{R}} \leq{} & -\beta_1|v||s|^{\frac{\alpha}{2-\alpha}} - \beta_2|v|^{\alpha+1} - \kappa_{\tilde{R}}\beta_3|v|^{\alpha}|s|^{\frac{2}{2-\alpha}} \\ & - \kappa_{\tilde{R}}\kappa_1 \alpha \mu_2 R^{-1}|s|^{\frac{6-2\alpha}{2-\alpha}}, \end{aligned} \tag{9.63}$$

where $\beta_1 = \mu_2|R_1 - R_1^{-1}| - \kappa_{\tilde{R}}\Theta$, $\beta_2 = (\mu_1 - M)R_1^{\alpha-1} - \kappa_{\tilde{R}}R_1\sqrt{2\tilde{R}}$, $\beta_3 = \mu_2 R_1^{-1}(1+\alpha) - \kappa_1\alpha(\mu_1 + M)R_1^{\alpha-1}\rho^{\frac{4-3\alpha}{2}}\sqrt{2\tilde{R}}$, (9.53) and the equality $|s|^3 = |s|^{\frac{2}{2-\alpha}}|s|^{\frac{4-3\alpha}{2-\alpha}}$ are utilised. It should be noted that the expressions $\beta_1 > 0$, $\beta_2 > 0$, $\beta_3 > 0$ hold true due to (9.50). Ignoring the negative semi-definite terms with β_1, β_3, (9.63) can be rewritten as follows:

$$\dot{V}_{\tilde{R}} \leq -\beta_2|v|^{\alpha+1} - \kappa_{\tilde{R}}\kappa_1\alpha\mu_2 R^{-1}|s|^{\frac{6-2\alpha}{2-\alpha}} \tag{9.64}$$

Furthermore, the following inequalities hold true within the compact set (9.46):

$$\begin{aligned} v^2 = |v|^2 = |v|^{\alpha+1}|v|^{1-\alpha} &\leq |v|^{\alpha+1}\left(\sqrt{2\tilde{R}}\right)^{1-\alpha} \\ \Rightarrow -|v|^{\alpha+1} &\leq -\frac{v^2}{\left(\sqrt{2\tilde{R}}\right)^{1-\alpha}} \end{aligned} \tag{9.65}$$

Hence, (9.64) can be simplified as follows:

$$\dot{V}_{\tilde{R}} \leq -c_{\tilde{R}}\left[|s|^{\frac{6-2\alpha}{2-\alpha}} + v^2\right] \tag{9.66}$$

where,

$$c_{\tilde{R}} = \min\left\{\frac{\beta_2}{\left(\sqrt{2\tilde{R}}\right)^{1-\alpha}},\quad \kappa_{\tilde{R}}\kappa_1\alpha\mu_2 R_1^{-1}\right\} > 0. \tag{9.67}$$

Case 1: $|s| \geq 1$: The following inequality holds true for $|s| \geq 1$:

$$\frac{6-2\alpha}{2-\alpha} \geq \frac{2}{2-\alpha} \Leftrightarrow |s|^{\frac{6-2\alpha}{2-\alpha}} \geq |s|^{\frac{2}{2-\alpha}} \tag{9.68}$$

Also, the following can be obtained from (9.56):

$$\frac{M_{\tilde{R}}}{2}\max\{1, \mu_2(2-\alpha)\}(|s|^{\frac{2}{2-\alpha}} + v^2) \geq V_{\tilde{R}}(s, v) \tag{9.69}$$

Hence, the following inequality is then obtained for $|s| \geq 1$ by combining (9.66), (9.68) and (9.69):

$$\dot{V}_{\tilde{R}} \leq -\bar{\kappa}_1 V_{\tilde{R}} \tag{9.70}$$

where, $\bar{\kappa}_1 = \frac{2c_{\tilde{R}}}{M_{\tilde{R}}\max\{1, \mu_2(2-\alpha)\}}$.

Case 2: $|s| < 1$: Noting that the following inequalities hold true for $|s| < 1$:

$$|s|^{\frac{6-2\alpha}{2-\alpha}} > |s|^{\frac{2\gamma}{2-\alpha}} \Leftrightarrow \frac{6-2\alpha}{2-\alpha} < \frac{2\gamma}{2-\alpha} \Leftrightarrow \gamma > 3-\alpha, \tag{9.71}$$

for some $\gamma > 3-\alpha$. As $3-\alpha < \frac{5}{2}$ always holds true due to $\alpha \in (\frac{1}{2}, 1)$, $\gamma \geq \frac{5}{2}$ is a valid choice. In the following, $\gamma = 3$ is chosen. It can be seen that the following equality holds:

$$\begin{aligned}\left(|s|^{\frac{2}{2-\alpha}} + \nu^2\right)^3 &= |s|^{\frac{6}{2-\alpha}} + 3|s|^{\frac{4}{2-\alpha}}\nu^2 + 3|s|^{\frac{2}{2-\alpha}}\nu^4 + \nu^6 \\ &\leq \max\{\rho^\alpha, K_1\}\left(|s|^{\frac{6-2\alpha}{2-\alpha}} + \nu^2\right)\end{aligned} \tag{9.72}$$

where the bounds (9.53) have been utilised resulting in the definition

$$K_1 = \max\left\{3\,\rho^2, 3\,\rho(2\tilde{R}), (2\tilde{R})^2\right\} > 0.$$

Note that the following can be obtained from (9.56):

$$\left(\frac{M_{\tilde{R}}}{2}\max\{1, \mu_2(2-\alpha)\}(|s|^{\frac{2}{2-\alpha}} + \nu^2)\right)^3 \geq \left(V_{\tilde{R}}(s,\nu)\right)^3 \tag{9.73}$$

Then, combining (9.66), (9.72) and (9.73):

$$\dot{V}_{\tilde{R}}(s,\nu) \leq -c_{\tilde{R}}\left(|s|^{\frac{6-2\alpha}{2-\alpha}} + \nu^2\right) \leq -\bar{\kappa}\left(V_{\tilde{R}}\right)^3 \tag{9.74}$$

where,

$$\bar{\kappa} = \frac{c_{\tilde{R}}}{\left(\frac{M_{\tilde{R}}}{2}\max\{1, \mu_2(2-\alpha)\}\right)^3 \max\{\rho^\alpha, K_1\}} > 0. \tag{9.75}$$

Hence, the desired uniform negative definiteness (9.47) is obtained by combining (9.70) and (9.74) as $W_{\tilde{R}}(s,\nu) = \min\left\{\bar{\kappa}_1 V_{\tilde{R}}, \bar{\kappa}\left(V_{\tilde{R}}\right)^3\right\}$.

3. Global Equiuniform Asymptotic Stability

Since the inequality (9.74) holds on the solutions of the uncertain system (9.28), initialized within the compact set (9.46), the decay of the function $V_{\tilde{R}}(s,\nu)$ can be found by considering the majorant solution $\nu(t)$ of $V_{\tilde{R}}$ as follows:

$$\dot{\nu}(t) = \begin{cases} -\bar{\kappa}_1\nu, & |s| \geq 1; \\ -\bar{\kappa}_2\nu^\gamma, & |s| < 1. \end{cases} \tag{9.76}$$

where, $\gamma > 3 - \alpha$ is introduced for generality. A more conservative decay than that in (9.76) can be computed. There are two possible sub-cases, namely, $\nu(t) \geq 1$ and $\nu(t) < 1$ for each of the cases $|s| \geq 1$ and $|s| < 1$. The following expressions hold true for a positive definite function $\nu(t)$ and a scalar $\gamma > 1$:

$$\begin{aligned} \nu(t)^\gamma \geq \nu(t) \Rightarrow -\nu(t)^\gamma \leq -\nu(t) \quad \text{if } \nu(t) \geq 1; \\ \nu(t)^\gamma \leq \nu(t) \Rightarrow -\nu(t) \leq -\nu(t)^\gamma \quad \text{if } \nu(t) < 1. \end{aligned} \tag{9.77}$$

Hence, the decay (9.76) is modified by utilising (9.77) independent of the magnitude of $|s|$ and dependent on $\nu(t)$ as follows:

$$\dot{\nu}(t) = \begin{cases} -\bar{\kappa}\nu, & \text{if } \nu(t) \geq 1; \\ -\bar{\kappa}\nu^\gamma, & \text{if } \nu(t) < 1. \end{cases} \tag{9.78}$$

where $\bar{\kappa} = \min\{\bar{\kappa}_1, \bar{\kappa}_2\} > 0$. The solution for the case $\nu(t) < 1$ can be integrated as follows:

$$\int_{\nu_0}^{\nu(t)} \frac{\mathrm{d}\nu(t)}{\nu^\gamma} = -\bar{\kappa} \int_{t_0}^{t} \mathrm{d}t \tag{9.79}$$

where $\nu_0 = \nu(t_0)$. The general solution of $\nu(t)$ of (9.78) can then be obtained by utilising (9.78) and by combining the solutions of both (9.70) and (9.79):

$$\nu(t) = \begin{cases} \nu(t_0)\, e^{-\bar{\kappa}(t-t_0)}, & \text{if } \nu(t) \geq 1; \\ \nu(t_1) \left(\frac{1}{\bar{\kappa}(t-t_1)(\gamma-1)\nu_0^{\gamma-1}+1} \right)^{\frac{1}{\gamma-1}}, & \text{if } \nu(t) < 1. \end{cases} \tag{9.80}$$

It can be easily seen that the solution $\nu(t) \to 0$ as $t \to \infty$ and that the decay rate depends on the gain parameters μ_1, μ_2, bound M on the disturbance ω and the system property R_1. On the compact set (9.46), the following inequality holds (see (9.54), (9.56)):

$$L_{\tilde{R}} V(s, \nu) \leq V_{\tilde{R}}(s, \nu) \leq M_{\tilde{R}} V(s, \nu) \tag{9.81}$$

for all $(s, \nu) \in D_{\tilde{R}}$ and positive constants $L_{\tilde{R}}, M_{\tilde{R}}$. The above inequalities (9.80) and (9.81) ensure that the globally radially unbounded function $V(x_1, x_2)$ decays asymptotically

$$V(s(t), \nu(t)) \leq \begin{cases} L_{\tilde{R}}^{-1} M_{\tilde{R}} \tilde{R} e^{-\bar{\kappa}(t-t_0)}, & \text{if } V_{\tilde{R}} \geq 1; \\ L_{\tilde{R}}^{-1} M_{\tilde{R}} \tilde{R} \left(\frac{1}{\bar{\kappa}(t-t_1)(\gamma-1)\nu_0^{\gamma-1}+1} \right)^{\frac{1}{\gamma-1}}, & \text{if } V_{\tilde{R}} < 1. \end{cases} \tag{9.82}$$

on the solutions of (9.28) uniformly in ω and the initial data, located within an arbitrarily large set (9.46). This proves that the uncertain system (9.28) is globally equiuniformly asymptotically stable around the origin $(s, v) = (0, 0)$.

4. Global Equiuniform Finite Time Stability.

Due to (9.29), the piece-wise continuous uncertainty $R_1^{\alpha-1}\omega(t)\,\mathrm{sign}(s)$ on the right hand side of the system (9.28) is locally uniformly bounded by $R_1^{\alpha-1}M|v|^\alpha$ whereas the remaining part of the feedback is globally homogeneous with homogeneity degree $q = -1$ with respect to dilation $r = (\frac{2-\alpha}{1-\alpha}, \frac{1}{1-\alpha})$. It remains to verify, however, whether the existing quasi-homogeneity result [14, Theorem 3.2] can be extended to the continuous case in question. Let the piece-wise continuous function $R_1^{\alpha-1}\omega(s, v, t)\,\mathrm{sign}(s) = R_1^{\alpha-1}\omega^c(s, v, t)\,\mathrm{sign}(s)$ be defined for an arbitrary $c \geq \max\{1, c_0\}$, where c_0 is lower homogeneity parameter, as follows:

$$\omega^c(s, v, t) = c^{q+r_2}\omega(c^{-r_1}s, c^{-r_2}v, c^q t) \tag{9.83}$$

where the right hand side represents a parameterised set of uncertainties. Then the following holds true:

$$|\omega^c(s, v, t)| \leq c^{q+r_2-\alpha r_2} M R^\alpha |v|^\alpha \tag{9.84}$$

It follows that all parameterised uncertainties represented by the right hand side of (9.83) are admissible in the sense of (9.29) if the inequality $c^{q+r_2-\alpha r_2} \leq 1$ holds true. From the definitions $r_2 = \frac{1}{1-\alpha}, q = -1$, it is obtained that $q + r_2 - \alpha r_2 = 0 \Rightarrow c^{q+r_2-\alpha r_2} \leq 1$. Appealing to [14, Definitions 2.9, 2.10], the solutions $s^c(t) = c^{r_1}s(c^q t), v^c(t) = c^{r_2}v(c^q t)$ are solutions of the system (9.28) with the piece-wise continuous function $\omega(s, v, t) = \omega^c(s, v, t)$. It follows that any solution of the differential equation (9.28) evolving within a homogeneity ball B_δ, generates a parameterised set of solutions $s^c(t), v^c(t)$ with the parameter c large enough. Hence, (9.28) is locally homogeneous of degree $q = -1$ with the dilation $(r_1, r_2) = (\frac{2-\alpha}{1-\alpha}, \frac{1}{1-\alpha})$. Thus, the globally equiuniformly asymptotically stable system (9.28) and in turn the original impact system (9.19), (9.24) are globally equiuniformly finite time stable according to [14, Theorem 3.1]. The proof is completed by noting that the proof of Theorem 9.2 for the case $\alpha = 0$ coincides with that appearing in [8, Theorem 2]. □

9.5 Conclusion

It can be seen from the results presented that setting $\alpha = 0$ is admissible by the main claims of this chapter. The resulting analysis produces a discontinuous controller and a non-smooth Lyapunov function given by (9.24) and (9.42) respectively. The Lyapunov analysis presented here then coincides with recent results in [8] as can be seen from (9.48) which produces $U(s, v) = sv$ as given in [8]. Hence, this work is a generalization of variable structure systems parameterised by α both in the sense of encompassing continuous and discontinuous right hand sides as well as in the sense of unifying the Lyapunov analysis when resets are present in the dynamics. A possible extension of the approach can be studied for the more general case of dynamical systems where the non-smooth transformation is applied to one or more generalized position variables [1, Chapter 7] for co-dimension one constraints. A related future direction is to establish homogeneity results and their connections to finite time stability for impact systems.

Acknowledgements This work was supported by the EPSRC under Grant EP/G053979/1.

References

1. Bernard, B.: Nonsmooth Impact Mechanics. Springer, London (1996)
2. Goebel, R., Sanfelice, R.G., Teel, A.: Hybrid dynamical systems. IEEE Control Syst. Mag. **29**(2), 28–93 (2009)
3. Chevallereau, C., Abba, G., Aoustin, Y., Plestan, F., Westervelt, E.R., Canudas-De-Wit, C., Grizzle, J.W.: Rabbit: a testbed for advanced control theory. IEEE Control Syst. Mag. **23**(5), 57–79 (2003)
4. Westervelt, E.R., Grizzle, J.W., Chevallereau, C., Choi, J.H., Morris, B.: Feedback Control of Dynamic Bipedal Robot Locomotion. CRC Press, Taylor and Francis Ltd., Boca Raton (2007)
5. Johansson, K.H., Egerstedt, M., Lygeros, J., Sastry, S.: On the regularization of Zeno hybrid automata. Syst. Control Lett. **38**(3), 141–150 (1999)
6. O'Toole, M.D., Navarro-López, E.M.: A hybrid automaton for a class of multi-contact rigid-body systems with friction and impacts. In: Proceedings of the 4th IFAC Conference on Analysis and Design of Hybrid Systems, pp. 298–306 (2012)
7. Bartolini, G., Pisanoa, A., Puntab, E., Usai, E.: A survey of applications of second-order sliding mode control to mechanical systems. Int. J. Control **76**(9), 875–892 (2003)
8. Oza, H.B., Orlov, Y.V., Spurgeon, S.K.: Finite time stabilization of a perturbed double integrator with unilateral constraints. Math. Comput. Simul. **95**, 200–212 (2014)
9. Kawski, M.: Homogeneity in control: Geometry and applications. In: Proceedings of the IEEE European Control Conference, pp. 2449–2457 (2015)
10. Oza, H.B., Orlov, Y.V., Spurgeon, S.K.: Continuous uniform finite time stabilization of planar controllable systems. SIAM J. Control Optim. **53**(3), 1154–1181 (2015)
11. Zhuravlev, V.F.: Equations of motion of mechanical systems with ideal onesided links. J. Appl. Math. Mech. **42**(5), 839–847 (1978)
12. Zhuravlev, V.F.: Analytical methods in the theory of vibro-impact systems. J. Appl. Math. Mech. **57**(2), 221–236 (1993)
13. Zhuravlev, V.F.: Application of nonsmooth transformations to analyze a vibroimpact duffing system. Int. Appl. Mech. **44**(10), 1173–1179 (2008)
14. Orlov, Y.: Finite time stability and robust control synthesis of uncertain switched systems. SIAM J. Control Optim. **43**(4), 1253–1271 (2004)
15. Filippov, A.F.: Differential Equations with Discontinuous Right-hand Sides. Springer (1988)
16. Stewart, D.E.: Rigid-body dynamics with friction and impact. SIAM Rev. **42**(1), 3–39 (2000)
17. Bhat, S.P., Bernstein, D.S.: Continuous finite-time stabilization of the translational and rotational double integrators. IEEE Trans. Autom. Control **43**(5), 678–682 (1998)
18. Orlov, Y., Aoustin, Y., Chevallereau, C.: Finite time stabilization of a perturbed double integrator-Part I: Continuous sliding mode-based output feedback synthesis. IEEE Trans. Autom. Control **56**(3), 614–618 (2010)

19. Haimo, V.T.: Finite time controllers. SIAM J. Control Optim. **24**(4), 760–770 (1984)
20. Teel, A.R., Zaccarian, L.: On "uniformity" in definitions of global asymptotic stability for time-varying nonlinear systems. Automatica **42**(12), 2219–2222 (2006)
21. Kalman, R.E., Bertram, J.E.: Control system analysis and design via the "Second Method" of Lyapunov, Part I: Continuous-time systems. J. Basic Eng. Trans. Asme. Ho **82**(2), 371–393 (1960)
22. Khalil, H.K.: Nonlinear Systems. Prentice Hall, New Jersey (2002)
23. Haddad, W.M., Nersesov, S.G., Du, L.: Finite-time stability for time-varying nonlinear dynamical systems. In: Proceedings of the American Control Conference, pp. 2449–2457 (2015)
24. Bhat, S., Bernstein, D.: Geometric homogeneity with applications to finite-time stability. Math. Control Signal Syst. **17**(2), 101–127 (2005)
25. Levant, A.: Homogeneity approach to high-order sliding mode design. Automatica **41**(5), 823–830 (2005)
26. Orlov, Y.V.: Discontinuous Systems. Springer, London (2009)
27. Bhat, S.P., Bernstein, D.S.: Finite-time stability of continuous autonomous systems. SIAM J. Control Optim. **38**(3), 751–766 (2000)
28. Grizzle, J.W., Abba, G., Plestan, F.: Asymptotically stable walking for biped robots: Analysis via systems with impulse effects. IEEE Trans. Autom. Control **46**(1), 51–64 (2001)
29. Bhat, S.P., Bernstein, D.S.: Finite-time stability of homogeneous systems. In: Proceedings of the American Control Conference, pp. 2513–2514 (1997)
30. Cai, C., Goebel, R., Teel, A.R.: Smooth Lyapunov functions for hybrid systems Part II: (Pre)Asymptotically stable compact sets. IEEE Trans. Autom. Control **53**(3), 734–748 (2008)
31. Goebel, R., Teel, A.R.: Lyapunov characterization of Zeno behavior in hybrid systems. In: Proceedings of the 47th IEEE Conference on Decision and Control, pp. 2752–2757 (2008)
32. Acho, A.L., Orlov, Y.: An invariance principle for discontinuous dynamic systems with applications to a coulumb friction oscillator. ASME J. Dyn. Syst. Meas. Control **122**, 687–690 (2000)

Chapter 10
Robustification of Cooperative Consensus Algorithms in Perturbed Multi-agents Systems

Alessandro Pilloni, Alessandro Pisano and Elio Usai

10.1 Introduction

Multi-Agent Systems (MAS) consists of a set of dynamical systems interconnected by a communication network. One of the most challenging and studied topics in this area consists in the coordination of the whole system in a cooperative, distributed way. This problem is commonly referred to as *consensus problem*, i.e., the problem of understanding how to force the states of the agents operating over a network to converge to (or "agree upon") a common decision (or "value"), while meeting the communication constraints.

When the objective is to distributedly achieve the agreement on some averaged quantity related to the local initialization of the network (such as the positions of a platoon of mobile systems, the sensed temperatures or humidity values on a wide area, or in general any set of local quantities taken from the field where the agents are deployed) this problem is commonly referred to as *average consensus*.

The consensus on an averaged information is a problem that finds applications in different fields, e.g., distributed computing [1, 2] and optimization [3–5], motion coordination [6, 7], sensor-fusion [8] and many more.

After the seminal work on decentralized algorithms solving the average consensus problem [9], more sophisticated protocols able to work under different network

A. Pilloni (✉) · A. Pisano · E. Usai
Department of Electrical and Electronic Engineering,
University of Cagliari, Piazza D'Armi, Cagliari, Italy
e-mail: alessandro.pilloni@diee.unica.it

A. Pisano
e-mail: pisano@diee.unica.it

E. Usai
e-mail: eusai@diee.unica.it

S. Li et al. (eds.), *Advances in Variable Structure Systems and Sliding Mode Control—Theory and Applications*, Studies in Systems, Decision and Control 115, DOI 10.1007/978-3-319-62896-7_10

topologies, and featuring different convergence properties, have been also proposed. Among them, in [10, 11] the so-called *ratio-consensus* was proposed to achieve average consensus in directed graphs. Then, nonlinear protocols able to fulfill the task in finite-time under undirected [12, 13] and time-varying network topologies [14] have been also presented in more recent times.

All these protocols suffer, however, from two significant problems. Firstly, the potential existence of a single "outlier agent", i.e., an agent whose initial state holds an abnormal value and that can arbitrarily affects the whole network behavior.

This issue has been firstly investigated in [15–17]. There, the main idea was to identify the misbehaving agents in a decentralized way and then, recover a correct network state after the outlier agent was removed. In spite of these remarkable attempts to design clever reconfigurable interaction rules, the average value remain, from a statistical point of view, highly sensitive against the presence of outlier agents [18].

More recently, the development of algorithms providing the distributed computation of different, more robust, statistical measures, such as the median value, has become an interesting research issue (see [19, 20] and the references therein).

The second problem to be taken into account is that, whenever the agents consist of physical systems, e.g. mobile robots, unmodeled dynamics or external perturbations may affect the local agents dynamics. It follows that the presence of these perturbations shifts the steady-equilibrium of the network with respect to the expected "nominal" one. To deal with this problem, robust sliding-mode concepts have been involved in the designing of communication protocols for consensus purposes. In particular, the problem of steering the system's states to a common steady-behavior in spite of the heterogeneous, unknown perturbations has been addressed, in [21–23] for network of first-order agents, and in [24–26] for network of second order agents, to cite a few. Although all those algorithms preserve the network synchronization in spite of disturbances, due to the exogenous nature of the disturbance signals, the steady equilibrium of the network cannot be predicted a-priori because of these perturbations [27]. In general, in the perturbed scenario the network equilibrium becomes a time-varying unpredictable trajectory.

In the current literature, the achievement of consensus on a pre-specified function (or value) of the initial conditions of the network, such as the average or the median value of the agents' initial conditions, in spite of heterogenous perturbations is studied in [28, 29].

Thus motivated, in this chapter we discuss two schemes, that allow to establish, respectively, the average-consensus, and, the median-consensus, with the agents dynamics being governed by perturbed integrators under undirected communications. Both these schemes are based on the Integral Sliding-Mode Control (ISMC) design paradigm. In contrast to the conventional sliding mode control (SMC) schemes, the constrained sliding motion under ISMC has the same dimension of the original state space. Thus, no order reduction is achieved throughout the sliding motion. However, since the parameters of the sliding surface can be chosen arbitrarily, e.g., such that the system trajectories will slide on the surface from the initial time instant on, the reaching phase is eliminated [30]. Some interesting applications of ISMC solution can be found in [31–33]. More recently, the ISMC is also employed in the contest of MAS in [34], to achieve the network synchronization in a leader-follower scenario. However, the distributed tracking problem is a completely different task with respect to make consensus on same a-priory unknown function of the system states. In fact, the leader profile represents the reference steady profile for the whole MAS and this value is known at least from some agent operating over the network. In contrast, in the average (or median) consensus problem, the network itself cooperatively estimates such an aggregated information, and this information will become the global equilibrium for the networked system.

Summarizing, in this chapter we present from a unified perspective the two approaches in [28, 29], thereby allowing the reader to more clearly understand the basic principles of the underlying common ideas and permitting their generalization to more involved scenarios and "robust consensus" problems.

The Chapter is organized as follows: the notation used throughout, along with some concepts from the Graph Theory are outlined in Sect. 10.2. The problem formulation and the proposed algorithms are presented in Sect. 10.3. Simulation results, supporting the proposed design, are given in Sect. 10.4. Finally, conclusions and perspectives for future investigations are collected in the final Sect. 10.5.

10.2 Notations and Mathematical Preliminaries

10.2.1 Mathematical Notation

Hereinafter the mathematical notation used throughout the paper is listed. We denote by $\mathbb{R}$ the set of real numbers. A vector $\boldsymbol{x} = [x_i] \in \mathbb{R}^n$ is viewed as a column vector, and x_i $(i = 1, 2, \ldots, n)$ represents the ith component of $\boldsymbol{x}$. The transpose of matrix $\boldsymbol{A} \in \mathbb{R}^{n\times m}$ and of a vector $\boldsymbol{x} \in \mathbb{R}^n$ are denoted as $\boldsymbol{A}'$ and $\boldsymbol{x}'$, respectively. The spectrum of eigenvalues of a square matrix $\boldsymbol{B} \in \mathbb{R}^{n\times n}$ is denoted by $eig\{\boldsymbol{B}\}$. The scalar product of two vectors with compatible dimension, respectively, $\boldsymbol{x}$, and, $\boldsymbol{y} \in \mathbb{R}^n$, is denoted by $\boldsymbol{x}'\boldsymbol{y}$. $\|\boldsymbol{x}\|_p$ denotes the p-norm of the vector $\boldsymbol{x}$. We write $\|\boldsymbol{x}\|_\infty$ to denote the max norm, $\|\boldsymbol{x}\|_\infty = \max_{1\le i\le n} |x_i|$. The all-one and all-zero vectors are denoted as $\mathbf{1}_n = [1, \ldots, 1]' \in \mathbb{R}^n$ and $\mathbf{0}_n = [0, \ldots, 0]' \in \mathbb{R}^n$. $\boldsymbol{I}_n \in \mathbb{R}^{n\times n}$ stands for the identity matrix. A block-diagonal matrix is denoted by $diag(\boldsymbol{D}_1, \ldots, \boldsymbol{D}_n)$ where $\boldsymbol{D}_1,\ldots,\boldsymbol{D}_n$ are square matrices with arbitrary dimensions. Let $\boldsymbol{\mathcal{S}} = \{s_1, \ldots, s_p\}$ be a set of p arbitrary elements, $card(\boldsymbol{\mathcal{S}}) = p$ denotes its cardinality. $\otimes$ denotes the Matrix Kronecker Product.

10.2.2 Graph Theory

Here, a short overview of the main concepts and essential properties related with the Graph Theory, which are employed and referred throughout the paper, is presented.

For further details the reader is referred to [35]. A weighted directed graph (or simply a *digraph*) $\mathcal{G}(\mathcal{V}, \mathcal{E}, \mathcal{A})$ consists of a *vertex set* $\mathcal{V} = \{1, \ldots, n\}$, and an *edge set* $\mathcal{E} \subseteq \{\mathcal{V} \times \mathcal{V}\}$ having cardinality $m = card(\mathcal{E})$, whereas $\mathcal{A} \in \mathbb{R}^{n\times n}_{\geq 0}$ denotes its *adjacency matrix*, with $a_{ij} = 1$ iff $(v_i, v_j) \in \mathcal{E}$. The *set of neighbors of the node i* is denoted by $\mathcal{N}_i = \{j \in \mathcal{V} : (v_i, v_j) \in \mathcal{E}\}$. A digraph is *undirected* (or simply a *graph*), if $(v, u) \in \mathcal{E}$ whenever $(u, v) \in \mathcal{E}$. A *path* is a sequence of vertices connected by edges. A digraph is *strongly connected* (resp., a graph is *connected*) if there is a path between any pair of vertices. The topology of graph $\mathcal{G}$ is encoded by the *Laplacian Matrix* $\mathcal{L} = [\ell_{ij}] \in \mathbb{R}^{n\times n}$, whose entries are defined as follows

$$\ell_{ij} := \begin{cases} \sum_{j=1, j\neq i}^{n} a_{ij}, & \text{if } i = j, \\ -a_{ij}, & \text{if } (i, j) \in \mathcal{E} \text{ and } i \neq j, \\ 0, & \text{otherwise.} \end{cases} \tag{10.1}$$

By construction, $\mathcal{L}$ is a positive semi-definite matrix and satisfies $\mathcal{L}\mathbf{1}_n = \mathbf{0}_n$. If $\mathcal{G}$ is a strongly connected (resp. connected) digraph (resp. graph), then 0 is a simple eigenvalue of $\mathcal{L}$. If $\mathcal{L} = \mathcal{L}'$ we can conclude that $\mathcal{G}$ is undirected. Let $\mathcal{G}$ be a graph with unitary weights, i.e. $a_{i,j} \in \{0, 1\}$, the *incidence matrix* $\mathcal{D} = [d_{ij}] \in \mathbb{R}^{n\times m}$ associated with $\mathcal{G}$ is a rectangular matrix whose entries satisfies

$$d_{ij} := \begin{cases} 1/\sqrt{2}, & \text{if arc } j = (i, k) \in \mathcal{E} \\ -1/\sqrt{2}, & \text{if arc } j = (k, i) \in \mathcal{E} \\ 0, & \text{otherwise.} \end{cases} \tag{10.2}$$

The incidence matrix $\mathcal{D}$ of an undirected graph $\mathcal{G}$ with unitary weight is strictly related with the Laplacian Matrix $\mathcal{L}$. In particular, the next relations result to be satisfied by construction

$$\mathcal{L} = \mathcal{D}\mathcal{D}' \tag{10.3}$$

$$\mathcal{D}'\mathbf{1}_n = \mathbf{0}_m \tag{10.4}$$

10.3 Problem Formulation and Main Results

10.3.1 Problem Formulation

Consider a network consisting of n agents communicating to each other over an undirected communication network whose topological constraints are encoded by a connected graph $\mathcal{G}$. Let each agent be governed by the following first-order perturbed dynamics

$$\dot{x}_i(t) = w_i(t) + u_i(t), \quad i \in \mathcal{V}, \tag{10.5}$$

where $x_i \in \mathbb{R}$ and $u_i \in \mathbb{R}$ are respectively the state and the control input of the ith agent, whereas $w_i \in \mathbb{R}$ is an unknown time-varying perturbation. Without loss of generality, agents are labeled in ascending order according to their initial conditions, i.e., $x_1(0) \leq x_2(0) \leq \cdots \leq x_n(0)$. The objective is to present two ISMC-based local interaction protocols such that, in spite of the exogenous time-varying perturbations and in accordance with the topological constraints, all agents states will eventually agree upon a common steady-state value denoted by $x^* \in \mathbb{R}$, i.e.,

$$\lim_{t\to\infty} x_i(t) = \lim_{t\to\infty} x_j(t) = x^* \qquad \forall\, i,\ j \in \mathcal{V},\ t \in \mathbb{R}$$

where $x^* \in \mathbb{R}$ corresponds, for the first scenario under analysis, to the average value of the agent's initial conditions, i.e.,

$$\bar{x} \equiv \frac{1}{n} \cdot \sum_{i=1}^{n} x_i(0) \tag{10.6}$$

whereas, for the second use case, it corresponds to the median value $\tilde{x}$ of the agents' initial conditions, i.e.,

$$\tilde{x} \equiv \begin{cases} \text{if } n \text{ is even } \tilde{x} \in \left[x_k(0), x_{k+1}(0)\right] \text{ with } \; k = \frac{n}{2}, \\ \text{if } n \text{ is odd} \qquad\quad \tilde{x} = x_k(0) \qquad \text{with } k = \frac{n+1}{2}. \end{cases} \tag{10.7}$$

In the following two subsections the two proposed algorithms will be presented and their convergence properties evaluated by employing Lyapunov tools. From now on, the next assumption are assumed to be in force.

Assumption 10.1 The network topology is described by a connected undirected graph $\mathcal{G}$ and the local perturbations $w_i(t)$ are supposed to fulfill the following inequality

$$\|w_i(t)\|_\infty \leq \Pi_i \leq \Pi,\ \forall\, i \in \mathcal{V},\ \forall\, t \geq 0. \tag{10.8}$$

where

$$\Pi = \max_{i\in\mathcal{V}}\{\Pi_i\} < \infty \tag{10.9}$$

is an a-priori known constant.

10.3.2 Robust ISMC-Based Average-Consensus Protocol

In this subsection we present a communication protocol able to estimate the averaged value of the agent's initial conditions given by (10.6), in spite of the presence of unknown, bounded perturbations affecting the local agents' dynamics. The proposed coordination algorithm meets the communication constraints given by $\mathcal{G}$, and it is designed as follows

$$u_i(t) = -\alpha \cdot \sum_{j \in \mathcal{N}_i} (x_i(t) - x_j(t)) - k \cdot \mathrm{SIGN}(x_i(t) + z_i(t))$$
$$\dot{z}_i(t) = \alpha \cdot \sum_{j \in \mathcal{N}_i} (x_i(t) - x_j(t)), \qquad z_i(0) = -x_i(0) \tag{10.10}$$

where the SIGN(·) operator denotes the discontinuous and multi-valued function

$$\mathrm{SIGN}(\mathfrak{S}) := \begin{cases} 1 & \text{if } \mathfrak{S} > 0 \\ [-1, 1] & \text{if } \mathfrak{S} = 0 \\ -1 & \text{if } \mathfrak{S} < 0 \end{cases} \quad \mathfrak{S} \in \mathbb{R} \tag{10.11}$$

Remark 10.1 Due to the discontinuous local interaction rule (10.10) (resp. (10.31)), and the possibly discontinuous nature of the perturbations $w_i(t)$ in (10.5) (which are supposed to be only uniformly bounded, according to Assumption 10.1), the resulting closed loop network dynamics (10.5), (10.10) (resp. (10.5), (10.31)) will be discontinuous as well. Thus, the resulting solution notion needs to be discussed and clarified. Following [36], we will understand these solutions in the so-called Filippov sense, i.e., as the solution of an appropriate differential inclusion the existence of which is guaranteed (owing on mild properties of the original discontinuous system which are fulfilled in the present scenario) and for which noticeable properties, such as absolute continuity, are in force. The reader is referred to [37] for a comprehensive account of the notions of solution for discontinuous dynamical systems.

We are now in position to state the first main result of this study.

Theorem 10.1 *Consider the collective dynamics given by* (10.5) *and let Assumption 10.1 be satisfied. If the tuning parameters of the local interaction rule* (10.10) *satisfy the next relations*

$$\alpha > 0 \ , \quad k > \Pi \tag{10.12}$$

then the multi-agent system (10.5), (10.10) *cooperatively estimates the average value of the agents initial conditions. According to the next relation*

$$\lim_{t \to \infty} \boldsymbol{x}(t) = \mathbf{1}_n \otimes \bar{x} \quad \forall \ i \in \mathcal{V} \tag{10.13}$$

where $\boldsymbol{x}(t) = [x_1, x_2, \ldots, x_n]' \in \mathbb{R}^n$ *denotes the vector of the agent's states and* $\bar{x}$ *is the average value* (10.6) *of the agent's initial conditions.*

Proof The proof is split into two consecutive steps. First, it will be demonstrated that, thanks to the discontinuous part of proposed protocol (10.10), the perturbed MAS (10.5), (10.10) features after a finite transient time a sliding motion while rejecting the perturbations $w_i(t)$. Then, by exploiting the concept of *equivalent control* [33], it will be shown that, during the sliding motion, the agents trajectories $x_i(t)$, $\forall\, i \in \mathcal{V}$, will tend to the desired consensus value $\bar{x}$ in (10.6).

Let us substitute (10.10) into (10.5). Then, a compact representation of the collective dynamics of the networked system can be provided, as follows

$$\begin{aligned} \dot{\boldsymbol{x}}(t) &= \boldsymbol{w}(t) - \alpha \cdot \mathcal{L}\boldsymbol{x}(t) - k \cdot \mathrm{SIGN}(\boldsymbol{x}(t) + \boldsymbol{z}(t)) \\ \dot{\boldsymbol{z}}(t) &= \alpha \cdot \mathcal{L}\boldsymbol{x}(t), \qquad\qquad \boldsymbol{z}(0) = -\boldsymbol{x}(0), \end{aligned} \tag{10.14}$$

where $\boldsymbol{x}(t) = [x_1, x_2, \ldots, x_n]'$, $\boldsymbol{z}(t) = [z_1, z_2, \ldots, z_n]'$, and, $\boldsymbol{w}(t) = [w_1, w_2, \ldots, w_n]'$. The operator SIGN$(\cdot)$ is now understood component-wise, i.e.,

$$\mathrm{SIGN}(\boldsymbol{x}) = \left[\mathrm{SIGN}(x_1), \cdots, \mathrm{SIGN}(x_n)\right]'. \tag{10.15}$$

Let $\boldsymbol{\sigma}(t) = [\sigma_1, \sigma_2, \ldots, \sigma_n]'$ be a vector of switching manifolds designed as follows

$$\boldsymbol{\sigma}(t) = \boldsymbol{x}(t) + \boldsymbol{z}(t). \tag{10.16}$$

Now, we can rewrite the collective dynamics (10.14) in the new $(\boldsymbol{x}, \boldsymbol{\sigma})$ coordinates as next

$$\dot{\boldsymbol{x}}(t) = \boldsymbol{w}(t) - \alpha \cdot \mathcal{L}\boldsymbol{x}(t) - k \cdot \mathrm{SIGN}(\boldsymbol{\sigma}(t)) \tag{10.17}$$

$$\dot{\boldsymbol{\sigma}}(t) = \boldsymbol{w}(t) - k \cdot \mathrm{SIGN}(\boldsymbol{\sigma}(t)), \qquad \boldsymbol{\sigma}(0) = \boldsymbol{0}_n. \tag{10.18}$$

Consider the candidate Lyapunov function

$$V(t) = \frac{1}{2}\|\boldsymbol{\sigma}(t)\|_2^2 \tag{10.19}$$

By evaluating $\dot{V}(t)$ along the trajectories of (10.18), and by invoking the Hölder's Inequality [39], it yields the following chain of inequalities

$$\begin{aligned} \dot{V}(t) &= \boldsymbol{\sigma}(t)'\dot{\boldsymbol{\sigma}}(t) \\ &= \boldsymbol{\sigma}(t)'\boldsymbol{w}(t) - k \cdot \textstyle\sum_{i=1}^n \sigma_i(t)\mathrm{SIGN}(\sigma_i(t)) \\ &\le \left|\boldsymbol{\sigma}(t)'\boldsymbol{w}(t)\right| - k \cdot \|\boldsymbol{\sigma}(t)\|_1 \\ &\le -(k - \|\boldsymbol{w}(t)\|_\infty) \cdot \|\boldsymbol{\sigma}(t)\|_1 \\ &\le -(k - \Pi) \cdot \|\boldsymbol{\sigma}(t)\|_1 \prec 0 \quad \forall \quad k > \Pi. \end{aligned} \tag{10.20}$$

Note that, when $\sigma_i(t) \neq 0$ for all $i \in \mathcal{V}$, the time derivative of (10.19) exists in the conventional sense. On the contrary, whenever any entry of $\boldsymbol{\sigma}(t)$ is zero, we need to consider the generalized time derivative of $V(t)$ which is set-valued according to the right-hand side of (10.20) and to the definition (10.11) of the multi-valued

SIGN operator [36]. On the other hand, whenever an element $\sigma_i(t)$ is zero for any interval of time of positive measure, this implies that its time derivative $\dot{\sigma}_i(t)$ is zero as well, during the same interval of time. This, allows us to substitute $\sigma_i(t)\dot{\sigma}_i(t) = 0$ in the first row of (10.20), whenever the argument of the discontinuous operator is zero. This reasoning derives from the closely related example studied in [36, Th. 3, pp. 157–158]). From that, it follows that, in the case under investigation, the generalized time derivative of the Lyapunov function in (10.20) is a singleton, every t, whereas its set-valued nature can safely be disregarded throughout the analysis during the isolated time instants where $\dot{V}(t)$ is set-valued.

Thanks to the appropriate selection of the initial condition for the auxiliary control variable in (10.10), i.e., $z_i(0) = -x_i(0)$, it follows from (10.18) and (10.19) that the system trajectory lies from the initial time $t = 0$ onto the sliding manifold $\boldsymbol{\sigma}(t) = \mathbf{0_n}$. Thus from (10.19), it yields that $V(0) = 0$. However, since the right-hand side of (10.20) is always non-positive, then the condition $V(t) = 0$ is maintained at all $\forall$ $t \geq 0$. As a consequence, the closed loop system exhibits a sliding motion along the manifold

$$\boldsymbol{\sigma}(t) = \mathbf{0_n} \tag{10.21}$$

from the initial time instant on regardless of the presence of the bounded perturbation $w_i(t)$ acting at the local level of each agent operating over the MAS.

By virtue of this, we are in a position to exploit the equivalent control method (see [38]) to study the corresponding sliding mode trajectories of system (10.17)–(10.18) subject to the constraint $\boldsymbol{\sigma}(t) = \mathbf{0_n}\ \forall\, t \geq 0$.

Particularly, it results that during the sliding motion, the discontinuous term SIGN($\boldsymbol{\sigma}(t)$) in the right-hand side of (10.17) can be replaced by the corresponding continuous “equivalent value” $\text{SIGN}_{eq}(\boldsymbol{\sigma}(t))$. Such “equivalent behavior” can be derived by simply imposing the right-hand side of (10.18) to be identically zero, i.e.,

$$\dot{\boldsymbol{\sigma}} = \mathbf{0}_n \quad \Rightarrow \quad \text{SIGN}_{eq}(\boldsymbol{\sigma}(t)) = \frac{1}{k} \cdot \boldsymbol{w}(t) \tag{10.22}$$

Then, by substituting $\text{SIGN}_{eq}(\boldsymbol{\sigma}(t))$ in (10.22) into the right-hand side of the motion Eq. (10.17), during the sliding motion, the following collective agents dynamic arises

$$\dot{\boldsymbol{x}}(t) = -\alpha \cdot \boldsymbol{\mathcal{L}}\boldsymbol{x}(t) \qquad t \geq 0 \tag{10.23}$$

Let us now define the following auxiliary variable

$$q(t) = \sum_{i=1}^{n} x_i(t) = \mathbf{1}_n' \boldsymbol{x}(t) \tag{10.24}$$

From (10.23), and due to the positive semi-definitiveness of $\mathcal{L} \succeq 0$, along with its zero column-sum property $\mathbf{1}_n' \mathcal{L} = \mathbf{0}_n$, it results that

$$\dot{q}(t) = -\alpha \cdot \mathbf{1}_n' \mathcal{L} \boldsymbol{x}(t) = 0 \tag{10.25}$$

Thus, let a be a constant real value, it follows that

$$q(t) = q(0) = \sum_{i=1}^{n} x_i(0) \quad \forall \quad t \geq 0 \tag{10.26}$$

and

$$\lim_{t \to \infty} q(t) = \lim_{t \to \infty} \mathbf{1}_n' \boldsymbol{x}(t) = a \cdot \mathbf{1}_n' \mathbf{1}_n = a \cdot n \tag{10.27}$$

Hence we obtain

$$\mathbf{1}_n' \boldsymbol{x}(0) = a \cdot n \tag{10.28}$$

and so

$$a = \left(\frac{\mathbf{1}_n' \boldsymbol{x}(0)}{\mathbf{1}_n' \mathbf{1}_n} \right) \cdot \mathbf{1}_n = \bar{x} = \frac{1}{n} \cdot \sum_{i=1}^{n} x_i(0) \tag{10.29}$$

As a consequence, we have

$$\lim_{t \to \infty} \boldsymbol{x}(t) = \mathbf{1}_n \otimes \bar{x} \tag{10.30}$$

that confirms the global asymptotical convergence of the state variables of the (10.23) towards the expected value $\bar{x}$ given by (10.6). This conclude the proof.

Remark 10.2 Since the chosen Lyapunov function (10.19) is smooth, and $\boldsymbol{\sigma}(t)$ in (10.16) (resp. (10.39)) is, by definition of Filippov solution, absolutely continuous, then $V(\boldsymbol{\sigma}(t))$ is absolutely continuous as well. In fact it is the composition of a smooth function with an absolutely continuous function. Furthermore, since the discontinuous terms in the right-hand side of (10.20) (resp. (10.42)) are vanishing with the same arguments of the sign function, i.e. $\boldsymbol{\sigma}(t)$. Thus, in spite of the potentially set-valued nature of $\dot{V}(t)$, that would have required special tools of non-smooth analysis (see e.g. [20]), here $\dot{V}(t)$ is uniquely defined in the conventional sense. As a consequence of that, neither special derivative notions nor advanced non-smooth analysis tools need to be invoked to support the achievement of condition (10.20) (resp. (10.43)).

Remark 10.3 It is worth to remark that finite-time convergence towards the expected averaged value of the agents' initial conditions, as well as, application to different distributed consensus-based computational problems (see e.g. [3]) can easily be derived by simple modification of the proposed algorithm (10.10). In particular, in accordance with the ISMC paradigm, the idea is to re-design (10.10) in a such way that the resulting "equivalent dynamic" during the sliding motion (e.g., see (10.23)) solves the problem under nominal unperturbed conditions.

10.3.3 Robust ISMC-Based Median-Consensus Protocol

Let us now discuss the second scenario of this chapter, which deals with the problem of estimating the median value of the agent's initial conditions given by (10.7), in spite of the presence of the bounded perturbations $w_i(t)$ corrupting the agents' dynamics (10.5). To solve this problem, we take inspiration from [20], where the Authors solve the median consensus problem in the unperturbed scenario (i.e., with $w_i(t) = 0\ \forall\ i \in \mathcal{V}$). Then, according to Remark 10.3, we design the local communication protocol $u_i(t) \in \mathbb{R}$ as follows

$$\begin{aligned} u_i(t) &= -y_i(t) - k \cdot \mathrm{SIGN}\left(x_i(t) + z_i(t)\right), \\ \dot{z}_i(t) &= y_i(t), \qquad\qquad z_i(0) = -x_i(0), \\ y_i(t) &= \alpha \cdot \mathrm{sign}\left(x_i(t) - x_i(0)\right) + \lambda \cdot \sum_{j \in \mathcal{N}_i} \mathrm{sign}\left(x_i(t) - x_j(t)\right), \end{aligned} \tag{10.31}$$

where the SIGN$(\cdot)$ operator is the multi-valued signum function given by (10.11), whereas, the sign$(\cdot)$ operator is the single-valued function

$$\mathrm{sign}\left(\mathfrak{S}\right) := \begin{cases} 1 & \text{if } \mathfrak{S} > 0, \\ 0 & \text{if } \mathfrak{S} = 0, \quad \mathfrak{S} \in \mathbb{R}. \\ -1 & \text{if } \mathfrak{S} < 0, \end{cases} \tag{10.32}$$

We now present the second main result of this study.

Theorem 10.2 *Consider the collective dynamics* (10.5) *and let Assumption 10.1 be satisfied. If the tuning parameters of the local interaction rules* (10.31) *satisfy the next relations*

$$\alpha > 0, \qquad \lambda > \alpha, \qquad k > \Pi, \tag{10.33}$$

then, the multi-agent system (10.5), (10.31) *cooperatively estimates the median value of the agents initial conditions after a finite time T. Thus, it follows that*

$$\lim_{t \to T} \boldsymbol{x}(t) = \mathbf{1}_n \otimes \tilde{x} \quad \forall\ i \in \mathcal{V}, \quad t \geq T \in \mathbb{R} \tag{10.34}$$

holds, where $\boldsymbol{x}(t) = [x_1, x_2, \ldots, x_n]' \in \mathbb{R}^n$ denotes the vector of the agent's states and $\bar{x}$ is the median value of the agent's initial conditions (10.7).

Proof With the same spirit of the Proof of Theorem 10.1, by substituting (10.31) into (10.5), a compact representation of the resulting perturbed closed-loop network dynamics is derived

$$\begin{aligned}\dot{\boldsymbol{x}}(t) =& \boldsymbol{w}(t) - \alpha \cdot \text{sign}\left(\boldsymbol{x}(t) - x(0)\right) - \lambda\sqrt{2} \cdot \mathcal{D}\text{sign}\left(\sqrt{2} \cdot \mathcal{D}'\boldsymbol{x}(t)\right) \\ &- k \cdot \text{SIGN}\left(\boldsymbol{x}(t) + z(t)\right), \\ \dot{z}(t) =& \alpha \cdot \text{sign}\left(\boldsymbol{x}(t) - x(0)\right) + \lambda \cdot \sqrt{2} \cdot \mathcal{D}\text{sign}\left(\sqrt{2} \cdot \mathcal{D}'\boldsymbol{x}(t)\right), \\ z(0) =& -\boldsymbol{x}(0),\end{aligned} \tag{10.35}$$

where $\boldsymbol{x} = [x_1, x_2, \dots x_n]'$, $\boldsymbol{z} = [z_1, z_2, \dots z_n]'$, and $\boldsymbol{w} = [w_1, w_2, \dots w_n]'$. $\boldsymbol{x}(0) = \{x_1(0), \dots, x_n(0)\}$ collects the agents' initial conditions. The discontinuous operators SIGN$(\cdot)$, and sign$(\cdot)$ are now generalized in terms of element-wise operators, i.e.,

$$\text{sign}(\boldsymbol{x}) = \left[\text{sign}(x_1), \cdots, \text{sign}(x_n)\right]' \tag{10.36}$$

$$\text{SIGN}(\boldsymbol{x}) = \left[\text{SIGN}(x_1), \cdots, \text{SIGN}(x_n)\right]' \tag{10.37}$$

Furthermore, it is worth to note that, according to the definition of the Incidence Matrix in (10.2), the following equivalence

$$\sqrt{2} \cdot \mathcal{D}\text{sign}\left(\sqrt{2} \cdot \mathcal{D}'\boldsymbol{x}(t)\right) = \begin{bmatrix} \sum_{j \in \mathcal{N}_1} \text{sign}(x_1(t) - x_j(t)) \\ \vdots \\ \sum_{j \in \mathcal{N}_n} \text{sign}(x_n(t) - x_j(t)) \end{bmatrix} \tag{10.38}$$

holds by construction. Then, by considering as a vector of local switching manifolds the same combination of states considered in (10.16), i.e.,

$$\boldsymbol{\sigma}(t) = \boldsymbol{x}(t) + \boldsymbol{z}(t), \tag{10.39}$$

we can rewrite the collective closed-loop dynamics (10.35) in the new set of coordinates $(\boldsymbol{x}, \boldsymbol{\sigma})$ as follows

$$\begin{aligned}\dot{\boldsymbol{x}}(t) =& \boldsymbol{w}(t) - \alpha \cdot \text{sign}\left(\boldsymbol{x}(t) - x(0)\right) \\ &- \lambda\sqrt{2} \cdot \mathcal{D}\text{sign}\left(\sqrt{2} \cdot \mathcal{D}'\boldsymbol{x}(t)\right) - k \cdot \text{SIGN}(\boldsymbol{\sigma}(t)),\end{aligned} \tag{10.40}$$

$$\dot{\boldsymbol{\sigma}}(t) = \boldsymbol{w}(t) - k \cdot \text{SIGN}\left(\boldsymbol{\sigma}(t)\right), \qquad \boldsymbol{\sigma}(0) = \boldsymbol{0}_n. \tag{10.41}$$

It should be stressed that, due to the proper selection of $\boldsymbol{z}(0) = -\boldsymbol{x}(0)$, which implies $\boldsymbol{\sigma}(0) = \boldsymbol{0}_n$ (see (10.39)), then $\boldsymbol{\sigma}(t)$ holds the zero value from the initial time instant $t = 0$. Let us now study the stability of the closed-loop dynamics (10.40)–(10.41). Following a similar treatment as in the Proof of Theorem 10.1, we choose as a candidate Lyapunov function the same function (10.19). By differentiating (10.19) along the trajectories of (10.40)–(10.41), by simple computations, and by making straightforward manipulations taking the advantages of the Hölder Inequality, it yields

$$\begin{aligned} \dot{V}(t) &\le \left|\boldsymbol{\sigma}(t)'\boldsymbol{w}(t)\right| - k \cdot \|\boldsymbol{\sigma}(t)\|_1 \\ &\le -(k - \|\boldsymbol{w}(t)\|_\infty) \cdot \|\boldsymbol{\sigma}(t)\|_1 \\ &\le -(k - \Pi)\|\boldsymbol{\sigma}(t)\|_1 \end{aligned} \tag{10.42}$$

Thus, due to the tuning constraints in (10.33), in particular $k > \Pi$, one readily verifies that

$$\dot{V}(t) \prec 0, \quad \forall \;\; t \ge 0. \tag{10.43}$$

According to the ISMC design paradigm, it can be seen that, from (10.41), it holds

$$V(0) = \frac{1}{2}\|\boldsymbol{\sigma}(0)\|_2^2 = 0 \tag{10.44}$$

then, by coupling (10.44) with (10.43), it follows that

$$V(t) = 0, \quad \forall t \ge 0. \tag{10.45}$$

which means that the reaching phase is eliminated, in spite of any bounded perturbation $\boldsymbol{w}(t) \in \mathbb{R}^n$. By exploiting the equivalent control method, we can now evaluate the state trajectories of the system along the sliding manifold $\sigma(t) = \mathbf{0}_n$. The equivalent value associated to the discontinuous term $\text{SIGN}(\boldsymbol{\sigma}(t))$ can be easily derived from (10.41) as follows

$$\dot{\boldsymbol{\sigma}} = \mathbf{0}_n \quad \Rightarrow \quad \text{SIGN}_{eq}(\boldsymbol{\sigma}(t)) = \frac{1}{k} \cdot \boldsymbol{w}(t). \tag{10.46}$$

By replacing the term $\text{SIGN}(\boldsymbol{\sigma}(t))$ in (10.40) with its equivalent value (10.46), it results that from the initial time instant $t = 0$, the network evolutions follow the next perturbation-free evolutions law

$$\dot{\boldsymbol{x}}(t) = -\alpha \cdot \text{sign}(\boldsymbol{x}(t) - \boldsymbol{x}(0)) - \lambda\sqrt{2} \cdot \boldsymbol{\mathcal{D}}\text{sign}(\sqrt{2} \cdot \boldsymbol{\mathcal{D}}'\boldsymbol{x}(t)). \tag{10.47}$$

A complete convergence analysis of the resulting dynamics (10.47) goes beyond the scope of this study, focused on the robustification of cooperative consensus algorithms. However, following [20, Theorem 3.3], and by selecting the coefficients $\lambda > \alpha$ as in (10.33), it results that the collective dynamic (10.47) exhibits a two step convergence process outlined as follows:

1. There exists a finite time instant $\tau_1 \in \mathbb{R}_{>0}$, with $\tau_1 < \infty$, such that

$$x_i(t) = c(t), \quad \forall \ i \in \mathcal{V}, \quad t \geq \tau_1, \tag{10.48}$$

where $c(t)$ is the common time-varying profile for the agents state variables.
2. There exists a finite time instant $\tau_2 \geq \tau_1$, $\tau_2 < \infty$, such that

$$c(t) = \tilde{x}, \quad \forall \ i \in \mathcal{V}, \quad t \geq \tau_2. \tag{10.49}$$

This concludes the proof. The reader is referred to [20] for a detailed non-smooth Lyapunov analysis of the discontinuous differential equation (10.47).

10.4 Simulations and Discussion

To demonstrate the efficacy of the proposed protocols (10.10) and (10.31), in this Section the performance of these algorithms is tested via computer simulations.

It is worth to remark that the presence of perturbation in (10.5), modifies the equilibrium of the network and thus any standard algorithm for estimating the averaged or the median value of the agents initial conditions will fail. The scenario under test consists in a connected network of $n = 9$ agents, whose topology is displayed in Fig. 10.1. Agents' dynamics are governed by (10.5). The disturbances are selected as exogenous, time-varying biased sinusoidal signals

$$w_i(t) = b_i + a_i \cos(2\pi f_i t) \tag{10.50}$$

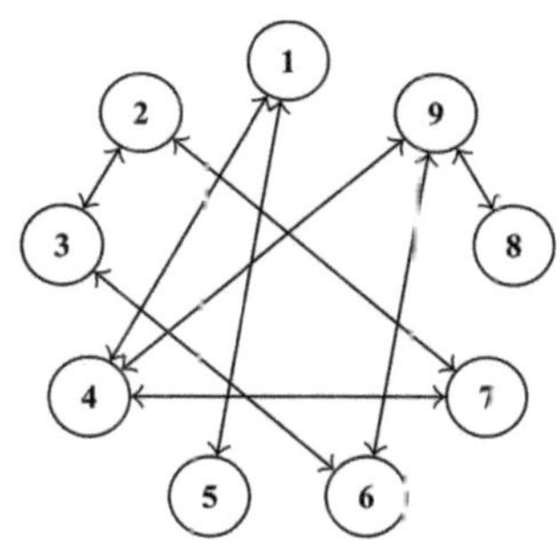

Fig. 10.1 Communication topology for a network of ten agents

with randomly selected coefficients

$$a_i \in [1, 9] \quad , \quad b_i \in [-4, 4] \quad , \quad f_i \in [1, 10]. \tag{10.51}$$

Let $\boldsymbol{w} = [w_1, \ldots, w_n]$ be the column-wise collection of the local perturbations, by simple computations, from (10.50), and (10.51), we can easily derive that $\|\boldsymbol{w}\|_\infty \leq \Pi = 13$.

The initial states of the agents for the following simulative tests are intentionally chosen as $x_i(0) = i$ for $i = 1, \ldots, 8$ and $x_9(0) = 100$. As it can be seen, the initial condition of the 9th agent is much bigger than that of the remaining agents in order to simulate an *outlier agent*. It follows that the average and median values of the agents' initial conditions, following (10.6) and (10.7) are $\bar{x} \approx 15.11$ $\tilde{x} = 5$, respectively.

As expected, it the median value is less sensitive to the presence of the 9th outlier agent with respect to the average value.

The set of parameters for the proposed algorithms are selected, according to Theorems 10.1 and 10.2, respectively, as $\alpha = 1$, $\lambda = 3$, and $k = 15$.

In Figs. 10.2, 10.3, 10.4 and 10.5, we can observe the performance of the reference algorithms proposed in [9], i.e.,

$$u_i = -\alpha \cdot \sum_{j \in \mathcal{N}_i} \text{sign}(x_i - x_j) \tag{10.52}$$

and [20], i.e.,

$$u_i = -\alpha \cdot \text{sign}(x_i - x_i(0)) - \lambda \cdot \sum_{j \in \mathcal{N}_i} \text{sign}(x_i - x_j) \tag{10.53}$$

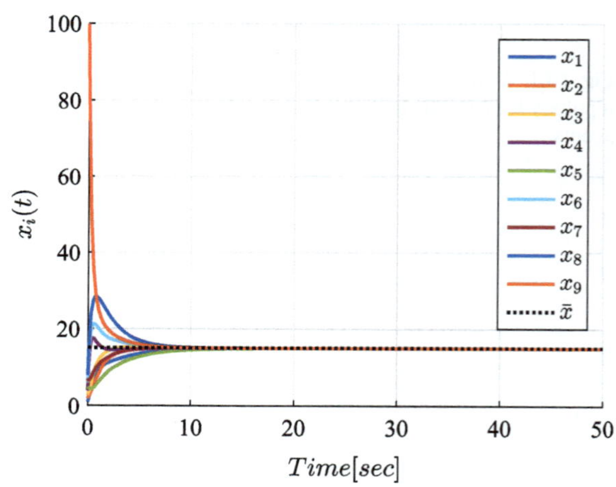

Fig. 10.2 Performance of the average consensus protocol (10.52) in the absence of perturbations (i.e., with $w_i(t) = 0 \; \forall \, i \in \mathcal{V}$)

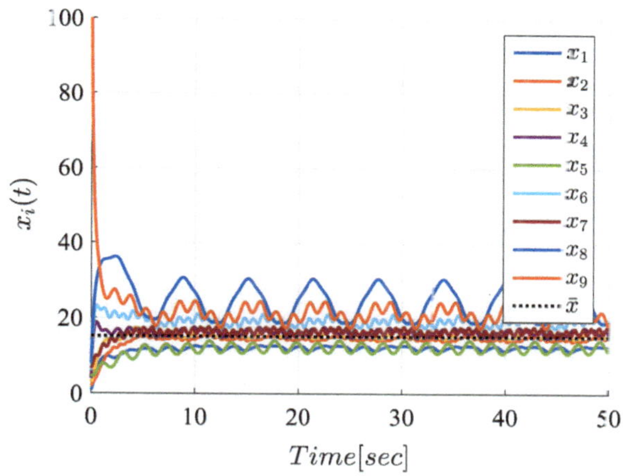

Fig. 10.3 Performance of the average consensus protocol (10.52) under the effect of the perturbations (10.50)–(10.51) (i.e., with $w_i(t) \neq 0 \; \forall \, i \in \mathcal{V}$)

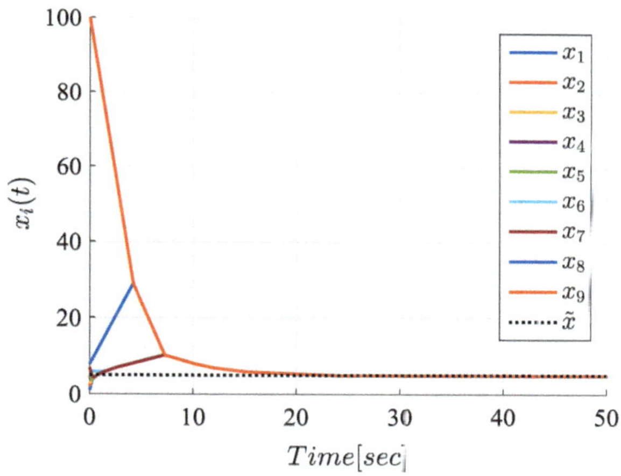

Fig. 10.4 Performance of the median consensus protocol (10.53) in the absence of perturbations (i.e., with $w_i(t) = 0 \; \forall \, i \in \mathcal{V}$)

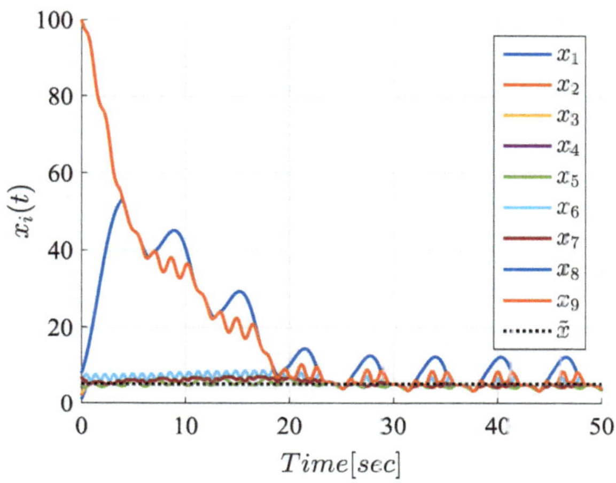

Fig. 10.5 Performance of the median consensus protocol (10.53) under the effect of the perturbations (10.50)–(10.51) (i.e., with $w_i(t) \neq 0 \; \forall \, i \in \mathcal{V}$)

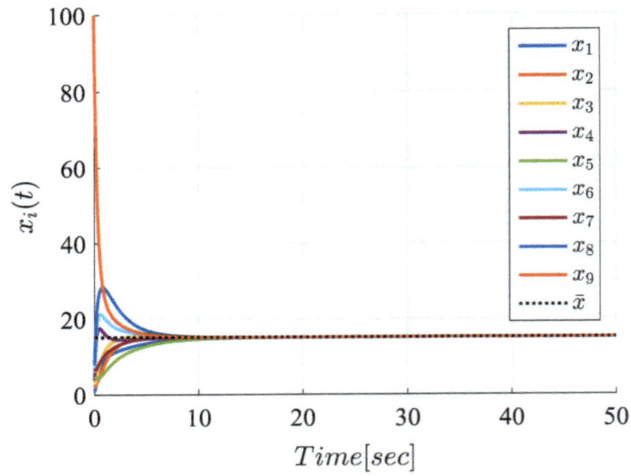

Fig. 10.6 Performance of the proposed ISMC-based average consensus protocol (10.10) under the effect of the perturbations (10.50)–(10.51) (i.e., with $w_i(t) \neq 0 \; \forall \, i \in \boldsymbol{\mathcal{V}}$)

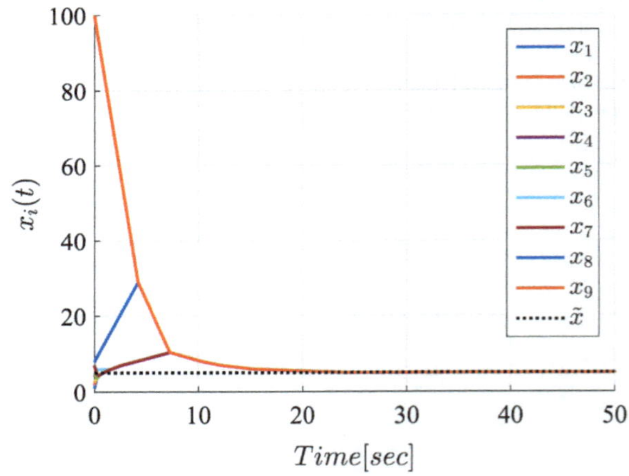

Fig. 10.7 Performance of the proposed ISMC-based median consensus protocol (10.31) under the effect of the perturbations (10.50)–(10.51) (i.e., with $w_i(t) \neq 0 \; \forall \, i \in \boldsymbol{\mathcal{V}}$)

in achieving average- and median-consensus, respectively in the ideal scenario, i.e., no perturbation (see Figs. 10.2, 10.4), and in the perturbed scenario, i.e. $w_i(t) \neq 0 \; \forall \, i \in \boldsymbol{\mathcal{V}}$ (see Figs. 10.3 and 10.5). It is apparent that these *standard* algorithms estimate correctly the expected averaged and median value of the networks's initialization only in the unperturbed case.

The temporal evolutions of the agent's states under the proposed ISMC-based protocol (10.10) and (10.31) in the presence of perturbations are depicted in Figs. 10.6, 10.7, respectively. It is evident that the proposed algorithms, in accordance with their respective convergence analysis, correctly estimate the average and median value of the agent's initial condition.

Particularly, it can be noted that, due to the intrinsic feature of the ISMC design paradigm of eliminating the reaching phase, the agents behavior depicted in Figs. 10.6

and 10.7 result to be identical to that obtained by using the standard algorithms (10.52) and (10.53) during the unperturbed (ideal) scenario depicted in Figs. 10.2 and 10.4.

10.4.1 Chattering Alleviation

Due to the discontinuous nature of the proposed local interaction rules (10.10) and (10.31), high-frequency chattering on the agents state variables arises in practical implementation. To alleviate this phenomenon, smooth approximations of the discontinuous sign functions are commonly used. Particularly, in the following the $\text{sign}(\cdot)$ and $\text{SIGN}(\cdot)$ functions in the proposed local interaction protocols (10.10) and (10.31) will be replaced by the sigmoidal smooth approximation

$$\text{Sig}_\mu(\mathfrak{S}) := \frac{\mathfrak{S}}{\mu + |\mathfrak{S}|} \quad \text{with} \quad \mathfrak{S} \in \mathbb{R}, \quad \mu \in \mathbb{R}_{>0}, \tag{10.54}$$

where the smoothing parameter $\mu > 0$ has to be be taken smaller and smaller to get a more and more accurate approximation of the original discontinuous functions.

It yields the following smooth approximations of the local interaction rules (10.10) and (10.31) are, respectively,

$$\begin{aligned} u_i(t) &= -y_i(t) - k \cdot \text{Sig}_\mu(x_i(t) + z_i(t)), \\ \dot{z}_i(t) &= y_i(t), \qquad\qquad z_i(0) = -x_i(0), \\ y_i(t) &= \alpha \cdot \sum_{j \in \mathcal{N}_i} (x_i(t) - x_j(t)). \end{aligned} \tag{10.55}$$

for the consensus on the averaged value of the agent's initial conditions, and

$$\begin{aligned} u_i(t) &= -y_i(t) - k \cdot \text{Sig}_\mu(x_i(t) + z_i(t)), \\ \dot{z}_i(t) &= y_i(t), \qquad\qquad z_i(0) = -x_i(0), \\ y_i(t) &= \alpha \cdot \text{Sig}_\mu(x_i(t) - x_i(0)) + \lambda \cdot \sum_{j \in \mathcal{N}_i} \text{Sig}_\mu(x_i(t) - x_j(t)). \end{aligned} \tag{10.56}$$

for the consensus on the resulting median value. Using the smooth approximation (10.54) with $\mu = 0.1$ (see Figs. 10.8 and 10.9) it is seen that a small vicinity of the average and median value, respectively, is attained for both the proposed algorithms.

Thanks to the approximation (10.54), the convergence towards the expected average/median value of the agents initial condition is preserved with accuracy that proportional to the deadzone layer μ. Thus, the chattering is alleviated at the expense of robustness and accuracy (see e.g. the bottom plots of Figs. 10.8 and 10.9). More effective chattering alleviation techniques were suggested in [30], however deeper

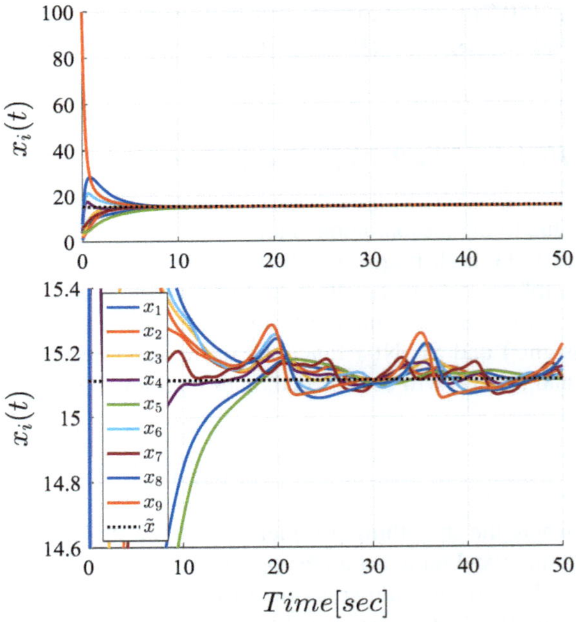

Fig. 10.8 Performance of the chattering-free consensus protocol (10.55) under the effect of the perturbations (10.50)–(10.51) (i.e., with $w_i(t) \neq 0 \; \forall \, i \in \boldsymbol{\mathcal{V}}$)

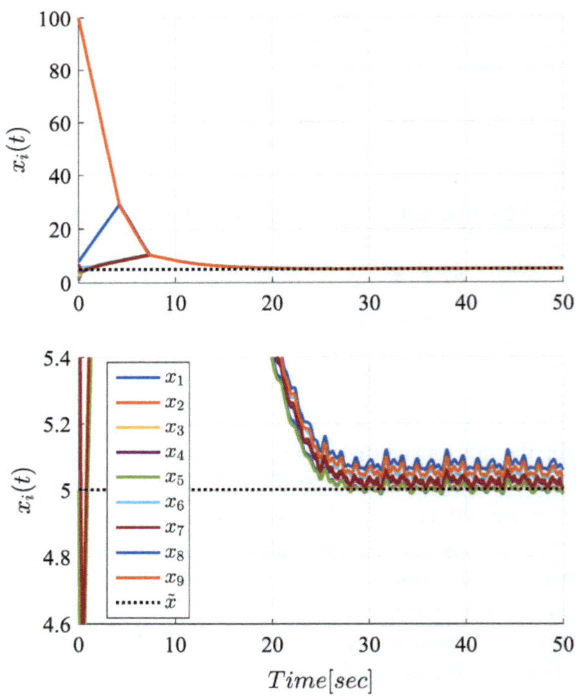

Fig. 10.9 Performance of the chattering-free consensus protocol (10.56) under the effect of the perturbations (10.50)–(10.51) (i.e., with $w_i(t) \neq 0 \; \forall \, i \in \boldsymbol{\mathcal{V}}$)

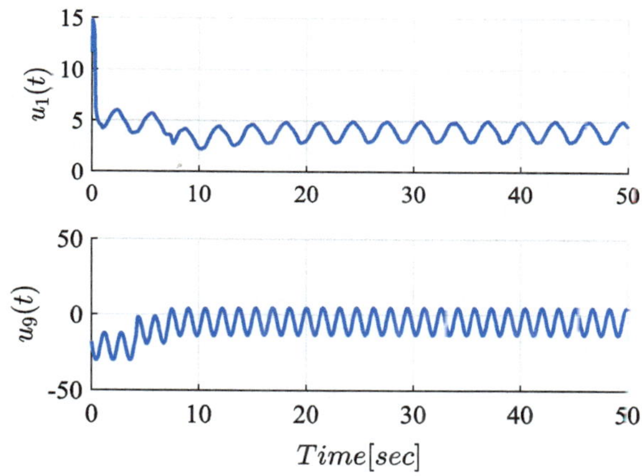

Fig. 10.10 Temporal evolution of the control signals $u_1(t)$ and $u_9(t)$ using the smooth approximation (10.56)

analysis into this topic goes beyond the aims of this study. Finally, in Fig. 10.10 the local control inputs $u_1(t)$ and $u_9(t)$ of protocol (10.56) are shown, and one can easily verify that these control actions are continuous signals.

10.5 Conclusions

In this chapter we have shown how the IMSC design paradigm can be usefully applied in the framework of MAS to allow the agents dynamics to be affected by unknown disturbances. Our solution, with respect the existing algorithms allows the distributed computation of quantities such as, e.g., the average or median value of the agents initial conditions, even when the network is perturbed. We show that a suitable redesign of the original "non robust" algorithm can be made, owing on integral sliding mode ingredients, allowing to restore the ideal performance (e.g., convergence to the median or average value). Further investigations will focus on the extension of this solution towards application that involve consensus as a prerequisite for the accomplishment of more sophisticated tasks in a distributed way, such as distributed multi-objective optimization for application involving network of high-order systems.

Acknowledgements This work was supported by the Region Sardinia (L.R.7/2007) under Grant agreement CRP-7733 ["Sviluppo, progettazione e realizzazione prototipale di sistemi di gestione e controllo ottimali per una Micro Smart Grid"].

References

1. Wang, J., Elia, N.: Control approach to distributed optimization. In: Proceedings of the 48th IEEE Conference on Communication, Control, and Computing, pp. 557–561 (2010)
2. Ugrinovskii, V.: Conditions for detectability in distributed consensus-based observer networks. IEEE Trans. Autom. Control **58**(10), 2659–2664 (2012)
3. Cortes, J.: Distributed algorithms for reaching consensus on general functions. Automatica **44**(3), 726–737 (2008)
4. Pilloni, A., Pisano, A., Franceschelli, M., Usai, E.: A discontinuous algorithm for distributed convex optimization. In: Proceedings of the 14th IEEE International Workshop on Variable Structure Systems, pp. 22–27 (2016)
5. Pilloni, V., Carta, A., Atzori, L.: Resource allocation using virtual objects in the internet of things: a qoi oriented consensus algorithm. In: Proceedings of the 19th IEEE International Conference on Innovations in Clouds, Internet and Network, pp. 82–87 (2016)
6. Lin, P., Ren, W., Song, Y., Farrell, J.: Distributed optimization with the consideration of adaptivity and finite-time convergence. In: Proceedings of the IEEE American Control Conference, pp. 3177–3182 (2014)
7. Rahili, S., Ren, W., Ghapani, S.: Distributed convex optimization of time-varying cost functions with swarm tracking behavior for continuous-time dynamics. In: Proceedings of the 54th IEEE Conference on Decision and Control, pp. 362–367 (2015)
8. Pilloni, V., Navaratnam, P., Vural, S., Atzori, L., Tafazolli, R.: Cooperative task assignment for distributed deployment of applications in wsns. In: Proceedings of the IEEE International Conference on Communications, pp. 2229–2234, (2013)
9. Olfati-Saber, R., Murray, R.: Consensus problems in networks of agents with switching topology and time-delays. IEEE Trans. Autom. Control **49**(9), 1520–1533 (2004)
10. Hadjicostis, C., Dominguez-Garcia, A., Vaidya, N.: Resilient average consensus in the presence of heterogeneous packet dropping links. In: Proceedings of the 54th IEEE Conference on Decision and Control, pp. 106–111 (2012)
11. Dominguez-Garcia, A., Hadjicostis, C.: Distributed strategies for average consensus in directed graphs. In: Proceedings of the 50th IEEE Conference on Decision and Control and European Control Conference, pp. 2124–2129 (2011)
12. Wang, L., Xiao, F.: Finite-time consensus problems for networks of dynamic agents. IEEE Trans. Autom. Control **55**(4), 950–955 (2010)
13. Menon, P., Edwards, C.: A discontinuous protocol design for finite-timeaverage consensus. In: Proceedings of the 54th IEEE Conference on Conference on Control Applications, pp. 2029–2034 (2010)
14. Rao, S., Ghose, D.: Sliding mode control-based algorithms for consensus in connected swarms. Int. J. Control **84**(9), 1477–1490 (2011)
15. Franceschelli, M., Egerstedt, M., Giua, A.: Motion probes for fault detection and recovery in networked control systems. In: Proceedings of the IEEE American Control Conference, pp. 4358–4363 (2008)
16. Guo, M., Dimarogonas, D., Johansson, K.: Distributed real-time fault detection and isolation for cooperative multi-agent systems. In: Proceedings of the IEEE American Control Conference, pp. 5270–5275 (2012)
17. LeBlanc, H., Zhang, H., Sundaram, S., Koutsoukos, X.: Consensus of multi-agent networks in the presence of adversaries using only local information. In: Proceedings of the 1st IEEE International Conference on High Confidence Networked Systems, pp. 1–10 (2012)
18. Hampel, F., Ronchetti, E., Rousseeuw, P., Stahel, W.: Robust Statistics: The Approach Based on Influence Functions. Wiley, New York (1986)
19. Franceschelli, M., Giua, A., Pisano, A.: Finite-time consensus on the median value by discontinuous control. In: Proceedings of the IEEE American Control Conference, pp. 946–951 (2014)
20. Franceschelli, M., Giua, A., Pisano, A.: Finite-time consensus on the median value with robustness properties. IEEE Trans. Autom. Control (2016). doi:10.1109/TAC.2016.2590602

21. Franceschelli, M., Pilloni, A., Pisano, A., Giua, A., Usai, E.: Finite-time consensus with disturbance attenuation for directed switching network topologies by discontinuous local interactions. In: Proceedings of the 52nd IEEE Conference on Decision and Control, pp. 2611–2616 (2013)
22. Davila, J.: Distributed tracking of first order systems using second-order sliding modes. In: Proceedings of the 19th IFAC World Congress, pp. 1392–1397 (2014)
23. Franceschelli, M., Pisano, A., Giua, A., Usai, E.: Finite-time consensus with disturbance rejection by discontinuous local interactions in directed graphs. IEEE Trans. Autom. Control **60**(4), 1133–1138 (2015)
24. Pilloni, A., Franceschelli, M., Pisano, A., Usai, E.: Recent advances in sliding-mode based consensus strategies. In: Proceedings of the 13th IEEE International Workshop on Variable Structure Systems, pp. 1–6 (2014)
25. Pilloni, A., Pisano, A., Franceschelli, M., Usai, E.: Finite-time consensus for a network of perturbed double integrators by second-order sliding mode technique. In: Proceedings of the 52nd IEEE Conference on Decision and Control, pp. 2145–2150 (2013)
26. Fu, J., Wang, J.: Robust finite-time containment control for high-order multi-agent systems with matched uncertainties under directed communication graphs. Int. J. Control **89**(6), 1137–1151 (2016)
27. Pilloni, A., Pisano, A., Orlov, Y., Usai, E.: Consensus-based control for a network of diffusion pdes with boundary local interaction. IEEE Trans. Autom. Control **61**(9), 2708–2713 (2016)
28. Pilloni, A., Pisano, A., Franceschelli, M., Usai, E.: Integral sliding modes for the robustification of consensus-based multi-agent based systems. In: Proceedings of the 14th IEEE International Workshop on Variable Structure Systems, pp. 222–227 (2016)
29. Pilloni, A., Pisano, A., Franceschelli, M., Usai, E.: Robust distributed consensus on the median value for networks of heterogeneously perturbed agents. In: Proceedings of the 55th IEEE Conference on Decision and Control, pp. 1–6 (2016)
30. Utkin, V., Shi, J.: Integral sliding mode in systems operating under uncertainty conditions. In: Proceedings of the 35th IEEE Conference on Decision and Control, pp. 4591–4596 (1996)
31. Bejarano, F., Fridman, L., Poznyak, A.: Output integral sliding mode for min-max optimization of multi-plant linear uncertain systems. IEEE Trans. Autom. Control **54**(11), 2611–2620 (2009)
32. Castanos, F., Hernandez, D., Fridman, L.: Integral sliding-mode control for linear time-invariant implicit systems. Automatica **50**(3), 971–975 (2014)
33. Gao, Q., Feng, G., Liu, L., Qiu, J., Wang, Y.: Robust control for stochastic t-s fuzzy systems via integral sliding-mode approach. IEEE Trans. Fuzzy Syst. **22**(4), 870–881 (2014)
34. Jiang, Y., Liu, J., Wang, S.: Robust integral sliding-mode consensus tracking for multi-agent systems with time-varying delay. Asian J. Control **18**(1), 224–235 (2016)
35. Godsil, C., Royle, G.: Algebraic Graph Theory. Cambridge University Press, Cambridge (1996)
36. Filippov, A.: Differential Equations with Discontinuous Righthand Sides. Kluwer Academic Publisher, London (1988)
37. Cortes, J.: Discontinuous dynamical systems. IEEE Control Syst. Mag. **28**(3), 36–73 (2008)
38. Utkin, V.: Sliding Modes in Control and Optimization. Springer, Berlin (1992)
39. Owen, D.: Handbook of mathematical functions with formulas. Technometrics **7**(1), 78–79 (1965)

Chapter 11
Finite-Time Consensus for Disturbed Multi-agent Systems with Unmeasured States via Nonsingular Terminal Sliding-Mode Control

Xiangyu Wang and Shihua Li

11.1 Introduction

During the past decade, distributed cooperative control has become a more and more active research field because of its broad applications and advantages compared with conventional centralized control [1–3]. Specifically, its typical applications include coordination control of multiple robots [4, 5], attitude alignment of multiple spacecrafts [6], and distributed filtering of sensor networks [7], to name just a few. Furthermore, its advantages lie in higher robustness, less communication cost and better efficiency [8].

Consensus is a basic issue in distributed cooperative control field, which aims to design appropriate protocols via neighboring information for the agents such that some of their state elements (e.g., attitudes or velocities) reach a common value [2]. On the consensus problem, there are usually two factors in agent dynamics which bring great challenge for the distributed control design, namely, the influence of disturbances (including both system uncertainties, unmodelling dynamics and external disturbances) and the lack of full-state information.

On the consensus problem of multi-agent systems with disturbances, several kinds of anti-disturbance control approaches have been proposed, such as sliding-mode control [3, 9, 10], adaptive control [11], H_∞ control [12], and output regulation theory [13]. A common point of these results is that they are focused on the

X. Wang · S. Li (✉)
School of Automation, Southeast University, Nanjing 210096,
Jiangsu, People's Republic of China
e-mail: lsh@seu.edu.cn

X. Wang
e-mail: w.x.y@seu.edu.cn

X. Wang · S. Li
Key Laboratory of Measurement and Control of Complex Systems of Engineering,
Ministry of Education, Nanjing 210096,
Jiangsu, People's Republic of China

S. Li et al. (eds.), *Advances in Variable Structure Systems and Sliding Mode Control—Theory and Applications*, Studies in Systems, Decision and Control 115, DOI 10.1007/978-3-319-62896-7_11

agent dynamics with matched disturbances, namely, the disturbances enter the agent dynamics through the same channels as the control inputs. Then, a question arises naturally, that is how to achieve consensus of multi-agent systems subject to mismatched disturbances? On one hand, the consensus problem with mismatched disturbances is important, because lots of practical multi-agent systems suffer from the adverse effects of such disturbances, e.g., multi-missile systems [14] and multi-hydraulic manipulator systems [15]. On the other hand, it is doubtless that the consensus problem in the presence of mismatched disturbances is more difficult, because the mismatched disturbances (which are in different channels from the control inputs) cannot be directly suppressed by feedback controllers [16, 17]. Inspired by both aspects, some researchers have been devoted to consensus with mismatched disturbances and a few control schemes have been presented, such as adaptive backstepping control for multi-agent systems subject to bounded mismatched disturbances [18] and output regulation theory for multi-agent systems with mismatched disturbances generated from exosystems [19].

From the point view of control patterns to cope with disturbances, all the aforementioned control methods belong to passive anti-disturbance control methods. In detail, these control methods realize disturbance rejection goal through feedback regulation on the tracking errors between the measured outputs and their reference signals. In this way, they are not prompt enough when facing strong disturbances and they usually achieve disturbance rejection robustly by sacrificing the nominal control performances.

To improve the above-mentioned drawbacks in disturbance rejection, an effective way is to utilize active anti-disturbance control, which is feedforward-feedback composite control rather than pure feedback control. Usually, the feedforward compensations are made based on direct measurements (e.g., by using sensors) or soft measurements (e.g., by developing disturbance estimators), and the baseline feedback control is performed on the basis of existing feedback control methods. Among active anti-disturbance control, there are two most popular approaches, namely, disturbance observer based control (DOBC) [20, 21] and active disturbance rejection control (ADRC) [22]. Specifically, DOBC uses disturbance observers to estimate the disturbances and ADRC utilizes extended-state observers to observe the disturbances. In dealing with disturbances, both DOBC and ADRC have faster responses and less conservativeness than passive anti-disturbance control. Motivated by these nice features, DOBC [23, 24] and ADRC [25] have been employed into consensus control design with disturbances. However, these results are still limited to consensus with matched disturbances. In [26, 27], DOBC schemes were constructed to solve asymptotic output consensus for higher-order multi-agent systems with mismatched disturbances.

It is worth noticing that most of the above reported results are obtained based on full-state information, that is, all the system state elements are assumed to be measurable. In practice, the systems usually contain unmeasured states, due to the absence of sensors or some other reasons. In such cases, to obtain the state information for control design, a common way is to design observers for reconstructing state information. Similar cases also happen in consensus control field. To solve the consensus problem without disturbances but with unknown states, some state observers were designed in [28, 29] and corresponding output feedback protocols were proposed. Nevertheless, if the disturbances are involved into the agent dynamics, the consensus control schemes proposed in [28, 29] do not work anymore. Not only the observer tracking error systems but also the whole closed-loop consensus error systems are not convergent or even not stable. To this end, it is desirable and meaningful to work on the consensus problem of multi-agent systems with both disturbances and unmeasured states.

In this chapter, the output consensus problem of leader-follower higher-order multi-agent systems with mismatched disturbances and unmeasured states are investigated. By combining a finite-time observer technique and the nonsingular terminal sliding-mode control method together, a feedforward-feedback composite consensus control scheme is developed. The developed distributed control scheme realizes finite-time output consensus for all the agents. The main contributions of this chapter are threefold. First of all, for the first time, the finite-time output consensus of higher-order multi-agent systems with both mismatched disturbances and unmeasured states is solved by using distributed active anti-disturbance control. Secondly, the results obtained in this chapter extend the applicable scope of the baseline feedback control part in distributed active anti-disturbance control from state feedback control to output feedback control. The last but not the least, the mismatched/matched disturbances considered in the systems are no longer limited to slow time-varying types and they are allowed to be faster time-varying or have higher-order forms.

The rest parts of this chapter are organized as follows. In Sect. 11.2, some useful preliminary knowledge and problem formulation are given. In Sect. 11.3, the detailed consensus control design and the main result are shown. Then in Sect. 11.4, a simulation example is exhibited to validate the proposed consensus control scheme. Finally, some conclusions of this chapter are drawn in Sect. 11.5.

11.2 Preliminaries and Problem Formulation

11.2.1 Notations

Denote $\mathbf{1}_n=[1,\ldots,1]^T\in\mathbb{R}^n$. Given a vector $x=[x_1,\ldots,x_n]^T\in\mathbb{R}^n$ and $\alpha\in\mathbb{R}$, denote $\mathrm{sig}^\alpha(x)=[\mathrm{sig}^\alpha(x_1),\ldots,\mathrm{sig}^\alpha(x_n)]^T$, especially, $\mathrm{sign}(x)=[\mathrm{sign}(x_1),\ldots,\mathrm{sign}(x_n)]^T$, where $\mathrm{sig}^\alpha(z)=|z|^\alpha\mathrm{sign}(z)$, $\forall z,\alpha\in\mathbb{R}$ and $\mathrm{sign}(\cdot)$ is the standard sign function. Moreover, denote the 1-norm, Euclidean norm and infinity norm of vector x as $\|x\|_1=\sum_{i=1}^n|x_i|$, $\|x\|_2=\sqrt{x^{\mathrm{T}}x}$ and $\|x\|_\infty=\max_{i=1,\ldots,n}\{|x_i|\}$, respectively. A fundamental property is that $\|x\|_2\le\|x\|_1\le\sqrt{n}\|x\|_2$, $\forall x\in\mathbb{R}^n$. Given a symmetric matrix $P\in\mathbb{R}^{n\times n}$, denote its eigenvalues as $\lambda_{\min}(P)=\lambda_1(P)\le\lambda_2(P)\le\cdots\le\lambda_n(P)=\lambda_{\max}(P)$, which are in non-decreasing order. Let I_n denote $n\times n$ identity matrix.

11.2.2 Some Lemmas and Definitions

Lemma 11.1 ([30]) *For any* $x_i \in \mathbb{R}, i = 1, \dots, n$ *and* $0 < q \leq 1$, *it holds that*

$$\left(\sum_{i=1}^{n} |x_i|\right)^q \leq \sum_{i=1}^{n} |x_i|^q.$$

Lemma 11.2 ([30]) *If* $c > 0, d > 0$ *and* $\gamma(x, y) > 0$ *is a real-valued function for* $x \in \mathbb{R}, y \in \mathbb{R}$, *then*

$$|x|^c |y|^d \leq \frac{c\gamma(x, y)|x|^{c+d}}{c+d} + \frac{d\gamma^{-c/d}(x, y)|y|^{c+d}}{c+d}.$$

Consider the following autonomous system

$$\dot{x} = f(x), \quad x \subseteq \mathbb{R}^n, \quad f(0) = 0, \tag{11.1}$$

where $f : D \to \mathbb{R}^n$ is continuous on an open neighborhood $D \subseteq \mathbb{R}^n$ of the origin.

Definition 11.1 ([31]) The origin is a finite-time convergent equilibrium of system (11.1) if there are an open neighbourhood $U \subseteq D$ of the origin and a function $T_x : U\backslash\{0\} \to (0, \infty)$, such that every solution trajectory $x(t, x_0)$ of system (11.1) starting from the initial point $x_0 \in U\backslash\{0\}$ is well-defined for $t \in [0, T_x(x_0))$, and $\lim_{t \to T_x(x_0)} x(t, x_0) = 0$. Here $T_x(x_0)$ is called as the convergence-time function (with respect to x_0). The origin is said to be a finite-time stable equilibrium if it is finite-time convergent and Lyapunov stable. If $U = D = \mathbb{R}^n$, the origin is said to be a globally finite-time stable equilibrium.

Lemma 11.3 ([32]) *Let* $c_1, \dots, c_n > 0$ *be such that the polynomial* $s^n + c_n s^{n-1} + \cdots + c_2 s + c_1$ *is Hurwitz, and consider*

$$\dot{x}_i = x_{i+1}, \ i = 1, \dots, n-1, \ \dot{x}_n = u. \tag{11.2}$$

There exists $\varepsilon \in (0, 1)$ *such that, for every* $\alpha \in (1 - \varepsilon, 1)$, *the origin is a globally finite-time stable equilibrium for system (11.2) under the feedback control law*

$$u = -c_1 \text{sig}^{\alpha_1}(x_1) - \cdots - c_n \text{sig}^{\alpha_n}(x_n), \tag{11.3}$$

where $\alpha_{i-1} = \frac{\alpha_i \alpha_{i+1}}{2\alpha_{i+1} - \alpha_i}, i = 2, \dots, n, \alpha_{n+1} = 1$ *and* $\alpha_n = \alpha$.

Consider the system

$$\begin{aligned} \dot{x}_k &= x_{k+1}, \ k = 1, \dots, m-1, \\ \dot{x}_m &= u + d, \end{aligned} \tag{11.4}$$

where $x_k, k = 1, \ldots, m$ are state elements, u is the control input, and $d(t)$ is the external disturbance.

Lemma 11.4 *If $d(t)$ is differentiable and $\dot{d}(t)$ has a known Lipschitz constant $L > 0$, and a nonlinear observer is designed as*

$$\begin{aligned}
\dot{z}_0 &= v_0,\ v_0 = -\lambda_0 L^{\frac{1}{m+1}} \text{sig}^{\frac{m}{m+1}}(z_0 - x_1) + z_1,\\
\dot{z}_l &= v_l,\ v_l = -\lambda_l L^{\frac{1}{m+1-l}} \text{sig}^{\frac{m-l}{m+1-l}}(z_l - v_{l-1}) + z_{l+1},\ l = 1, \ldots, m-2,\\
\dot{z}_{m-1} &= v_{m-1} + u,\ v_{m-1} = -\lambda_{m-1} L^{\frac{1}{2}} \text{sig}^{\frac{1}{2}}(z_{m-1} - v_{m-2}) + z_m,\\
\dot{z}_m &= -\lambda_m L \text{sign}(z_m - v_{m-1}),
\end{aligned} \tag{11.5}$$

where $\lambda_0, \ldots, \lambda_m$ are positive gains, $z_0 = \hat{x}_1$ is the estimate of x_1, $z_l = \hat{x}_{l+1}, l = 1, \ldots, m-2$ are the estimates of x_{l+1}, and $z_m = \hat{d}$ is the estimate of d. Then by choosing appropriate gains $\lambda_0, \ldots, \lambda_m$, observer (11.5) *is finite-time convergent.*

Proof Denote observation errors as $w_0 = x_1 - z_0, \ldots, w_{m-1} = x_m - z_{m-1}, w_m = d - \hat{d}$. Then the observation error dynamics are

$$\begin{aligned}
\dot{w}_0 &= -\lambda_0 L^{\frac{1}{m}} \text{sig}^{\frac{m}{m+1}}(w_0) + w_1\\
\dot{w}_l &= -\lambda_l L^{\frac{1}{m-l}} \text{sig}^{\frac{m-l}{m+1-l}}(w_l - \dot{w}_{l-1}) + w_{l+1},\ l = 1, \ldots, m-1\\
\dot{w}_m &= -\lambda_m L \text{sign}(w_m - \dot{w}_m) + \dot{d}.
\end{aligned} \tag{11.6}$$

According to [33], there are appropriate gains $\lambda_0, \ldots, \lambda_m$ such that observation error dynamics (11.6) are finite-time stable.

Remark 11.1 Observer gains $\lambda_0, \ldots, \lambda_m$ can be chosen recursively and only λ_0 needs to be assigned in advance [33].

11.2.3 Graph Theory Notions

Let $G = (\mathscr{V}, \mathscr{E}, \mathscr{A})$ be a directed graph, where $\mathscr{V} = \{1, \ldots, n\}$ is the node set, $\mathscr{E} \subseteq \mathscr{V} \times \mathscr{V}$ is the edge set and $\mathscr{A} = [a_{ij}] \in \mathbb{R}^{n \times n}$ is the weighted adjacency matrix of G. For $\mathscr{A}$, $a_{ij} > 0$ if $(j, i) \in \mathscr{E}$ while $a_{ij} = 0$ otherwise. Moreover, $a_{ii} = 0, \forall i \in \mathscr{V}$. The neighbor set of node i is denoted as $N_i = \{j \in \mathscr{V} | (j, i) \in \mathscr{E}\}$. An undirected graph G is defined such that $(j, i) \in \mathscr{E} \Leftrightarrow (i, j) \in \mathscr{E}$ and $a_{ij} = a_{ji}$. The Laplacian matrix of the graph G is defined as $\mathscr{L} = [l_{ij}] \in \mathbb{R}^{n \times n}$, where $l_{ii} = \sum_{j=1}^{n} a_{ij}$ and $l_{ij} = -a_{ij}$ for $i \neq j$. In a directed graph, a directed path is an edge sequence as $(k_1, k_2), (k_2, k_3), \ldots, k_i \in \mathscr{V}$. If there is a directed edge from node i to node j, then node i is the parent and node j is the child. A directed tree is a graph, where every node has and only has one parent except for the root node, which has no parent, and the root has a directed path to every other node. A directed spanning tree of a graph is a directed tree formed by graph edges connecting all the nodes of the graph. A graph has a directed spanning tree if there exists at least one node having a directed path to all the other nodes.

For the leader-follower multi-agent systems, the leader is denoted by node 0 and the followers are denoted by nodes $1, \ldots, n$. The undirected communication topology graph of the followers is denoted by $G = (\mathscr{V}, \mathscr{E}, \mathscr{A})$ and its Laplacian matrix is denoted by $\mathscr{L}$. The communication topology graph for all the agents including the leader is denoted as $\bar{G} = (\bar{\mathscr{V}}, \bar{\mathscr{E}}, \bar{\mathscr{A}})$ with the node set $\bar{\mathscr{V}} = \mathscr{V} \cup \{0\}$. The communication between the leader and a follower is unidirectional from the leader to the follower. The edge weights between the leader and the followers are denoted by $b_i, i \in \mathscr{V}$. If the i-th follower is connected with the leader, then $b_i > 0$, otherwise, $b_i = 0$. Matrices $\mathscr{B} = \text{diag}\{b_1, \ldots, b_n\}$ and $\bar{\mathscr{L}} = \mathscr{L} + \mathscr{B} \in \mathbb{R}^{n \times n}$ are called as the leader adjacency and Laplacian matrices of the graph $\bar{G}$, respectively.

Assumption 11.1 For a leader-follower multi-agent system, the communication topology graph $\bar{G}$ contains at least one directed spanning tree.

Lemma 11.5 ([34]) *For a leader-follower multi-agent system, if Assumption 11.1 holds, then the Laplacian matrix $\bar{\mathscr{L}}$ is positive definite.*

11.2.4 Problem Formulation

The agent dynamics to be studied in this chapter are ($i \in \mathscr{V}$)

$$\begin{aligned} \dot{x}_{i,k} &= x_{i,k+1} + d_{i,k}(t), \; k = 1, \ldots, m-1, \\ \dot{x}_{i,m} &= u_i + d_{i,m}(t), \end{aligned} \tag{11.7}$$

where $x_{i,1}, \ldots, x_{i,m}$ are state elements, $y_i = x_{i,1}$ is output, u_i is control input to be designed, and $d_{i,1}(t), \ldots, d_{i,m-1}(t)$ and $d_m(t)$ are mismatched and matched disturbances, respectively. In system (11.7), only the output y_i is measurable and the other state elements $x_{i,2}, \ldots, x_{i,m}$ are unknown. The leader dynamics are

$$\begin{aligned} \dot{x}_{0,k} &= x_{0,k+1}, \; k = 1, \ldots, m-1, \\ \dot{x}_{0,m} &= u_0, \end{aligned} \tag{11.8}$$

where $x_{0,1}, \ldots, x_{0,m}$ are leader state elements, $y_0 = x_{0,1}$ is output, and u_0 is the predefined control input, which is independent from the followers.

The objective of this chapter is to design finite-time consensus protocols through output feedback for leader-follower multi-agent system (11.7) and (11.8) such that the outputs of all the $n+1$ agents reach consensus in finite time.

11.3 Control Design

The consensus control design is composed of two parts, namely, finite-time observer design and composite finite-time consensus protocols design.

11.3.1 Finite-Time Observer Design

The motivation to design an observer for each follower is natural, because in order to achieve the output consensus goal, information exchanges are needed among the neighboring agents. However, for the agents described by (11.7), only the outputs are available and can be directly used for transferring, which is not enough to realize consensus. Under this circumstance, it is necessary to design an observer for each follower to estimate the unknown state elements and the disturbances.

Assumption 11.2 The disturbance $d_{i,l}, i \in \mathscr{V}, l = 1, \ldots, m$ in system (11.7) is $(m-l+1)$ times differentiable and $d_{i,l}^{(m-l+1)}$ has a bounded Lipschitz constant $L_{i,l}$.

Remark 11.2 Assumption 11.2 is motivated by [33, 35]. Under this assumption, the result of Lemma 11.4 can be used to design observers for the followers. In practice, several kinds of disturbances satisfy Assumption 11.2, such as constant disturbances, ramp disturbances, high-order disturbances, and sinusoidal disturbances.

Denote $\chi_{i,1}=x_{i,1}, \chi_{i,k+1}=x_{i,k+1} + \sum_{l=1}^{k} d_{i,l}^{(k-l)}, k=1, \ldots, m-1, i \in \mathscr{V}$. Under Assumption 11.2, system (11.7) can be rewritten as

$$
\begin{aligned}
\dot{\chi}_{i,k} &= \chi_{i,k+1}, \ k = 1, \ldots, m-1, \\
\dot{\chi}_{i,m} &= u_i + d_i,
\end{aligned} \tag{11.9}
$$

where $d_i = \sum_{l=1}^{m} d_{i,l}^{(m-l)}$. The derivative $\dot{d}_i$ of the lumped disturbance d_i has a Lipschitz constant $L_i = \sum_{l=1}^{m} L_{i,l}$. For system (11.9), to estimate the unknown state elements and the disturbances, the following observer is designed

$$
\begin{aligned}
\dot{z}_{i,0} &= v_{i,0}, \ v_{i,0} = -\lambda_{i,0} L_i^{\frac{1}{m+1}} \operatorname{sig}^{\frac{m}{m+1}}(z_{i,0} - x_{i,1}) + z_{i,1}, \\
\dot{z}_{i,l} &= v_{i,l}, \ v_{i,l} = -\lambda_{i,l} L_i^{\frac{1}{m+1-l}} \operatorname{sig}^{\frac{m-l}{m+1-l}}(z_{i,l} - v_{i,l-1}) + z_{i,l+1}, \ l = 1, \ldots, m-2, \\
\dot{z}_{i,m-1} &= v_{i,m-1} + u_i, \ v_{i,m-1} = -\lambda_{i,m-1} L_i^{\frac{1}{2}} \operatorname{sig}^{\frac{1}{2}}(z_{i,m-1} - v_{i,m-2}) + z_{i,m}, \\
\dot{z}_{i,m} &= -\lambda_{i,m} L_i \operatorname{sign}(z_{i,m} - v_{i,m-1}),
\end{aligned} \tag{11.10}
$$

where $z_{i,0} = \hat{\chi}_{i,1}$ is the estimate of $\chi_{i,1}$, $z_{i,l} = \hat{\chi}_{i,l+1}, l = 1, \ldots, m-1$ are the estimates of $\chi_{i,l+1}$, and $z_{i,m} = \hat{d}_i$ is the estimate of d_i. Denote observation errors as $w_{i,l} = \chi_{i,l+1} - \hat{\chi}_{i,l+1}, l = 0, \ldots, m-1, w_{i,m} = d_i - \hat{d}_i$. Then the error dynamics are

$$\begin{aligned}
\dot{w}_{i,0} &= -\lambda_{i,0} L_i^{\frac{1}{m}} \text{sig}^{\frac{m}{m+1}}(w_{i,0}) + w_{i,1} \\
\dot{w}_{i,l} &= -\lambda_{i,l} L_i^{\frac{1}{m-l}} \text{sig}^{\frac{m-l}{m+1-l}}(w_{i,l} - \dot{w}_{i,l-1}) + w_{i,l+1}, \; l = 1, \ldots, m-1 \\
\dot{w}_{i,m} &= -\lambda_{i,m} L_i \text{sign}(w_{i,m} - \dot{w}_{i,m-1}) + \dot{d}_i. \qquad (11.11)
\end{aligned}$$

Proposition 11.1 *By choosing appropriate gains* $\lambda_{i,0}, \ldots, \lambda_{i,m}$, *observer* (11.10) *is finite-time convergent, namely, there is a finite time instant* T_o *such that* $w_{i,0}(t) = \ldots = w_{i,m}(t) = 0, i \in \mathscr{V}, \forall t \geq T_o$.

Proof By Lemma 11.4, the result of Proposition 11.1 follows directly. Thus, the proof is omitted.

Remark 11.3 For a follower described by (11.7), both its state elements (expect for the output) and the disturbances are unknown, and they are in the same mismatched/matched channels. Consequently, it is very difficult or even impossible to separately estimate the unknown state elements and the disturbances in the same channels by using observers. In such a case, observation of the combination variables of the unknown state elements and the disturbances, e.g., $\chi_{i,k}$, is feasible and efficient. This is the essential reason to construct an observer as (11.10) for each follower to estimate the combination variables of unknown state elements and the disturbances.

11.3.2 Consensus Protocol Design

Denote $\eta_k=[\hat{\chi}_{1,k}, \ldots, \hat{\chi}_{n,k}]^T, k=1, \ldots, m,\; u = [u_1, \ldots, u_n]^T, d = [d_1, \ldots, d_n]^T$, $\hat{d} = [\hat{d}_1, \ldots, \hat{d}_n]^T$ and $W_l = [\chi_{1,l+1}, \ldots, \chi_{n,l+1}]^T - \eta_{l+1}, l = 0, \ldots, m-1, W_m = d - \hat{d}$. Denote the consensus tracking errors as $\mathscr{E}_k = [e_{1,k}, \ldots, e_{n,k}]^T$, where

$$e_{i,k} = \sum_{j=1}^{n} a_{ij}(\hat{\chi}_{i,k} - \hat{\chi}_{j,k}) + b_i(\hat{\chi}_{i,k} - x_{0,k}),\; k = 1, \ldots, m,$$

Then $\mathscr{E}_k = \bar{\mathscr{L}}\eta_k - \mathscr{B}\mathbf{1}_n x_{0,k}, k = 1, \ldots, m$. Because $\mathscr{L}\mathbf{1}_n = 0$ and $\bar{\mathscr{L}} = \mathscr{L} + \mathscr{B}$, then $\bar{\mathscr{L}}\mathbf{1}_n x_{0,k} = \mathscr{B}\mathbf{1}_n x_{0,k}, k = 1, \ldots, m$. As a result, $\mathscr{E}_k = \bar{\mathscr{L}}\eta_k - \bar{\mathscr{L}}\mathbf{1}_n x_{0,k}, k = 1, \ldots, m$. From (11.7) and (11.8), the consensus tracking error dynamics are described as

$$\begin{aligned}
\dot{\mathscr{E}}_k &= \mathscr{E}_{k+1} + \tilde{W}_{k-1},\; k = 1, \ldots, m-1, \\
\dot{\mathscr{E}}_m &= \bar{\mathscr{L}}u + \bar{\mathscr{L}}(d - \dot{W}_{m-1} - \mathbf{1}_n u_0) \qquad (11.12)
\end{aligned}$$

where $\tilde{W}_{k-1} = \bar{\mathscr{L}}(W_k - \dot{W}_{k-1}), k = 1, \ldots, m-1$.

Assumption 11.3 The state estimates $\hat{\chi}_{i,k}, i \in \mathscr{V}, k = 1, \ldots, m$ and disturbance estimate $\hat{d}_i$ generated from observer (11.9) can be exchanged between neighboring agents, namely, each agent can receive such information sent from its neighbors.

Assumption 11.4 The leader state elements $x_{0,k}, k = 1, \ldots, m$ are bounded in any finite time and there exists $\bar{u}_0$ such that $\|u_0(t)\|_2 \leq \bar{u}_0, \forall t \geq 0$.

The main result of this chapter is given in the following theorem.

Theorem 11.1 *For leader-follower multi-agent system* (11.7) *and* (11.8), *under Assumptions 11.1–11.4, the consensus protocol for each follower is designed as*

$$u_i = -k_i \mathrm{sign}\left(\sum_{j=1}^{n} a_{ij}(s_i - s_j) + b_i s_i\right) - \hat{d}_i,\ i \in \mathscr{V}, \tag{11.13}$$

where $k_i \geq \max_{j \in \mathscr{V}} \left[b_j \bar{u}_0 + \left|\sum_{k=1}^{M} c_{j,k} \mathrm{sig}^{\alpha_{j,k}}(e_{j,k})\right| + \delta_j\right] / (\lambda_{\min}(\bar{\mathscr{L}})/\sqrt{n}), \delta_i > 0$, $\hat{d}_i$ *is the estimate of the disturbance* d_i *generated from observer* (11.10), *and the sliding-mode surface* s_i *is designed as*

$$s_i = e_{i,m} + \int_0^t \left[\sum_{k=1}^{m} c_{i,k} \mathrm{sig}^{\alpha_{i,k}}(e_{i,k})\right] \mathrm{d}\tau,\ i \in \mathscr{V}, \tag{11.14}$$

where $\alpha_{i,k-1} = \frac{\alpha_{i,k}\alpha_{i,k+1}}{2\alpha_{i,k+1} - \alpha_{i,k}}, k = 2, \ldots, m, \alpha_{i,m+1} = 1, \alpha_{i,m} = \alpha_{i,0} \in (1 - \varepsilon_i, 1), \varepsilon_i \in (0, 1)$, *and* $c_{i,k} > 0$ *satisfy that polynomial* $\lambda^m + c_{i,m}\lambda^{m-1} + \cdots + c_{i,2}\lambda + c_{i,1}$ *is Hurwitz. Then there are* $\varepsilon_i \in (0, 1)$ *such that under protocol* (11.13), *the agents' outputs reach consensus in finite time, namely,* $y_i - y_0 \to 0, \forall i \in \mathscr{V}$ *in finite time.*

Proof Denote $S = [s_1, \ldots, s_n]^T$. From (11.14), it follows that

$$S = \mathscr{E}_m + \int_0^t \left(\sum_{k=1}^{m} C_k \zeta_k\right) \mathrm{d}\tau, \tag{11.15}$$

where $C_k = \mathrm{diag}\{c_{1,k}, \ldots, c_{n,k}\}, \zeta_k = [\mathrm{sig}^{\alpha_{1,k}}(e_{1,k}), \ldots, \mathrm{sig}^{\alpha_{n,k}}(e_{n,k})]^T$. Based on (11.13), the control input vector u can be written as

$$u = -K \mathrm{sign}(\bar{\mathscr{L}} S) - \hat{d}, \tag{11.16}$$

where $K = \mathrm{diag}\{k_1, \ldots, k_n\}$. By substituting (11.16), the derivative of S along system (11.12) satisfies

$$\begin{aligned}\dot{S} &= \dot{\mathscr{E}}_m + \sum_{k=1}^{m} C_k\zeta_k \\ &= \bar{\mathscr{L}}\left(-K\,\mathrm{sign}(\bar{\mathscr{L}}S) + W_m - \dot{W}_{m-1} - \mathbf{1}_n u_0\right) + \sum_{k=1}^{m} C_k\zeta_k. \end{aligned} \tag{11.17}$$

Simultaneously, system (11.12) can be rewritten as

$$\begin{aligned}\dot{\mathscr{E}}_k &= \mathscr{E}_{k+1} + \tilde{W}_{k-1},\ k = 1, \ldots, m-1, \\ \dot{\mathscr{E}}_m &= -\sum_{k=1}^{m} C_k\zeta_k + \dot{S}. \end{aligned} \tag{11.18}$$

The following proof consists of two steps: the first step is to prove state boundedness of closed-loop system (11.18) in $t \in [0, T_o)$, where T_o is the finite convergence time of observer (11.9), and the second step is to prove global finite-time stability of the closed-loop system (11.18).

Step 1 State boundedness in $[0, T_o)$ of the closed-loop system (11.18)

Denote a function $V_b = \frac{1}{2}S^T S + \frac{1}{2}\sum_{k=1}^{m} \mathscr{E}_k^T \mathscr{E}_k$. Then the derivative of V_b along system (11.18) satisfies

$$\begin{aligned}\dot{V}_b &= S^T\dot{S} + \sum_{k=1}^{m} \mathscr{E}_k^T \dot{\mathscr{E}}_k \\ &= -\, S^T\bar{\mathscr{L}}K\,\mathrm{sign}(\bar{\mathscr{L}}S) + S^T\bar{\mathscr{L}}(W_m - \dot{W}_{m-1} - \mathbf{1}_n u_0) + S^T\sum_{k=1}^{m} C_k\zeta_k \\ &\quad + \sum_{k=1}^{m-1} \mathscr{E}_k^T(\mathscr{E}_{k+1} + \tilde{W}_{k-1}) + \mathscr{E}_m^T\bar{\mathscr{L}}(-K\,\mathrm{sign}(\bar{\mathscr{L}}S) + W_m - \dot{W}_{m-1} - \mathbf{1}_n u_0). \end{aligned} \tag{11.19}$$

From the basic properties of 1-norm and Euclidean norm, it follows that

$$-S^T\bar{\mathscr{L}}K\,\mathrm{sign}(\bar{\mathscr{L}}S) \le -\frac{\lambda_{\min}(\bar{\mathscr{L}})}{\sqrt{n}}\underline{k}\sum_{i=1}^{n}|s_i| \le 0, \tag{11.20}$$

where $\underline{k} = \min_{i\in\mathscr{V}}\{k_i\}$. It is straightforward to obtain that

$$S^T\bar{\mathscr{L}}(W_m - \dot{W}_{m-1} - \mathbf{1}_n u_0) \le (\|W_m\|_2 + \|\dot{W}_{m-1}\|_2 + \sqrt{n}\bar{u}_0)\|\bar{\mathscr{L}}\|_2\|S\|_2, \tag{11.21}$$

$$\sum_{k=1}^{m-1} \mathscr{E}_k^T(\mathscr{E}_{k+1} + \tilde{W}_{k-1}) \le \sum_{k=1}^{m-1}\|\tilde{W}_{k-1}\|_2\sum_{k=1}^{m}\|\mathscr{E}_k\|_2 + \sum_{k=1}^{m}\|\mathscr{E}_k\|_2^2, \tag{11.22}$$

$$\mathscr{E}_m^T \bar{\mathscr{L}}(-K\,\mathrm{sign}(\bar{\mathscr{L}}S) + W_m - \dot{W}_{m-1} - \mathbf{1}_n u_0) \le \left[\sqrt{n}(\bar{k} + \bar{u}_0) + \|W_m\|_2 + \|\dot{W}_{m-1}\|_2\right] \times \|\bar{\mathscr{L}}\|_2 \|\mathscr{E}_m\|_2, \tag{11.23}$$

where $\bar{k} = \max_{i\in\mathscr{V}}\{k_i\}$. Since $S^T \sum_{k=1}^m C_k \zeta_k = \sum_{i=1}^n \sum_{k=1}^m c_{i,k} s_i \mathrm{sig}^{\alpha_{i,k}}(e_{i,k})$, by Lemma 11.2, it holds that

$$\begin{aligned} s_i \mathrm{sig}^{\alpha_{i,k}}(e_{i,k}) &\le \frac{|s_i|^{1+\alpha_{i,k}}}{1+\alpha_{i,k}} + \frac{\alpha_{i,k}|e_{i,k}|^{1+\alpha_{i,k}}}{1+\alpha_{i,k}} \\ &\le \frac{s_i^2}{2} + \frac{\alpha_{i,k} e_{i,k}^2}{2} + \frac{1-\alpha_{i,k}}{2}. \end{aligned}$$

Then it can be verified that

$$\begin{aligned} S^T \sum_{k=1}^m C_k \zeta_k &\le \sum_{i=1}^n \left(\sum_{k=1}^m \frac{c_{i,k}}{2} \right) s_i^2 + \sum_{k=1}^m \left(\sum_{i=1}^n \frac{c_{i,k}\alpha_{i,k}}{2} e_{i,k}^2 \right) + \sum_{k=1}^m \sum_{i=1}^n \frac{c_{i,k}(1-\alpha_{i,k})}{2} \\ &\le \sigma_1 \|S\|_2^2 + \sigma_2 \sum_{k=1}^m \|\mathscr{E}_k\|_2^2 + \sum_{k=1}^m \sum_{i=1}^n \frac{c_{i,k}(1-\alpha_{i,k})}{2}, \end{aligned} \tag{11.24}$$

where $\sigma_1 = \max_{i\in\mathscr{V}} \left\{ \sum_{k=1}^m \frac{c_{i,k}}{2} \right\}$ and $\sigma_2 = \max_{i\in\mathscr{V}, k=1,\ldots,m} \left\{ \frac{c_{i,k}\alpha_{i,k}}{2} \right\}$. Substituting (11.20)–(11.24) into (11.19) yields

$$\begin{aligned} \dot{V}_b \le &\left\{ \sum_{k=1}^{m-1} \|\tilde{W}_{k-1}\|_2 + \left[\sqrt{n}(\bar{k} + \bar{u}_0) + \|W_m\|_2 + \|\dot{W}_{m-1}\|_2\right] \|\bar{\mathscr{L}}\|_2 \right\} \\ &\times \left(\|S\|_2 + \sum_{k=1}^m \|\mathscr{E}_k\|_2 \right) + (1+\sigma_2) \sum_{k=1}^m \|\mathscr{E}_k\|_2^2 + \sigma_1 \|S\|_2^2 + \sum_{k=1}^m \sum_{i=1}^n \frac{c_{i,k}(1-\alpha_{i,k})}{2}. \end{aligned} \tag{11.25}$$

By Lemma 11.2, the following inequalities hold

$$\|S\|_2 \le \frac{1+\|S\|_2^2}{2}, \ \|\mathscr{E}_k\| \le \frac{1+\|\mathscr{E}_k\|_2^2}{2}.$$

Then it follows from (11.25) that

$$\dot{V}_b \le \sigma_3 V_b + \sigma_4, \tag{11.26}$$

where

$$\sigma_3 = \sum_{k=1}^{m-1} \|\tilde{W}_{k-1}\|_2 + \left[\sqrt{n}(\bar{k} + \bar{u}_0) + \|W_m\|_2 + \|\dot{W}_{m-1}\|_2\right] \|\bar{\mathscr{L}}\|_2 + 2\sigma_1 + 2\sigma_2 + 2,$$

$$\sigma_4 = \frac{m+1}{2}\left\{\sum_{k=1}^{m-1} \|\tilde{W}_{k-1}\|_2 + \left[\sqrt{n}(\bar{k} + \bar{u}_0) + \|W_m\|_2 + \|\dot{W}_{m-1}\|_2\right] \|\bar{\mathscr{L}}\|_2\right\} + \sum_{k=1}^{m}\sum_{i=1}^{n} \frac{c_{i,k}(1-\alpha_{i,k})}{2}.$$

Since observer (11.9) is finite-time convergent, $W_m(t)$, $\dot{W}_{m-1}(t)$ and $\tilde{W}_{k-1}(t), k = 1, \ldots, m-1$ are all bounded $\forall t \in [0, +\infty)$. Therefore, both $S(t)$ and $\mathscr{E}_k(t), k = 1, \ldots, m$ are bounded $\forall t \in [0, T_o)$.

Step 2 Global finite-time stability of the closed-loop system (11.18)

When $t \geq T_o$, by Proposition 11.1, $W_0(t) = 0, \ldots, W_m(t) = 0$, which means that $\tilde{W}_{k-1}(t) = 0, k = 1, \ldots, m-1$ and $\dot{W}_{m-1}(t) = 0$. Take the Lyapunov function as $V_S = \frac{1}{2}S^T S$. From (11.17), it can be obtained that

$$\begin{aligned}
\dot{V}_S &= S^T \bar{\mathscr{L}}(-K\,\mathrm{sign}(\bar{\mathscr{L}}S) - \mathbf{1}_n u_0) + S^T \sum_{k=1}^{m} C_k \zeta_k \\
&\leq -\sum_{i=1}^{n}\left[\frac{\lambda_{\min}(\bar{\mathscr{L}})}{\sqrt{n}}\underline{k} - b_i \bar{u}_0 - \left|\sum_{k=1}^{m} c_{i,k}\mathrm{sig}^{\alpha_{i,k}}(e_{i,k})\right|\right]|s_i| \\
&\leq -\underline{\delta}\sum_{i=1}^{n}|s_i| \leq -\sqrt{2}\underline{\delta}(V_S)^{\frac{1}{2}},
\end{aligned} \tag{11.27}$$

where $\underline{\delta} = \min_{i\in\mathscr{V}}\{\delta_i\}$. Hence, the agents' states reach the sliding-mode surface $S = 0$ in finite time $T_S \geq T_o$. On $S = 0$, from (11.18), the equivalent dynamics are

$$\begin{aligned}
\dot{\mathscr{E}}_k &= \mathscr{E}_{k+1},\ k = 1, \ldots, m-1, \\
\dot{\mathscr{E}}_m &= -\sum_{k=1}^{m} C_k \zeta_k,
\end{aligned} \tag{11.28}$$

which can also be written as

$$\begin{aligned}
\dot{e}_{i,k} &= e_{i,k+1},\ k = 1, \ldots, m-1, \\
\dot{e}_{i,m} &= -\sum_{k=1}^{m} c_{i,k}\mathrm{sig}^{\alpha_{i,k}}(e_{i,k}).
\end{aligned} \tag{11.29}$$

By Lemma 11.3, there are $\varepsilon_i \in (0, 1)$ such that $\mathscr{E}_1$ converges to zero in finite time. By noting that $\hat{\chi}_{i,1}(t) = x_{i,1}(t) = y_i(t), i \in \mathscr{V}, \forall t \geq T_o$, then $y_i - y_0, i \in \mathscr{V}$ converge to zero in finite time. This completes the proof.

Remark 11.4 When $t \geq T_o$, by Proposition 11.1, it holds that $W_k(t) = 0$, $\dot{W}_k(t) = 0, k = 0, \ldots, m-1$, $W_m(t) = 0$. On one hand, it follows from (11.17) that

$$\dot{S} = -\bar{\mathscr{L}}K\mathrm{sign}(\bar{\mathscr{L}}S) - \bar{\mathscr{L}}\mathbf{1}_n u_0 + \sum_{k=1}^{m} C_k\zeta_k, \tag{11.30}$$

which means that $\dot{S}$ is discontinuous and then the solution of the closed-loop system (11.30) would be understood in the sense of Filippov. On the other hand, it has been proved above that $S(t) \equiv 0, \forall t \geq T_S \geq T_o$, namely, for any $t_2 > t_1 \geq T_S$, the integration of $\dot{S}(t)$ on $[t_1, t_2]$ is $\int_{t_1}^{t_2} \dot{S}(t)\mathrm{d}t = 0$. In this way, $\dot{S}$ has an infinite switching frequency and a zero average. Then it is naturally filtered out and does not bring effects on the stability of the integration-based closed-loop system. As a result, the equivalent dynamics (11.28) are obtained and Theorem 11.1 follows.

Remark 11.5 The proposed consensus protocol (11.13) is a feedforward-feedback composite protocol. For the i-th follower, it utilizes the agent's and its neighbors' state estimation information and the estimate of its disturbances. In detail, the switching law is $u_i^s = -k_i\mathrm{sign}\Big(\sum_{j=1}^{n} a_{ij}(s_i - s_j) + b_i s_i\Big)$, which guarantees that the agents' states converge to the sliding-mode surface $S = [s_1, \ldots, s_n]^T$ in finite time.

The block diagram of the closed-loop system under the proposed distributed control scheme in Theorem 11.1 is shown in Fig. 11.1.

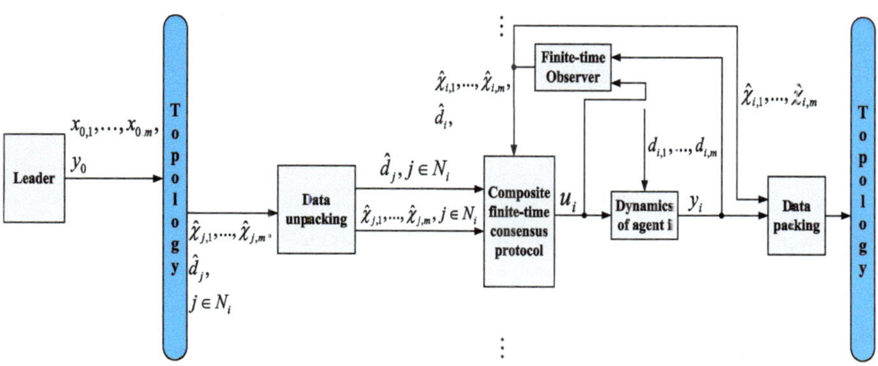

Fig. 11.1 Block diagram of the closed-loop multi-agent system under the proposed distributed control scheme

11.4 Numerical Simulations

In this section, to illustrate the effectiveness of the proposed consensus control scheme in Theorem 11.1, a simulation example is given on a leader-follower multi-agent system with 4 followers. The dynamics of the i-th ($i = 1, 2, 3, 4$) follower are

$$\begin{aligned} \dot{x}_{i,1} &= x_{i,2} + d_{i,1}, \\ \dot{x}_{i,2} &= x_{i,3} + d_{i,2}, \\ \dot{x}_{i,3} &= u_i + d_{i,3}, \\ y_i &= x_{i,1}. \end{aligned} \tag{11.31}$$

The leader dynamics are

$$\dot{x}_{0,1} = x_{0,2},\ \dot{x}_{0,2} = x_{0,3},\ \dot{x}_{0,3} = u_0,\ y_0 = x_{0,1}. \tag{11.32}$$

The communication topology of the agents is shown in Fig. 11.2.

The motion function of the leader is

$$x_{0,1}(t) = \begin{cases} -\frac{(t-6)^4}{324} + 10, & 0\text{ s} \leq t \leq 6\text{ s}. \\ 10, & t > 6\text{ s}. \end{cases} \tag{11.33}$$

It is not difficult to obtain that $\bar{u}_0 = 0.5$. The initial states of the followers are

$$\begin{aligned} &\text{follower 1}: \ x_{1,1}(0) = 3,\ x_{1,2}(0) = 0,\ x_{1,3}(0) = 0, \\ &\text{follower 2}: \ x_{2,1}(0) = 2,\ x_{2,2}(0) = 0,\ x_{2,3}(0) = 0, \\ &\text{follower 3}: \ x_{3,1}(0) = 6,\ x_{3,2}(0) = 0,\ x_{3,3}(0) = 0, \\ &\text{follower 4}: \ x_{4,1}(0) = 7,\ x_{4,2}(0) = 0,\ x_{4,3}(0) = 0. \end{aligned}$$

The disturbances are imposed on the agents at $t = 15$ s. For $t \geq 15$ s, the disturbances are

$$\begin{aligned} \text{follower 1}: \ & d_{1,1}(t) = 2\sin(1.5(t-15)),\ d_{1,2}(t) = -1.2(t-15) + 3, \\ & d_{1,3}(t) = \sin(2(t-15)), \\ \text{follower 2}: \ & d_{2,1}(t) = 0.2(t-15)^2 + (t-15),\ d_{2,2}(t) = 1.5(t-15),\ d_{2,3}(t) = -5, \end{aligned}$$

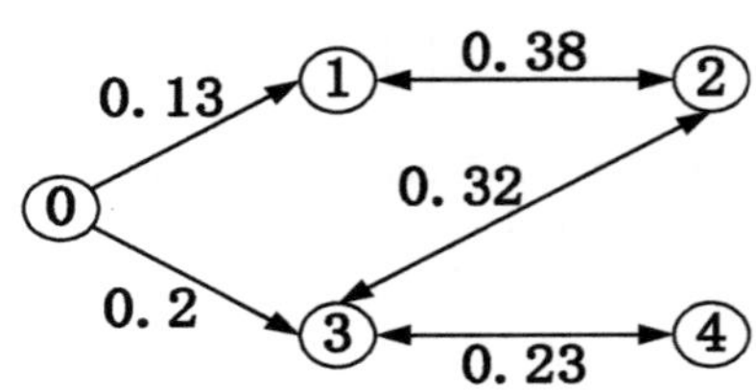

Fig. 11.2 The communication topology (0 denotes the leader and $\mathscr{V} = \{1, 2, 3, 4\}$)

$$\text{follower } 3: \; d_{3,1}(t) = 1 + 2(t-15), \; d_{3,2}(t) = \sin(4(t-15)), \; d_{3,3}(t) = 0.5(t-15),$$
$$\text{follower } 4: \; d_{4,1}(t) = 1.2(t-15)^2 - 2(t-15) + 3, \; d_{4,2}(t) = 2(t-15) - 2,$$
$$d_{4,3}(t) = 2.$$

According to the sufficient conditions given in Theorem 11.1, the control parameters of protocol (11.13) are chosen as

$$\text{follower } 1: \; \delta_1 = 0.04, \; c_{1,1} = 5, \; c_{1,2} = 9, \; c_{1,3} = 6, \; \alpha_{1,0} = \frac{7}{10},$$
$$\text{follower } 2: \; \delta_2 = 0.038, \; c_{2,1} = 4, \; c_{2,2} = 9, \; c_{2,3} = 5, \; \alpha_{2,0} = \frac{4}{5},$$
$$\text{follower } 3: \; \delta_3 = 0.045, \; c_{3,1} = 4, \; c_{3,2} = 8, \; c_{3,3} = 5, \; \alpha_{3,0} = \frac{3}{4},$$
$$\text{follower } 4: \; \delta_4 = 0.05, \; c_{4,1} = 5, \; c_{4,2} = 8, \; c_{4,3} = 6, \; \alpha_{4,0} = \frac{3}{5}.$$

The parameters of observer (11.10) are selected as

$$\text{follower } 1: \; L_1 = 18, \; \lambda_{1,0} = 12, \; \lambda_{1,1} = 25, \; \lambda_{1,2} = 9, \; \lambda_{1,3} = 26,$$
$$\text{follower } 2: \; L_2 = 13, \; \lambda_{2,0} = 9, \; \lambda_{2,1} = 10, \; \lambda_{2,2} = 8, \; \lambda_{2,3} = 7,$$
$$\text{follower } 3: \; L_3 = 19, \; \lambda_{3,0} = 11, \; \lambda_{3,1} = 12, \; \lambda_{3,2} = 9, \; \lambda_{3,3} = 15,$$
$$\text{follower } 4: \; L_4 = 17, \; \lambda_{4,0} = 12, \; \lambda_{4,1} = 10, \; \lambda_{4,2} = 11, \; \lambda_{4,3} = 15.$$

In simulations, the time step length is set as 0.01 s. The simulation results are exhibited in Figs. 11.3, 11.4 and 11.5. It can be seen from Fig. 11.4 that observer (11.10) provides accurate estimates of the unmeasured states and the disturbances in finite time. From Fig. 11.3, on one hand, in the absence of disturbances (namely, when $t \in [0\text{ s}, 15\text{ s})$), all the followers' outputs track the leader's output in finite time. On the other hand, when the disturbances are imposed on the followers at $t = 15$ s, the followers' outputs are suddenly pushed away from the leader's output, but after a few seconds, they come back and their outputs accurately track the leader's output once again. This nice feature mainly stems from the designed observer (11.10), because the observer offers the state and disturbance estimates, which are organically and effectively used to handle the effects caused by the unmeasured state elements and disturbances.

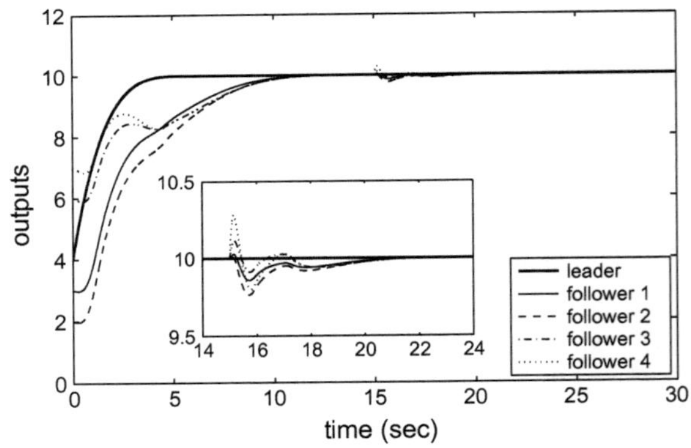

Fig. 11.3 Response curves of the agents outputs under the consensus protocol (11.13)

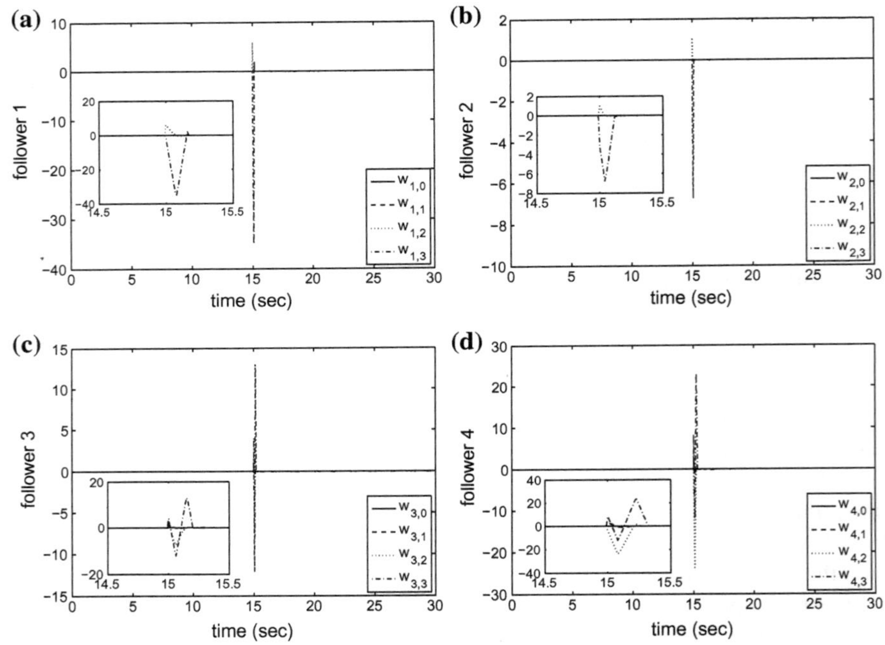

Fig. 11.4 Observation errors of observer (11.10). **a** Observation errors of follower 1. **b** Observation errors of follower 2. **c** Observation errors of follower 3. **d** Observation errors of follower 4

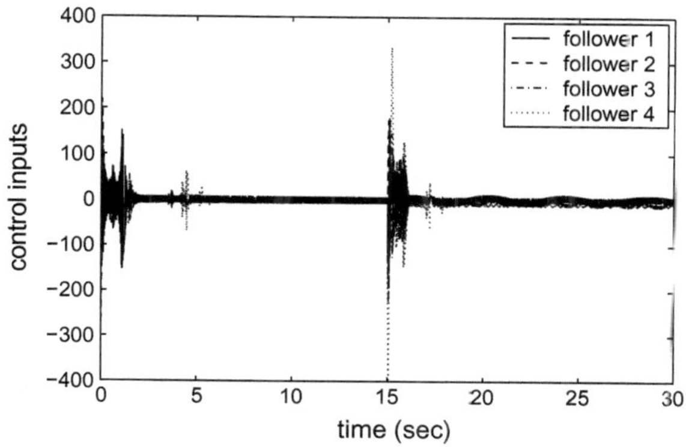

Fig. 11.5 Time histories of the agents inputs in the form of the consensus protocol (11.13)

11.5 Conclusions

In this chapter, a solution has been worked out for the finite-time output consensus problem of leader-follower higher-order multi-agent systems subject to both unmeasured states and mismatched disturbances. The main control design is composed of two stages, namely, finite-time observer design and nonsingular terminal sliding-mode consensus protocol design. In the first stage, a finite-time observer has been constructed for each follower to estimate the unmeasured state elements and the disturbances in finite time. In the second stage, nonlinear integral-type nonsingular terminal sliding-mode surfaces have been designed for the followers, where the state and disturbance estimates have been distributedly embedded. Then, based on such surfaces, composite sliding-mode consensus protocols have been proposed, which realize finite-time output consensus for all the agents. The most distinguished feature of the proposed composite consensus control scheme lies in that it achieves finite-time output consensus in the presence of both unmeasured states and fast time-varying mismatched disturbances.

Acknowledgements This work was supported by the National Natural Science Foundation of China under Grants 61503078 and 61473080, the Natural Science Foundation of Jiangsu Province under Grant BK20150626, the China Postdoctoral Science Foundation Funded Project under Grants 2015M570398 and 2016T90407, the Fundamental Research Funds for the Central Universities under Grant 2242016K41013, and the Priority Academic Program Development of Jiangsu Higher Education Institutions.

References

1. Ren, W., Beard, R.W.: Consensus seeking in multiagent systems under dynamically changing interaction topologies. IEEE Trans. Autom. Control **50**(5), 655–661 (2005)
2. Olfati-Saber, R., Murray, R.M.: Consensus problems in networks of agents with switching topology and time-delays. IEEE Trans. Autom. Control **49**(9), 1520–1533 (2004)

3. Pilloni, A., Pisano, A., Orlov, Y., Usai, E.: Consensus-based control for a network of diffusion PDEs with boundary local interaction. IEEE Trans. Autom. Control **61**(9), 2708–2713 (2016)
4. Bullo, F., Corters, J., Martinez, S.: Distributed Control of Robotic Networks: A Mathematical Approach to Motion Coordination Algorithms. Princeton University Press, Princeton (2009)
5. Li, S., Wang, X.: Finite-time consensus and collision avoidance control algorithms for multiple AUVs. Automatica **49**(11), 3359–3367 (2013)
6. Du, H., Li, S., Qian, C.: Finite-time attitude tracking control of spacecraft with application to attitude synchronization. IEEE Trans. Autom. Control **56**(11), 2711–2717 (2011)
7. Liu, Q., Wang, Z., He, X., Zhou, D.H.: Event-based recursive distributed filtering over wireless sensor networks. IEEE Trans. Autom. Control **60**(9), 2470–2475 (2015)
8. Wang, X., Li, S., Shi, P.: Distributed finite-time containment control for double-integrator multiagent systems. IEEE Trans. Cybern. **44**(9), 1518–1528 (2014)
9. Khoo, S., Xie, L., Man, Z.: Robust finite-time consensus tracking algorithm for multirobot systems. IEEE/ASME Trans. Mechatron. **14**(2), 219–228 (2009)
10. Pilloni, A., Pisano, A., Franceschelli, M., Usai, E.: Integral sliding modes for the robustification of consensus-based multi-agent based systems. In: Proceedings of the 14th International Workshop on Variable Structure Systems, pp. 222–227 (2016)
11. Li, Z., Duan, Z., Lewis, F.L.: Distributed robust consensus control of multi-agent systems with heterogeneous matching uncertainties. Automatica **50**(3), 883–889 (2014)
12. Li, P., Qin, K., Shi, M.: Distributed robust H_∞ rotating consensus control for directed networks of second-order agents with mixed uncertainties and timedelay. Neurocomputing **148**, 332–339 (2015)
13. Tang, Y., Hong, Y., Wang, X.: Distributed output regulation for a class of nonlinear multi-agent systems with unknown-input leaders. Automatica **62**, 154–160 (2015)
14. Chen, W.-H.: Nonlinear disturbance observer-enhanced dynamic inversion control of missiles. J. Guid. Control Dyn. **26**(1), 61–166 (2003)
15. Zeng, H., Sepehri, N.: Nonlinear position control of cooperative hydraulic manipulators handling unknown payloads. Int. J. Control **78**(3), 196–207 (2005)
16. de Loza, A.F., Cieslak, J., Henry, D., Zolghadri, A., Fridman, L.: Output tracking of systems subjected to perturbations and a class of actuator faults based on HOSM observation and identification. Automatica **59**, 200–205 (2015)
17. Estrada, A., Fridman, L., Iriarte, R.: Combined backstepping and HOSM control design for a class of nonlinear MIMO systems. Int. J. Robust Nonlinear Control **27**(4), 566–581 (2017)
18. Shen, Q., Shi, P.: Distributed command filtered backstepping consensus tracking control of nonlinear multiple-agent systems in strict feedback form. Automatica **53**, 120–124 (2015)
19. Su, Y., Huang, J.: Cooperative global robust output regulation for nonlinear uncertain multi-agent systems in lower triangular form. IEEE Trans. Autom. Control **60**(9), 2378–2389 (2015)
20. Chen, W.-H.: Disturbance observer based control for nonlinear system. IEEE/ASME Trans. Mechatron. **9**(4), 706–710 (2004)
21. Li, S., Yang, J., Chen, W.-H., Chen, X.: Disturbance Observer-Based Control: Methods and Applications. CRC Press, Boca Raton (2014)
22. Han, J.: From PID to active disturbance rejection control. IEEE Trans. Ind. Electron. **56**(3), 900–906 (2009)
23. Yang, H., Zhang, Z., Zhang, S.: Consensus of second-order multiagent systems with exogenous disturbances. Int. J. Robust Nonlinear Control **21**(9), 945–956 (2011)
24. Ding, Z.: Consensus disturbance rejection with disturbance observers. IEEE Trans. Ind. Electron. **62**(9), 5829–5837 (2015)
25. Cao, W., Zhang, J., Ren, W.: Leader-follower consensus of linear multi-agent systems with unknown external disturbances. Syst. Control Lett. **82**, 64–70 (2015)
26. Wang, X., Li, S., Lam, J.: Distributed active anti-disturbance output consensus algorithms for higher-order multi-agent systems with mismatched disturbances. Automatica **74**, 30–37 (2016)
27. Li, G., Wang, X., Li, S.: Distributed composite output consensus protocols of higher-order multi-agent systems subject to mismatched disturbances. IET Control Theory Appl. **11**(8), 1162–1172 (2017)

28. Zhou, B., Xu, C., Duan, G.: Distributed and truncated reduced-order observer based output feedback consensus of multi-agent systems. IEEE Trans. Autom. Control **59**(8), 2264–2270 (2014)
29. Zhang, H., Feng, G., Yan, H., Chen, Q.: Observer-based output feedback event-triggered control for consensus of multi-agent systems. IEEE Trans. Ind. Electron. **61**(9), 4884–4894 (2014)
30. Huang, X., Lin, W., Yang, B.: Global finite-time stabilization of a class of uncertain nonlinear systems. Automatica **41**(5), 881–888 (2005)
31. Bhat, S.P., Bernsteinm, D.S.: Finite-time stability of continuous autonomous systems. SIAM J. Control Optim. **38**(3), 751–766 (2000)
32. Bhat, S.P., Bernstein, D.S.: Geometric homogeneity with applications to finite-time stability. Math. Contr., Signals Syst. **17**(2), 101–127 (2005)
33. Levant, A.: Higher-order sliding modes, differentiation and output-feedback control. Int. J. Control **76**(9–10), 924–941 (2003)
34. Hong, Y., Hu, J., Gao, L.: Tracking control for multi-agent consensus with an active leader and variable topology. Automatica **42**(7), 1177–1182 (2006)
35. Shtessel, Y.B., Shkolnikov, I.A., Levant, A.: Smooth second-order sliding modes: missile guidance application. Automatica **43**(8), 1470–1476 (2007)

Chapter 12
Discrete Event-Triggered Sliding Mode Control

Abhisek K. Behera and Bijnan Bandyopadhyay

12.1 Introduction

Sliding mode control (SMC) is a robust control technique that achieves desired control objective for a dynamical system by forcing the state trajectory onto a predesigned (sliding) manifold by the action of a discontinuous control law [1, 2]. The motion on this manifold is often described as *sliding mode* for the dynamical system. Generally, there are two phases in SMC, one is to bring the trajectory to the sliding manifold and is known as *reaching phase*, and the other one is to force the trajectory to remain on this manifold which is called *sliding phase* [3]. This sliding phase must occur in finite-time only in a system to ensure the existence of sliding mode. In this phase, the system becomes unaffected by the external disturbances and thus robustness of the system is achieved [4]. The challenging task is to implement the discontinuous control law due to limitations of actuators. So, the implementation of SMC laws have drawn attention since last three decades.

The most important research in this direction has lead to so called discrete-time sliding mode (DTSM). Broadly, in two ways DTSM is investigated namely, direct discretization of continuous-time control laws [5] and the design of control law from the discrete-time plant [6–14]. However, in both these cases, the system trajectory remain bounded in the vicinity of sliding manifold. The band around this manifold explicitly depends on sampling interval, system parameters and disturbance bound. Many improvements have been suggested to reduce the size of this band resulting in numerical chattering. Further, it may be noted that in all these cases the control is implemented in periodic manner.

A.K. Behera · B. Bandyopadhyay (✉)
Interdisciplinary Programme in Systems and Control Engineering,
Indian Institute of Technology Bombay, Mumbai 400076, India
e-mail: bijnan@ee.iitb.ac.in

A.K. Behera
e-mail: abhisek@sc.iitb.ac.in

S. Li et al. (eds.), *Advances in Variable Structure Systems and Sliding Mode Control—Theory and Applications*, Studies in Systems, Decision and Control 115, DOI 10.1007/978-3-319-62896-7_12

Recently, a new control strategy has been proposed in literature, event-triggered control, where the control signal is applied to the plant whenever it is required [15–17]. In this technique, the state trajectory is continuously monitored and when it crosses certain threshold value from its previous sampled value, the control signal is updated. In most of the time, the periodic updating of control signal is not required if the state trajectory changes barely. By avoiding the periodic control execution, the computational burden is also reduced on the processor. So, event-triggered control has become a more promising control implementation strategy in almost every resource constrained systems [18–24]. To obtain robust performance of the system, event-triggered SMC is designed that not only uses resources optimally but also ensures robustness of the system during sliding mode [25–35].

Like other event-triggered control, in event-triggered SMC the state measurements are continuously done for evaluating the triggering instant. To realize this, the sophisticated sensors may be needed for measuring states continuously which may incur an additional cost. Many variants of event-triggered strategy have been proposed to avoid the continuous state measurement. One is the self-triggering strategy where triggering instant is determined from the previous sampled value while maintaining system stability [36–41]. So, the extra hardware circuit is also not needed to evaluate the event. However, the price is paid in terms of reduced inter event time (the time between any two consecutive triggering instants). Nevertheless, this is more useful way of implementing event-triggered control. The self-triggered SMC is also proposed in [28].

Another school of thought is to evaluate the event periodically rather than continuously. This strategy is more practical than the self-triggered control since the triggering scheme is designed based on the available sensor measurements. In this case, the cost of sensors may also be reduced as only periodic measurements are used in the control design. The sensors send the sampled state information at periodic intervals and that is used for the event-triggered control design. So, in this chapter, the periodic evaluation of triggering rule is carried out by using the sampled information. To analyse the closed loop system stability, we use discrete-time approach to design SMC using event-triggering strategy. In literature, a similar design strategy has been proposed called periodic event-triggered control [42, 43]. In this case also, the event is evaluated at periodic time intervals and updated whenever it is violated. However, event-triggered design of SMC is not discussed in literature.

12.2 System Description

Consider a single-input single output (SISO) linear time-invariant (LTI) system as given below

$$\dot{x} = Ax + B(u + d), \qquad x \in \mathbb{R}^n, \quad x_0 = x(0). \tag{12.1}$$

The system and input matrices are of appropriate dimensions. The disturbance, $d(t)$, is acting on the system through the input channel and is assumed to be bounded for all time. So, the following assumption holds.

Assumption 12.1 There exists a constant d_0 such that $\sup_{t\geq 0} |d(t)| \leq d_0$. In this chapter, it is further assumed that d_0 is a known constant.

Here, we discuss periodic evaluation of event-triggering rule such that the system (12.1) is stable. To realize this approach, we use discrete-time approach to design the SMC. Hence, we consider the discrete-time model of the system (12.1). In this approach, it is considered that the disturbance does not change much in between chosen sampling interval, so it can be assumed to be constant for a given sampling interval. The zero-order hold (ZOH) discretization for some sampling interval, h, can be obtained as

$$x(k+1) = \Phi x(k) + \Gamma(u(k) + d(k)) \tag{12.2}$$

where

$$\Phi = \mathrm{e}^{Ah} \quad \text{and} \quad \Gamma = \int_0^h \mathrm{e}^{As}\,\mathrm{d}s\, B.$$

It is apparent from the above that the discrete-time model is not in regular form. However, it can be transformed into regular form by some nonsingular transformation $z = Tx$. Thus, the system (12.2) can be transformed into regular form as

$$z(k+1) = \overline{\Phi} z(k) + \overline{\Gamma}(u(k) + d(k)) \tag{12.3}$$

where $\overline{\Phi} = T\Phi T^{-1}$ and $\overline{\Gamma} = T\Gamma$. This system can also be written in the (regular) form as

$$z_1(k+1) = \overline{\Phi}_{11} z_1(k) + \overline{\Phi}_{12} z_2(k) \tag{12.4}$$

$$z_2(k+1) = \overline{\Phi}_{21} z_1(k) + \overline{\Phi}_{22} z_2(k) + \overline{\Gamma}_2(u(k) + d(k)) \tag{12.5}$$

where $z_1 \in \mathbb{R}^{n-1}$ and $z_2 \in \mathbb{R}$. In the subsequent discussions of this chapter, we henceforth use either the system (12.3) or (12.4), (12.5) to design event-triggered discrete-time SMC. First, we present the design of discrete-time SMC for the system (12.3).

12.3 Discrete-Time Sliding Mode Control

In this section, we introduce DTSM control for LTI system. We present the design of DTSM control using the reaching law proposed in [9] and the similar is used in the design of event-triggered SMC.

Consider the sliding variable $s = c^\top z$ and the corresponding sliding manifold given by

$$\mathscr{S} = \left\{z \in \mathbb{R}^n : s = c^\top z = 0\right\}. \tag{12.6}$$

The vector $c = \left[c_1^\top \; 1\right]^\top$ is deigned such that sliding motion dynamics is stable. To see this, during sliding the system trajectory is constrained to the manifold $\mathscr{S}$ and hence $z_2 = -c_1^\top z_1$. So, the sliding motion dynamics can be given as

$$z_1(k+1) = \left(\overline{\Phi}_{11} - \overline{\Phi}_{12} c_1^\top\right) z_1(k) \tag{12.7}$$

$$z_2(k) = -c_1^\top z_1(k). \tag{12.8}$$

If the pair $(\overline{\Phi}, \overline{\Gamma})$ is controllable then so is the pair $(\overline{\Phi}_{11}, \overline{\Phi}_{12})$. This implies there exist a gain $c_1^\top$ such that $\overline{\Phi}_{11} - \overline{\Phi}_{12} c_1^\top$ is Schur[1] stable. Then, the stability of closed loop system follows immediately. So, the main objective is to bring the sliding mode in the system, the stability of the system can be assured.

The following reaching law is considered in the design of DTSM control in this chapter and is given as

$$s(k+1) = \tilde{d}(k) \tag{12.9}$$

where $\tilde{d}(k) = c^\top \overline{\Gamma} d(k)$. It may be noted that in a disturbance free case, i.e., $d(k) = 0$ for all $k \in \mathbb{Z}_{\geq 0}$, the system trajectory can be reached to the sliding manifold using the reaching law $s(k+1) = 0$. This ensures in one sampling step the sliding trajectory reaches the sliding manifold and then stays there for all time. However, in presence of disturbance the system trajectory remains in the vicinity of manifold. This is defined as quasi sliding mode (QSM) and the corresponding band is called QSM band. A formal definition of QSM is given below.

Definition 12.1 (*Quasi Sliding Mode*) The system (12.3) is said to be in QSM if for some $\varepsilon > 0$ there exist $\bar{k} \geq 0$ such that sliding trajectory $s(k) = c^\top z(k)$ satisfies

$$|s(k)| \leq \varepsilon$$

for all $k \geq \bar{k}$. The constant $\varepsilon > 0$ is the size of QSM band.

The control law can be designed as follows from the sliding variable dynamics given by

$$s(k+1) = c^\top \overline{\Phi} z(k) + c^\top \overline{\Gamma} u(k) + c^\top \overline{\Gamma} d(k).$$

---

[1] A matrix is said to be *Schur* stable if all the eigenvalues of this matrix are located within an unit disk in complex plane.

The control law that brings sliding motion in the system in the vicinity of sliding manifold is given as

$$u(k) = -(c^\top \overline{\Gamma})^{-1} c^\top \overline{\Phi} z(k). \tag{12.10}$$

It may be noted that this control brings the trajectory to the vicinity of sliding manifold in one discrete-time step. Once the trajectory reaches there, it remains bounded within the band of size $\left|c^\top B\right| d_0$. This is the QSM band for the system (12.3). However, in bringing the trajectory to QSM band in one discrete-time step, a large control effort may be required. But, in practice only limited control magnitude can be provided to any plant due to practical limitations of actuator. Due to this, in this case, it may not be possible to bring the system trajectory to the QSM band. But, the convergence of system trajectory to sliding manifold can be shown within a region in the state space.

For a saturated actuator with saturation constraint u_0, the control signal may be given as

$$u(k) = \begin{cases} u(k) & \text{if } \ |u(k)| < u_0, \\ u_0 \dfrac{u(k)}{|u(k)|} & \text{if } \ |u(k)| \geq u_0. \end{cases} \tag{12.11}$$

Now it is obvious that if the control signal is not saturated in some domain then the sliding mode will be enforced in the system. Otherwise, the system stability may not be ensured. However, we can find a domain of attraction for the system such that sliding mode can be ensured even if the actuator is saturated. This can be shown as follows.

Let the actuator is saturated, i.e., $|u(k)| \geq u_0$. Then, sliding variable dynamics with the saturated control can be given below as proposed in [9].

$$\begin{aligned}
|s(k+1)| &= \left| c^\top \overline{\Phi} z(k) + c^\top \overline{\Gamma} u_0 \frac{u(k)}{|u(k)|} + c^\top \overline{\Gamma} d(k) \right| \\
&\leq \left| c^\top \overline{\Phi} z(k) \right| \left(1 - \frac{u_0}{|u(k)|} \right) + \left| c^\top \overline{\Gamma} \right| d_0 \\
&= \left| c^\top \overline{\Phi} z(k) \right| - u_0 \frac{\left| c^\top \overline{\Phi} z(k) \right|}{\left| (c^\top \overline{\Gamma})^{-1} \right| \left| c^\top \overline{\Phi} z(k) \right|} + \left| c^\top \overline{\Gamma} \right| d_0 \\
&= \left| \left(c^\top \overline{\Phi} - c^\top \right) z(k) + s(k) \right| - \frac{u_0}{\left| (c^\top \overline{\Gamma})^{-1} \right|} - \left| c^\top \overline{\Gamma} \right| d_0 \\
&\leq |s(k)| + \left| \left(c^\top \overline{\Phi} - c^\top \right) z(k) \right| - \frac{u_0}{\left| (c^\top \overline{\Gamma})^{-1} \right|} + \left| c^\top \overline{\Gamma} \right| d_0 \\
&\leq |s(k)|
\end{aligned}$$

if $\left| (c^\top \overline{\Gamma})^{-1} \right| \left| \left(c^\top \overline{\Phi} - c^\top \right) z(k) \right| + d_0 \leq u_0$. Hence, the domain of attraction (DOA) for the saturated control signal may be given below as

$$\Omega_1 := \left\{ z \in \mathbb{R}^n : \left|(c^\top \overline{\Gamma})^{-1}\right| \left|\left(c^\top \overline{\Phi} - c^\top\right) z\right| + d_0 \le u_0 \right\}.$$

This shows that QSM occurs in the system in the region Ω_1 even if the control magnitude is saturated. Once the QSM occurs in the system, the system trajectory remain bounded in the vicinity of sliding manifold. However, in the disturbance free case, the DOA may be given as

$$\left\{ z \in \mathbb{R}^n : \left|(c^\top \overline{\Gamma})^{-1}\right| \left|\left(c^\top \overline{\Phi} - c^\top\right) z\right| \le u_0 \right\}.$$

From this we observe that the presence of disturbance restricts the DOA than that of disturbance free case.

12.4 Event-Triggered Sliding Mode Control

In this section, we present the design of event-triggered SMC using similar reaching law as discussed in the previous section. Here the control is not updated periodically but at aperiodic instants that is multiple of sampling interval. In this case, also the system trajectory remain bounded within a band called practical QSM band which is defined below.

Definition 12.2 (*Practical Quasi Sliding Mode*) The system (12.3) is said to be in practical QSM if the system is in QSM with band size greater than the QSM band (ε). That means for some $\varepsilon_1 > \varepsilon > 0$ there exist $\bar{k} > 0$ such that the sliding trajectory $s(k) = c^\top z(k)$ satisfies

$$|s(k)| \le \varepsilon_1$$

for all $k \ge \bar{k}$. The constant $\varepsilon_1 > 0$ is the size of practical QSM band.

From this it implies that practical QSM band is larger than the QSM band. In event-triggered strategy, the control law is implemented whenever certain conditions are satisfied. In this chapter, we evaluate the triggering rule at discrete instants and apply the control when it is violated. So, the control is updated to the plant at multiples of sampling interval only. We propose here event-triggering strategy for the control law (12.10). The event-triggered DTSM control is given as

$$u(k) = -(c^\top \overline{\Gamma})^{-1} c^\top \overline{\Phi} z(k_i) \tag{12.12}$$

for all $k \in [k_i, k_{i+1})$. Here, k_i denotes the triggering instant at which the control signal is updated and it is equal to the some multiples of sampling interval. During the time interval $[k_i, k_{i+1})$, the control signal is held constant. Now, we define $e(k) = x(k_i) - x(k)$ as the error introduced in the system due to aperiodic implementation of discrete control law (12.10).

The sufficient condition for existence of practical QSM for the system (12.3) and is given in the following Theorem.

Theorem 12.1 *Consider the system* (12.3) *and the control law* (12.12). *Let* $\alpha > 0$ *be given such that*

$$\left|c^{\top}\overline{\Phi}e(k)\right| < \alpha \tag{12.13}$$

holds for all $k \in \mathbb{Z}_{\geq 0}$. *Then, practical QSM occurs in the system in the finite number of discrete time steps.*

Proof Consider the sliding variable dynamics with the control law (12.12) in the interval $[k_i, k_{i+1})$ as given below

$$\begin{aligned} s(k+1) &= c^{\top}\overline{\Phi}z(k) + c^{\top}\overline{\Gamma}u(k) + c^{\top}\overline{\Gamma}d(k) \\ &= -c^{\top}\overline{\Phi}e(k) + c^{\top}\overline{\Gamma}d(k). \end{aligned}$$

Using (12.13) in the above relation, it gives

$$s(k+1) < \alpha + \left|c^{\top}\overline{\Gamma}\right| d_0.$$

It is to be noted that in this discrete event-triggered SMC the system trajectory reaches the practical QSM band in the first sampling instant, $k_0 + \tau$, as the control is updated at $k_0 = 0$. Define $\alpha_1 = \alpha + \left|c^{\top}\overline{\Gamma}\right| d_0$. However, for all discrete time instants $k \in [k_i, k_{i+1})$ with $i \geq 1$, the system trajectory remains bounded as

$$\left\{z \in \mathbb{R}^n : |s| = \left|c^{\top}z\right| < \alpha_1\right\}.$$

It is concluded that in one discrete step, the trajectory reaches the practical QSM band and remains there for all time. Hence, practical sliding mode in the system is established and this completes the proof.

Remark 12.1 In the above theorem, we see that in the proposed discrete event-triggered SMC the practical QSM band is reached in one discrete-time step similar to the case of periodic control implementation. But, it remains bounded within the practical QSM band for all subsequent triggering intervals.

One important consequence to the above Theorem is given below.

Corollary 12.1 *Consider the system* (12.3) *with* $d(k) = 0$ *for all* $k \in \mathbb{Z}_{\geq 0}$. *Let* (12.13) *holds for all* $k \in \mathbb{Z}_{\geq 0}$ *for some* $\alpha > 0$. *Then, practical QSM occurs in the system with a band given by*

$$\left\{z \in \mathbb{R}^n : |s| = \left|c^{\top}z\right| < \alpha\right\}.$$

This shows that in discrete event-triggered SMC, the system trajectories remain bounded in the vicinity of sliding manifold even in the disturbance free case. However, in periodic control implementation, the trajectory reaches the manifold exactly due to the reaching law $s(k+1) = 0$. That means, in discrete event-triggered SMC there is always a QSM band around the sliding manifold in disturbance free case. Moreover, this band size can be designed apriori by suitable value of α.

The relation (12.13) is very essential for the existence of practical QSM in the system. This relation must be ensured for all sampling period. So, this is used as triggering mechanism for the discrete event-triggered SMC and is given by

$$k_{i+1} = \left\{k > k_i : \|c\| \left\|\overline{\Phi}\right\| \|e(k)\| \geq \sigma\alpha\right\} \tag{12.14}$$

for some $\sigma \in (0, 1)$. This triggering maintains the relation $\|c\| \left\|\overline{\Phi}\right\| \|e(k)\| < \sigma\alpha$ for all $k \in \mathbb{Z}_{\geq 0}$. So, the stability of system is guaranteed.

In this case also, since the trajectory reaches practical QSM band in one triggering instant, large control effort may also be required. So, the actuator can be saturated and limited control signal is applied to the plant instead of actual control signal. This may lead to system instability in some cases. However, there always exist a DOA that ensures system stability in spite of saturated actuator. In the next, we provide an estimate of DOA in case of event-triggered discrete-time SMC.

Let the control signal is saturated at the triggering instant k_i, i.e., $|u(k_i)| \geq u_0$. Then, DOA can be obtained from the sliding variable dynamics in the time interval $[k_i, k_{i+1})$ as given below

$$\begin{aligned}
s(k+1) &= c^\top\overline{\Phi}z(k) + c^\top\overline{\Gamma}u_0\frac{u(k_i)}{|u(k_i)|} + c^\top\overline{\Gamma}d(k) \\
&= c^\top\overline{\Phi}z(k) - c^\top\overline{\Phi}z(k_i)\frac{u_0}{|u(k_i)|} + c^\top\overline{\Gamma}d(k) \\
&= c^\top\overline{\Phi}z(k) - c^\top\overline{\Phi}z(k_i) + c^\top\overline{\Phi}z(k_i)\left(1 - \frac{u_0}{|u(k_i)|}\right) + c^\top\overline{\Gamma}d(k) \\
&\leq \left|c^\top\overline{\Phi}e(k)\right| + \left|c^\top\overline{\Phi}z(k_i)\right|\left(1 - \frac{u_0}{|u(k_i)|}\right) + \left|c^\top\overline{\Gamma}\right| d_0.
\end{aligned}$$

Here, we are interested in the DOA where the system stability is always ensured. So, using the relation (12.13) in the above, one obtains

$$\begin{aligned}
s(k+1) &< \alpha_1 + \left|c^\top\overline{\Phi}z(k_i)\right| - u_0\frac{\left|c^\top\overline{\Phi}z(k_i)\right|}{|u(k_i)|} \\
&\leq \alpha_1 + \left|c^\top\overline{\Phi}z(k_i) - c^\top z(k) + c^\top z(k)\right| - u_0\frac{\left|c^\top\overline{\Phi}z(k_i)\right|}{\left|(c^\top\overline{\Gamma})^{-1}\right|\left|c^\top\overline{\Phi}z(k_i)\right|} \\
&< |s(k)| + \alpha_1 + \alpha + \left|\left(c^\top\overline{\Phi} - c^\top\right)z(k)\right| - \frac{u_0}{\left|(c^\top\overline{\Gamma})^{-1}\right|}.
\end{aligned}$$

So, $|s(k+1)| < |s(k)|$ if

$$\left|(c^\top \overline{\Gamma})^{-1}\right| \left(\alpha_2 + \left|\left(c^\top \overline{\Phi} - c^\top\right) z(k)\right|\right) \le u_0$$

where $\alpha_2 = 2\alpha + \left|c^\top \overline{\Gamma}\right| d_0$. So, DOA can be given for the discrete event-triggered SMC as

$$\Omega_2 := \left\{z \in \mathbb{R}^n : \left|(c^\top \overline{\Gamma})^{-1}\right| \left(\alpha_2 + \left|\left(c^\top \overline{\Phi} - c^\top\right) z\right|\right) \le u_0\right\}.$$

This shows that practical QSM can be ensured with saturated control signal if trajectory starts within Ω_2. Hence, system stability can be established.

It may be mentioned here that in disturbance free case, the DOA for the sliding trajectory to reach the band in event-triggering strategy is obtained by setting $d_0 = 0$ as

$$\left\{z \in \mathbb{R}^n : \left|(c^\top \overline{\Gamma})^{-1}\right| \left(2\alpha + \left|\left(c^\top \overline{\Phi} - c^\top\right) z\right|\right) \le u_0\right\}.$$

We see that in the disturbance free case DOA depends on the event parameter α which further restricts the DOA.

12.5 Simulation Results

Consider an LTI system

$$\dot{x} = \begin{bmatrix} 0 & 1 \\ 4 & 5 \end{bmatrix} x + \begin{bmatrix} 0 \\ 1 \end{bmatrix} (u + d).$$

We shall design the discrete event-triggered SMC for the above system. Select $\tau = 0.1$ as the sampling period for the system. The ZOH discrete-time model is obtained as

$$x(k-1) = \begin{bmatrix} 1.024 & 0.1306 \\ 0.5224 & 1.677 \end{bmatrix} x(k) + \begin{bmatrix} 0.005969 \\ 0.1306 \end{bmatrix} (u(k) + d(k)).$$

Note that ZOH model is not in regular form while the continuous-time plant model is in regular form. We define a nonsingular transformation as given below

$$T = \begin{bmatrix} 1.0000 & -0.0457 \\ 0 & 1.0000 \end{bmatrix}$$

that transforms the ZOH model into regular form. Then, the regular form is obtained using this nonsingular transformation as

$$x(k+1) = \begin{bmatrix} 1 & 0.09967 \\ 0.5224 & 1.701 \end{bmatrix} x(k) + \begin{bmatrix} 0 \\ 0.1306 \end{bmatrix} (u(k) + d(k)).$$

The sliding surface parameter c is chosen such that reduced order system has eigenvalue at 0.9 and is given as $\begin{bmatrix}1.0033 & 1\end{bmatrix}^\top$. The event parameters α and σ are selected as 0.1 and 0.85, respectively. The objective is to bring the trajectory to the practical QSM band in finite number of discrete steps. First we consider unsaturated case and then the actuator saturation case. The initial condition is taken as $z(0) = \begin{bmatrix}1 & 2\end{bmatrix}^\top$.

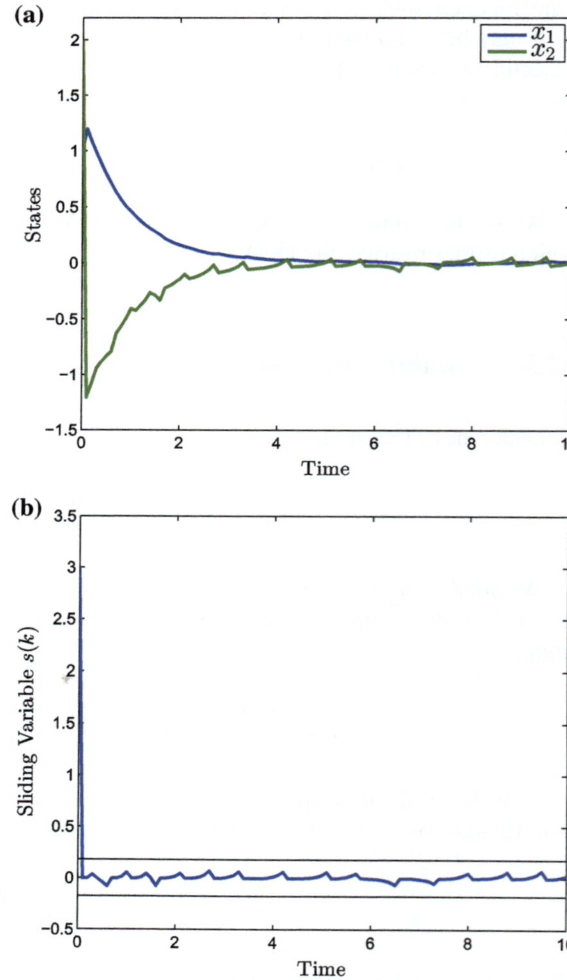

Fig. 12.1 System trajectories of discrete event-triggered SMC without actuator saturation, **a** State trajectories with time and **b** Sliding trajectory with time

12.5.1 *Without Saturation*

In this case, the event-triggered DTSM control that brings the system trajectory to the vicinity of sliding manifold is given as

$$u(k) = -\left[0.1993\ 0.2352\right] z(k_i)$$

for all $k \in [k_i, k_{i+1})$. This control signal is directly applied to the plant whenever event is triggered. The response of the system is shown in Fig. 12.1. The system trajectory reaches the practical QSM band in one step as shown in Fig. 12.1b. Once it reaches the band, it remains within the QSM band as claimed in earlier section.

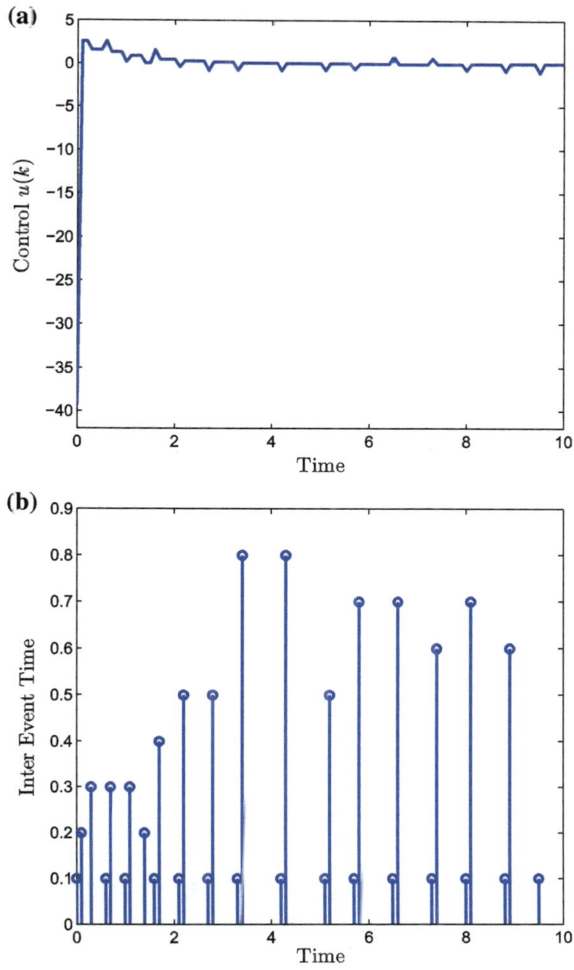

Fig. 12.2 Performance of discrete event-triggered SMC without actuator saturation, **a** Event-triggered SMC signal and **b** Inter event time versus time

The corresponding state trajectories are plotted in Fig. 12.1a. The state trajectories also converge to origin and remain ultimately bounded.

From Fig. 12.2a, it is observed that a large control effort is required to bring the system trajectory to the practical QSM band in one triggering instant. The control signal during initial time as high as 40 units but it significantly reduced to a small value as the state is forced to the vicinity of sliding manifold. Figure 12.2b shows the plot of variation of inter event time with time. At initial time, the control is updated, so the inter event time is found to be 0.1 unit of time. But, immediately after that the event is not triggered at sampling intervals that leads to no update in the control signal. The relation between Fig. 12.2a and b shows that if the inter event time is increased, no changes in the control action is seen. So, if event parameter is properly

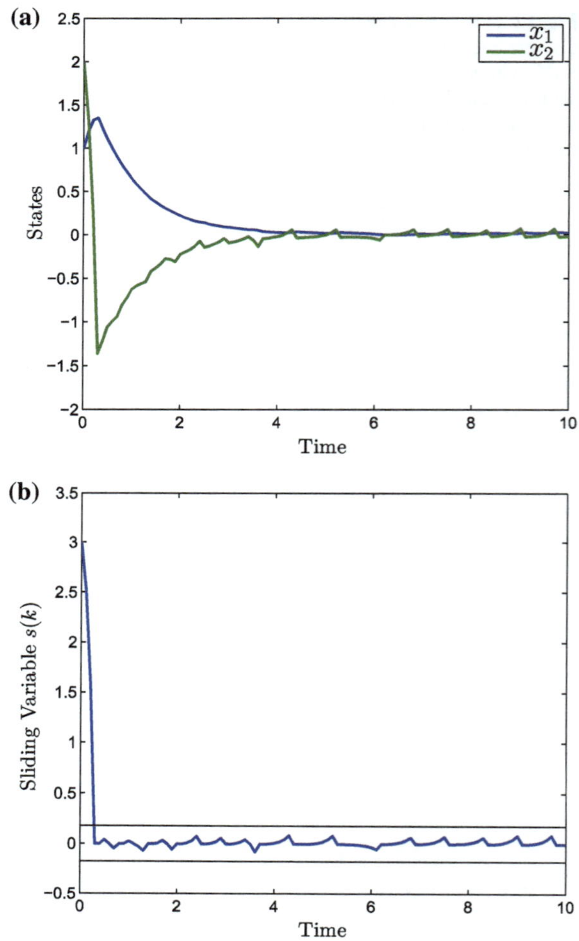

Fig. 12.3 System trajectories of discrete event-triggered SMC with actuator saturation, **a** State trajectories with time and **b** Sliding trajectory with time

chosen the inter event time can be increased further subject to some satisfactory system performance.

12.5.2 With Saturation

In this case actuator saturation is considered in the design of discrete event-triggered SMC. To demonstrate the discussions of the previous section, consider the saturation limit for the actuator as $u_0 = 20$. That means the saturated control signal is applied

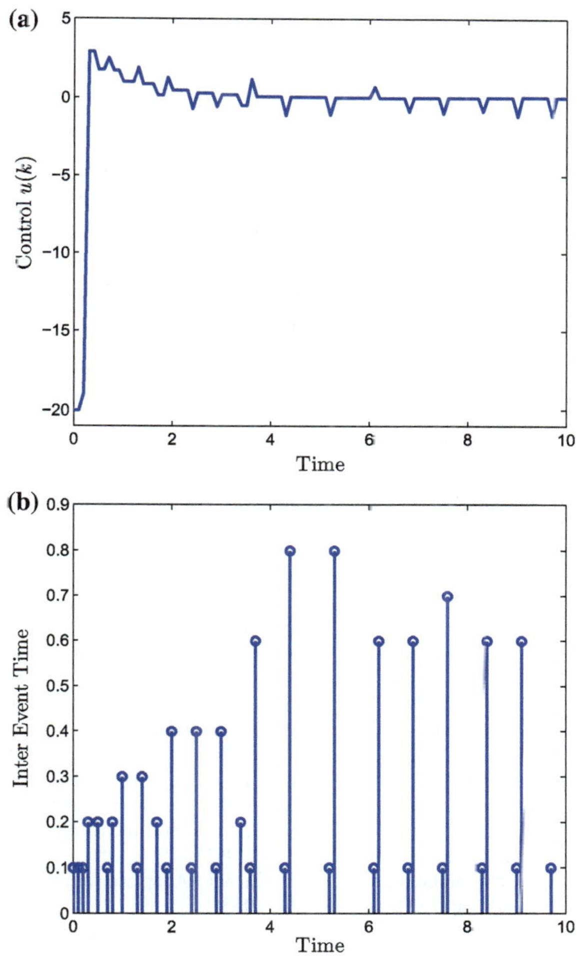

Fig. 12.4 Performance of discrete event-triggered SMC with actuator saturation, **a** Event-triggered SMC signal and **b** Inter event time versus time

to the plant instead of actual control signal whenever the control signal magnitude is larger than the saturation limit of the actuator.

It is verified that the initial condition $z(0)$ is located within the set Ω_2. This implies that even if the actuator is saturated the system trajectory would reach the practical QSM band in finite number of steps. In Fig. 12.3b, it is seen that the system trajectory does not reach the manifold in one triggering interval as the control signal crosses the saturation limit. But, the control signal is reduced as the sliding trajectory decreases to the practical QSM band at 0.3 s and remain bounded henceforth. The state trajectory is plotted in Fig. 12.3a which also shows the stability of the closed loop system and similarly the saturated control signal is plotted in Fig. 12.4a. The inter event time is shown in Fig. 12.4b. We observe that the inter event time is increased to same value in both the cases considered in this simulation study.

12.6 Conclusion

In this chapter, a discrete event-triggered SMC is proposed that evaluates the event periodically. So, continuous monitoring of state trajectory is avoided. In this case, the control signal is updated whenever event is triggered at one of these periodic instants only. It is also seen that in this triggering mechanism a positive lower bound for triggering mechanism is always guaranteed. Finally, simulation results are presented to demonstrate the performance of the system.

References

1. Utkin, V.I.: Variable structure systems with sliding modes. IEEE Trans. Autom. Control **22**(2), 212–222 (1977)
2. Utkin, V.I., Gulnder, J., Shi, J.: Sliding Mode Control in Electromechanical Systems. CRC Press, Taylor and Francis Group (1999)
3. Edwards, C., Spurgeon, S.K.: Sliding Mode Control: Theory and Applications. CRC Press, Taylor and Francis Group (1998)
4. Draženović, D.: The invariance conditions in variable structure systems. Automatica **5**(3), 287–295 (1969)
5. Galias, Z., Yu, X.: Euler's discretization of single input sliding-mode control systems. IEEE Trans. Autom. Control **52**(9), 1726–1730 (2007)
6. Sarpturk, S.Z., Istefanopulus, I., Kaynak, O.: On the stability of discrete-time sliding mode control systems. IEEE Trans. Autom. Control **32**(10), 930–932 (1987)
7. Furuta, K.: Sliding mode control of a discrete system. Syst. Control Lett. **14**(2), 145–152 (1990)
8. Gao, W., Wang, Y., Homaifa, A.: Discrete-time variable structure control systems. IEEE Trans. Ind. Electron. **42**(2), 117–122 (1995)
9. Bartolini, G., Ferrara, A., Utkin, V.I.: Adaptive sliding mode control in discrete-time systems. Automatica **31**(5), 769–773 (1995)
10. Bartoszewicz, A.: Discrete-time quasi-sliding-mode control strategies. IEEE Trans. Ind. Electron. **45**(4), 633–637 (1998)
11. Chakrabarty, S., Bandyopadhyay, B.: A generalized reaching law for discrete time sliding mode control. Automatica **52**, 83–86 (2015)

12. Janardhanan, S., Bandyopadhyay, B.: Output feedback sliding-mode control for uncertain systems using fast output sampling technique. IEEE Trans. Ind. Electron. **53**(5), 1677–1682 (2006)
13. Behera, A.K., Bandyopadhyay, B.: Steady-state behaviour of discretized terminal sliding mode. Automatica **54**, 176–181 (2015)
14. Bandyopadhyay, B., Janardhanan, S.: Discrete-Time Sliding Mode Control- A Multirate Output Feedback Approach. Lecture Notes in Control and Information Sciences. Springer, Berlin (2005)
15. Årźen, K.-E.: A simple event-based PID controller. In: Proceedings of 14th IFAC World Congress, pp. 423–428 (1999)
16. Åström, K.J., Bernhardsson, B.: Comparison of Riemann and Lebesgue sampling for first order stochastic systems. In: Proceedings of 41st IEEE Conference on Decision and Control, pp. 2011–2016 (2002)
17. Tabuada, P.: Event-triggered real-time scheduling of stabilizing control tasks. IEEE Trans. Autom. Control **52**(9), 1680–1685 (2007)
18. Heemels, W.P.M.H., Johansson, K.H., Tabuada, P.: An introduction to event-triggered and self-triggered control. In: Proceedings of 51st IEEE Conference on Decision and Control, pp. 3270–3285 (2010)
19. Lunze, J., Lehmann, D.: A state-feedback approach to event-based control. Automatica **46**(1), 211–215 (2010)
20. Tallapragada, P., Chopra, N.: On event triggered tracking for nonlinear systems. IEEE Trans. Autom. Control **58**(9), 2343–2348 (2013)
21. Yue, D., Tian, E., Han, Q.-L.: A delay system method for designing event-triggered controllers of networked control systems. IEEE Trans. Autom. Control **58**(2), 475–481 (2013)
22. Heemels, W.P.M.H., Sandee, J.H., Boscho, P.P.J.V.: Analysis of event-driven controllers for linear systems. Int. J. Control **81**(4), 571–590 (2008)
23. Borger, D.P., Heemels, W.P.M.H.: Event-separation properties of event-triggered control systems. IEEE Trans. Autom. Control **59**(10), 2644–2656 (2014)
24. Girard, A.: Dynamic triggering mechanisms for event-triggered control. IEEE Trans. Autom. Control **60**(7), 1992–1997 (2015)
25. Behera, A.K., Bandyopadhyay, B.: Event based robust stabilization of linear systems. In: Proceedings of 40th Annual Conference of the IEEE Industrial Electronics Society, pp. 133–138 (2014)
26. Ferrara, A., Incremona, G.P., Magini, L.: Model-based event-triggered robust MPC/ISM. In: Proceedings of 13th Eurpean Control Conference, pp. 2931–2936 (2014)
27. Cucuzzella, M., Incremona, G.P., Ferrara, A.: Event-triggered sliding mode control algorithms for a class of uncertain nonlinear systems: experimental assessment. In: Proceedings of American Control Conference, pp. 6549–6554 (2016)
28. Behera, A.K., Bandyopadhyay, B.: Self-triggering-based sliding-mode control for linear systems. IET Contr. Theory Appl. **9**(17), 2541–2547 (2015)
29. Behera, A.K., Bandyopadhyay, B.: Event based sliding mode control with quantized measurement. In: Proceedings of International Workshop on Recent Advances in Sliding Modes, pp. 1–6 (2015)
30. Behera, A.K., Bandyopadhyay, B., Xavier, N., Kamal, S.: Event-triggered sliding mode control for robust stabilization of linear multivarible systems. In: Yu, X., Efe, M.Ö. (eds.) Recent Advances in Sliding Modes: From Control to Intelligent Mechatronics. Studies in Systems, Decision and Control, vol. 24, pp. 155–175. Springer International Publishing, Cham (2015)
31. Behera, A.K., Bandyopadhyay, B.: Event-triggered sliding mode control for a class of nonlinear systems. Int. J. Control **89**(9), 1916–1931 (2016)
32. Behera, A.K., Bandyopadhyay, B.: Robust sliding mode control: an event-triggering approach. IEEE Trans. Circuits Syst.-II: Express Briefs **64**(2), 146–150 (2017)
33. Behera, A.K., Bandyopadhyay, B.: Decentralized event-triggered sliding mode control. In: Proceedings of 10th Asian Control Conference, pp. 1–5 (2015)
34. Behera, A.K., Bandyopadhyay, B., Reger, J.: Discrete event-triggered sliding mode control with fast output sampling feedback. In: Proceedings of 14th International Workshop on Variable Structure Systems, pp. 148–153 (2016)

35. Kumari, K., Bandyopadhyay, B., Behera, A.K., Reger, J.: Event-triggered sliding mode control for delta opeartor systems. In: Proceedings of 42nd Annual Conference of the IEEE Industrial Electronics Society, pp. 148–153 (2016)
36. Wang, X., Lemmon, M.D.: Self-triggered feedback control systems with finite-gain $\mathscr{L}_2$ stability. IEEE Trans. Autom. Control **54**(3), 452–467 (2009)
37. Wang, X., Lemmon, M.D.: Self-triggering under state-independent disturbances. IEEE Trans. Autom. Control **55**(6), 1494–1500 (2010)
38. Anta, A., Tabuada, P.: To sample or not to sample: self-triggered control for nonlinear systems. IEEE Trans. Autom. Control **55**(9), 2030–2042 (2010)
39. Almieda, J., Silvestre, C., Pascoal, A.M.: Self-triggered output feedback controller of linear plants in the presence of unkonwn disturbances. IEEE Trans. Autom. Control **59**(11), 3040–3045 (2014)
40. Mazo, M.J., Anta, A., Tabuada, P.: An ISS self-triggered implementation of linear controllers. Automatica **46**(8), 1310–1314 (2010)
41. Gommans, T., Antunes, D., Donkers, T., Tabuada, P., Heemels, M.: Self-triggered linear quadratic control. Automatica **50**(4), 1279–1287 (2014)
42. Heemels, W.P.M.H., Donkers, M.C.F., Teel, A.R.: Periodic event-triggered control for linear systems. IEEE Trans. Autom. Control **58**(4), 847–861 (2013)
43. Postoyan, R., Anta, A., Heemels, W.P.M.H., Tabuada, P., Nešić, D.: Periodic event-triggered control for nonlinear systems. In: Proceedings of 52nd IEEE Conference on Decision and Control, pp. 7397–7402 (2013)

Chapter 13
Fault Tolerant Control Using Integral Sliding Modes

Christopher Edwards, Halim Alwi and Mirza Tariq Hamayun

13.1 Introduction

Sliding mode control schemes have a number of interesting properties, which fuelled the continued research interest in this field since the 1960s. Perhaps the key property is its inherent insensitivity (at least theoretically) to so-called matched uncertainty – i.e. uncertainty acting in the channel of the input control signals [9, 23, 28]. In conventional first order sliding modes, the order of the original open-loop dynamical system is reduced by an amount equal to the number of input control signals, and the associated reduced order dynamics are determined by the specific choice of sliding surface/switching function – which forms a key part of the design process. Many different paradigms for the design of linear sliding surfaces for uncertain linear systems have been developed, and this area of research is quite mature [9, 23]. In conventional sliding modes the closed-loop behaviour can be split into two well-defined distinguishable phases: (a) the pre-sliding phase in which the controller drives the system states towards the sliding surface prior to achieving a sliding mode; (b) the reduced order sliding motion that occurs once the surface is attained and the states are forced to evolve along this surface. Note that the insensitivity (robustness) properties only appear once the sliding mode has been achieved.

Integral sliding modes (ISM) were first discussed in the late 1980s and early 1990s [19, 27]. Fundamentally the key distinction between integral sliding modes and conventional first order sliding modes is that during sliding, ISM systems retain

C. Edwards (✉) · H. Alwi
College of Engineering and Physical Sciences, University of Exeter, Exeter, UK
e-mail: c.edwards@exeter.ac.uk

H. Alwi
e-mail: h.alwi@exeter.ac.uk

M.T. Hamayun
COMSATS Institute of Information Technology, Lahore Campus, Lahore, Pakistan
e-mail: mhamayun@ciitlahore.edu.pk

S. Li et al. (eds.), *Advances in Variable Structure Systems and Sliding Mode Control—Theory and Applications*, Studies in Systems, Decision and Control 115, DOI 10.1007/978-3-319-62896-7_13

the order of the original open-loop system, although the property of insensitivity to matched uncertainty remains. Several other interesting and useful differences from conventional sliding modes are observed – the key one perhaps being that there is no reaching phase and sliding begins at time $t = 0$. Key subsequent work has refined these ideas to combat unmatched uncertainty [8], investigated output feedback formulations [5, 15] and extended these ideas to nonlinear systems [21].

This chapter focuses on the exploitation of these ideas for the development of fault tolerant controllers– i.e. control laws which retain an acceptable level of performance in the face of actuator or sensor faults [3, 6]. It is intuitively clear that certain classes of actuator faults can be modelled as matched uncertainty. Consequently, for a class of faults which merely deteriorate the capabilities of the actuators and in which some level of functionality is retained post-fault, sliding mode controllers appear to offer de-facto fault tolerance. This chapter explores these ideas for a wider class of over-actuated systems and allows the possibility of total failure of certain actuators. Two distinct classes of problems are considered: firstly a fault tolerant ISM controller is designed for a class of over-actuated linear systems; and secondly an ISM scheme is retrofitted to an existing feedback control scheme for a class of over-actuated linear systems with the objective of retaining as close to nominal performance as possible, in the face of actuator faults and failures. This chapter focusses on the situation in which full state information is available: work considering the output case appears in [15, 20].

13.2 ISM Control of Uncertain Linear Systems

Consider initially the uncertain linear system given by

$$\dot{x}(t) = Ax(t) + Bu(t) + f(t, x) \tag{13.1}$$

where $A \in \mathbb{R}^{n \times n}$ and $B \in \mathbb{R}^{n \times m}$ are known, and the pair (A, B) is controllable. In this section it is assumed that B has full column rank.[1] Here $f(t, x)$ is unknown and represents lumped uncertainty in the model of the system. For the purpose of control law design it is assumed that the states $x(t)$ are measured and are available. Assume initially that $f(t, x)$ satisfies the matching condition

$$f(t, x) = B\xi(t, x) \tag{13.2}$$

where $\xi(t, x)$ is unknown but worst case norm bounded by a known function $\bar{\xi}(t, x)$. A so-called integral sliding mode surface for the uncertain system in (13.1) is

[1]This is usually the case in 'typical' control problems and can be interpreted as the control inputs being distinct and non-redundant. However the situation where this is not the case will be discussed at length later in the chapter, since actuator redundancy is very beneficial in safety critical systems requiring fault tolerant control.

$$\sigma(t,x) = Gx(t) - Gx(0) - \int_0^t G(A+BK)x(\tau)\mathrm{d}\tau \tag{13.3}$$

where $G \in \mathbb{R}^{m\times n}$ is a fixed design matrix chosen to ensure that $\det(GB) \neq 0$ and the matrix $K \in \mathbb{R}^{m\times n}$ is chosen so that $(A+BK)$ is stable. The existence of K is guaranteed from the assumption that (A, B) is a controllable pair. The switching variable $\sigma : \mathbb{R}_+ \times \mathbb{R}^n \mapsto \mathbb{R}^m$ in (13.3) has the property that $\sigma(0, x(0)) = 0$ for any initial condition $x(0)$. The objective is to design a control law so that $\sigma \equiv 0$ for all time; i.e. to ensure a sliding mode is enforced on

$$\mathscr{S} = \{x \in \mathbb{R}^n \ : \ \sigma(t,x) = 0\} \tag{13.4}$$

A controller will now be described which ensures the condition $\sigma^T\dot{\sigma} \leq -\eta\|\sigma\|$ holds for all time. Such a condition is described in the sliding mode literature as the η-reachability condition [9, 23]. Normally this is sufficient to ensure that a sliding motion is attained in finite time and maintained for all subsequent time. Here, by design the states are already on the sliding surface at time $t = 0$ because the expression in (13.3) is constructed to make $\sigma(0, x(0)) = 0$ for any initial condition $x(0)$. Taking the derivative of (13.3) it follows

$$\begin{aligned}\dot{\sigma} &= G\big(Ax + B(u+\xi)\big) - G(A+BK)x(t)\\ &= GB(u + \xi - Kx)\end{aligned} \tag{13.5}$$

Choose the control law as

$$u(t) = Kx(t) - \rho(t,x)(GB)^{-1}\frac{\sigma(t)}{\|\sigma(t)\|} \quad \text{for } \sigma \neq 0 \tag{13.6}$$

where the modulation gain $\rho(t,x) = \|GB\|\bar{\xi}(t,x) + \eta$ (which is realistic since it is assumed the upper-bound on the uncertainty represented by $\bar{\xi}(t,x)$ is known). Then substituting for u in (13.5) yields

$$\begin{aligned}\sigma^T\dot{\sigma} &= \sigma^T\left(GB\xi(t,x) - \rho(t,x)\frac{\sigma(t)}{\|\sigma(t)\|}\right)\\ &\leq \|\sigma\|\|GB\xi(t,x)\| - \rho(t,x)\|\sigma(t)\|\\ &\leq -\eta\|\sigma(t)\|\end{aligned} \tag{13.7}$$

by choice of $\rho(t,x)$. Note during the sliding motion, the equivalent control [9, 28] necessary to maintain a sliding motion on $\mathscr{S}$, obtained by formally solving for the control signal when $\dot{\sigma} = 0$ in (13.5), is given by

$$u_{eq}(t) = Kx(t) - \xi(t,x)$$

Substituting $u_{eq}(t)$ in (13.1) means sliding is governed by

$$\begin{aligned}\dot{x}(t) &= Ax(t) + B(u_{eq} + \xi(t,x)) \\ &= (A + BK)x(t)\end{aligned} \tag{13.8}$$

which is stable by choice of K. It is clear from (13.8) that the sliding motion is independent of the uncertainty. This is the classical property of sliding modes: viz, the ability to completely reject the effect of matched uncertainty. It is also interesting to note that the dynamics in (13.8) are independent of the choice of G.

Now consider the situation where $f(t,x)$ is not matched to the input channels. In this case the uncertainty can be written in the form

$$f(t,x) = B\xi(t,x) + B^{\perp}\phi(t,x) \tag{13.9}$$

where $\xi(t,x)$ and $\phi(t,x)$ encapsulate the uncertainty, and $B^{\perp} \in \mathbb{R}^{n\times(n-m)}$ is a full column rank matrix with the property that $B^T B^{\perp} = 0$. (This is often termed the annihilator of the matrix B and has the property that $\det\begin{bmatrix} B & B^{\perp}\end{bmatrix} \neq 0$.) Thus it is clear that any $f(t,x)$ can be 'factored' into the form given in (13.9). In a situation where there exists unmatched uncertainty

$$\dot{\sigma} = GB\big(u + \xi(t,x) - Kx(t)\big) + GB^{\perp}\phi(t,x) \tag{13.10}$$

Suppose a sliding motion can be maintained on $\mathscr{S}$. In this case the equivalent control obtained from equating $\sigma = \dot{\sigma} = 0$ in (13.10) is given by

$$u_{eq} = K(x) - \xi(t,x) - (GB)^{-1}GB^{\perp}\phi(t,x) \tag{13.11}$$

and so from (13.1), (13.9) and (13.11) the sliding motion is governed by

$$\dot{x}(t) = (A + BK)x(t) + \big(I - B(GB)^{-1}G\big)B^{\perp}\phi(t,x) \tag{13.12}$$

Notice the impact of $\phi(t,x)$ is not rejected and in fact the scalar $\|I - B(GB)^{-1}G\|$ is indicative of the amplification of the effects of the disturbance/uncertainty $B^{\perp}\phi(t,x)$. If G is selected as $G = (B^T B)^{-1}B^T$ then

$$(I - B(GB)^{-1}G) = (I - B(B^T B)^{-1}B^T) \tag{13.13}$$

Notice that the projection operator $(I - B(B^T B)^{-1}B^T)$ in (13.13) is symmetric and idempotent since

$$\big(I - B(B^T B)^{-1}B^T\big)^2 = (I - B(B^T B)^{-1}B^T)$$

These properties of symmetry and idempotency imply that $\|I - B(B^T B)^{-1}B^T\| = 1$. As argued in [8], in general, for any given G, the norm $\|I - B(GB)^{-1}G\| \geq 1$ and so the choice $G = (B^T B)^{-1}B^T$ is an optimal one in the sense of non-amplification of the unmatched uncertainty. Note in this case G is a left pseudo-inverse of B.

13.3 Applications to Fault Tolerant Control

Now consider a nominal LTI system with actuator faults/failures represented as

$$\dot{x}(t) = Ax(t) + BW(t)u(t) \tag{13.14}$$

where $W(t) = diag\{w_1(t), \ldots, w_m(t)\}$ is a diagonal matrix in which the scalars $w_1(t), \ldots, w_m(t)$ model the effectiveness level of the actuators. If $w_i(t) = 1$, it means that the ith actuator has no fault and is working perfectly, whereas if $1 > w_i(t) > 0$, an actuator fault is present. If $w_i(t) = 0$, the ith actuator has completely failed and the control input component u_i does not affect the dynamics.

Assume only l actuators are needed to give the requisite closed-loop performance i.e. the system in (13.14) is over-actuated. The remaining $m - l$ actuators can then be used to induce fault tolerance. Assume an estimate of $W(t)$ is available[2] and written as $\widehat{W}(t) = diag\{\hat{w}_1(t), \ldots, \hat{w}_m(t)\}$, where the scalars $0 \leq \hat{w}_i(t) \leq 1$. This will subsequently be used in the control law. However invariably the estimate $\widehat{W}(t)$ will not be a perfect copy of $W(t)$, and here it is assumed to satisfy

$$W(t) = (I - \Delta(t))\widehat{W}(t) \tag{13.15}$$

where $\Delta(t) = diag\{\delta_1(t), \ldots, \delta_m(t)\}$ and the scalars $\delta_1(t), \ldots, \delta_m(t)$ are unknown. To cope with total failures of certain actuators, a control allocation structure will be incorporated [17, 22]. The approach proposed in [1, 14] is to first reorder the states so that the input distribution matrix in (13.14) can be partitioned as

$$B = \begin{bmatrix} B_1 \\ B_2 \end{bmatrix} \tag{13.16}$$

where $B_1 \in \mathbb{R}^{(n-l)\times m}$, $B_2 \in \mathbb{R}^{l\times m}$ and the latter matrix is of rank $l < m$. Furthermore as a result of reordering the states it is assumed that the elements of B_2 have large magnitude compared to $||B_1||$. Once this re-ordering has taken place, scale the last l states so that $B_2 B_2^T = I_l$. Now define a virtual control input as

$$\nu(t) := B_2 u(t) \tag{13.17}$$

If the physical control signal is chosen as

$$u(t) = B_2^{\dagger}(t)\nu(t) \tag{13.18}$$

where $B_2^{\dagger}(t) \in \mathbb{R}^{m\times l}$ is any right pseudo-inverse of the matrix B_2, then $u(t)$ from (13.18) clearly satisfies the condition in (13.17). One possible choice of $B_2^{\dagger}(t)$ is

---

[2]The information need to compute $\widehat{W}$ can be supplied from a fault reconstruction scheme [3] or by using extended Kalman filters [29]. Furthermore on many modern fly-by-wire aircraft sensors are present to measure the actual surface deflections which can be used to create estimates of $\widehat{W}$.

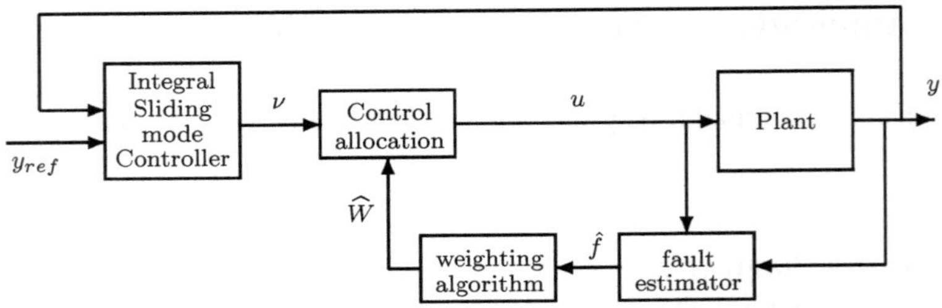

Fig. 13.1 Schematic of the overall control strategy

$$B_2^{\dagger}(t) = \widehat{W}(t) B_2^T (B_2 \widehat{W}(t) B_2^T)^{-1} \tag{13.19}$$

assuming $\det(B_2 \widehat{W}(t) B_2^T) \neq 0$. The overall control structure is given in Fig. 13.1.

Note from (13.19) that if $\hat{w}_i \to 0$ then $u_i(t) \to 0$ and the control signals are redistributed to the 'healthy' actuators. Define

$$\mathscr{W} = \{(\hat{w}_1, \ldots, \hat{w}_m) \; : \; \det(B_2 \widehat{W} B_2^T) \neq 0\} \tag{13.20}$$

Because $l < m$, it is possible that $\det(B_2 \widehat{W} B_2^T) \neq 0$ even if up to $m - l$ of the entries $\hat{w}_i(t) = 0$ in the matrix $\widehat{W}(t)$. In other words, up to $m - l$ can totally fail and yet $\det(B_2 \widehat{W} B_2^T) \neq 0$. Define

$$\hat{\nu}(t) := (B_2 \widehat{W}^2(t) B_2^T)(B_2 \widehat{W}(t) B_2^T)^{-1} \nu(t) \tag{13.21}$$

then the linear system with respect to the virtual input $\hat{\nu}(t)$ can be written as

$$\dot{x}(t) = Ax(t) + \underbrace{\begin{bmatrix} B_1(I - \Delta(t)) B_2^+(t) \\ B_2(I - \Delta(t)) B_2^+(t) \end{bmatrix}}_{\widehat{B}(t)} \hat{\nu}(t) \tag{13.22}$$

where

$$B_2^+(t) := \widehat{W}^2(t) B_2^T (B_2 \widehat{W}^2(t) B_2^T)^{-1} \tag{13.23}$$

Notice that $B_2^+(t)$ is a right pseudo-inverse of B_2 if $\widehat{W}(t) \in \mathscr{W}$. Using the properties of pseudo-inverses in [24], as argued in [1], there exists a scalar γ_o such that

$$\|B_2^+(t)\| = \|\widehat{W}^2(t) B_2^T (B_2 \widehat{W}^2(t) B_2^T)^{-1}\| < \gamma_o \tag{13.24}$$

for all $(\hat{w}_1(t), \ldots, \hat{w}_m(t)) \in \mathscr{W}$. In the case when the estimates of the efficiency are perfect (i.e. $\Delta(t) = 0$), and when the system is fault-free (i.e. $\widehat{W}(t) = I$), the system in (13.22) becomes

$$\dot{x}(t) = Ax(t) + \underbrace{\begin{bmatrix} B_1 B_2^T \\ I_l \end{bmatrix}}_{B_\nu} \nu(t) \tag{13.25}$$

since $B_2^+(t)|_{\widehat{W}(t)=I} = B_2^T$. Equation (13.25) will be used to *design* the control scheme. Suppose that by design of the partition in (13.16), the pair (A, B_ν) associated with (13.25) is controllable, then there exists a state feedback controller $\nu(t) = -F_\nu x(t)$, such that the nominal system

$$\dot{x}(t) = (A - B_\nu F_\nu)x(t) \tag{13.26}$$

possesses an appropriate closed loop response. Then define

$$\sigma(t, x) := Gx(t) - Gx(t_0) - G\int_0^t \big(A - B_\nu F_\nu\big)x(\tau)\mathrm{d}\tau \tag{13.27}$$

where $G \in \mathbb{R}^{l\times n}$ is design freedom. As in [14] choose

$$G := B_2(B^T B)^{-1} B^T \tag{13.28}$$

This choice of G has the property that $GB_\nu = I_l$. The derivative of Eq. (13.27) is

$$\dot{\sigma}(t) = G\dot{x}(t) - GAx(t) + GB_\nu F_\nu x(t) \tag{13.29}$$

where $\dot{x}(t)$ is given by (13.22). Consequently the equivalent control whilst sliding is

$$\hat{\nu}_{eq}(t) = -(G\widehat{B}(t))^{-1} F_\nu x(t) \tag{13.30}$$

Substituting (13.30) into equation (13.22) and adding and subtracting $B_\nu F_\nu x(t)$ yields

$$\dot{x}(t) = (A - B_\nu F_\nu)x(t) + \big(B_\nu - \widehat{B}(t)(G\widehat{B}(t))^{-1}\big) F_\nu x(t) \tag{13.31}$$

where B_ν is defined in (13.25) and $\widehat{B}(t)$ in (13.22). Using G from (13.28), further manipulation of Eq. (13.31) yields

$$\dot{x}(t) = (A - B_\nu F_\nu)x(t) + \tilde{B}\tilde{\Phi}(t)F_\nu x(t) \tag{13.32}$$

where

$$\tilde{\Phi}(t) := B_1 B_2^T - B_1(I - \Delta(t))B_2^+(t)(B_2(I - \Delta(t))B_2^+(t))^{-1} \tag{13.33}$$

and

$$\tilde{B} := \begin{bmatrix} I_{n-l} \\ 0 \end{bmatrix} \tag{13.34}$$

Remark 13.1 Notice in the case of perfect knowledge of the actuator efficiency (i.e. $\triangle(t) = 0$), and when there are no faults in the system (i.e. $\widehat{W}(t) = I$), the matrices $\widehat{B}|_{\widehat{W}(t)=I} = B_\nu$ and $B_2^+(t)|_{\widehat{W}(t)=I} = B_2^T$. Then using the fact that $GB_\nu = I$, Eq. (13.32) becomes

$$\dot{x}(t) = (A - B_\nu F_\nu)x(t) \tag{13.35}$$

However in the presence of faults/failures, the sliding motion is governed by

$$\dot{x}(t) = (A - B_\nu F_\nu + \tilde{B}\tilde{\Phi}(t)F_\nu)x(t) \tag{13.36}$$

which needs to be proved to be stable. To help achieve this goal define a transfer function matrix $\tilde{G}(s) = F_\nu(sI - (A - B_\nu F_\nu))^{-1}\tilde{B}$ and let

$$\gamma_2 = ||\tilde{G}(s)||_\infty \tag{13.37}$$

Then the following can be proved:

Proposition 13.1 ([14]) *Assume the effectiveness gain estimate $\widehat{W}(t)$ is sufficiently accurate so that $\triangle_{max}\gamma_o < 1$ holds, where γ_o is defined in (13.24) and $\|\triangle(t)\| < \triangle_{max}$. Then during fault/failure conditions, for any $(\hat{w}_1(t), \ldots, \hat{w}_m(t)) \in \mathscr{W}$, the reduced order sliding motion is stable provided*

$$\frac{\gamma_2\,\gamma_3(1 + \gamma_o)}{1 - \triangle_{max}\gamma_o} < 1 \tag{13.38}$$

where $\gamma_o > ||B_2^+(t)||$, $\gamma_3 = ||B_1||$ and γ_2 is from (13.37). □

A control structure to guarantee sliding is maintained is

$$\hat{\nu}(t) = \hat{\nu}_l(t) + \hat{\nu}_n(t) \tag{13.39}$$

where $\hat{\nu}_l(t) := -F_\nu x(t)$ and

$$\hat{\nu}_n(t) := \begin{cases} -\rho(t,x)\frac{\sigma(x,t)}{||\sigma(x,t)||} & \text{if } \sigma(t) \neq 0 \\ 0 & \text{otherwise} \end{cases} \tag{13.40}$$

where $\rho(t, x)$ is a scalar modulation function. Then provided $\triangle_{max} < \frac{1}{\gamma_o}$, where γ_o is defined in (13.24), if

$$\rho(t,x) = \frac{\triangle_{max}\gamma_o\|\hat{\nu}_l\| + \eta}{1 - \triangle_{max}\gamma_o} \tag{13.41}$$

where η is a positive scalar, the control law proposed in (13.39) guarantees sliding on $\mathscr{S}$ in (13.27). Finally, using Eqs. (13.18), (13.19) and (13.21), it follows that the physical control law is given by

$$u(t) = -\widehat{W}(t)B_2^T(B_2\widehat{W}^2(t)B_2^T)^{-1}\left(F_v x(t) + \rho\frac{\sigma(t)}{||\sigma(t)||}\right) \tag{13.42}$$

Note the design freedom F_v can be selected to ensure the conditions of Proposition 13.1 are fulfilled.

13.3.1 An Example

The problem of controlling the lateral axis of a large transport aircraft [10] will be used to demonstrate the effectiveness of the proposed scheme. The states are roll rate, yaw rate, side-slip angle and roll angle. The controlled outputs are side-slip angle and roll angle (which means $l = 2$). The available control surfaces are anti-symmetric aileron deflection (rad), rudder deflection (rad) and differential aggregated engine pressure ratios (i.e. $m = 3$).

In the simulations which follow the aircraft undertakes a turning manoeuvre in which the reference command requests a change in ϕ to 25° during the period of time 60–90 s. A 0° reference command is applied to β throughout. Figures 13.2 and

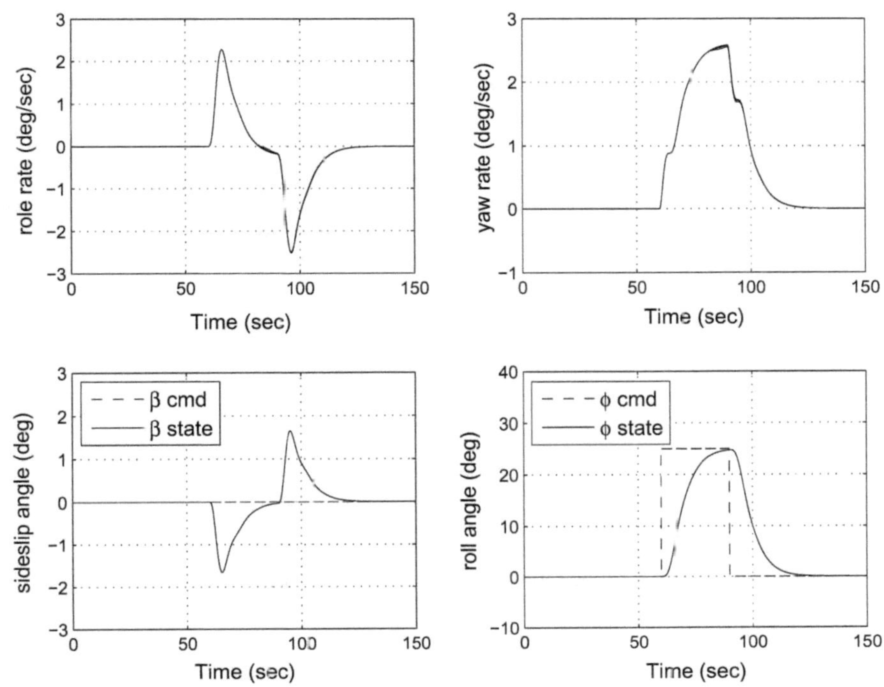

Fig. 13.2 Aileron-fault: plant states

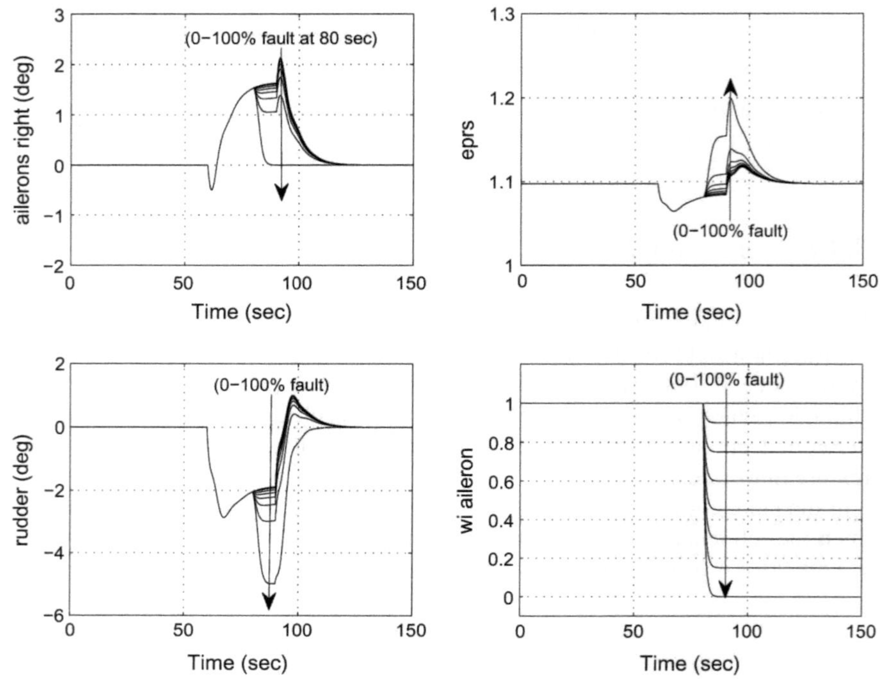

Fig. 13.3 Aileron-fault: actuator deflections

13.3 show various levels of aileron faults (from 0–100%) each occurring at 80 s in 15% increments. It can be seen that the CA systematically redistributes the control signals to the rudder and the engines, while maintaining the same level of tracking performance as in the fault-free condition.

This section has demonstrated the development of an ISM based fault tolerant control law with design parameters F_ν (and G). In the next section it will be assumed that fault tolerance needs to be induced *in an existing feedback loop*, and the designer must incorporate an existing choice of feedback gain.

13.4 Retrofitting for FTC

Consider once again the system in (13.14) and partition the input distribution matrix as

$$B = \begin{bmatrix} B_o & B_s \end{bmatrix} \tag{13.43}$$

where $B_o \in \mathbb{R}^{n\times l}$, $B_s \in \mathbb{R}^{n\times m-l}$ and l represents the number of primary actuators. Here B_o is the input distribution matrix associated with the primary actuators, whilst

B_s is associated with secondary actuators (which impart redundancy). Assume B_o has full column rank and let $T_o \in \mathbb{R}^{n\times n}$ be an orthogonal matrix associated with a coordinate change such that

$$T_o B_o = \begin{bmatrix} 0 \\ B_{21} \end{bmatrix} \tag{13.44}$$

where $B_{21} \in \mathbb{R}^{l\times l}$ (and $\det(B_{21}) \neq 0$). Therefore in suitable coordinates the system from (13.14) has a distribution matrix

$$B = \begin{bmatrix} 0 & B_{12} \\ B_{21} & B_{22} \end{bmatrix} \tag{13.45}$$

where $B_{22} \in \mathbb{R}^{l\times(m-l)}$. As before, scale the last l states to ensure that $B_{21}^T B_{21} = I_l$. Assume the system (A, B_o) is controllable and a state feedback controller, based only on the primary actuators, has been designed. Specifically assume a state feedback gain matrix $F_o \in \mathbb{R}^{l\times n}$ has been designed a-priori so that

$$\dot{x}(t) = (A + B_o F_o)x(t) \tag{13.46}$$

has a suitable dynamic response. A control allocation scheme will now be *retrofitted* to the existing controller (Fig. 13.4) without disturbing the existing feedback loop.

The physical control signals sent to the actuators are

$$u(t) = N(t)\nu(t) \tag{13.47}$$

where $\nu(t) \in \mathbb{R}^l$ is the virtual control effort. Now partition the effectiveness matrix from equation (13.14) as $W(t) = diag\{W_1(t), W_2(t)\}$ where the sub-matrices

Fig. 13.4 Schematic of the overall control strategy

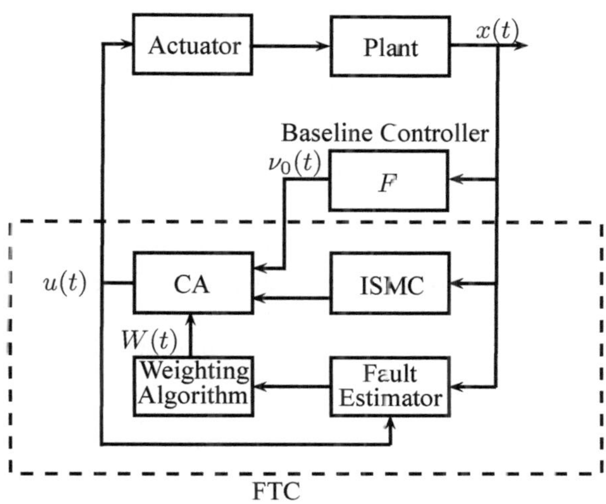

$W_1(t) = diag\{w_1(t), \ldots, w_l(t)\}$ and $W_2 = diag\{w_{l+1}(t), \ldots, w_m(t)\}$. A suitable choice of allocation matrix is

$$N(t) = \begin{bmatrix} I_l \\ N_2(t)(I_l - W_1(t)) \end{bmatrix} \tag{13.48}$$

where

$$N_2(t) = B_{22}^T B_{21}(B_{21}^T B_{22} W_2(t) B_{22}^T B_{21})^{-1} \tag{13.49}$$

In order that $\det(B_{21}^T B_{22} W_2(t) B_{22}^T B_{21})^{-1} \neq 0$, because by construction $\det(B_{21}) \neq 0$, this constraint is equivalent to $\det(B_{22} W_2(t) B_{22}^T) \neq 0$. This imposes a limitation on the number of elements of W_2 that can become zero and therefore limits the number of total failures in the *secondary actuators* that can be accommodated.

With respect to the virtual control, after some manipulation

$$\dot{x}(t) = Ax(t) + \underbrace{\begin{bmatrix} B_{12} W_2 N_2 (I_l - W_1) \\ B_{21} \end{bmatrix}}_{B_w} \nu(t) \tag{13.50}$$

and in a fault-free situation when $W = I$, Eq. (13.50) becomes

$$\dot{x}(t) = Ax(t) + \underbrace{\begin{bmatrix} 0 \\ B_{21} \end{bmatrix}}_{B_o} \nu(t) \tag{13.51}$$

Consequently if $\nu(t) = F_o x(t)$, then the nominal baseline performance is achieved. Furthermore when $W = I$, substituting in (13.47) and (13.48), it follows

$$u(t) = \begin{bmatrix} F_o x(t) \\ 0 \end{bmatrix}$$

and only the primary actuators are used. Choose once again

$$\sigma(x, t) := Gx(t) - Gx(t_0) - G \int_0^t (A + B_o F_o) x(\tau) \mathrm{d}\tau \tag{13.52}$$

and this time define $G := B_o^T$. With this choice of G it follows $GB_o = I_l$ and $GB_w = I_l$. It can be shown that the sliding motion is given by

$$\dot{x}(t) = (A + B_o F_o) x(t) + \tilde{B} \Phi(t) F_o x(t) \tag{13.53}$$

where

$$\tilde{B} := \begin{bmatrix} B_{12} \\ 0 \end{bmatrix} \tag{13.54}$$

and $\Phi(t) := W_2 N_2 (I_l - W_1)$. It is clear that during fault-free conditions (i.e. when $W = I$), $\Phi(t) = 0$, and Eq. (13.53) becomes (13.46), which is stable by design. However the sliding motion in (13.53) depends on the matrix $W(t)$, and a stability analysis needs to be carried out to ensure closed-loop stability for different faults and failures. Define

$$\gamma_2 = \|\tilde{G}(s)\|_\infty \tag{13.55}$$

where $\tilde{G}(s) := F_o(sI - (A + B_o F_o))^{-1}\tilde{B}$. It can be verified that $W_2 N_2$ is a pseudo-inverse for $B_{21}^T B_{22}$, and so using arguments similar to those in [1], the properties of the pseudo-inverse proved in [24] ensures $\|W_2 N_2\| < \gamma_1$ for some γ_1 provided $\det(B_{22} W_2 B_{22}^T) \neq 0$. Define a scalar γ_1^* (which is guaranteed to exist) to be the smallest number satisfying

$$\|\Phi(t)\| < \gamma_1^* \tag{13.56}$$

Proposition 13.2 ([16]) *During faults or failures, for any combination of* $0 < w_i \leq 1$, *the closed loop system will be stable if* $\gamma_2 \gamma_1^* < 1$ □

Again this can be extended to the case when $W(t)$ is not known and is replaced by $\widehat{W}(t)$ in the control law. The proposed integral sliding mode control law, which depends on the nominal system (13.51) is defined as

$$\nu(t) = \nu_l(t) + \nu_n(t) \tag{13.57}$$

where $\nu_l(t) = F_o x(t)$ and

$$\nu_n(t) = -\rho \frac{\sigma(x,t)}{\|\sigma(x,t)\|} \quad \text{for } \sigma(x,t) \neq 0 \tag{13.58}$$

where ρ is a scalar gain to enforce the sliding motion. Finally the physical control law is given by

$$u(t) = \begin{bmatrix} I_l \\ \widehat{N}_2(t)(I_l - \widehat{W}_1(t)) \end{bmatrix} \left(F_o x(t) - \rho \frac{\sigma(x,t)}{\|\sigma(x,t)\|} \right) \tag{13.59}$$

where $\widehat{N}_2(t)$ is obtained from replacing W by $\widehat{W}$ in N_2.

Proposition 13.3 ([16]) *Suppose that the condition*

$$(1 + \gamma_3 \gamma_1^*)\Delta_{max} < 1 \tag{13.60}$$

holds where $\gamma_3 = \|B_{22}\|$. *Then during fault/failure conditions, including failure of all the primary actuators and for any* $\hat{w}_{l+1}(t), \ldots, \hat{w}_m(t) \in \mathcal{W}_o$ *where*

$$\mathcal{W}_o = \{(\hat{w}_{l+1}(t), \ldots, \hat{w}_m(t)) \; : \; \det(B_{22}\widehat{W}_2 B_{22}^T) \neq 0\}$$

the closed loop system will be stable if:

$$\frac{\gamma_2 \gamma_1^*(1+\Delta_{max})}{1-(1+\gamma_3 \gamma_1^*)\Delta_{max}} < 1 \tag{13.61}$$

where γ_2 is defined in (13.55). □

13.4.1 Example: Fault Tolerant Control Law for Yaw Damping

Consider a yaw damper stability-augmentation system for the lateral equations of motion of an aircraft [12]. Here the baseline control law F_o for the nominal system (13.51) has been a-priori designed using eigenstructure assignment. The states comprise the washout filter state, roll angle, side slip angle, yaw rate and roll rate. The available control inputs are aileron deflections, spoiler deflections (left: 1–4, 5 and right: 8, 9–12), rudder deflection, and individual engine thrusts scaled by 10^5. For details see [16]. The results presented here are based on a 'real world' model of a B747-100/200 aircraft which has been used as the basis for the GARTEUR AG16 benchmark [10]. Figures 13.5 and 13.6, show the states and actuators deflections

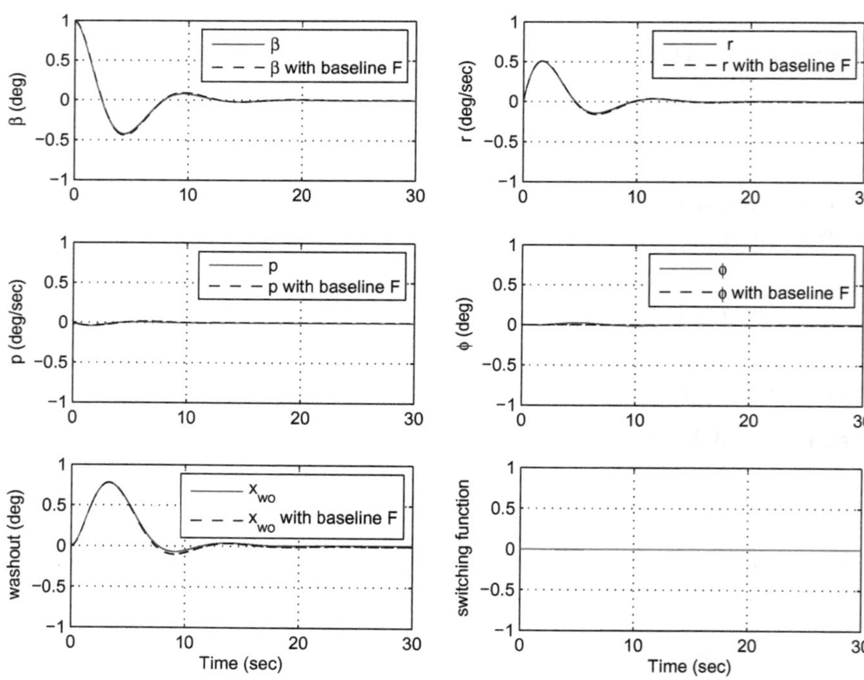

Fig. 13.5 Primary failure: system states vs ideal states with baseline F_o

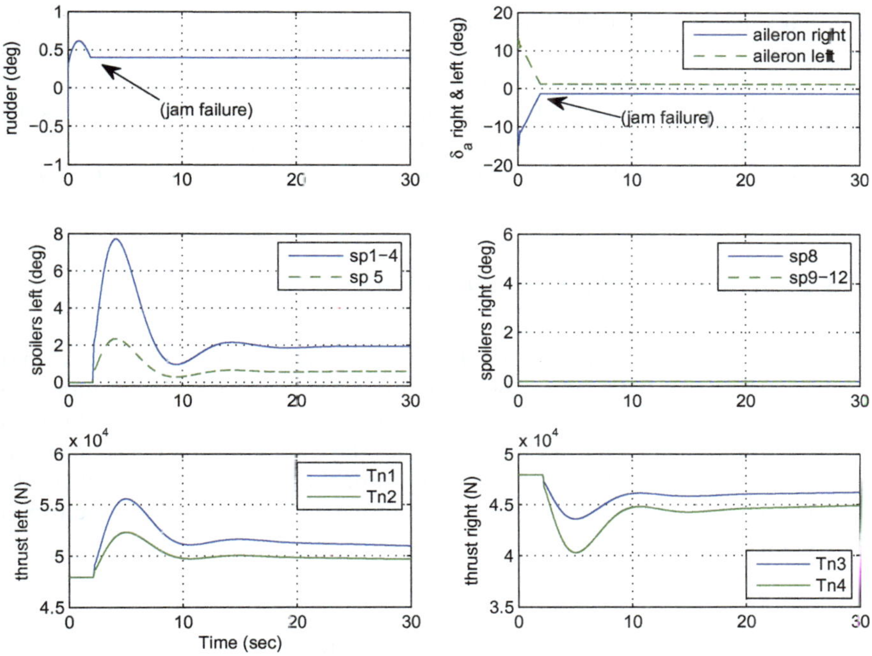

Fig. 13.6 Primary failure: actuators deflections

when both the primary actuators become stuck at some offset positions. Figure 13.5 shows that in this extreme failure case nominal performance is maintained.

13.5 RECOVER and the SIMONA Research Simulator

The design frameworks presented in the earlier sections has demonstrated that the combination of the robustness properties of ISMC, and the ability of CA to redistribute control signals, can create a single controller that has the ability to handle both nominal fault-free and fault/failure conditions without changing the structure of the controller. The designs described previously were however based on a single LTI system obtained through linearization about a certain operating point. In order to deal with the challenge of maintaining performance across a wide range of operating conditions, recent years have seen the emergence of linear parameter varying (LPV) based SMC design. LPV-based SMC design provides a natural extension of many of the existing linear time invariant (LTI) based designs, and this can be seen as an added advantage for industrial designers.

Despite the recent development in LPV-based SMC, there is still a lack of implementation results for these schemes to allow real-time assessment. Some of the recent

SMC work that has considered real-time aerospace implementations (especially on a flight simulator) appears in [2]. However, the results in [2] are based on LTI designs. The remainder of this chapter will showcase the implementation of an LPV integral sliding mode scheme, evaluated by a pilot, on a motion simulator. This shows that the proposed scheme can be implemented in real-time, and feedback from the pilot provides an evaluation of the controller from a pilot's operational point of view.

13.5.1 RECOVER Benchmark Model

The model considered for design, simulation and as the basis for the flight simulator is called RECOVER (REconfigurable COntrol for Vehicle Emergency Return). The RECOVER model represents a *high fidelity nonlinear model of a B747-100/200 large transport aircraft*, and it has been used as a test-bed and evaluation benchmark for the GARTEUR FM-AG16 project [10] (a consortium of academics, research institutions and industry), which investigated state-of-the-art FTC schemes for aerospace applications.

13.5.2 SIMONA Research Simulator (SRS)

The SRS is a realistic 6 degree-of-freedom (DOF) motion flight simulator at the International Research Institute for Simulation, Motion and Navigation (SIMONA) at Delft University of Technology, The Netherlands.

The SRS has a typical commercial aircraft cockpit with two side-by-side pilot seats, and typical pilot controls. A mode control panel (MCP) allows the pilot to give auto-pilot commands, with a primary flight display (PFD) and a navigation display (ND). An engine indication and crew alerting system (EICAS) are also presented on electronic displays. The latter is augmented with control surface deflections to allow the pilot to monitor the controller's actions. The SRS has an outside virtual world projection with a $180 \times 40^\circ$ field of view, coupled with 6 large electro-hydraulic motion actuators (configured in a hexapod configuration) (Figs. 13.7 and 13.8).

The motion cueing algorithm provides the pilot with a high level of immersion. The SRS has a modular software and hardware architecture allowing the inclusion of a wide variety of vehicle models, controllers and displays. The custom middleware software DUECA (Delft University Environment for Communication and Activation) handles the real-time scheduling, computer synchronization, and data transport. The custom-built motion and visualization system and its modular structure have made the SRS a powerful tool for the research of human-machine interaction, handling qualities [11, 25] as well as the study of FTC schemes [10, 26]. In this work, the SRS has been configured to represent the RECOVER B747-100/200 aircraft with an outside virtual world representation of the area around Amsterdam-Schiphol airport.

Fig. 13.7 SIMONA research simulator

Fig. 13.8 SIMONA research simulator cockpit configuration

13.6 Design

In this chapter, only the longitudinal axis of the B747-100/200 aircraft will be considered for the design of the LPV ISM FTC scheme. However the lateral controller from earlier work will also be implemented to provide complete longitudinal, lateral and directional control of the aircraft. The design is based on the longitudinal LPV plant from [18]. Table 13.1 shows the details of the LPV model. For design purposes, the state $\bar{h}_e$ has been removed and the remaining states from [18] have been reordered as $[\bar{\theta}, \bar{\alpha}, \bar{V}_{tas}, \bar{q}]^{\mathrm{T}}$.

As in the RECOVER benchmark model, it will be assumed that the elevators and the stabilizer are potentially prone to faults/failures. The LPV plant subject to actuator faults or failures is represented by

$$\dot{\bar{x}}(t) = A(\rho)\bar{x}(t) + B(\rho)W(t)u(t) \tag{13.62}$$

where $W = \text{diag}\{w_1(t), w_2(t), w_3(t)\}$ represents the effectiveness levels of the elevators, stabilizer and thrust respectively. The scheduling parameter ρ (defined in

Table 13.1 Longitudinal LPV model from [18]. (The bar $(\bar{\cdot})$ represent the deviation from its trim.)

States\ctrl surf	Description	Trim value
$\bar{\alpha}$	Angle of attack	1.05°
$\bar{q}$	Pitch rate	0 (°/s)
$\bar{V}_{tas}$	True air speed	227.02 (m/s)
$\bar{\theta}$	Pitch angle	1.05 (°)
$\bar{h}_e$	Altitude	7000 (m)
$\bar{\delta}_e, \bar{\delta}_s$	Elevator, stabilizer	0.163, 0.590 (°)
$\bar{T}_n$	Total engine thrust	42291 (N)
LPV parameters	Description	Range
$[\rho_1, \rho_2, \ldots, \rho_7]$	$[\bar{\alpha}, \bar{V}_{tas}, \bar{V}_{tas}\bar{\alpha}, \bar{V}_{tas}^2, \ldots, \bar{V}_{tas}^2\bar{\alpha}, \bar{V}_{tas}^3, \bar{V}_{tas}^4]$	V_{tas} : [150, 250] m/s α : [−2, 8]°

Table 13.1) is assumed to be available and to lie in a bounded compact set $\Omega \subset \mathbb{R}^7$. For control purposes, all the states $\bar{x}(t)$ are assumed to be available. In (13.62), $A(\rho) \in \mathbb{R}^{4\times4}$, $B(\rho) \in \mathbb{R}^{4\times3}$ are the time varying system and input matrices respectively where

$$A(\rho) = A_0 + \sum_{i=1}^{7} A_i \rho_i \quad \text{and} \quad B(\rho) = B_0 + \sum_{i=1}^{7} B_i \rho_i \tag{13.63}$$

For further details of the LPV matrices see [4, 18]. For design purposes, the input distribution matrix $B(\rho)$ can be written as

$$\begin{aligned} B(\rho) &= \underbrace{\begin{bmatrix} 0 & 0\,0 \\ 0.01 & 0\,0 \\ \hline 0 & 1\,0 \\ 0 & 0\,1 \end{bmatrix}}_{B_f} \underbrace{\begin{bmatrix} 100b_{31}(\rho) & 100b_{32}(\rho) & 100b_{33}(\rho) \\ 0 & 0 & b_{23}(\rho) \\ b_{41}(\rho) & b_{42}(\rho) & b_{43}(\rho) \end{bmatrix}}_{E(\rho)} \\ &= \begin{bmatrix} B_1 \\ \hline B_2 \end{bmatrix} E(\rho) \end{aligned} \tag{13.64}$$

Note that (13.64) represents a factorisation of $B(\rho)$ into a fixed and varying component – which will be exploited during the design process. Also note that the term B_1 (the top two rows of B_f) in (13.64) is smaller compared to the B_2 term. This will be exploited during the design of the ISM controller.

The control objectives are to manipulate the flight path angle (FPA) ($\bar{\gamma} = \bar{\theta} - \bar{\alpha}$) and speed $\bar{V}_{tas}$, and therefore the controlled output $\bar{y}_c$ is defined as

$$\bar{y}_c(t) = \underbrace{\begin{bmatrix} 1 & -1 & 0 & 0 \\ 0 & 0 & 1 & 0 \end{bmatrix}}_{C_c} \bar{x}(t) = \begin{bmatrix} \bar{\gamma} \\ \bar{V}_{tas} \end{bmatrix} \tag{13.65}$$

To include tracking of FPA and speed, the following integral action states [3] are incorporated

$$\dot{x}_r(t) = r(t) - C_c\bar{x}(t) \tag{13.66}$$

Here $r(t)$ is the command to be tracked, and C_c is the controlled output distribution matrix defined in (13.65). The ISM control design will be based on the following augmented system:

$$\dot{x}_a(t) = A_a(\rho)x_a(t) + B_{\nu_a}\nu(t) + B_r r(t) \tag{13.67}$$

where the augmented states $x_a(t) = \left[\, x_r^{\mathrm{T}}(t)\ \bar{x}^{\mathrm{T}}(t)\,\right]^{\mathrm{T}}$ and

$$A_a(\rho) := \begin{bmatrix} 0 & -C_c \\ 0 & A(\rho) \end{bmatrix}, \quad B_{\nu_a} := \begin{bmatrix} 0 \\ B_\nu \end{bmatrix}, \quad B_r = \begin{bmatrix} I_l \\ 0 \end{bmatrix}$$

In the above (using the fact that from (13.64) $B_2B_2^{\mathrm{T}} = I_2$)

$$B_\nu := \begin{bmatrix} B_1B_2^{\mathrm{T}} \\ I_l \end{bmatrix}$$

For the design of the ISM controller, the following sliding surface is considered

$$\mathscr{S} = \{x_a \in \mathbb{R}^6 : \sigma_a(x_a, \rho, t) = 0\}$$

where $\sigma_a \in \mathbb{R}^2$ is the ISM switching function for the augmented system given by

$$\sigma_a(\cdot) := G_a\,(x_a(t) - x_a(0)) - G_a\int_0^t \big(A_a(\rho) - B_{\nu_a}F_a\big)x_a(\tau)d\tau \tag{13.68}$$

In (13.68) the matrix F_a must be chosen to make $\big(A_a(\rho) - B_{\nu_a}F_a\big)x_a(t)$ stable for all $\rho \in \Omega$. (Details of the design of F_a will be discussed later). In (13.68) the design freedom G_a has been chosen as

$$G_a := B_2(B_{f_a}^{\mathrm{T}}B_{f_a})^{-1}B_{f_a}^{\mathrm{T}} = \begin{bmatrix} 0 & 0 & 0 & 0 & 1 & 0 \\ 0 & 0 & 0 & 0 & 0 & 1 \end{bmatrix} \tag{13.69}$$

where

$$B_{f_a} = \begin{bmatrix} 0_{2\times 3} \\ B_f \end{bmatrix}$$

and B_2 represents the bottom two rows of B_f defined in (13.64).

The physical control law is based on the augmented system from (13.67) and is given by

$$u(t) = -E(\rho)^{-1}B_2^{\mathrm{T}}\left(B_2E(\rho)\widehat{W}(t)(E(\rho))^{-1}B_2^{\mathrm{T}}\right)^{-1} \cdot\left(F_a x_a(t) + \kappa(t)\frac{\sigma_a}{\|\sigma_a\|}\right) \tag{13.70}$$

for $\sigma_a \neq 0$ where $\kappa(t)$ represents an adapting gain and $\widehat{W}(t)$ represents an estimate of the actual actuator efficiency $W(t)$. It is assumed the estimate of $W(t)$ satisfies the form in (13.15). Due to this imperfection, an adaptive scheme has been considered for the positive scalar modulation function $\kappa(t)$ in (13.70). This has the advantage of allowing the modulation gain to be small in fault-free situations, but also to become sufficiently large when faults are present. Here specifically $\kappa(t)$ is given by

$$\kappa(t) = \|F_a\|\|x_a(t)\|\bar{\kappa}(t) + \eta \tag{13.71}$$

where η is a positive design scalar and the positive adaptation gain $\bar{\kappa}(t)$ evolves according to

$$\dot{\bar{\kappa}}(t) = -\beta\bar{\kappa}(t) + \gamma\varepsilon_0\|F_a\|\|x_a(t)\|\|\sigma_a(t)\| \tag{13.72}$$

In this chapter, the design scalars η, β, γ and ε_0 have been chosen as $\eta = 1$, $\beta = 1$, $\gamma = 0.01$ and $\varepsilon_0 = 0.01$ respectively. Note that the adaptation scheme in (13.71)–(13.72) is different from the one in [13, 14].

The tracking requirements are to decouple the FPA and true air speed V_{tas} dynamics, and to provide demand tracking with fault-free settling times of 20 and 45 s respectively. During faults/failures to any of the actuators, a slower degraded settling time response of 30 s for the FPA is acceptable, while the settling time for V_{tas} should remain unchanged. Based on this requirement, the fixed matrix F_a from (13.68) and (13.70) has been obtained by using LMIs to achieve two main objectives.

The first objective is to ensure the required level of nominal performance for all values of ρ using an LQR based design. The second objective is to ensure the following stability condition

$$\gamma_0\gamma_1\left(1 + \frac{c}{\sqrt{\varepsilon}}\right) < 1 \tag{13.73}$$

is satisfied for all permissible faults/failures which belong to the set

$$\mathscr{W}_\varepsilon = \{(w_1, w_3, w_3) \in [0\ 1] \times [0\ 1] \times [0\ 1]\ :\ (G_a B_w(\rho))^{\mathrm{T}}(G_a B_w(\rho)) > \varepsilon I\} \tag{13.74}$$

where ε is a small positive scalar satisfying $0 < \varepsilon \ll 1$ and

$$G_a B_w(\rho) = B_2E(\rho)W(t)(E(\rho))^{-1}B_2^{\mathrm{T}} \tag{13.75}$$

The set $\mathscr{W}_\varepsilon$ represents the class of actuator faults/failures for which stability of the sliding motion can be maintained. In (13.73), $c = \max_{\rho \in \Omega} \|E(\rho)\| \, |(E(\rho))^{-1}|$ represents the worst case condition number of $E(\rho)$ from (13.64) and $\gamma_1 = \|B_1\|$. The positive scalar γ_0 is the $\mathscr{L}_2$ gain associated with the nominal fault-free closed loop sliding motion. Here both the LQR performance objective and the closed loop stability condition Bounded Real Lemma (BRL) have been formulated as LMIs. Since the LPV system considered here is affine, the LMI formulation can be represented by a polytopic system and the LMIs are solved for all the vertices of the polytope to synthesize the state feedback gain F_a (see [13] for further details).

In this chapter the design matrices associated with the LQR cost function have been chosen as $Q = \text{diag}\{1.1, 0.04, 1, 1, 0.03, 5\}$, and $R = \text{diag}\{0.007, 1.1\}$ respectively, where the first two states in the Q matrix are the integral action states associated with the augmented system in (13.67). The fixed state feedback matrix F_a resulting from the LMIs is given by

$$F_a = \begin{bmatrix} -1.1161 & -2.3532 & -10.3807 & 3.8107 & 3.7409 & -1.3623 \\ -0.9891 & 0.0177 & 9.6902 & -4.9097 & -0.0222 & 3.3779 \end{bmatrix}$$

Here $\varepsilon = 0.28$ has been chosen since it is assumed that the engines are fault-free (it is the only available actuator for V_{tas} tracking) while the elevator is considered as the primary control surface for FPA tracking and the horizontal stabilizer acts as redundancy. To satisfy the closed-loop stability condition in (13.73), γ_0 associated with (13.73) should satisfy $\gamma_0 < \frac{\sqrt{\varepsilon}}{\gamma_1(\sqrt{\varepsilon}+c)} = 14.8588$. This has been obtained using a numerical search. From the LMIs, $\gamma_0 = 11.0000$ has been obtained during the design process for F_a and the stability condition in (13.73) is satisfied.

For the SRS implementation, the discontinuous term in (13.70) has been smoothed to ensure appropriate control signals are sent to the actuators. This is achieved by using the sigmoidal approximation [3] given by $\frac{\sigma_a}{\|\sigma_a\|+\delta}$ where δ is a small positive scalar chosen as $\delta = 0.01$. This sigmoidal approximation also allows extra design freedom especially when a fault/failure has occurred.

Note that the control law in (13.70) requires the actuator effectiveness level W. Since the actual actuator deflection is assumed to be available, the effectiveness of the actuator can be estimated by comparing the actual actuator deflection with the commanded signals from the controller. This is not an unrealistic assumption in modern aircraft systems [7].

13.7 SRS Implementation

Before being implemented on the SRS, the controller was initially developed and tested in a desktop simulation environment using the RECOVER software. This was carried out under Matlab R2006b, the original version supported by the RECOVER model, using a fixed time solver ODE4 with a time step of 0.01 s.

For implementation in the SRS, the same input/output interface to the SRS as used in the GARTEUR FM-AG16 evaluation campaign has been considered. The SIMULINK block of the ISM controller was converted into C code using the Real-Time Workshop utility in Matlab. This C code was then implemented on the SRS flight control computer, which is a PC with an Intel(R) Xeon(R) 3.07 GHz processor. The ISM controller has a relatively low computational load allowing the controller to run in real-time using a 10 ms per step size.

Figure 13.9 shows the high level overview of the controller-SRS interconnection which has been implemented. From Fig. 13.9, the ISM controller provides an inner-loop FPA and speed tracking capability which can be directly controlled by the pilot through the MCP dials in the centre of the cockpit. To emulate an actual aircraft autopilot, outer-loop PID based control of altitude has been used to provide command signals for the inner-loop FPA ISM controller. The gains associated with the outer-loop PID altitude control have been selected as $K_p = 0.1$, $K_i = 0.07$ and $k_d = 0.1$.

The FTC scheme was later flown and evaluated by an experienced commercial pilot, using the SRS, under several scenarios including fault-free and actuator failure cases. Although the controller is designed based on an LPV model from [18], the results obtained during the SRS evaluation (offline and piloted) involve the *full high fidelity nonlinear aircraft model*.

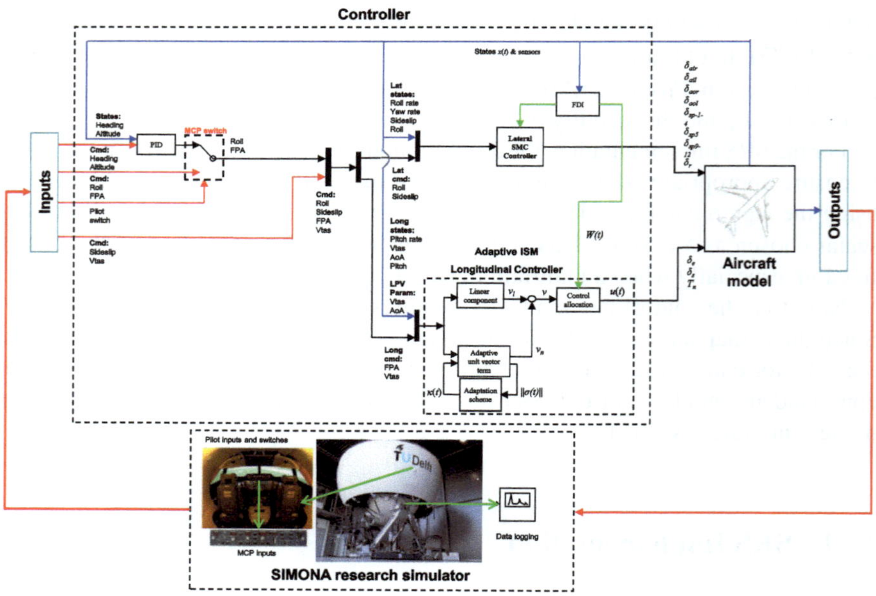

Fig. 13.9 Controller interconnection

13.8 SRS Offline Evaluation Results

The results presented in this section are based on the offline evaluation results which have been obtained using the software architecture DUECA. The offline tests conducted in DUECA represent the initial evaluations of the controller with real-time scheduling to ensure synchronization between all the computers in the network comprising the SRS. A series of tests was carried out before the controller was evaluated by the pilot on the SRS with the motion system turned on. Apart from testing the real-time synchronization of the controller, the offline tests also provide a valuable preliminary evaluation of the controller performance under the same (consistent) conditions and command inputs for the fault-free case and when faults/failures occur. Despite the fact that the controller is designed based on the LPV model from [18] which is obtained from a trim condition at 7000 m as described in Table 13.1, the controller has been tested at the trim condition [5.53°, 0.0017 °/s, 134 m/s, 5.53°, 600 m] with an input trim [2°, −1.59°, 45,568 N] with an initial mass of 317,000 kg and with the flaps fully retracted. This represents one of the trim conditions used for the GARTEUR FM-AG16 benchmark problem and it is different to the trim conditions of the model in [18].

The overall manoeuvre considered during the SRS offline evaluation is given in Fig. 13.10. This covers a wide range of the flight envelope and includes significant longitudinal (speed and altitude) and lateral (roll) deviations from the trim points to highlight the efficacy of the proposed scheme – especially when dealing with faults

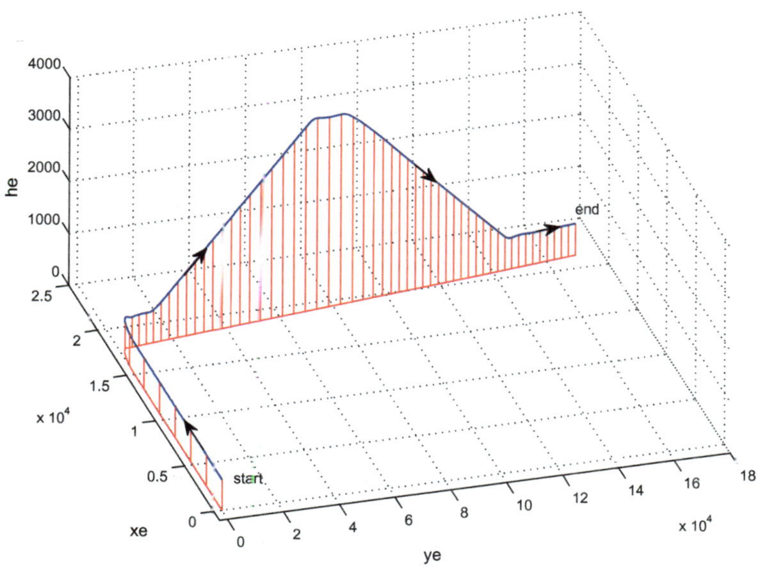

Fig. 13.10 Offline evaluation: trajectory

and failures. In this chapter two benchmark failure scenarios will be considered [10]: one is an elevator jam and the other is a stabilizer runaway. For consistency, the same heading, altitude and speed (recorded) commands will be used for all the fault-free and failure cases, while the failures are set to occur at approximately 260 s.

13.8.1 Elevator Jam

Figure 13.11 shows a comparison between the fault-free case and a scenario when an elevator jams at 260 s. The effect of the jammed elevator can be seen in the plot of the control surface deflections (Fig. 13.11b) where from 260 s onwards, the elevator does not move. The difference between the fault-free case and the failure case can clearly be seen. Despite the elevator jam, there is no visible difference in terms of the flight path angle γ and speed V_{tas} tracking performance (the lines visually overlap), which highlights the efficacy of the proposed scheme. There is also no visible difference in terms of the altitude change between the failure and the fault-free case. From Fig. 13.11c, it can be seen that immediately after the failure at 260 s, the elevator effectiveness level drops to 0%. The plot of the norm of the switching function $\|\sigma_a(t)\|$ in Fig. 13.11d also shows no visible difference between the fault-free and the failure case. Finally, the plot of the adaptive gain in Fig. 13.11d shows the variation of $\kappa(t)$ in (13.71). Again, there is no visible difference in terms of the adaptive gain between the fault-free and the failure case. Note that in the fault-free case, the variations in the adaptive gain in Fig. 13.11d are due to the combination of variations in $\|\sigma_a(t)\|$ and $\|x_a(t)\|$ as described in (13.72).

13.8.2 Stabilizer Runaway

Figure 13.12 shows the results for the case when a stabilizer runaway occurs. The effect of the stabilizer runaway can be seen in the control surface plot in Fig. 13.12b where the stabilizer moves at maximum rate to a maximum position of 3°. The effect of control reallocation can be seen as the elevator moves to around 8° immediately after the failure occurs at 260 s. Despite the stabilizer runaway, there is no visible difference in terms of tracking performance between the fault-free and the failure case as shown in Fig. 13.12a. The plot of the norm of the switching function $\|\sigma_a(t)\|$ in Fig. 13.12d shows the difference between the fault-free and the failure case. Here it can be seen that the norm for the failure case is slightly higher than for the fault-free case immediately after the failure at 260 s, but is still relatively small. Also note that a stabilizer runaway fault is considered as a 'catastrophic failure case' in the benchmark problem [10]. Finally, the plot of the adaptive gain in Fig. 13.12d shows the variation of $\kappa(t)$ in (13.71) and indicates there is a slight difference in terms of the adaptive gain between the fault-free and the failure case, starting from the time when the failure occurs.

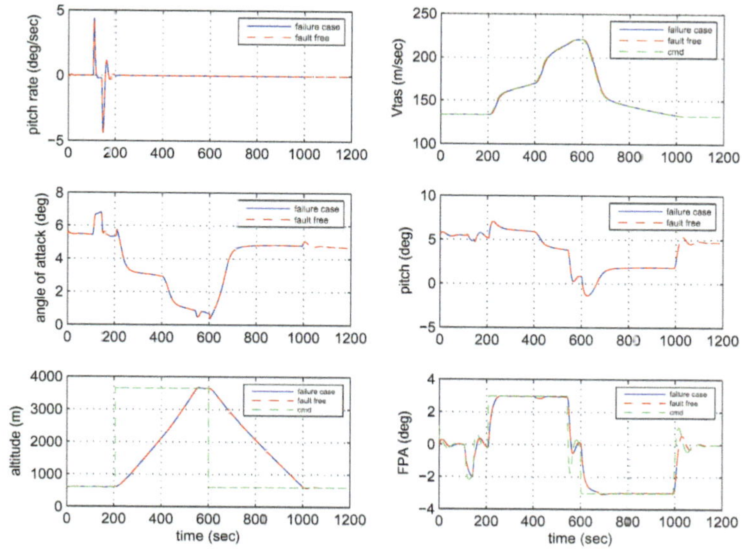

(a) states

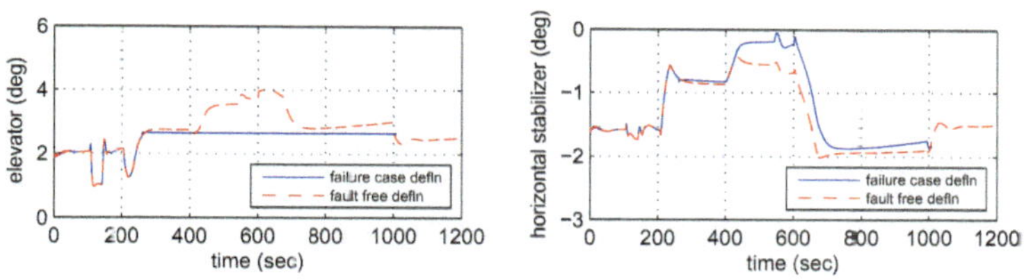

(b) control surfaces

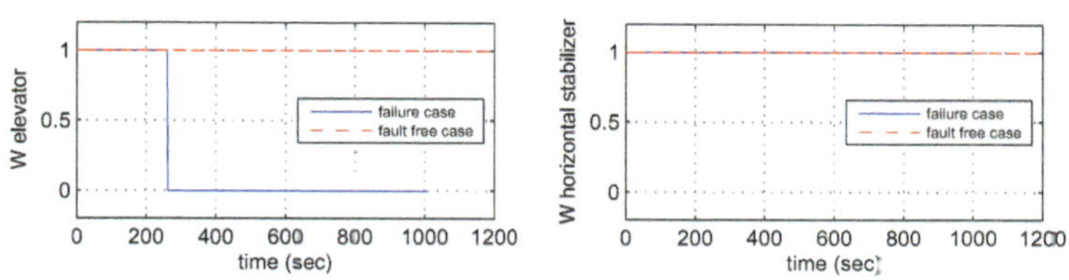

(c) control surface effectiveness

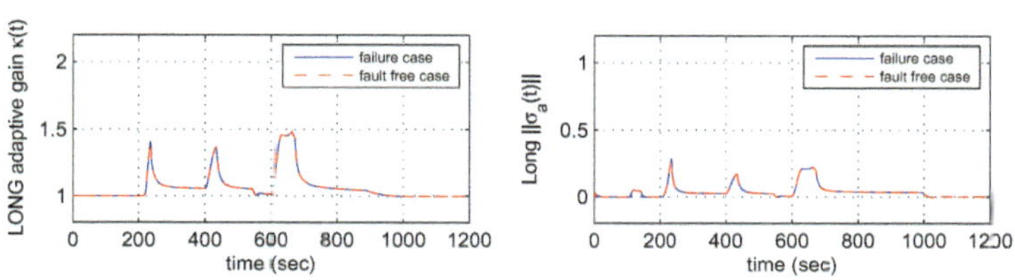

(d) switching function and adaptive gain

Fig. 13.11 Offline evaluation: elevator jam

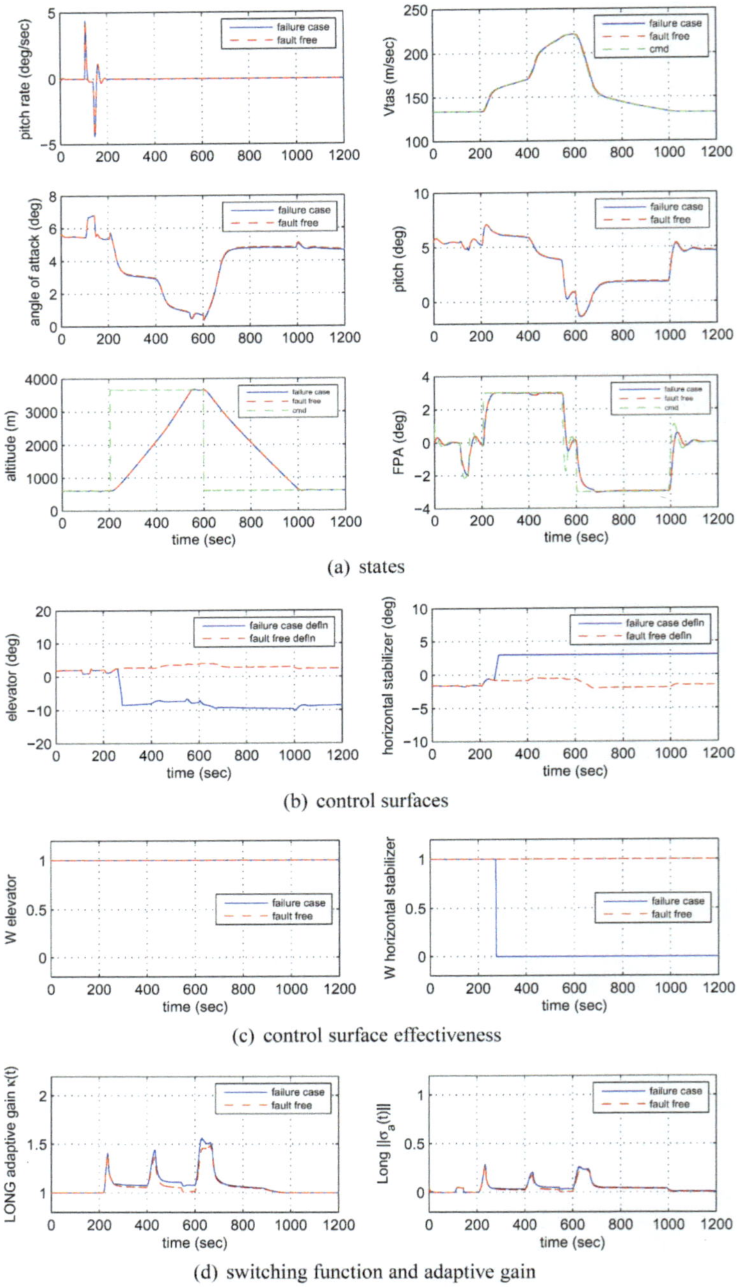

Fig. 13.12 Offline evaluation: stabilizer runaway

13.9 SRS Piloted Evaluation Results

The results presented in this section are based on the piloted evaluation results on the SRS. The pilot has more than 9000 flying hours and served as a commercial air transport captain for the Boeing 757-200 and Boeing 767-300. Valuable feedback from the pilot, on the proposed controller from a practical and operational point of view, will be discussed at the end of the chapter.

In this section, two sets of test results will be presented. As shown in Figs. 13.14, 13.15 and 13.16, the first set of results is associated with the perfect estimation of actuator efficiency W, conducted under fault-free, elevator jam and stabilizer runaway scenarios. These results are the actual pilot evaluation results. The second set of results (also plotted on top of Figs. 13.14, 13.15 and 13.16 - i.e. overlapped) is associated with a 10% imperfection in the estimate of actuator efficiency $\widehat{W}$. These results are based on the 'offline' simulation where the pilot inner loop FPA and speed commands, saved from the logged data from the SRS, have been used to drive desktop simulations. These 'offline' results with imperfect estimation of $\widehat{W}$ were not tested during the piloted evaluation due to implementation issues associated with additional hardware required for the activation of the imperfection. Despite being tested using desktop simulations, the results with imperfect $\widehat{W}$ estimation provide a unique opportunity for an exact comparison with the case of perfect W knowledge.

Figure 13.13 shows three different trajectories of the aircraft during the pilot evaluation associated with the fault-free, elevator jam and stabilizer runaway. For consistency and to allow comparison between the three different cases, the same sequence of manoeuvres were conducted by the pilot. Note that even though same sequence

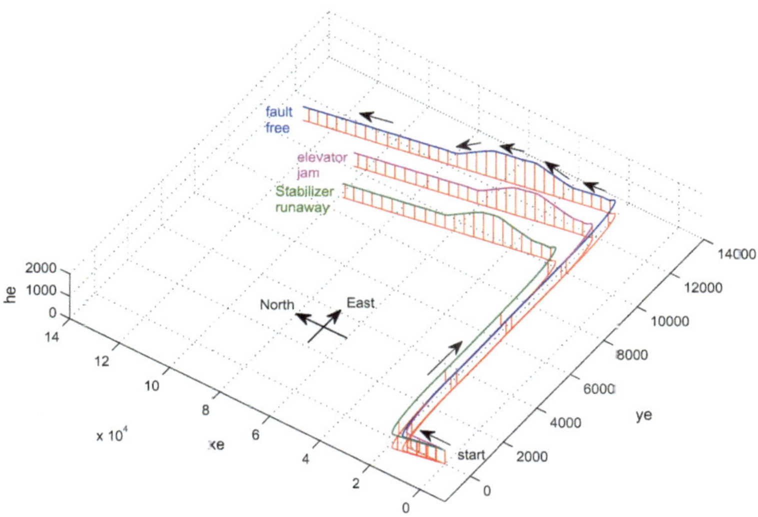

Fig. 13.13 Pilot evaluation: trajectory

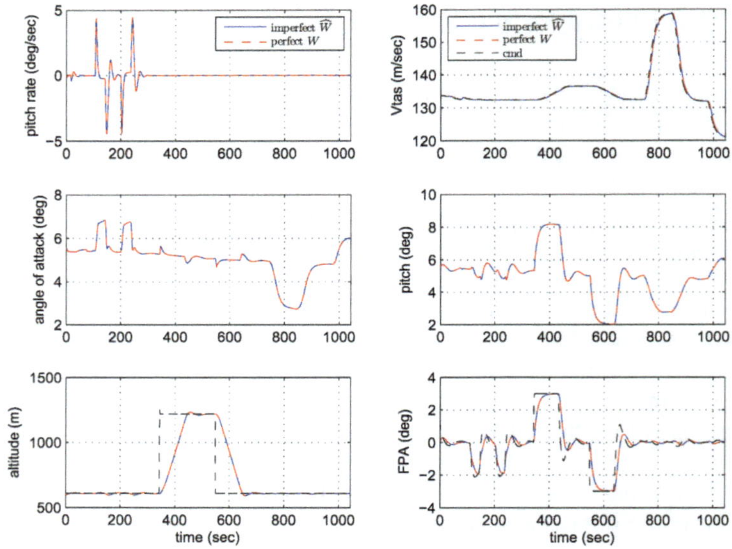

(a) Longitudinal states

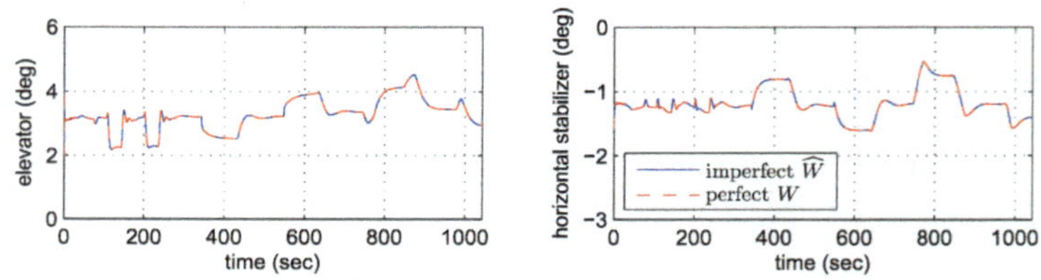

(b) Longitudinal control surfaces

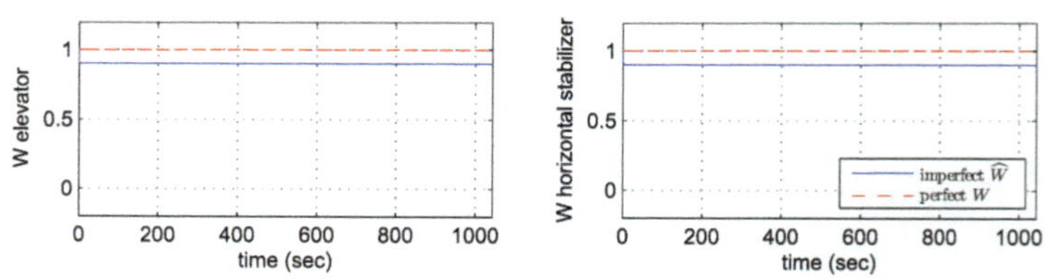

(c) control surface effectiveness

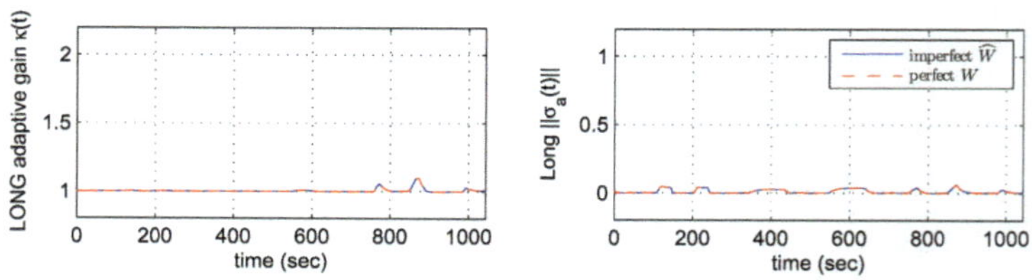

(d) switching function and adaptive gain

Fig. 13.14 Pilot evaluation - fault-free with perfect and imperfect $\widehat{W}$

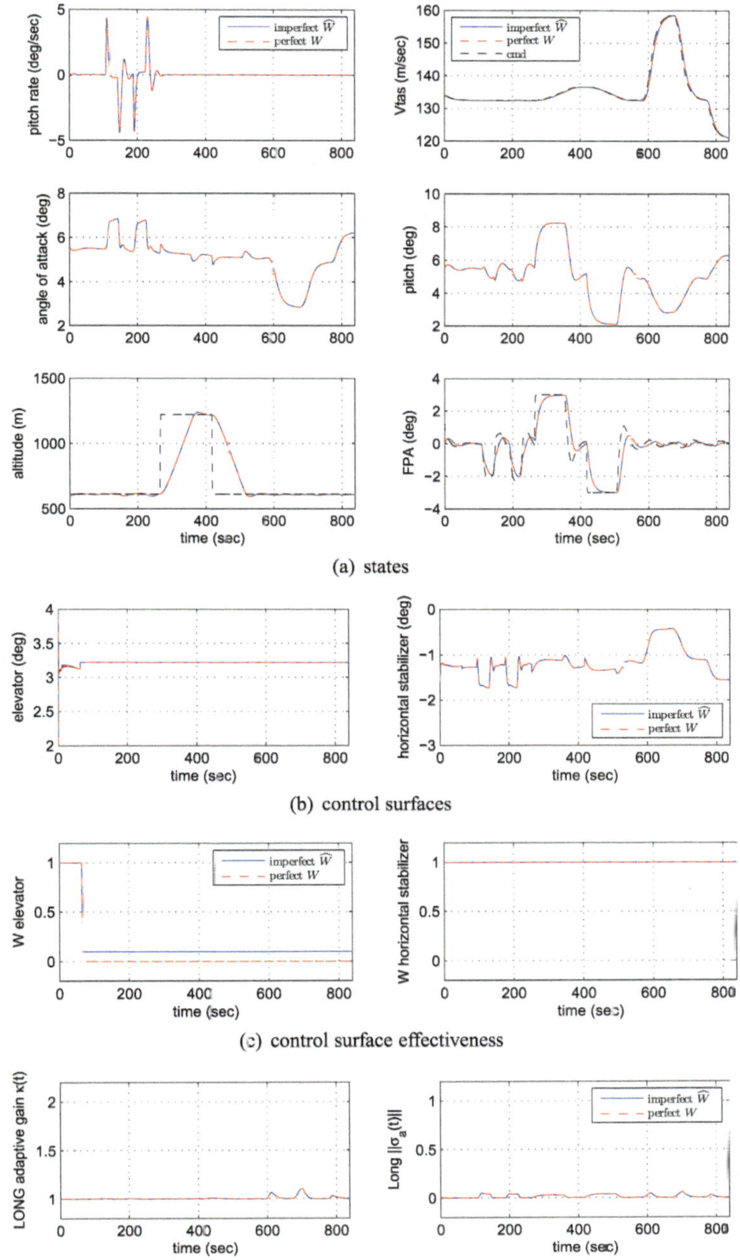

(a) states

(b) control surfaces

(c) control surface effectiveness

(d) switching function and adaptive gain

Fig. 13.15 Pilot evaluation - elevator jam with perfect and imperfect $\widehat{W}$

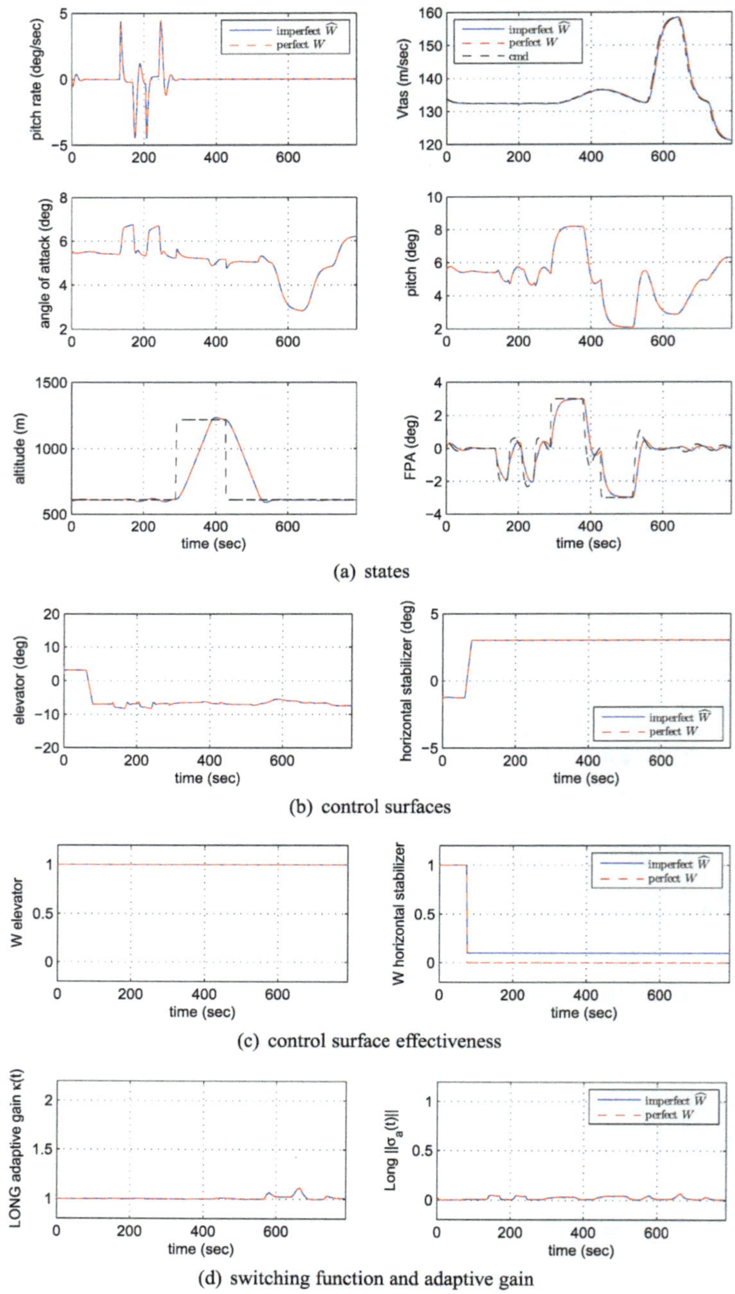

(a) states

(b) control surfaces

(c) control surface effectiveness

(d) switching function and adaptive gain

Fig. 13.16 Pilot evaluation: stabilizer runaway with perfect and imperfect $\widehat{W}$

of manoeuvres were carried out, the final manoeuvres shown in Fig. 13.13 are not exactly the same as compared to the ones in Fig. 13.10 which use recorded commands to reproduce the manoeuvres exactly. The difference in Fig. 13.13 is due to the different time each sequence was executed as the pilot manually changes the heading, altitude and speed commands from the dial on the MCP. It has to be noted however that despite the different execution times, the magnitude of each command is the same.

13.9.1 Pilot Evaluation: Fault-Free

Figure 13.14 shows the piloted evaluation results of the proposed controller during fault-free conditions. Figure 13.14 also shows the comparison between perfect W and imperfect $\widehat{W}$ in fault-free conditions. Figure 13.14c shows the nominal case when $W = 1$ and the case of imperfect estimation of effectiveness level when $\widehat{W} = 0.9$. From Fig. 13.14a, it can be seen that there are no visible differences between the states associated with the perfect and imperfect levels of actuator effectiveness estimation. This highlights the ability of the proposed controller to handle 10% imperfections in the actuator effectiveness levels. In both perfect and imperfect situations, the controller provides good altitude, FPA and speed tracking. Figure 13.14b shows no visible difference between perfect and imperfect estimation in the deflections of the elevator and stabilizer during fault-free conditions. Finally Fig. 13.14d shows the nominal variations in the sliding mode switching function σ_a and the adaptive gain due to changes in operating condition for the fault-free case. Again there are no visible differences between the perfect and imperfect case.

13.9.2 Pilot Evaluation: Elevator Jam

Figure 13.15 shows the piloted evaluation for the case of an elevator jam. The effect of the elevator jam can be seen in Fig. 13.15b where the elevator locks in place from approximately 63 s onwards. After the elevator jam, the stabilizer becomes more active compared to the fault-free case in Fig. 13.14b. Figure 13.15 also shows the comparison between the perfect and imperfect estimates of the effectiveness level in the event of an elevator jam. The imperfect estimation of $\widehat{W}$ can be seen in Fig. 13.15c. The actual (perfect) elevator effectiveness in the event of a jam is $W = 0$. However, in the imperfect case, $\widehat{W} = 0.1$. Despite this imperfection, it can be seen from Fig. 13.15a that there are no visible differences in the states and tracking performance. In fact, there is also no visible difference when compared to the fault-free case in Fig. 13.14a.

13.9.3 Pilot Evaluation: Stabilizer Runaway

Figure 13.16 shows the piloted evaluation results in the event of a stabilizer runaway at approximately 74 s. The effect of the stabilizer runaway can be seen in Fig. 13.16b where the stabilizer moves at its maximum rate to its maximum deflection of 3°. After this point, it can be seen that the elevator becomes more active and moves to $-10°$ in order to compensate for the failed stabilizer. Figure 13.16c shows the comparison between the perfect $W = 0$ and imperfect $\widehat{W} = 0.1$ case. Despite the stabilizer runaway and imperfect estimation of the stabilizer effectiveness level, there is no visible difference in the states and tracking performance as seen in Fig. 13.16a. In fact, there is also no visible difference compared to the fault-free case in Fig. 13.14a. Finally Fig. 13.16d shows that sliding is being maintained because $\|\sigma_a(t)\|$ is close to zero and the adaptive gain remains low, despite the challenging and critical failure of a stabilizer runaway [10].

13.9.4 Piloted Evaluation: Pilot Feedback

General feedback from the pilot and the SRS researcher indicates that similar performance was observed on all three scenarios being evaluated. In fact the pilot was unable to notice a meaningful difference between the fault-free and the elevator and stabilizer failure cases, and reported no transient behaviour when the failure occurs. Furthermore the SRS researcher had to double check that the failure did occur and the pilot only noticed that the failure had occurred when looking at the control surface deflections on the SRS EICAS display at the centre of the cockpit.

13.10 Discussion

The offline and piloted evaluations show very similar tracking performance in the nominal and failure conditions, which demonstrates the capability of the controller. In the piloted evaluation, the controller was tested over a large part of the flight envelope and in a realistic environment. Apart from demonstrating that the controller can be implemented and flown in a real-time simulator, this evaluation also provided feedback on its closed-loop behaviour from an operational standpoint. The pilot comments confirmed very similar performance in both nominal and failure conditions, but also indicated where this performance should be tuned to be more in line with typical operational requirements. By tuning the outer loops of the controller, it is expected that a second version will be able to fulfil all relevant performance criteria.

The results in Sect. 13.9 also provide a comparison between the piloted evaluation results based on perfect knowledge of W and also imperfect estimation of $\widehat{W}$. This comparison provides a measure of performance of the proposed controller when estimation of the actuator effectiveness level is not exactly known. In fact, as shown in Figs. 13.14, 13.15 and 13.16 that there are no visible differences in the performance of the aircraft between perfect W and imperfect $\widehat{W}$, even in the presence of elevator and stabilizer failures. This indicates that the pilot will not see any difference in terms of performance. Therefore despite being conducted on a desktop simulation, the observation and comments given for the piloted evaluation also apply to the desktop simulation with imperfect $\widehat{W}$.

13.11 Conclusions

The chapter has considered integral sliding modes and how they can be employed in the context of fault tolerant control. Two distinct classes of problems were considered: firstly a fault tolerant ISM controller was designed for a class of faults in an over-actuated linear system; and secondly an ISM scheme was retrofitted to an existing feedback control scheme for an over-actuated uncertain linear system with the objective of retaining the pre-existing nominal performance in the face of faults and failures. Aerospace examples have been used throughout to demonstrate the efficacy of the approach. In order to handle wider variations in the flight conditions, the controller ideas have been extended using LPV ideas. These controllers have been implemented and tested on the SIMONA research flight simulator. One of the highlights of the piloted evaluation was the fact that the pilot was unable to see any discernible difference between the fault-free and the failure cases whilst flying the aircraft. Furthermore there was no transient observed when the elevator jammed and when a more catastrophic stabilizer runaway was tested. The evaluation results highlight the potential of the proposed scheme for actual real-time industrial implementation.

Acknowledgements The work presented here involves contributions from a number of people not directly involved with the preparation of this chapter. In particular we must gratefully acknowledge the contributions of O. Stroosma and Prof J.A. Mulder from Delft University of Technology for their continued support in terms of allowing us to use the SIMONA research simulator.

References

1. Alwi, H., Edwards, C.: Fault tolerant control using sliding modes with on-line control allocation. Automatica **44**(7), 1859–1865 (2008)
2. Alwi, H., Edwards, C., Stroosma, O., Mulder, J.A.: Evaluation of a sliding mode fault-tolerant controller for the El-Al incident. J. Guid. Control Dyn. **33**(3), 677–694 (2010)
3. Alwi, H., Edwards, C., Tan, C.P.: Fault Detection and Fault-tolerant Control Using Sliding Mode. Springer, Berlin (2011)
4. Alwi, H., Edwards, C., Marcos, A.: Fault reconstruction using a LPV sliding mode observer for a class of LPV systems. J. Frankl. Inst. **349**(2), 510–530 (2012)
5. Bejarano, F., Fridman, L., Poznyak, A.: Output integral sliding mode control based on algebraic hierarchical observer. Int. J. Control **80**(3), 443–453 (2007)

6. Blanke, M., Kinnaert, M., Lunze, J., Staroswiecki, M.: Diagnosis and Fault-Tolerant Control. Springer, Berlin (2006)
7. Brière, D., Traverse, P.: Airbus A320/A330/A340 electrical flight controls-a family of fault-tolerant systems. In: Proceedings of the 23rd International Symposium on Fault-Tolerant Computing, pp. 616–623 (1993)
8. Castanos, F., Fridman, L.: Analysis and design of integral sliding manifolds for systems with unmatched perturbations. IEEE Trans. Autom. Control **51**(5), 853–858 (2006)
9. Edwards, C., Spurgeon, S.K.: Sliding Mode Control: Theory and Applications. CRC Press, Boca Raton (1998)
10. Edwards, C., Lombaerts, T., Smaili, H.: Fault Tolerant Flight Control: A Benchmark Challenge. Lecture Notes in Control and Information Sciences. Springer, Berlin (2010)
11. Field, E.J., Pinney, T.R., (René) van Paassen, M.M., Stroosma, O., Rivers, R.A.: Effects of implementation variations on the results of piloted simulator handling qualities evaluations. In: Proceedings of AIAA Modeling and Simulation Technologies Conference (2004)
12. Franklin, G.F., Powell, J.D., Emami-Naeini, A.: Feedback Control of Dynamic Systems. Prentice Hall, Upper Saddle River (2002)
13. Hamayun, M.T., Alwi, H., Edwards, C.: An LPV fault tolerant control scheme using integral sliding modes. In: Proceedings of the 51th IEEE Conference on Decision and Control, pp. 1840–1845 (2012)
14. Hamayun, M.T., Edwards, C., Alwi, H.: Design and analysis of an integral sliding mode fault tolerant control scheme. IEEE Trans. Autom. Control **57**(7), 1783–1789 (2012)
15. Hamayun, M.T., Edwards, C., Alwi, H.: A fault tolerant control allocation scheme with output integral sliding mode. Automatica **49**(6), 1830–1837 (2013)
16. Hamayun, M.T., Edwards, C., Alwi, H.: Augmentation scheme for fault-tolerant control using integral sliding modes. IEEE Trans. Control Syst. Technol. **22**(1), 307–313 (2014)
17. Härkegård, O., Glad, S.T.: Resolving actuator redundancy-optimal control vs. control allocation. Automatica **41**(1), 137–144 (2005)
18. Khong, T.H., Shin, J.: Robustness analysis of integrated LPV-FDI filters and LTI-FTC system for a transport aircraft. In: Proceedings of AIAA Guidance, Navigation and Control Conference and Exhibit (2007)
19. Matthews, G., DeCarlo, R.A.: Decentralized tracking for a class of interconnected nonlinear systems using variable structure control. Automatica **24**(2), 187–193 (1988)
20. Rios, H., Kamal, S., Fridman, L., Zolghadri, A.: Fault tolerant control alloaction via continuous integral sliding modes: a HOSM-observer approach. Automatica **51**, 318–325 (2015)
21. Rubagotti, M., Estrada, A., Castanos, F., Ferrara, A., Fridman, L.: Integral sliding mode control for nonlinear systems with matched and unmatched perturbations. IEEE Trans. Autom. Control **56**(11), 2699–2704 (2011)
22. Shtessel, Y., Buffington, J., Banda, S.: Multiple time scale flight control using re-configurable sliding modes. J. Guid. Control Dyn. **22**(6), 873–883 (1999)
23. Shtessel, Y., Edwards, C., Fridman, L., Levant, A.: Sliding Mode Control and Observation. Birkhauser, New York (2013)
24. Stewart, G.W.: On scaled projections and pseudoinverses. Linear Algebr. Appl. **112**, 189–193 (1989)
25. Stroosma, O., van Paassen, M.M., Mulder, M.: Using the SIMONA research simulator for human-machine interaction research. In: Proceedings of the AIAA Modeling and Simulation Technologies Conference (2003)
26. Stroosma, O., Smaili, H., Lombaerts, T., Mulder, J.A.: Piloted simulator evaluation of new fault-tolerant flight control algorithms for reconstructed accident scenarios. In: Proceedings of the AIAA Modeling and Simulation Technologies Conference (2008)
27. Utkin, V.I., Shi, J.: Integral sliding mode in systems operating under uncertainty conditions. In: Proceedings of the 35th IEEE Conference on Decision and Control, pp. 4591–4596 (1996)
28. Utkin, V.I.: Sliding Modes in Control Optimization. Springer, Berlin (1992)
29. Zhang, Y.M., Jiang, J.: Active fault-tolerant control system design against partial actuator failures. IEE Proc. Control Theory Appl. **149**(1), 95–104 (2002)

Part III
Applications of VSS/SMC to Real Time Systems

Chapter 14
Speed Control of Induction Motor Servo Drives Using Terminal Sliding-Mode Controller

Yong Feng, Minghao Zhou, Fengling Han and Xinghuo Yu

14.1 Introduction

The induction motor (IM) is one of the most common electrical motor used in most applications. This motor runs at a speed less than its synchronous speed, therefore it is also called as asynchronous motor. The synchronous speed is the speed of rotation of the magnetic field in a rotary machine and it depends upon the frequency and number poles of the IM. The IM has been extensively used in many practical applications due to its simply construction, lower repair and maintenance costs, high reliability and relatively low manufacturing cost, etc [1]. With the development of power electronics, electrical technique and control theories, IMs have been able to be used in high-performance servo systems, such as speed servo systems, even position servo systems.

Three methods can be used for the control of IMs: the scalar control, the direct torque control (DTC) and the field oriented control (FOC). The former method is very simple method for controlling the speed of IM compared to the vector control which is more complex. The latter two methods can be utilized to implement the high-performance IM servo systems.

Y. Feng (✉) · M. Zhou
Department of Electrical Engineering, Harbin Institute of Technology,
Harbin 150001, People's Republic of China
e-mail: yfeng@hit.edu.cn

M. Zhou
e-mail: zhouminghao@aliyun.com

F. Han
School of Science, RMIT University, Melbourne, VIC 3001, Australia
e-mail: fengling.han@rmit.edu.au

X. Yu
Research & Innovation Portfolio, RMIT University, Melbourne, VIC 3001, Australia
e-mail: x.yu@rmit.edu.au

S. Li et al. (eds.), *Advances in Variable Structure Systems and Sliding Mode Control—Theory and Applications*, Studies in Systems, Decision and Control 115, DOI 10.1007/978-3-319-62896-7_14

In DTC-based IM servo systems both the stator flux and the torque are regulated respectively using the bang-bang control strategies. This control method may lead to the torque ripple. If the IM runs at low speed, its performances will become poorer, and the speed range will be limited.

FOC is widely used in high performance control of IM servo systems. Since the torque and flux of an IM are decoupled using FOC, the IM systems can yield faster dynamic response and lower steady-state error. The mathematical model of an IM in a three-dimensional stationary reference frame (abc) can be converted into a model in a two-dimensional rotating reference frame (dq) using the Clarke-Park transformation. The d-axis current in the stator represents the rotor flux and the q-axis current represents the torque. Therefore, the decoupled rotor flux and the torque of an IM can be separately controlled like as a decoupled excited DC motor. Consequently it is possible to achieve good steady-state and dynamic performances of IMs [2, 3].

However, an accurate information on both the magnitude and the angular position of the rotor flux are needed by FOC for the transformation between the rotating and the stationary reference frames. There are two main methods for obtaining the magnitude and the angular position of the rotor flux, the direct measurements or the indirect estimation. The former needs special sensors, therefore it is difficult in practical applications. The latter is popular and widely used [4]. It applies the measurements of the stator currents, stator voltages and the motor speed into some estimation algorithms to estimate the magnitude and the angular position of the rotor flux. A lot of estimation methods for the rotor flux have been proposed, such as Luenberger observer-based methods [5, 6], model reference methods [7, 8], Kalman filter-based methods [9], and neural networks [10].

The high performance control of IMs is a challenge due to multi-variable, strong coupling, and nonlinearities in the model of IMs [11, 12]. So far, a lot of control methods have been proposed to improve the robustness and dynamical performances of IMs, such as neural network control, fuzzy control, optimal control, adaptive control and sliding-mode control [13, 14]. Sliding-mode control has attractive advantages compared to other control methods, such as low sensitivity to the system parameter variations and strong robustness to external disturbances [15, 16]. However, the chattering phenomena limit the practical applications of conventional sliding-mode control [17]. In this chapter, a nonsingular terminal sliding-mode control (NTSM) method is applied for IM velocity servo systems. To implement the FOC of IMs, an NTSM observer is designed in the chapter to estimate the rotor flux of IMs with equivalent smooth control signals [18, 19]. Additionally, the speed sensorless technology is also utilized in this chapter, afterwards an NTSM observer is utilized in the FOC system of IMs to estimate the speed instead of practical sensors.The simulations have been carried out to validate the applied method.

14.2 Mathematical Model of Induction Motor

An accurate mathematical model of IMs is the basic factor for the implementation of high-performance servo systems of IMs, especially for the FOC algorithms. For simplicity of the analysis some assumptions for IMs can be described as follows:

(1) the effect of magnetic saturation is neglected.

(2) the three-phase windings have the same structure and the fringe effect is neglected.

(3) the slots effect is ignored.

(4) the iron core loss is not taken into account.

14.2.1 Mathematical Model of IM in Three-Dimensional Stationary Coordinate (abc)

The mathematical model of IMs is usually composed of the voltage, flux, and motion equations. Based on the assumptions above, the voltage equations of the IMs in three-dimensional stationary coordinate (*abc*) are below:

$$\begin{bmatrix} u_{sa} \\ u_{sb} \\ u_{sc} \\ u_{ra} \\ u_{rb} \\ u_{rc} \end{bmatrix} = \begin{bmatrix} R_s & 0 & 0 & 0 & 0 & 0 \\ 0 & R_s & 0 & 0 & 0 & 0 \\ 0 & 0 & R_s & 0 & 0 & 0 \\ 0 & 0 & 0 & R_r & 0 & 0 \\ 0 & 0 & 0 & 0 & R_r & 0 \\ 0 & 0 & 0 & 0 & 0 & R_r \end{bmatrix} \begin{bmatrix} i_{sa} \\ i_{sb} \\ i_{sc} \\ i_{ra} \\ i_{rb} \\ i_{rc} \end{bmatrix} + \begin{bmatrix} \dot{\psi}_{sa} \\ \dot{\psi}_{sb} \\ \dot{\psi}_{sc} \\ \dot{\psi}_{ra} \\ \dot{\psi}_{rb} \\ \dot{\psi}_{rc} \end{bmatrix} \quad (14.1)$$

where u_{sa}, u_{sb} and u_{sc} are three stator voltages in *abc* axes;

u_{ra}, u_{rb} and u_{rc} are three rotor voltages in *abc* axes;

i_{sa}, i_{sb} and i_{sc} are three stator currents in *abc* axes;

i_{ra}, i_{rb} and i_{rc} are three rotor currents in *abc* axes;

ψ_{sa}, ψ_{sb} and ψ_{sc} are three stator fluxes in *abc* axes;

ψ_{ra}, ψ_{rb} and ψ_{rc} are three rotor fluxes in *abc* axes;

R_s is the stator resistance;

R_r is the rotor resistance.

Therefore the flux equations of an IM in the three-dimensional stationary coordinate (*abc*) include both the stator and rotor flux equations and describe the relationships between the currents and the fluxes in the stator and rotor. They are given by the following equations:

$$\begin{bmatrix} \psi_s \\ \psi_r \end{bmatrix} = \begin{bmatrix} L_{ss} & L_{sr} \\ L_{rs} & L_{rr} \end{bmatrix} \begin{bmatrix} i_s \\ i_r \end{bmatrix} \quad (14.2)$$

with

$$L_{ss} = \begin{bmatrix} L_{sm} + L_{sl} & -\frac{1}{2}L_{sm} & -\frac{1}{2}L_{sm} \\ -\frac{1}{2}L_{sm} & L_{sm} + L_{sl} & -\frac{1}{2}L_{sm} \\ -\frac{1}{2}L_{sm} & -\frac{1}{2}L_{sm} & L_{sm} + L_{sl} \end{bmatrix}, L_{rr} = \begin{bmatrix} L_{sm} + L_{rl} & -\frac{1}{2}L_{sm} & -\frac{1}{2}L_{sm} \\ -\frac{1}{2}L_{sm} & L_{sm} + L_{rl} & -\frac{1}{2}L_{sm} \\ -\frac{1}{2}L_{sm} & -\frac{1}{2}L_{sm} & L_{sm} + L_{rl} \end{bmatrix},$$

$$L_{sr} = L_{rs}^T = L_{sm} \begin{bmatrix} \cos\theta & \cos(\theta - 120°) & \cos(\theta + 120°) \\ \cos(\theta + 120°) & \cos\theta & \cos(\theta - 120°) \\ \cos(\theta - 120°) & \cos(\theta + 120°) & \cos\theta \end{bmatrix}$$

where ψ_s and ψ_r are the stator flux phase and the rotor flux phase in *abc* axes respectively, by $\psi_s = [\psi_{sa}, \psi_{sb}, \psi_{sc}]^T$, $\psi_r = [\psi_{ra}, \psi_{rb}, \psi_{rc}]^T$

$\boldsymbol{i}_s$ and $\boldsymbol{i}_r$ are the stator current vector and the rotor current vector in abc axes respectively, by $\boldsymbol{i}_s = [\boldsymbol{i}_{sa}, \boldsymbol{i}_{sb}, \boldsymbol{i}_{sc}]^T$, $\boldsymbol{i}_r = [\boldsymbol{i}_{ra}, \boldsymbol{i}_{rb}, \boldsymbol{i}_{rc}]^T$;

L_{sl} and L_{rl} are the stator and rotor leakage inductance respectively;

L_{sm} and L_{rm} are the mutual inductance between the stator and rotor windings;

θ is the electric angle between the stator and rotor windings.

The torque equation of IMs can be given as follows:

$$\begin{aligned} T_e = n_p L_{sm}[(i_{sa}i_{ra} + i_{sb}i_{rb} + i_{sc}i_{rc}) \sin\theta + (i_{sa}i_{ra} + i_{sb}i_{rb} + i_{sc}i_{rc}) \sin(\theta + 120°) \\ + (i_{sa}i_{ra} + i_{sb}i_{rb} + i_{sc}i_{rc}) \sin(\theta - 120°)] \end{aligned} \quad (14.3)$$

where n_p is the number of pole pairs.

Finally the mechanical equation of an IM can be described as follows:

$$T_e = T_L + \frac{J}{n_p}\dot{\omega} \quad (14.4)$$

where ω is the electric angular velocity of the rotor;

J is the inertia of the motor;

T_L is the load torque.

14.2.2 Mathematical Model of IM in Two-Dimensional Stationary Coordinate ($\alpha\beta$)

Based on Clarke transformation, the mathematical model of IMs in a three-dimensional stationary coordinate (*abc*) can be converted into a two-dimensional stationary coordinate ($\alpha\beta$). The Clarke transformation is given by the following equation:

$$\boldsymbol{C}_{3/2} = \sqrt{\frac{2}{3}} \begin{bmatrix} 1 & -\frac{1}{2} & -\frac{1}{2} \\ 0 & \frac{\sqrt{3}}{2} & \frac{\sqrt{3}}{2} \end{bmatrix} \quad (14.5)$$

Subsequently, the voltage equations of the IMs in a two-dimensional stationary coordinate ($\alpha\beta$) can be described as follow:

$$\begin{cases} u_{s\alpha} = R_s i_{s\alpha} + L_s \dot{i}_{s\alpha} + L_m \dot{i}_{r\alpha} \\ u_{s\beta} = R_s i_{s\beta} + L_s \dot{i}_{s\beta} + L_m \dot{i}_{rb} \\ u_{r\alpha} = R_r i_{r\alpha} + L_r \dot{i}_{r\alpha} + L_m \dot{i}_{r\alpha} + \omega(L_m i_{s\beta} + L_r i_{r\beta}) \\ u_{r\beta} = R_r i_{r\beta} + L_r \dot{i}_{r\beta} + L_m \dot{i}_{s\beta} + \omega(\mathrm{L}_m i_{s\alpha} + L_r i_{r\alpha}) \end{cases} \tag{14.6}$$

where $u_{s\alpha}$ and $u_{s\beta}$ are two stator voltages in $\alpha\beta$ axes;
$u_{r\alpha}$ and $u_{r\beta}$ are two rotor voltages in $\alpha\beta$ axes;
$i_{s\alpha}$ and $i_{s\beta}$ are two stator currents in $\alpha\beta$ axes;
$i_{r\alpha}$ and $i_{r\beta}$ are two rotor currents in $\alpha\beta$ axes;
L_s, L_r, and L_m the stator inductance, the rotor inductance and the mutual inductance between the stator and rotor windings, which are described by

$$L_s = L_m + L_{ls},\ L_r = L_m + L_{lr},\ L_m = \frac{3}{2} L_{ms}$$

Then the flux equations of IMs in a two-dimensional stationary coordinate ($\alpha\beta$) can be described as follow:

$$\begin{cases} \psi_{s\alpha} = L_s i_{s\alpha} + L_m i_{r\alpha} \\ \psi_{s\beta} = L_s i_{s\beta} + L_m i_{r\beta} \\ \psi_{r\alpha} = L_m i_{s\alpha} + L_r i_{r\alpha} \\ \psi_{r\beta} = L_m i_{s\beta} + L_r i_{r\beta} \end{cases} \tag{14.7}$$

where $\psi_{s\alpha}$ and $\psi_{s\beta}$ are the stator fluxes in $\alpha\beta$ axes respectively;
$\psi_{r\alpha}$ and $\psi_{r\beta}$ are the rotor fluxes in axes respectively.

The torque equation of IMs in a two-dimensional stationary coordinate ($\alpha\beta$) is described as follows:

$$T_e = n_p L_m (i_{s\beta} i_{r\alpha} - i_{s\alpha} i_{r\beta}) \tag{14.8}$$

14.2.3 *Mathematical Model of IMs in Two-Dimensional Rotating Coordinate (dq)*

The FOC strategy can decouple the d-axis and q-axis currents in the stator of an IM applying the Park transformation, and make the rotor flux and torque of the IM controlled separately like as decoupled excited DC motors. The relationship between the two-dimensional stationary and rotating coordinates can be shown in Fig. 14.1, where θ_e represents the electric angle between the d-axis and the α-axis. The speed of the two-dimensional rotating coordinate equals to ω_1 which is the synchronous electric angular velocity of the IM.

The Park transformation is given by

$$\boldsymbol{C}_{2s/2r} = \begin{bmatrix} \cos\theta_e & \sin\theta_e \\ -\sin\theta_e & \cos\theta_e \end{bmatrix} \tag{14.9}$$

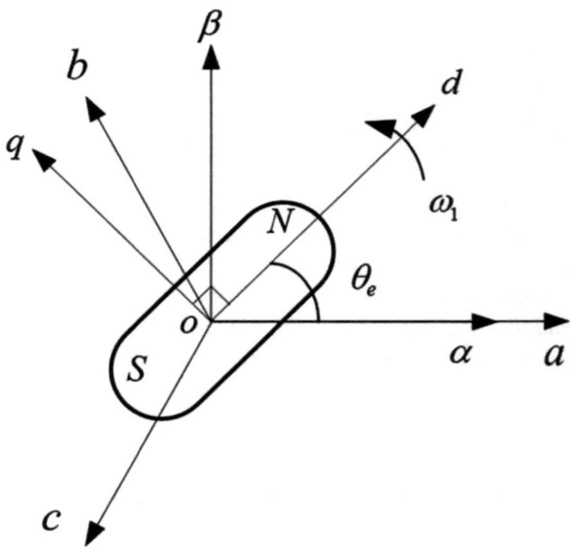

Fig. 14.1 $\alpha\beta$ and dq coordinates

Then the voltage equations of the IMs in a two-dimensional rotating coordinate (dq) can be described as follow:

$$\begin{cases} u_{sd} = R_s i_{sd} + \dot{\psi}_{sd} - \omega_1 \psi_{sq} \\ u_{sq} = R_s i_{sq} + \dot{\psi}_{sq} + \omega_1 \psi_{sd} \\ u_{rd} = R_r i_{rd} + \dot{\psi}_{rd} - \omega_s \psi_{rq} \\ u_{rq} = R_r i_{rq} + \dot{\psi}_{rq} - \omega_s \psi_{rd} \end{cases} \tag{14.10}$$

where u_{sd} and u_{sq} are two stator voltages in dq axes;
u_{rd} and u_{rq} are two rotor voltages in dq axes;
i_{sd} and i_{sq} are the stator currents in d- and q- axes;
i_{rd} and i_{rq} are the rotor currents in d- and q- axes;
ψ_{sd} and ψ_{sq} are the stator fluxes in d- and q- axes;
ψ_{rd} and ψ_{rq} are the rotor fluxes in d- and q- axes;
ω_s is the slip angle velocity;
ω_1 is the synchronous angular velocity.

The flux equations of the IMs in a two-dimensional rotating coordinate system (dq) can be described as follow:

$$\begin{cases} \psi_{sd} = L_s i_{sd} + L_m i_{rd} \\ \psi_{sq} = L_s i_{sq} + L_m i_{rq} \\ \psi_{rd} = L_m i_{sd} + L_r i_{rq} \\ \psi_{rq} = L_m i_{sq} + L_r i_{rq} \end{cases} \tag{14.11}$$

The torque equation of IMs in a two-dimensional rotating coordinate (dq) is given by

$$T_e = \frac{n_p L_m}{L_r}(i_{sq}\psi_{rd} - i_{sd}\psi_{rq}) \tag{14.12}$$

Summarizing, the mathematical model of IMs in a three-dimensional stationary coordinate (abc) is described by Eqs.(14.1)–(14.4), which is further transferred to the model in a two-dimensional stationary coordinate $(\alpha\beta)$, as shown as in Eqs. (14.6), (14.7) and (14.8), by the Clark transformation (14.5). Finally, the model of IMs in a two-dimensional rotating coordinate (dq) is obtained in Eqs. (14.4), (14.10), (14.11) and (14.12) by the Park transformation (14.9).

14.3 Field Oriented Control System

For squirrel cage IMs, the rotor voltages are

$$u_{rd} = u_{rq} = 0 \tag{14.13}$$

The mathematical model of the IMs in the FOC systems can be finally described in dq axes as follows by using Eqs. (14.4), and (14.10)–(14.13):

$$\begin{cases} \dot{i}_{sd} = \xi \frac{1}{T_r}\psi_{rd} + \xi\omega\psi_{rq} - \lambda i_{sd} + \omega_1 i_{sq} + K u_{sd} \\ \dot{i}_{sq} = -\xi\omega\psi_{rd} + \xi\frac{1}{T_r}\psi_{rq} - \omega_1 i_{sd} - \lambda i_{sq} + K u_{sq} \\ \dot{\psi}_{rd} = -\frac{1}{T_r}\psi_{rd} + (\omega_1 - \omega)\psi_{rq} + \frac{L_m}{T_r} i_{sd} \\ \dot{\psi}_{rq} = -(\omega_1 - \omega)\psi_{rd} - \frac{1}{T_r}\psi_{rq} + \frac{L_m}{T_r} i_{sq} \\ \dot{\omega} = \frac{n_p^2 L_m}{J L_r}\left(i_{sq}\psi_{rd} - i_{sd}\psi_{rq}\right) - \frac{n_p}{J} T_L \end{cases} \tag{14.14}$$

where $\sigma = 1 - L_m^2/L_s L_r$ is the leakage coefficient; $T_r = L_r/R_r$ is the rotor time constant; and $K = 1/\sigma L_s$, $\xi = K(L_m/L_r)$, $\lambda = K(R_s + L_m^2/T_r L_r)$.

Based on the FOC strategy, the d-axis is orientated in accord with the axis of the rotor flux. Therefore d- and q- axis rotor flux can be determined as follows

$$\begin{cases} \psi_{rd} = \psi_r \\ \psi_{rq} = 0 \end{cases} \tag{14.15}$$

where ψ_r is the rotor flux.

The FOC of IMs is based on their mathematical model (14.14) and the field orientated principle (14.15). It can be seen that the control of IMs is decoupled into the rotor flux control and torque control. Then the FOC-based speed servo system of IMs can be built using four nonsingular terminal sliding-mode controllers in

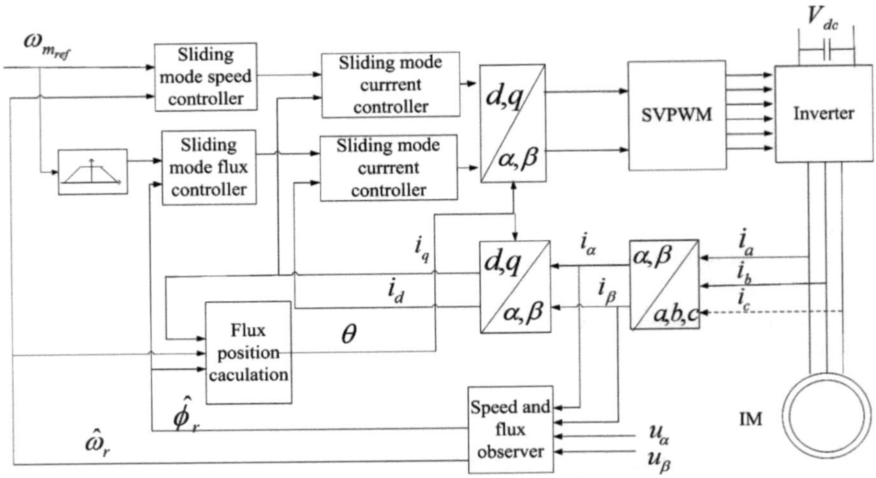

Fig. 14.2 An FOC-based speed closed-loop control system of IMs

the speed-, flux-, d-axis current- and q-axis current-loops respectively, as shown in Fig. 14.2.

In Fig. 14.2, $\omega_{m_{ref}}$ is the required speed of the IM. The outputs of the speed and flux controllers are the references of the stator currents in d- and q-axes respectively. To implement high-performance control of the FOC position servo system of IMs, four full-order sliding-mode controllers will be designed in the next section.

14.4 NTSM Controllers for IM Servo System

14.4.1 Speed Controller

The motion equation of the IMs can be obtained from the model of the IMs (14.14) as follows:

$$\dot{\omega}_m = \frac{n_p L_m}{J L_r} i_{sq} \psi_r - \frac{1}{J} T_L \tag{14.16}$$

where $\omega_m = \omega / n_p$ is the mechanical angular velocity of the rotor.

Defining the desired mechanical velocity of the motor as $\omega_{m_{ref}}$, which should be smooth enough up to the second order time derivative, the error between the actual velocity and the given velocity e_ω is:

$$e_\omega = \omega_{m_{ref}} - \omega_m \tag{14.17}$$

Then the speed error dynamics can be obtained as follows using Eqs.(14.16) and (14.17):

$$\dot{e}_\omega = \dot{\omega}_{m_{ref}} - \dot{\omega}_m = \dot{\omega}_{m_{ref}} - \frac{n_p L_m}{J L_r}\psi_r i_{sq} + \frac{1}{J}T_L \tag{14.18}$$

A NTSM manifold [7] is designed as follows:

$$s_\omega = e_\omega + \gamma_1 \dot{e}_\omega^{p_1/q_1} \tag{14.19}$$

where $\gamma_1 > 0$; p_1 and q_1 are odd, and $1 < p_1/q_1 < 2$.

Theorem 14.1 *The NTSM surface (14.19) and the following control assure that finite-time convergence of the speed error dynamics (14.18):*

$$i_{sq_{ref}} = i_{sq_{eq}} + i_{sq_n} \tag{14.20}$$

$$i_{sq_{eq}} = \frac{J L_r}{n_p L_m \psi_r}\dot{\omega}_{m_{ref}} \tag{14.21}$$

$$\dot{i}_{sq_n} + T i_{sq_n} = v_\omega \tag{14.22}$$

$$v_\omega = \frac{J L_r}{n_p L_m \psi_r}\left(\frac{s_\omega \dot{e}_\omega^{p_1/q_1-1}}{\left|s_\omega \dot{e}_\omega^{p_1/q_1-1}\right|^2}|s_\omega|\left|\dot{e}_\omega^{p_1/q_1-1}\right|(k_1+\eta_1) + \frac{q_1}{\gamma_1 p_1}\dot{e}_\omega^{2-p_1/q_1}\right) \tag{14.23}$$

where $k_1 > 0$, $\eta_1 > 0$ are design parameters, and $k_1 > \left(\left|\frac{1}{J}\dot{T}_L\right| + \left|\frac{n_p L_m \psi_r}{J L_r} T i_{sq_n}\right|\right)$.

Proof The following Lyapunov function is considered $V_\omega(t) = 0.5{s_\omega}^2(t)$. We have:

$$\dot{V}_\omega(t) = s_\omega(t)\dot{s}_\omega(t) = s_\omega\left[\dot{e}_\omega + \frac{\gamma_1 p_1}{q_1}\dot{e}_\omega^{p_1/q_1-1}\ddot{e}_\omega\right]$$

$$= \frac{s_\omega \gamma_1 p_1}{q_1}\dot{e}_\omega^{p_1/q_1-1}\left[-\frac{n_p L_m}{J L_r}\psi_r \dot{i}_{sq_n} + \frac{1}{J}\dot{T}_L + \frac{q_1}{\gamma_1 p_1}\dot{e}_\omega^{2-p_1/q_1}\right]$$

$$= \frac{s_\omega \gamma_1 p_1}{q_1}\dot{e}_\omega^{p_1/q_1-1}\left[-\frac{n_p L_m \psi_r}{J L_r}v_\omega + \frac{n_p L_m \psi_r}{J L_r}T i_{sq_n} + \frac{1}{J}\dot{T}_L - \frac{q_1}{\gamma_1 p_1}\dot{e}_\omega^{2-p_1/q_1}\right]$$

$$= \frac{s_\omega \gamma_1 p_1}{q_1} \dot{e}_\omega^{p_1/q_1-1} \left[-\frac{s_\omega \dot{e}_\omega^{p_1/q_1-1}}{\left|s_\omega \dot{e}_\omega^{p_1/q_1-1}\right|^2} |s_\omega| \left|\dot{e}_\omega^{p_1/q_1-1}\right| (k_1+\eta_1) + \frac{n_p L_m \psi_r}{J L_r} T i_{sq_n} + \frac{1}{J}\dot{T}_L \right]$$

$$= \frac{\gamma_1 p_1}{q_1} \left[-|s_\omega| \left|\dot{e}_\omega^{p_1/q_1-1}\right| (k_1+\eta_1) + s_\omega \dot{e}_\omega^{p_1/q_1-1} \left(\frac{1}{J}\dot{T}_L + \frac{n_p L_m \psi_r}{J L_r} T i_{sq_{ref}} \right) \right]$$

$$\le \frac{\gamma_1 p_1}{q_1} |s_\omega| \left|\dot{e}_\omega^{p_1/q_1-1}\right| \left[-(k_1+\eta_1) + \left|\frac{1}{J}\dot{T}_L\right| + \left|\frac{n_p L_m \psi_r}{J L_r} T i_{sq_{ref}}\right| \right]$$

$$\le \frac{\gamma_1 p_1}{q_1} \eta_1 \left|\dot{e}_\omega^{p_1/q_1-1}\right| |s_\omega|$$

since $k_1 > \left(\left|\dot{T}_L/J\right| + \left|(n_p L_m \psi_r / J L_r) T i_{sq_{ref}}\right| \right)$, therefore

$$\dot{V}_\omega(t) \le -\gamma_1 (p_1/q_1) \eta_1 \left|\dot{e}_\omega^{p_1/q_1-1}\right| |s_\omega|| < 0 \quad \text{for } |s_\omega| \ne 0$$

which means that the speed error dynamics (14.18) can reach the sliding-mode surface in finite time, and then both e_ω and $\dot{e}_\omega$ can converge to zero within infinite time. This completes the proof.

14.4.2 Rotor Flux Controller Design

Define the desired rotor flux as $\psi_{r_{ref}}$ = sconst, the error between the actual rotor flux and the desired flux is e_ψ:

$$e_\psi = \psi_{r_{ref}} - \psi_r \tag{14.24}$$

The rotor flux error system can be obtained as follows according to the mathematical model (14.14):

$$\dot{e}_\psi = -\dot{\psi}_r = \frac{1}{T_r}\psi_r - \frac{L_m}{T_r} i_{sd} \tag{14.25}$$

The sliding-mode surface s_ψ is designed as the follows:

$$s_\psi = e_\psi + \gamma_2 \dot{e}_\psi^{p_2/q_2} \tag{14.26}$$

where $\gamma_2 > 0$, p_2, q_2 are odd, and $1 < p_2/q_2 < 2$.

Theorem 14.2 *The NTSM surface (14.26) and the following control assure the finite-time convergence of the rotor flux error system (14.25):*

$$i_{sd_{ref}} = i_{sd_{eq}} + i_{sd_n} \tag{14.27}$$

$$i_{sd_{eq}} = \frac{\psi_r}{L_m} \tag{14.28}$$

$$\dot{i}_{sd_n} + T i_{sd_n} = v_\psi \tag{14.29}$$

$$v_\psi = \frac{J L_r}{n_p L_m \psi_r} \left(\frac{s_\psi \dot{e}_\psi^{p_2/q_2-1}}{\left| s_\psi \dot{e}_\psi^{p_2/q_2-1} \right|^2} |s_\psi| \left| \dot{e}_\psi^{p_2/q_2-1} \right| (k_2 + \eta_2) + \frac{q_2}{\gamma_2 p_2} \dot{e}_\psi^{2-p_2/q_2} \right) \tag{14.30}$$

where $k_2 > 0$, $\eta_2 > 0$ *is the design parameter.*

Proof This follows straightforwardly from Theorem 14.1.

14.4.3 *q*-axis Current Controller Design

Define the error between the desired current in q-axis ($i_{sq_{ref}}$) and the actual current in q-axis (i_{sq}) as follows:

$$e_{sq} = i_{sq_{ref}} - i_{sq} \tag{14.31}$$

The q-axis current error system can be obtained as the follow according to the mathematical model (14.14):

$$\dot{e}_{sq} = \dot{i}_{sq_{ref}} - \dot{i}_{sq} = \dot{i}_{sq_{ref}} + \xi\omega\psi_r + \omega_1 i_{sd} + \lambda i_{sq} - K u_{sq} \tag{14.32}$$

A NTSM manifold s_{sq} is designed as the follow:

$$s_{sq} = e_{sq} + \gamma_3 \dot{e}_{sq}^{p_3/q_3} \tag{14.33}$$

where $\gamma_3 > 0$, p_3, q_3 are odd, and $1 < p_3/q_3 < 2$.

Theorem 14.3 *The NTSM surface (14.33) and the following control assure the finite-time convergence of the q-axis current error system (14.32):*

$$u_{sq} = u_{sq_{eq}} + u_{sq_n} \tag{14.34}$$

$$u_{sq_{eq}} = \left(\dot{i}_{sq_{ref}} + \xi\omega\psi_r + \omega_1 i_{sd} + \lambda i_{sq}\right)/K \tag{14.35}$$

$$\dot{u}_{sq_n} + T u_{sq_n} = v_{sq} \tag{14.36}$$

$$v_{sq} = \frac{1}{K}\left(\frac{s_{sq}\dot{e}_{sq}^{p_3/q_3-1}}{\left|s_{sq}\dot{e}_{sq}^{p_3/q_3-1}\right|^2}\left|s_{sq}\right|\left|\dot{e}_{sq}^{p_3/q_3-1}\right|(k_3+\eta_3) + \frac{q_3}{\gamma_3 p_3}\dot{e}_{sq}^{2-p_3/q_3}\right) \tag{14.37}$$

where $k_3 > 0$, $\eta_3 > 0$ are design parameters.

Proof This follows straightforwardly from Theorem 14.1 as well.

14.4.4 d-axis Current Controller Design

Define the error between the desired and the actual d-axis current as the follow:

$$e_{sd} = i_{sd_{ref}} - i_{sd} \tag{14.38}$$

The d-axis current error system can be obtained as follows according to the mathematical model (14.14):

$$\dot{e}_{sd} = i_{sd_{ref}} - \xi\frac{1}{T_r}\psi_r + \lambda i_{sd} - \omega_1 i_{sq} - K u_{sd} \tag{14.39}$$

A NTSM manifold is designed as the follow:

$$s_{sd} = e_{sd} + \gamma_4 \dot{e}_{sd}^{p_4/q_4} \tag{14.40}$$

where $\gamma_4 > 0$, p_4 and q_4 are odd, and $1 < p_4/q_4 < 2$.

Theorem 14.4 *The NTSM surface (14.40) and the following control assure the finite-time convergence of the d-axis current error system (14.39):*

$$u_{sd} = u_{sd_{eq}} + u_{sd_n} \tag{14.41}$$

$$u_{sdeq} = \left(\dot{i}_{sd_{ref}} - \xi\frac{1}{T_r}\psi_r + \lambda i_{sd} - \omega_1 i_{sq}\right)/K \tag{14.42}$$

$$\dot{u}_{sd_n} + T u_{sd_n} = v_{sd} \tag{14.43}$$

$$v_{sd} = \frac{1}{K}\left(\frac{s_{sd}\dot{e}_{sd}^{p_4/q_4-1}}{\left|s_{sd}\dot{e}_{sd}^{p_4/q_4-1}\right|^2}\left|s_{sd}\right|\left|\dot{e}_{sd}^{p_4/q_4-1}\right|(k_4+\eta_4) + \frac{q_4}{\gamma_4 p_4}\dot{e}_{sd}^{2-p_4/q_4}\right) \tag{14.44}$$

where $k_4 > 0$, $\eta_4 > 0$ are parameters to design.

Proof This follows straightforwardly from Theorem 14.1.

14.5 Numerical Simulation Test

Some simulations are carried out for an IM control system to validate the applied NTSM controllers in MATLAB-Simulink. The parameters of the IM are give as follows:

$P_N = 1.1\,kW$, $I_N = 2.8\,A$, $U_N = 380\,V$, $f_N = 50\,Hz$, $n_p = 2$, $R_s = 5.9\,\Omega$, $R_r = 5.6\,\Omega$, $L_r = 37.94\,mH$, $L_s = 0.58\,H$, $L_m = 0.55\,H$, $J = 0.021\,kg \cdot m^2$, $\psi_{ref} = 0.7\,Wb$.

And the NTSM controllers are designed with the following parameters:

$p_1 = 5$, $q_1 = 3$, $\gamma_1 = 50$, $k_1 = 5000$, $T_1 = 10$; $p_2 = 5$, $q_2 = 3$, $\gamma_2 = 30$, $k_2 = 3000$, $T_2 = 10$; $p_3 = 7, q_3 = 5, \gamma_3 = 20, k_3 = 6000, T_3 = 5$; $p_4 = 13, q_4 = 11, \gamma_4 = 10$, $k_4 = 3000$, $T_4 = 5$.

The desired speed is $20sin(4t)$ rpm and the desired rotor flux is $0.7Wb$.

The motor speed and the its tracking error are displayed in Fig. 14.3. For clear observation, the desired speed is shifted a little bit manually. It can be seen that the motor speed can track its reference fast and accurately. The rotor flux and its tracking error are shown in Fig. 14.4. It is clear that the rotor flux can track its reference value reference. The desired flux is shifted a little as well. The stator currents are displayed in Fig. 14.5. Both of the two currents track their references quickly and accurately. The control signals of the NTSM current controllers are shown in Fig. 14.6. It can be seen that the two control signals are smooth, which means that the chattering is attenuated by the applied new high-order NTSM method, and better performances of IM systems can be obtained.

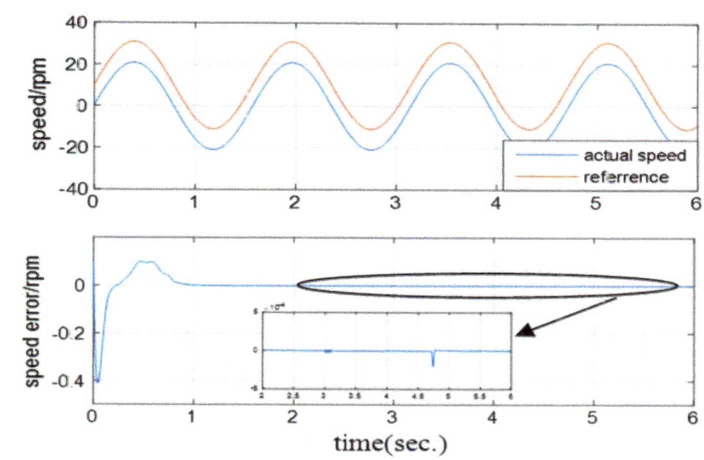

Fig. 14.3 The motor speed and its tracking error

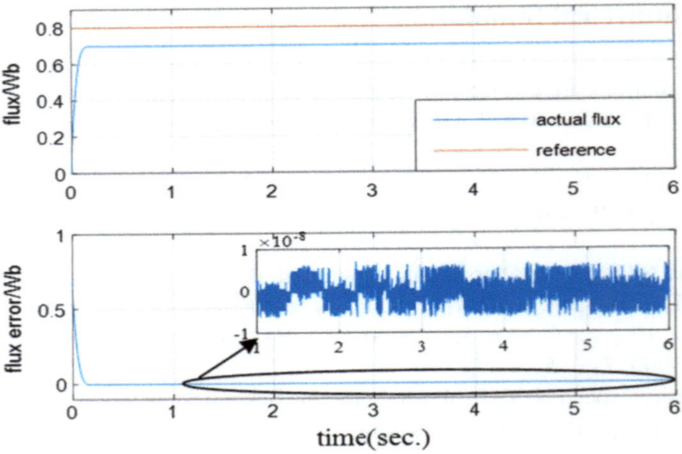

Fig. 14.4 The rotor flux and its tracking error

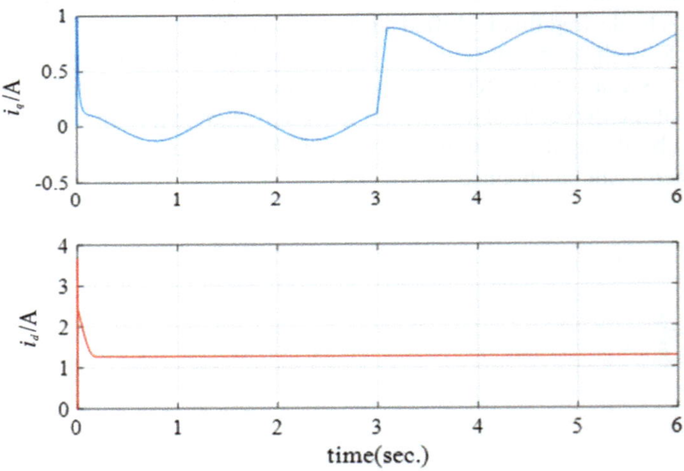

Fig. 14.5 The stator currents in dq

14.6 Conclusion

This chapter has introduced a nonsingular terminal sliding-mode control method for IM velocity servo systems. The NTSM controllers are applied into the speed, flux and current closed loop of the FOC-based IM velocity servo systems. The designed NTSM control law can suppress the chattering which exists in conventional sliding-mode control. The results of simulation have proved that the applied method is corrective and effective.

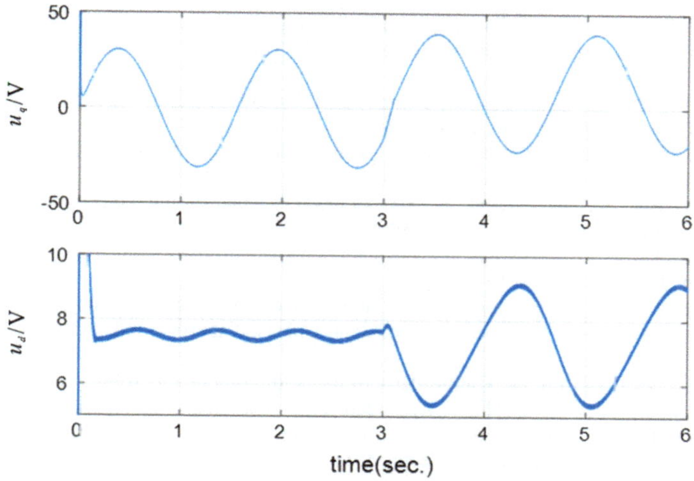

Fig. 14.6 The control signals of the NTSM current controllers

Acknowledgements This work was supported by the National Natural Science Foundation of China under Grant 61673132.

References

1. Zhou, M., Feng, Y., Yu, X.: High-order terminal sliding-mode observer for speed estimation of induction motors. In: Proceedings of the 8th IEEE Conference on Industrial Electronics and Applications, pp. 1045–1048 (2013)
2. Feng, Y., Zhou, M., Shi, H., Yu, X.: Flux estimation of induction motors using high-order terminal sliding-mode observer. In: Proceedings of 10th World Congress on Intelligent Control and Automation, pp. 1860–1863 (2012)
3. Feng, Y., Zheng, J., Yu, X., Truong, N.V.: Hybrid terminal sliding-mode observer design method for a permanent-magnet synchronous motor control system. IEEE Trans. Ind. Electron. **56**(9), 3424–3431 (2009)
4. Feng, Y., Zhou, M., Yu, X.: Sliding-Mode observer based flux estimation of induction motors. In: Proceedings of 5th International Conference on Intelligent Robotics and Applications, pp. 530–539 (2012)
5. Song, J., Lee, K.-B., Song, J.-H., Choy, I., Kim, K.-B.: Sensorless vector control of induction motor using a novel reduced-order extended Luenberger observer. In: Proceedings of Conference Record of the 2000 IEEE Industry Applications Conference, pp. 1828–1834 (2000)
6. Feng, Y., Yu, X., Han, F.: On nonsingular terminal sliding-mode control of nonlinear systems. Automatica **46**(6), 1715–1722 (2013)
7. Feng, Y., Yu, X., Man, Z.: Non-singular adaptive terminal sliding mode control of rigid manipulators. Automatica **38**(12), 2159–2167 (2002)
8. Camara, H.T., Carati, E.G., Hey, H.L., Pinheiro, H., Pinheiro, J.R., Grundling, H.A.: Speed and position servo for induction motor using robust model reference adaptive control. In: Proceedings of the 28th Annual Conference of the Industrial Electronics Society, pp. 1721–1727 (2002)

9. Feng, Y., Yu, X., Han, F.: High-order terminal sliding-mode observer for parameter estimation of a permanent-magnet synchronous motor. IEEE Trans. Ind. Electron. **60**(10), 4272–4280 (2013)
10. Lin, F.J., Wai, R.J.: Adaptive fuzzy-neural-network control for induction spindle motor drive. IEEE Trans. Energy Convers. **17**(4), 507–513 (2002)
11. Feng, Y., Yu, X., Zheng, X.: Second-order terminal sliding mode control of input-delay systems. Asian J. Control **8**(1), 12–20 (2006)
12. Feng, Y., Han, X., Wang, Y., Yu, X.: Second-order terminal sliding mode control of uncertain multivariable systems. Int. J. Control **80**(6), 856–862 (2007)
13. Feng, Y., Bao, S., Yu, X.: Design method of non-singular terminal sliding mode control systems. Control Decis. **17**(2), 194–198 (2002)
14. Zhou, M., Feng, Y.: High-order terminal sliding-mode control of uncertain systems with mismatched disturbance. In: Proceedings of 32nd Chinese Control Conference, pp. 3190–3193 (2013)
15. Zheng, J., Feng, Y., Lu, Q.: High-order terminal sliding-mode control for permanent magnet synchronous motor. Contr. Theory Appl. **26**(6), 697–700 (2009)
16. Wang, Y., Feng, Y., Lu, Q.: Design of free-chattering sliding mode control systems for permanent magnet synchronous motor. Electr. Mach. Control **2**(5), 514–519 (2008)
17. Feng, Y., Bao, S., Yu, X.: Inverse dynamics nonsingular terminal sliding mode control of two-link flexible manipulators. Int. J. Robot. Autom. **19**(2), 91–102 (2004)
18. Feng, Y., Yu, X., Man, Z.: Non-singular terminal sliding mode control and its application for robot manipulators. In: Proceedings of IEEE International Symposium on Circuits and Systems, pp. 545–548 (2001)
19. Zheng, X., Feng, Y., Bao, S.: Terminal sliding mode decomposed control of multivariable linear uncertain systems. Contr. Theory Appl. **21**(1), 11–16 (2004)

Chapter 15
Sliding Modes Control in Vehicle Longitudinal Dynamics Control

Antonella Ferrara and Gian Paolo Incremona

15.1 Introduction

The use of automatic control in ground vehicles has grown significantly over the last two decades. The main control problems considered in the literature and studied in depth, providing several solutions, are engine control, driveline control and vehicle dynamics control [31]. The problems related to vehicle dynamics control can be distinguished in longitudinal, lateral, vertical and roll dynamics control problems [57].

Several studies have been devoted to the design of traction/braking control systems, starting since the 80–90s [15]. This represents the basis to design the so-called Anti-block Breaking Systems (ABS) and belongs to the class of longitudinal control dynamics problems. It consists in enhancing the wheeled vehicle with *all terrain* capabilities in order to prevent loss of traction of the driving wheels.

As for the lateral dynamics control, this is realized by making the yaw rate of the vehicle track, at any time instant and for any road condition, the suitable reference signal. The yaw rate control allows for oversteering and understeering prevention [25, 26] and represents the basis to design the so-called Electronic Stability Program (ESP).

Vertical dynamics control includes instead active/semi-active suspension control, allowing for filtration of road irregularities for a better ride quality and car handling. In particular, active suspensions use separate actuators which can exert an independent force on the suspension. Semi-active ones can only change the viscous damping coefficient of the shock absorber.

A. Ferrara · G.P. Incremona (✉)
Dipartimento di Ingegneria Industriale e dell'Informazione, University of Pavia,
Via Ferrata 3, 27100 Pavia, Italy
e-mail: gp.incremona@gmail.com

A. Ferrara
e-mail: antonella.ferrara@unipv.it

S. Li et al. (eds.), *Advances in Variable Structure Systems and Sliding Mode Control—Theory and Applications*, Studies in Systems, Decision and Control 115, DOI 10.1007/978-3-319-62896-7_15

Finally, anti-roll-over control systems allows for road holding, vertical comfort and is the basis to design the on-hand systems meant to protect the vehicle from excessive variation of the roll angle [24]. For instance, in some commercial systems, if the system assesses the roll over risk, engine torque is lowered and some braking force is applied to one or more wheels to counteract the roll-over tendency.

Automotive subsystems present substantial nonlinearities and their models are subject to significant uncertainties and external disturbances [24]. Therefore, Sliding Mode Control (SMC) is a very appropriate and effective methodology to cope with this kind of systems [16, 58, 59]. In fact, SMC solutions have been developed since the 90 s and many proposals, among the more recent ones, have been successfully tested on prototypes [2, 12, 15, 23, 27, 29, 32].

The implementation of a feedback control requires several quantities to be available to the controller at any sampling time instant (e.g., vehicle velocity, wheel-slip, tire-road friction coefficient). Some of them are directly provided by sensors which are available on the vehicle. As for traction/braking control and lateral control, wheel speed sensors are fundamental. They consist of two hall effect sensors, a rare earth magnet and an appropriate evaluation electronics. Accelerometers/gyroscopic devices are typically used for yaw rate or roll rate measurements, while low-noise linear potentiometers are applied as suspension movement monitoring sensors. In contrast, there are quantities difficult to measure such as the longitudinal vehicle velocity and the tire-road friction coefficient. In the literature, the velocity of the vehicle has been estimated through Kalman filters, fuzzy logic, non-linear observers or custom digital filtering of the vehicle acceleration in combination with wheel speed measurements. As for the estimation of the tire-road friction coefficient, the research is still ongoing. Acoustic sensors are used to gather information on road condition by registering the acoustic waves emitted by the tires. Tire-road deformation sensors, or accelerometer glued inside the tire to estimate tire-road forces are used for evaluating the tire-road coefficient.

The problem of estimation of these quantities has been studied in recent years, also by means of sliding mode observers. In [56] a first order sliding mode observer has been designed to estimate the wheel rotational speed, which is in fact a measurable variable. The idea was to exploit the equivalent control of the observer to estimate the vehicle acceleration. This variable can then be employed in existing algorithms for vehicle speed estimation. Note that, a direct integration of vehicle acceleration cannot be performed on experimental data due to measurement noise and sensors nonidealities, such as biases and drifts, that cannot be perfectly compensated for by calibration procedures. Further, a tire-road friction estimator was proposed, which requires only the knowledge of the estimated vehicle acceleration. Other examples of use of sliding mode observers in the automotive field can be found in [4, 17, 30, 40, 43, 45, 53, 54].

In all the problems of vehicle dynamics control, the measured or estimated controlled variables are passed on to an electronic control unit (ECU) where the control algorithm is embedded. Conventional control schemes are typically of closed loop type and mainly based on PID controllers, often with feedforward components to enhance their performance. In the literature, also robust controllers relying on sliding mode generation have been proposed [1, 18–20, 28, 44, 46, 55, 61]. Rather recently, the assessment of some of these proposals through experimental tests has been performed with satisfactory results [25].

This chapter focuses on the use of SMC to solve traction control and vehicle platooning control problems. Traction control is an advanced driver assistance system which increases vehicle drivability, that is the degree of smoothness and steadiness of acceleration of an automotive vehicle, especially in difficult weather conditions allowing anti-skid braking and anti-spin acceleration.

The traction force produced by a vehicle is strongly influenced by road conditions. This is why it is necessary to design a robust traction force controller taking into account the time-varying tire/road interaction. As a further requirement, the designed control law has to prevent the generation of vibrations which could increase the discomfort. The major design requirement for the controllers is to make the wheel slip ratio follow a desired value, while guaranteeing that the control is stabilizing.

Platooning of vehicles can be useful in contexts where the air drag reduction can optimize the fuel consumption and improve safety, such as in heavy trucks formations in highways [39]. Even vehicle platooning can be realized through the solution of a traction control problem [34] and therefore by applying SMC techniques [21].

The present chapter is organized as follows. The considered vehicle model is introduced and the traction control problem is first discussed. A slip ratio SMC scheme is described and simulation results on a realistic scenario are illustrated. Then, the vehicle platooning control problem is described and a SMC scheme is proposed based on a Suboptimal Second Order Sliding Mode (SSOSM) control. It is also assessed in simulation, putting into evidence its capability of enhancing the vehicle platooning even in presence of modelling uncertainties and disturbances, typical of a real scenario.

Note that the present chapter is mainly based on [3, 4, 47] and the plenary talk given by A. Ferrara at the 14th International Workshop on Variable Structure System (VSS 2016).

15.2 The Vehicle Model

In order to suitably design a control algorithm for a vehicle subsystem, one has to specify which model of the vehicle is appropriate to capture the relevant dynamical effects. In the literature [24, 42], on the basis of the control problem of interest, different models were developed. The most simplified is the 2-degrees of freedom (DoF) model in which lateral and yaw motions are considered, and the presence of longitudinal accelerations and slip effects is assumed negligible. The latter require a 3-DoF model, while, if braking forces on handling maneuvers, wheel spin and rotational effects are needed, a 5-DoF model has to be used. A model with 8-DoF is instead appropriate when rolling motions and no symmetry is assumed between right and left sides of the vehicle. Typically, in the context of traction control design,

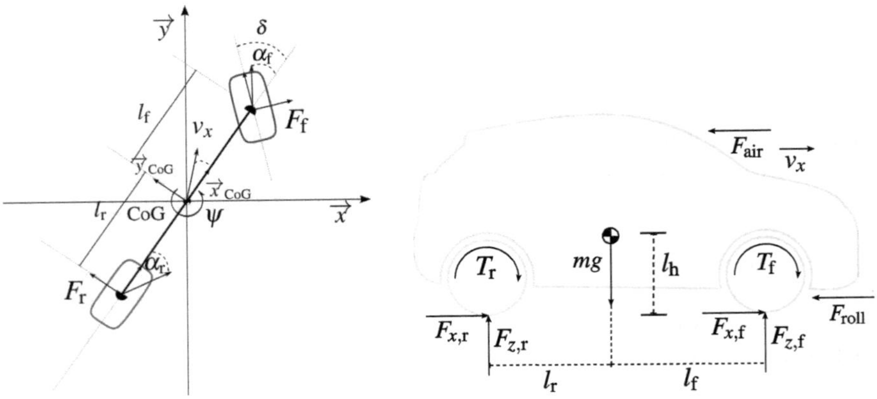

(a) single-track model (b) vehicle parameter

Fig. 15.1 Vehicle model. **a** Single-track model. **b** Vehicle parameters

the so-called bicycle single-track model [24] is adopted (see Fig. 15.1 where also the vehicle parameters are reported).

More specifically, the bicycle single-track model adopted in this chapter describes the longitudinal dynamics of the vehicle through the following equations

$$\begin{cases} m\dot{v}_x = 2\left[F_{x,\mathrm{f}}(\lambda_\mathrm{f}) + F_{x,\mathrm{r}}(\lambda_\mathrm{r})\right] - F_{\mathrm{loss}_i}(v) & (15.1\mathrm{a}) \\ J_\mathrm{f}\dot{\omega}_\mathrm{f} = T_\mathrm{f} - r_\mathrm{f} F_{x,\mathrm{f}}(\lambda_\mathrm{f}) & (15.1\mathrm{b}) \\ J_\mathrm{r}\dot{\omega}_\mathrm{r} = T_\mathrm{r} - r_\mathrm{r} F_{x,\mathrm{r}}(\lambda_\mathrm{r}) & (15.1\mathrm{c}) \\ F_\mathrm{loss}(v_x) = F_\mathrm{air}(v_x) + F_\mathrm{roll} = c_x v_x^2 \mathrm{sign}(v_x) + f_\mathrm{roll} mg & (15.1\mathrm{d}) \\ F_{x,\mathrm{f}} = \mu_\mathrm{f}(\lambda_\mathrm{f}) F_{z,\mathrm{f}} & (15.1\mathrm{e}) \\ F_{x,\mathrm{r}} = \mu_\mathrm{r}(\lambda_\mathrm{r}) F_{z,\mathrm{r}} & (15.1\mathrm{f}) \\ F_{z,\mathrm{f}} = \dfrac{l_\mathrm{r} mg - l_\mathrm{h} m\dot{v}_x}{2(l_\mathrm{f} + l_\mathrm{r})} & (15.1\mathrm{g}) \\ F_{z,\mathrm{r}} = \dfrac{l_\mathrm{f} mg + l_\mathrm{h} m\dot{v}_x}{2(l_\mathrm{f} + l_\mathrm{r})} & (15.1\mathrm{h}) \end{cases}$$

where v_x is the longitudinal velocity, $\omega = [\omega_\mathrm{f},\ \omega_\mathrm{r}]^T$ is the wheel angular velocity vector, $T = [T_\mathrm{f},\ T_\mathrm{r}]^T$ is the input torque acting on the wheels, F_x is the traction force on a wheel, F_z is the normal force on a wheel, F_air is the air drag, and F_roll is the rolling resistance. Note that the subscripts "f" and "r" stand for "front" and "rear", respectively. Moreover, m is the mass, c_x is the longitudinal wind drag coefficient, f_roll is the rolling resistance coefficient, J is the wheel moment of inertia, $\mu_\iota \in [0,\ 1]$ is the tire-road friction coefficient, $\iota = \{\mathrm{f},\ \mathrm{r}\}$, $r = [r_\mathrm{f},\ r_\mathrm{r}]^T$ is the wheels radius, l_f is the distance from the front axle to the center of gravity (CoG), l_r is the distance from the CoG to the rear axle, and l_h is the vertical distance to the CoG.

15.3 The Traction Force Control Problem

System (15.1) is expressed as a function of the slip ratio $\lambda = [\lambda_f, \lambda_r]^T$ which is defined as follows

$$\lambda_\iota = \frac{\omega_\iota r_\iota - v_x}{\max(\omega_\iota r_\iota, v_x)}, \quad \iota = \{f, r\} . \tag{15.2}$$

More precisely, two different cases can be distinguished. In case of acceleration, i.e., $\omega_\iota r_\iota > v_x$ and $\omega_\iota \neq 0$, the slip ratio results in being

$$\lambda_{a,\iota} = \frac{\omega_\iota r_\iota - v_x}{\omega_\iota r_\iota} , \tag{15.3}$$

with a corresponding slip ratio dynamics given by

$$\dot{\lambda}_{a,\iota} = -\frac{\dot{v}_x}{r_\iota \omega_\iota} - \frac{v_x F_{x,\iota}}{J_\iota \omega_\iota^2} + \frac{v_x}{J_\iota r_\iota \omega_\iota^2} T_\iota . \tag{15.4}$$

Analogously, in case of breaking, i.e., $\omega_\iota r_\iota < v_x$ and $v_x \neq 0$, one has that

$$\lambda_{b,\iota} = \frac{\omega_\iota r_\iota - v_x}{v_x} , \tag{15.5}$$

with the slip ratio expressible as

$$\dot{\lambda}_{b,\iota} = -\frac{r_\iota \omega_\iota \dot{v}_x}{v_x^2} - \frac{r_\iota^2 F_{x,\iota}}{J_\iota v_x} + \frac{v_x}{J_\iota r_\iota \omega_\iota^2} T_\iota . \tag{15.6}$$

Note that the subscripts "a" and "b" stand for "acceleration" and "breaking", respectively.

In the considered model (15.1)–(15.6), roll and yaw moments, lateral and vertical motions, brake, throttle, steering actuators and manifold dynamics are neglected. Note that, although taking into account more DoF in the model may increase its accuracy, it makes the controller design more complex, so that the control designer chooses the simplest model sufficient to describe the dynamical aspects of interest. As for the c_x coefficient, in order to verify the effectiveness of the proposed control schemes in presence of uncertainties, this is assumed to vary over time. Also the mass of the vehicle and the tire-road friction coefficient are time-varying and represent the unmatched uncertainties affecting the system. Moreover, relying on the control variables T_f and T_r, the total braking torque T_{brake}, and the engine torque exerted at the level of the driving shaft T_{shaft} can be calculated. For instance, for a front-wheel-driven-car, these torques can be computed as

$$\begin{cases} T_f = 0.5 T_{shaft} - 0.3 T_{brake} \\ T_r = -0.2 T_{brake} . \end{cases} \tag{15.7}$$

Table 15.1 Tire/road interaction models published in the literature [37]

Year	Model name	Properties	Features
	Piecewise linear model	Empirical	Cannot accurately fit curves
			Easy to identify
1993	Burckhardt model	Semi-empirical	Cannot accurately fit curves
			Has some revised formula
			Easy to identify
1994	Rill model	Semi-empirical	Easy to identify
1987	Magic formula	Semi-empirical	Cannot accurately fit curves
			Has some revised formula
			Can employ different factors
1977	Dahl model	Analytical	Can describe Coulomb friction
			Can get smooth transition around zero velocity
1991	Bliman–Sorine model	Analytical	Can capture the Stribeck effect
1995	LuGre model	Analytical	Can combine pre-sliding and sliding

T_{shaft} and T_{brake} are the reference signals for the throttle angle controller and for the brake controller, respectively [35].

Finally, as for the traction force, this also depends on the tire/road interaction which depends in turn on the road conditions. In the literature several models have been developed (see, for instance, those reported in Table 15.1). In this chapter we refer to the so-called "Magic Formula" by Bakker-Pacejka, which models the traction force as a function of the slip ratio, parametrized by the tire-road friction coefficient μ, as follows

$$F_x = \mu \cdot f_t(\lambda, F_z) \ . \tag{15.8}$$

The coefficient μ depends on the road conditions and the traction force results in being described by a family of plots, each of them corresponding to a different road condition (see Fig. 15.2a). It is apparent that the estimation of the road friction coefficient results in being a very determinant issue in traction control.

15.3.1 Fastest Acceleration/Deceleration Control (FADC) Problem

The classical way to design traction control is to solve the so-called Fastest Acceleration/Deceleration Control (FADC) problem. More specifically, the control objective for the FADC problem is to maximize the generated traction force. To this end, also shown in Fig. 15.2b, the desired slip ratio λ_d must be chosen as the abscissa of the extremal value of the $\lambda - F_x$ curve corresponding to the current road conditions identified by the estimated μ.

Thus, the FADC scheme must accomplish the following task:

- to estimate on-line the tire-road friction coefficient μ to identify the current $\lambda - F_z$ curve given the road condition;
- to calculate the desired slip ratio λ_d as the abscissa of the extremal value of the actual $\lambda - F_x$ curve;
- to design a control law that makes the current slip ratio λ track the desired value λ_d.

Making reference to the single-track model, this procedure must be followed for the front and rear wheel axles. Note that, the original control objective (i.e., that of solving the FADC problem) has been transformed into a slip ratio control problem, the objective of which is to steer the slip rate error, σ_ι, $\iota = \{f, r\}$, to zero, i.e.,

$$\sigma_\iota = \lambda_{\iota_d} - \lambda_\iota = 0 . \quad (15.9)$$

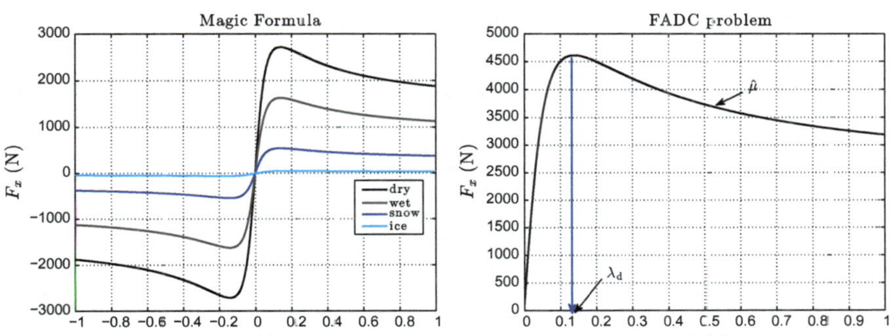

(a) $\lambda - F_x$ curves as a function of μ (b) Determination of λ_d

Fig. 15.2 Bakker-Pacejka model. **a** Dependence of μ on road conditions. **b** Desired reference slip ratio λ_d

15.4 Design of Sliding Mode Slip Controller

In this section two SMC algorithms will be discussed to solve the aforementioned traction control problem taking into account the vehicle model (15.1)–(15.6). SMC is particularly effective in case of systems that are affected by hard uncertainties. More specifically, SMC is able to make the controlled systems robust in front of the so-called matched uncertainties [16, 58], i.e., uncertainties acting on the same channel of the control variable. Moreover, the controlled system present an "order reduction" and the desired stability properties can be "assigned" to the controlled system in sliding mode by suitably selecting the so-called sliding manifold. So, in order to apply the SMC methodology to solve the considered traction control problem, the so-called sliding variable needs to be designed. In the following subsection, some preliminaries on the design of SMC for the considered vehicle systems will be recalled.

15.4.1 Preliminaries on Sliding Mode Control Design

Let ρ be the relative degree of the system, i.e., the minimum order of the time derivative of the sliding variable, $\sigma_\iota^{(\rho)}$, in which the control input T_ι explicitly appears. Now, compute the first and the second time derivative of the sliding variable, so that, by posing $\xi_{1,\iota} = \sigma_\iota$ and $\xi_{2,\iota} = \dot{\sigma}_\iota$, the so-called auxiliary system can be written as

$$\begin{cases} \dot{\xi}_{1,\iota}(t) &= \xi_{2,\iota}(t) \\ \dot{\xi}_{2,\iota}(t) &= f_{\varsigma,\iota}(t) + g_{\varsigma,\iota}(t)w(t), \ \ \varsigma = \{\mathrm{a}, \mathrm{b}\}, \ \iota = \{\mathrm{f}, \mathrm{r}\} \ , \end{cases} \tag{15.10}$$

where $w(t) = \dot{T}_\iota$ is the auxiliary control variable, while the function $f_{\varsigma,\iota}(\cdot)$ and $g_{\varsigma,\iota}(\cdot)$ are

$$\begin{aligned} f_{\mathrm{a},\iota}(t) &= \ddot{\lambda}_{\mathrm{d},\iota} + \frac{\ddot{v}_x}{r_\iota\omega_\iota} - \frac{\dot{v}_x\dot{\omega}_\iota}{r_\iota\omega_\iota^2} + \frac{\dot{v}_x F_{x,\iota}}{J_\iota\omega_\iota^3} + \frac{\dot{v}_x \dot{F}_{x,\iota}}{J_\iota\omega_\iota^2} + \\ &\quad - \frac{2v_x F_{x,\iota}\dot{\omega}_\iota}{J_\iota^2\omega_\iota^3} - \frac{v_x T_\iota \dot{\omega}_\iota}{J_\iota^2 r_\iota \omega_\iota^3} - \frac{\dot{v}_x T_\iota}{J_\iota^2 r_\iota \omega_\iota^2} \\ f_{\mathrm{b},\iota}(t) &= \ddot{\lambda}_{\mathrm{d},\iota} + \frac{r_\iota\ddot{\omega}_\iota}{v_x} + \frac{r_\iota\omega_\iota\ddot{v}_x}{v_x^2} + \frac{r_\iota^2\dot{F}_{x,\iota}}{J_\iota v_x} - \frac{r_\iota^2 F_{x,\iota}\dot{v}_x}{J_\iota v_x^2} + \\ &\quad - \frac{2r_\iota\omega_\iota\dot{v}_x}{v_x^2} - \frac{v_x T_\iota\dot{\omega}_\iota}{J_\iota^2 r_\iota\omega_\iota^3} - \frac{\dot{v}_x T_\iota}{J_\iota^2 r_\iota\omega_\iota^2} \\ g_{\varsigma,\iota}(t) &= -\frac{v_x}{J_\iota r_\iota \omega_\iota^2} \ . \end{aligned} \tag{15.11}$$

Since velocities are assumed to be always positive and physical limits exist, such as the limit characteristic curves of the torques which the engine can transfer to the wheels, it is assumed that functions $f_{\varsigma,\iota}(\cdot)$ and $g_{\varsigma,\iota}(\cdot)$ are bounded, i.e.,

$$|f_{\varsigma,\iota}(t)| \leq F \tag{15.12}$$

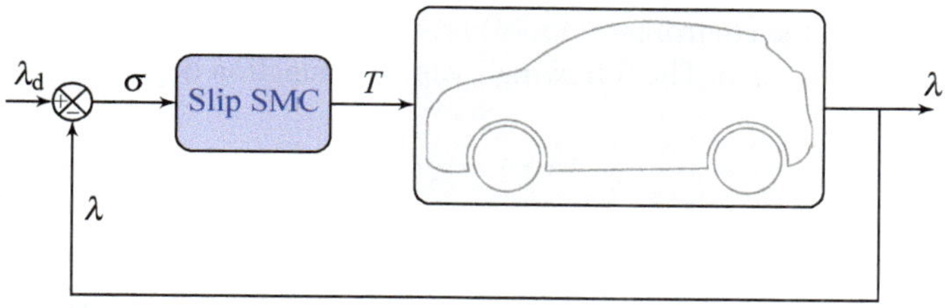

Fig. 15.3 SMC slip control scheme

$$-G_{\max} \leq g_{\varsigma,\iota}(t) \leq -G_{\min} < 0 \quad (15.13)$$

where F, $G_{\min}$ and $G_{\max}$ are positive constants, which in practical cases can be estimated and are therefore assumed known. Note that, these bounds depend on quantities which are typically provided by the manufacturer of the vehicle. In alternative, the bounds are estimated relying on data collection and analysis, through a trial-and-error procedure.

At this point, we are in a position to introduce the sliding mode control strategies, making reference to the proposed slip control scheme illustrated in Fig. 15.3. In the following subsections the design of a First Order Sliding Mode (FOSM) control [58], and of a Suboptimal Second Order Sliding Mode (SSOSM) control [5, 6, 8–10] are illustrated.

15.4.2 First Order Sliding Mode (FOSM) Control

The first strategy discussed in this chapter is the classical FOSM control [58]. Given the choice of the sliding variable (15.9), the relative degree results in being $\rho = 1$ so that a FOSM naturally applies. The control law in this case is

$$T_\iota(t) = -U_{\iota,\max}\text{sign}(\sigma_\iota(t)), \ \iota = \{\text{f}, \text{r}\} \quad (15.14)$$

where the control parameter $U_{\iota,\max}$ is a positive constant chosen so as to fulfill the so-called "η-reachability condition", i.e.,

$$\sigma_\iota(t)\dot{\sigma}_\iota(t) \leq -\eta|\sigma_\iota(t)| \quad (15.15)$$

with η being a positive constant depending on the bound on the uncertain terms and on $U_{\iota,\max}$. The "reachability problem" is effectively the sufficient condition in order to guarantee that an ideal sliding motion takes place. The sliding surface must be at least locally attractive and the trajectories of $\sigma_\iota(t)$ must be directed towards it, i.e., in some domain $\Omega \subset \mathbb{R}^n$ it yields

$$\begin{cases} \lim_{\sigma_\iota \to 0^+} \dot{\sigma}_\iota < 0 \\ \lim_{\sigma_\iota \to 0^-} \dot{\sigma}_\iota > 0 \, . \end{cases} \quad (15.16)$$

As proved in [58], the control law (15.14) is able to guarantee the finite time convergence of the error to zero. The following Lyapunov function depending on σ_ι can be selected

$$V(\sigma_\iota) = \frac{1}{2}\sigma_\iota^2 . \tag{15.17}$$

In order to compute the "reaching time" t_r [50], in which the sliding variable becomes equal to zero, deriving (15.17) with respect to time and integrating from 0 to t_r, one has

$$\int_0^{t_r} \frac{1}{2}\frac{d}{dt}\sigma_\iota^2 \le -\int_0^{t_r} \eta|\sigma_\iota| . \tag{15.18}$$

Finally, it holds that

$$|\sigma_\iota(t_r)| - |\sigma_\iota(0)| \le -\eta t_r , \tag{15.19}$$

and the time necessary to reach the sliding surface, $\sigma_\iota(t_r) = 0$, can be calculated as

$$t_r \le \frac{|\sigma_\iota(0)|}{\eta} . \tag{15.20}$$

However, the main difficulty in applying this approach to solve the slip control problem is the discontinuity of the control variable which can cause the so-called chattering phenomenon [11, 22, 36], which may be hardly acceptable in practice. In the next section this will be better clarified in view of simulation results.

15.4.3 Suboptimal Second Order Sliding Mode (SSOSM) Control

Among the solution proposed in the literature to attenuate the chattering phenomenon, Higher Order Sliding Mode (HOSM) control [7] may represent an effective solution [52]. The SSOSM control discussed in this chapter is a particular case of HOSM control (see, for instance, [7, 14] for other second order sliding mode algorithms). Given the auxiliary system (15.10), in which the relative degree is artificially increased by introducing the auxiliary control variable w and bounds (15.12)–(15.13) hold, according to a "Bang-Bang" control principle, the following "switching line" (see Fig. 15.4a) is considered

$$\xi_{1,\iota}(t) + \frac{1}{2}\frac{\xi_{2,\iota}(t)|\xi_{2,\iota}(t)|}{W_{\iota,\max}} = 0 \tag{15.21}$$

where $W_{\iota,\max}$ is the positive control gain. Then, the control law can be expressed as

$$T_\iota(t) = -\int_{t_0}^{t} \alpha_\iota W_{\iota,\max}\,\mathrm{sign}\left(\xi_{1,\iota}(\zeta) - \tfrac{1}{2}\xi_{\max}\right) d\zeta , \tag{15.22}$$

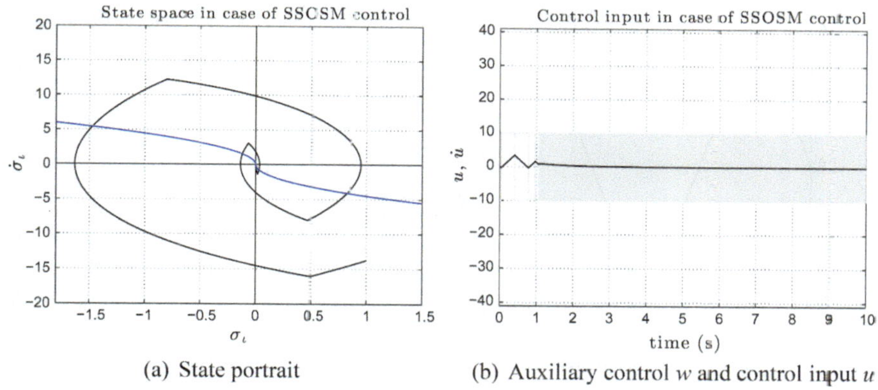

Fig. 15.4 SSOSM control. **a** Auxiliary system state portrait in case of SSOSM control. **b** Auxiliary discontinuous control and continuous control input

where $\xi_{\max}$ is the last local minimum or maximum of the sliding variable, while the control parameters $\alpha_\iota = \alpha_\iota^*$ and $W_{\iota,\max}$ are chosen such that

$$W_{\iota,\max} > \max\left(\frac{F}{\alpha_\iota^* G_{\min}}; \frac{4F}{3G_{\min} - \alpha_\iota^* G_{\max}}\right) \quad (15.23)$$

$$\alpha_\iota^* \in (0, 1] \cap \left(0, \frac{3G_{\min}}{G_{\max}}\right). \quad (15.24)$$

Note that, the SSOSM algorithm requires the control $w(t) = \dot{T}_\iota(t)$ to be discontinuous. Yet, the control actually fed into the plant is continuous (see Fig. 15.4b), which is highly appreciable in case of mechanical plants.

Moreover, in [6] it has been proved that, under constraints (15.23), the convergence of the auxiliary system trajectory to the origin takes place in a finite time. Observing the state portrait of the auxiliary system on the phase plane illustrated in Fig. 15.4a, it is apparent that the control law (15.22) implies a contraction property of the extremal values of the sliding variable so that the slip rate error and its first time derivative are steered to zero. More specifically, a sequence of states with coordinates $(\xi_{\max_i}, 0)$ featuring the following contraction

$$|\xi_{\max_{i+1}}| < |\xi_{\max_i}|, \ i = 1, 2, \cdots \quad (15.25)$$

is generated and the convergence of the system trajectory to the origin of the plane takes place in a finite time.

Moreover, an important advantage of the SSOSM control algorithm is that the knowledge of the first time derivative of the sliding variable is not required, but only the computation of its extremal values, for instance through the methods described in [9].

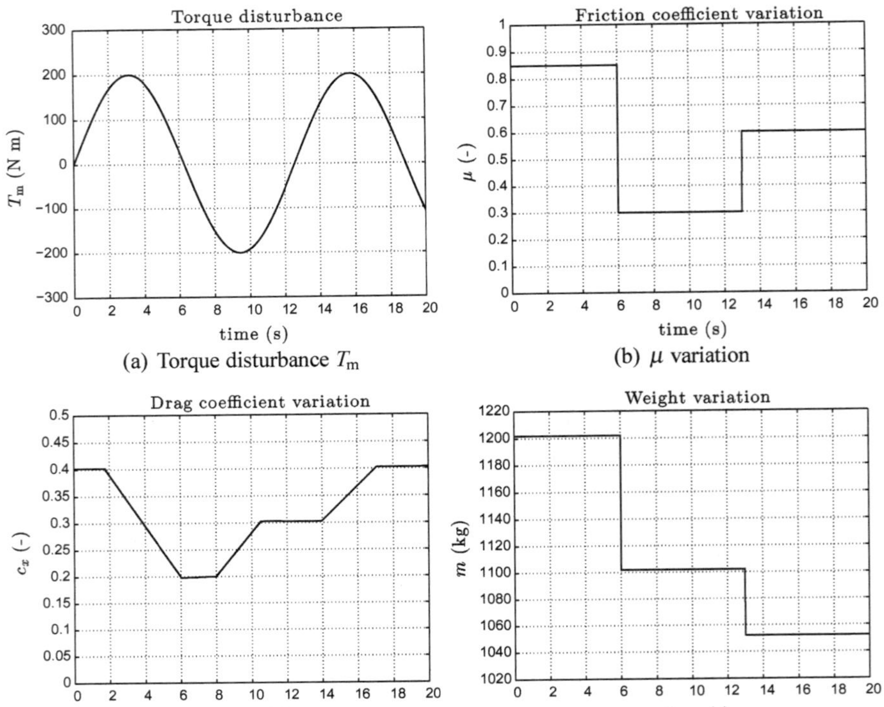

(a) Torque disturbance T_{m} (b) μ variation (c) c_x variation (d) mass m variation

Fig. 15.5 Matched and parameters uncertainties affecting the vehicle. **a** Sinusoidal torque disturbance. **b** Step friction coefficient variation. **c** Ramp drag coefficient variation. **d** Step vehicle weight variation

15.5 Simulation Results on Traction Control

In this section the sliding mode controllers previously presented are evaluated in a step response test, where the wheel-slip is regulated to a fixed value.

A PI controller is also used in the same test for the sake of comparison. It is assumed that the maximum traction/braking force is obtained for $\lambda_{\mathrm{d},\iota} = \pm 0.2$. Note that the sign corresponds to the sign of the driver's torque demand, which is considered as a matched disturbance.

The presented controllers are evaluated on a more realistic model of the vehicle dynamics, taking into account two different conditions:

1. matched uncertainties;
2. matched and parameters uncertainties.

Matched disturbances T_{m} are always included, as their rejection is the main advantage of using sliding mode controllers. The sinusoidal torques induce an acceleration

Table 15.2 Vehicle parameters considered in simulation

Parameter	Value
g	$9.81\,\mathrm{ms}^{-2}$
$m(0)$	$1202\,\mathrm{kg}$
J_{f}	$1.07\,\mathrm{kg\,m}^2$
J_{r}	$1.07\,\mathrm{kg\,m}^2$
c_x	0.4
f_{roll}	0.013
r_{f}	$0.32\,\mathrm{m}$
r_{r}	$0.32\,\mathrm{m}$
l_{f}	$1.15\,\mathrm{m}$
l_{r}	$1.45\,\mathrm{m}$
l_{h}	$0.65\,\mathrm{m}$
$v_x(0)$	$30\,\mathrm{ms}^{-1}$
$\lambda_{\mathrm{f}}(0)$	0.01
$\lambda_{\mathrm{r}}(0)$	0.01

or a braking phase on each axle, depending on the torque sign. With the purpose of showing the controllers robustness against unmatched disturbances, time varying uncertainties on vehicle mass, drag coefficient and road friction coefficient are taken into account, as shown in Fig. 15.5. The vehicle performs an acceleration maneuver for 20 s starting from velocity of $30\,\mathrm{ms}^{-1}$. All the other simulation data are reported in Table 15.2. Since results on front and rear axles are almost identical and the only differences affecting the respective dynamics are the geometric parameters introduced in Eqs. (15.1g) and (15.1h), only the results relevant to the front axle are shown. All controllers parameters have been tuned in order to obtain an adequate response for each algorithm.

In Fig. 15.6 the step response of the controlled systems obtained including in the loop the considered controllers is shown, together with the slip rate error, the front torque and the corresponding shaft torque. It can be observed that FOSM has strong oscillations, which do not vanish at all. The PI controller shows the highest overshoot and longest settling time, but nevertheless they converge to the reference value at steady state. In Fig. 15.6, it is also possible to appreciate the main advantage of using a SSOSM control algorithm in terms of torques fed into the plant. As expected, the SSOSM control signal is smooth and comparable with that obtainable through the PI controller, while the FOSM is discontinuous and scarcely applicable in practice. It can be seen that both FOSM and SSOSM controllers have a good performance versus the matched uncertainties, as it is confirmed by the Root Mean Square (RMS) error values reported in Table 15.3 (e_{RMS}). Moreover, Table 15.4 reports also the values of the control effort, E_{c}, for all controllers. A graphic rendering of the performance indices is instead reported in Fig. 15.7. The performance indices reported in Tables 15.3 and 15.4 help us assessing the performance of the different controllers considered in this chapter. The FOSM control is the most aggressive solution and surely the

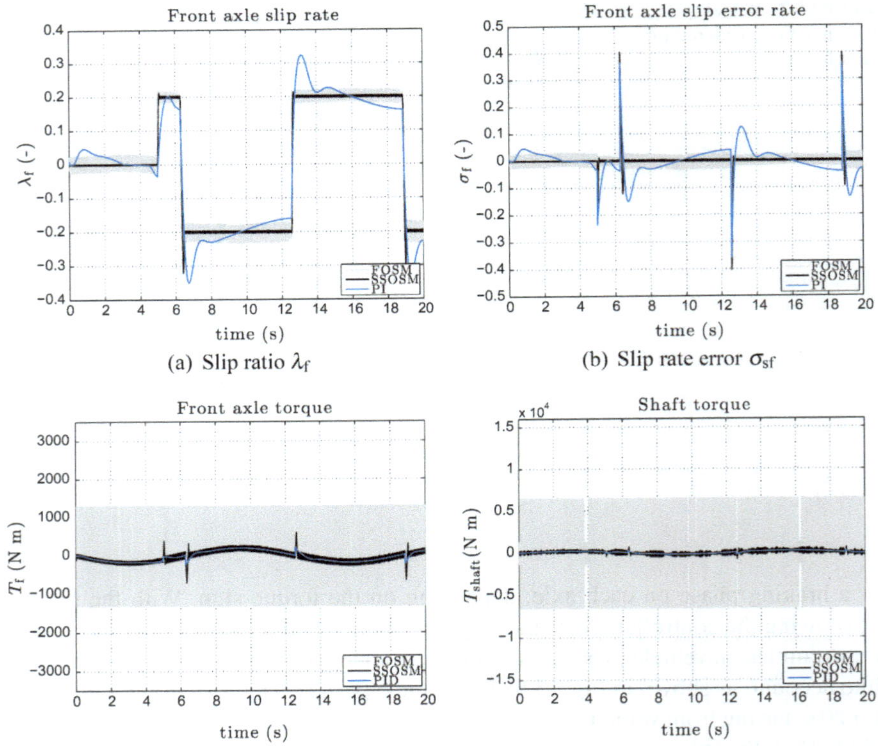

(a) Slip ratio λ_f (b) Slip rate error σ_{sf} (c) Torque T_f (d) Shaft torque T_{shaft}

Fig. 15.6 Step response of the vehicle in presence of matched and parametric uncertainties when FOSM, SSOSM and PI are used. **a** Front slip ratio. **b** Slip rate error. **c** Front torque. **d** Shaft torque

Table 15.3 RMS slip rate error

Test	Axle	FOSM	SSOSM	PI
1	front	0.0217	0.0364	0.0534
	rear	0.0215	0.0359	0.0540
2	front	0.0216	0.0365	0.0553
	rear	0.0221	0.0357	0.0554

most effective in zeroing the slip rate error. It ensures the minimum e_{RMS} in all the considered conditions, while at the same time it has a control signal, which is larger than those of all other controllers by one order of magnitude. This fact could make such a controller a non feasible solution, once the actuators limitations and dynamics are taken into account. Instead, SSOSM represents a valid solution, able to guarantee good tracking results with control actions really usable even in field implementations.

Table 15.4 RMS value of control signal

Test	Axle	FOSM	SSOSM	PI
1	front	1300	146.1	133.9
	rear	1300	150.4	138.2
2	front	1300	150.8	138.4
	rear	1300	153.4	140.9

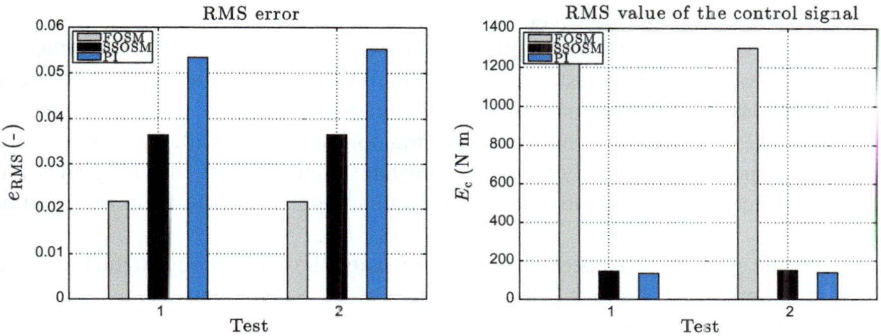

(a) RMS error e_{RMS} (b) RMS value E_c

Fig. 15.7 Graphic rendering of the computed performance indices for the front axle. **a** RMS error. **b** RMS value of the control signal

15.6 Control of a Platoon of Vehicles

The previous traction control strategies could be exploited also in formation control. A line formation of wheeled vehicles is called "platoon". Several well-known studies have focused on the impact of platooning on traffic flow regularization and road safety enhancement [13, 33, 41]. More recently, some works have been devoted to the evaluation of the advantage of vehicles platooning in terms of fuel consumption, this especially in the case of freight transportation [39].

As for traction control, the performances of a "platooning control system" are strongly influenced by road conditions. This implies that the "platooning control problem" has to be solved in a robust way. As a further requirement, also in this case, the designed control law has to prevent the generation of vibrations induced by the controller, as well as guarantee the so-called "string stability" [62].

Note that, in this framework, the so-called Cruise Control (CC) system is a control strategy devoted to track the desired velocity imposed by the driver, while the Adaptive CC (ACC) is a control strategy based on the velocity of the preceding vehicle and on the measure of the relative distance between the vehicle and its predecessor. Thus, the latter is the strategy used by all the vehicles of a platoon, apart from the leader vehicle which is controlled via a standard CC. Note that, standard devices, such as radars, can be adopted in order to capture the position of the preceding vehicle, which is assumed measurable in the following.

The main objective that an ACC vehicle has to attempt is to maintain a desired spacing, called "spacing policy", with respect to the preceding vehicle [47]. Typically, the spacing policy is a constant distance or a function of the vehicle velocity or of the relative velocity between the controlled vehicle and the preceding one. Furthermore, the spacing policy, according to which the control law is designed, plays a fundamental role in the determination of the vehicle safety, string stability, traffic flow stability and traffic flow capacity.

In the literature, several spacing policies have been proposed [47]. The simplest choice is a constant safety distance with a suitable communication system in order to provide the needed information about the leading vehicle of the platoon to the followers [49]. However, the most used spacing policy is the so-called Constant Time Gap (CTG) policy. Differently from the constant spacing policy, in CTG policy the safety distance tracking can be obtained without any inter-vehicular communication [48].

As for the stability properties guaranteed by using the CTG policy, in the literature there are conflicting opinions [38, 51]. More precisely, the controversy was about the stability for the traffic flow (under certain conditions) within which the platoon of vehicles moves. The claim was, in particular, that an unattenuated upstream propagation of disturbances may occur when a density perturbation is introduced into the traffic flow [60]. In fact, the work [51] shows that the traffic flow obtained on a highway is unstable when all vehicles on the highway use the CTG policy. In contrast, [38] states that the consequent traffic flow obtained with the CTG spacing policy is stable.

The focus of this chapter is on longitudinal control of vehicles in platoon, such that the spacing error between vehicles is steered to zero. The previously discussed SSOSM control algorithm is conveniently used to this end, as clarified in the following, relying on a spacing policy of CTG type. Note that string stability considerations are not included in this chapter, but they can be easily made using standard argumentations (see for instance [62]).

15.7 The Vehicle String Model

Consider a wirelessly interconnected vehicle platoon as schematically represented in Fig. 15.8. Note that, in the following, the subscript i will indicate the ith vehicle of the considered platoon, and it will be assumed that each vehicle can be adequately modeled by system (15.1). The considered platooning problem consists in making each vehicle of the string follow its predecessor while maintaining a pre-specified safety distance. Solving the platooning problem, even in presence of possible changes of road conditions, requires to re-formulate the problem as a traction control problem and take again into account the tire/road interaction model.

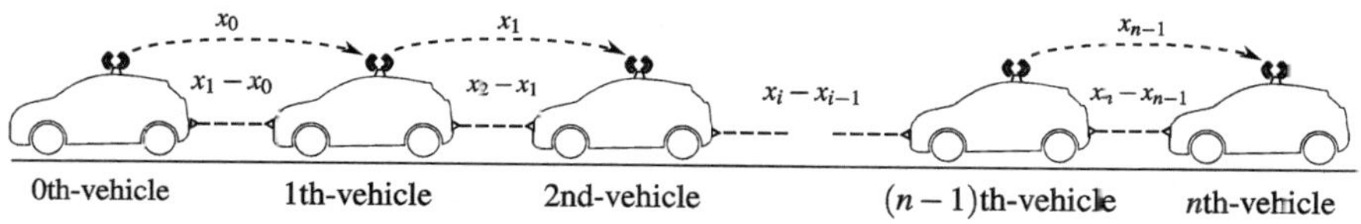

Fig. 15.8 Schematic representation of a string of n followers and the leader

In this chapter, as mentioned, the safety distance is determined according to the CTG policy, as it is frequently done in ACC design. More specifically, such a policy is given by

$$S_i(v_{x_i}(t)) = L + hv_{x_i}(t) \tag{15.26}$$

where S_i denotes the desired inter-vehicle spacing of the ith vehicle, $i = 1, 2, \ldots, n$, L is a constant value depending on the vehicles length, h is the so-called headway time, v_{x_i} is the velocity of the ith vehicle. The corresponding spacing error is given by

$$\delta_i(t) = \varepsilon_i(t) + L + hv_{x_i}(t) \tag{15.27}$$

where $\varepsilon_i = x_i - x_{i-1}$ is the relative distance between the ith vehicle and its predecessor. Then, assume that the longitudinal dynamics of the ith vehicle can be approximated by the following first order dynamics

$$\tau_a \ddot{v}_{x_i} + \dot{v}_{x_i} = \dot{v}_{x_{i_d}} \tag{15.28}$$

where τ_a is the time constant characterizing the actuator dynamics, while $\dot{v}_{x_{i_d}}$ is the desired value of acceleration to be attained in order to make the vehicle platoon string stable, i.e., to steer to zero the spacing error in steady-state [62].

15.8 The Proposed Control Scheme

The control scheme designed in this chapter is shown in Fig. 15.9. It consists of two nested control loops. As for the outer loop, on the basis of the spacing error, the controller determines the desired traction force. Relying on this latter and on the current tire/road friction coefficient, the desired slip ratios are calculated. These are used as references for the inner control loop. This latter has the objective of attaining the desired slip ratios by acting on the torques at the front and rear wheels, as illustrated in the previous sections.

15.8.1 Sliding Mode Longitudinal Controller

The longitudinal external controller has the role to generate the reference value of the slip ratio for the inner loop, while regulating to zero the spacing error. Analogously

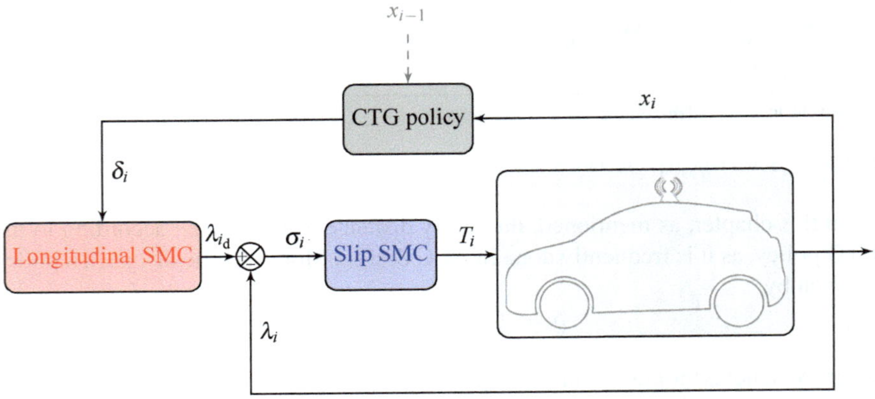

Fig. 15.9 SMC scheme for vehicles platoon

to the inner loop, the sliding variable for the ith vehicle is selected as the spacing error (15.27). In this second case, the effective input of the system is an acceleration expressed as $\dot{v}_{x_{i_d}}$ so that it is possible to verify that the natural relative degree of the system is equal to 1. Specifically, by defining the auxiliary variable $\xi_{1_i} = \delta_i$ one has

$$\begin{cases} \dot{\xi}_{1_i}(t) = \xi_{2_i}(t) \\ \dot{\xi}_{2_i}(t) = \varphi_i(t) + w_i(t) \end{cases} \tag{15.29}$$

where $w_i = h\ddot{v}_{x_i}$ is the so-called auxiliary control, while function $\varphi_i = \dot{v}_{x_i} - \dot{v}_{x_{i-1}}$ takes into account the difference of the velocity between two subsequent vehicles which is limited for physical reasons, so that it is reasonable to consider

$$|\varphi_i(\cdot)| \leq \Phi_i \ , \tag{15.30}$$

Φ_i being a positive constant assumed to be known. Note that, only the knowledge of the upperbound is necessary to design the control law. The control laws, which is proposed to steer $\xi_{1_i}(t)$ and ξ_{2_i} to zero in a finite time in spite of the uncertainties, in this second case, can be expressed as follows

$$\dot{v}_{x_{i_d}}(t) = -\frac{1}{h}\int_{t_0}^{t} W_{i_{\max}} \text{sign}\left(\xi_{1_i}(t) - \tfrac{1}{2}\xi_{\max_i}\right) \tag{15.31}$$

with $\xi_{\max_i}$ the local minima and maxima of the sliding variable, and control gain such that

$$W_{i_{\max}} > 2\Phi_i \tag{15.32}$$

in order to enforce the sliding mode. As a consequence, the spacing error between two subsequent vehicles vanishes in a finite time, i.e.,

$$\delta_i(t) = \varepsilon_i(t) + L + hv_{x_i}(t) = 0 \,. \tag{15.33}$$

On the basis of the auxiliary signal $w_i(t)$, the desired traction force $F_{x,\iota_{i\mathrm{d}}}$ (i.e., the actual control signal) can be determined as

$$F_{x,\mathrm{f}_{i\mathrm{d}}} + F_{x,\mathrm{r}_{i\mathrm{d}}} = \frac{1}{2}\left(m_i \dot{v}_{x_{i\mathrm{d}}} + F_{\mathrm{loss}_i}\right) \tag{15.34}$$

where $\dot{v}_{x_{i\mathrm{d}}}$ as in (15.31). Note that, as usual in this context, it is assumed that the force distribution is described by

$$\frac{F_{z,\mathrm{f}_i}}{F_{z,\mathrm{r}_i}} = \frac{l_{\mathrm{r}_i} + l_{\mathrm{h}_i}\left(\frac{F_{\mathrm{loss}_i}}{m_i g} + \mu_i\right)}{l_{\mathrm{f}_i} - l_{\mathrm{h}_i}\left(\frac{F_{\mathrm{loss}_i}}{m_i g} - \mu_i\right)} \,. \tag{15.35}$$

Furthermore, from (15.8), in the acceleration case, the following relationship should be maintaned

$$F_{x,\mathrm{f}_i} < \mu_i F_{z,\mathrm{f}_i} \tag{15.36}$$

$$F_{x,\mathrm{r}_i} < \mu_i F_{z,\mathrm{r}_i} \,. \tag{15.37}$$

By substituting (15.1g) and (15.1h) in (15.36) and (15.37), one obtains

$$F_{x,\mathrm{f}_i} < \mu_i \frac{l_{\mathrm{r}_i} m_i g - 2 l_{h_i} F_{x,\mathrm{r}_i} + l_{\mathrm{h}_i} F_{\mathrm{loss}_i}}{2(l_{\mathrm{f}_i} + l_{\mathrm{r}_i} + \mu_i l_{\mathrm{h}_i})} \tag{15.38}$$

$$F_{x,\mathrm{r}_i} < \mu_i \frac{l_{r_i} m_i g + 2 l_{\mathrm{h}_i} F_{x,\mathrm{f}_i} + l_{\mathrm{h}_i} F_{\mathrm{loss}_i}}{2(l_{\mathrm{f}_i} + l_{\mathrm{r}_i} - \mu_i l_{\mathrm{h}_i})} \,. \tag{15.39}$$

Note that, the optimal tire force distribution to achieve the best acceleration response (15.35) corresponds to the intersection point of the two boundary lines of (15.38) and (15.39).

Finally, combining (15.34) and (15.35), it is possible to compute the desired value of longitudinal forces $F_{x,\mathrm{f}_{i\mathrm{d}}}$ and $F_{x,\mathrm{r}_{i\mathrm{d}}}$. The latter are the forces necessary to steer the spacing error to zero. Moreover, these values are needed to find the slip ratio reference for the inner loop of the ith vehicle control system, $\lambda_{\iota_{\mathrm{d}}}$, $\iota = \{\mathrm{f}, \mathrm{r}\}$. In particular, considering the Magic Formula by Bakker-Pacejka, the desired slip ratio $\lambda_{\iota_{\mathrm{d}}}$ is selecetd as the abscissa of the $\lambda - F_x$ curve corresponding to the computed value of desired longitudinal force $F_{x,\iota_{i\mathrm{d}}}$ and to the current road condition identified by the estimated μ (see Fig. 15.10). Note that, in order to avoid stability problems, the slip ratio reference is constrained to lie inside a "stability zone" (see Fig. 15.10b) of the $\lambda - F_x$ curve. Future developments will be devoted to the analysis of the string stability in order to prove that the spacing errors are guaranteed not to amplify as they propagate towards the tail of the vehicle string.

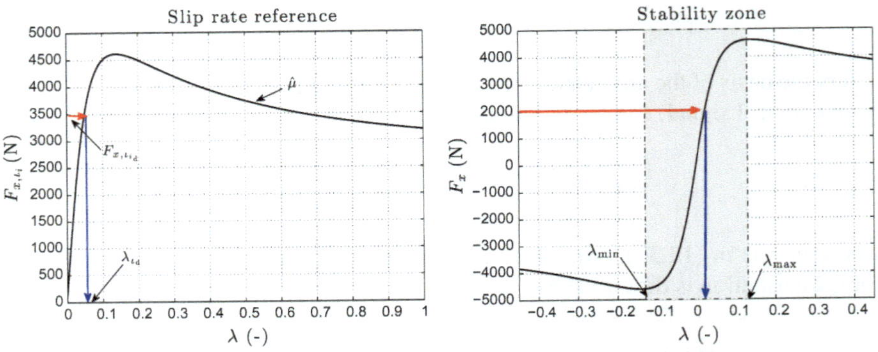

(a) Determination of λ_{ι_d} (b) Stability zone

Fig. 15.10 The slip rate reference for the inner loop of the ith vehicle control system. **a** Determination of λ_{ι_d}, $\iota = \{f, r\}$. **b** Stability zone

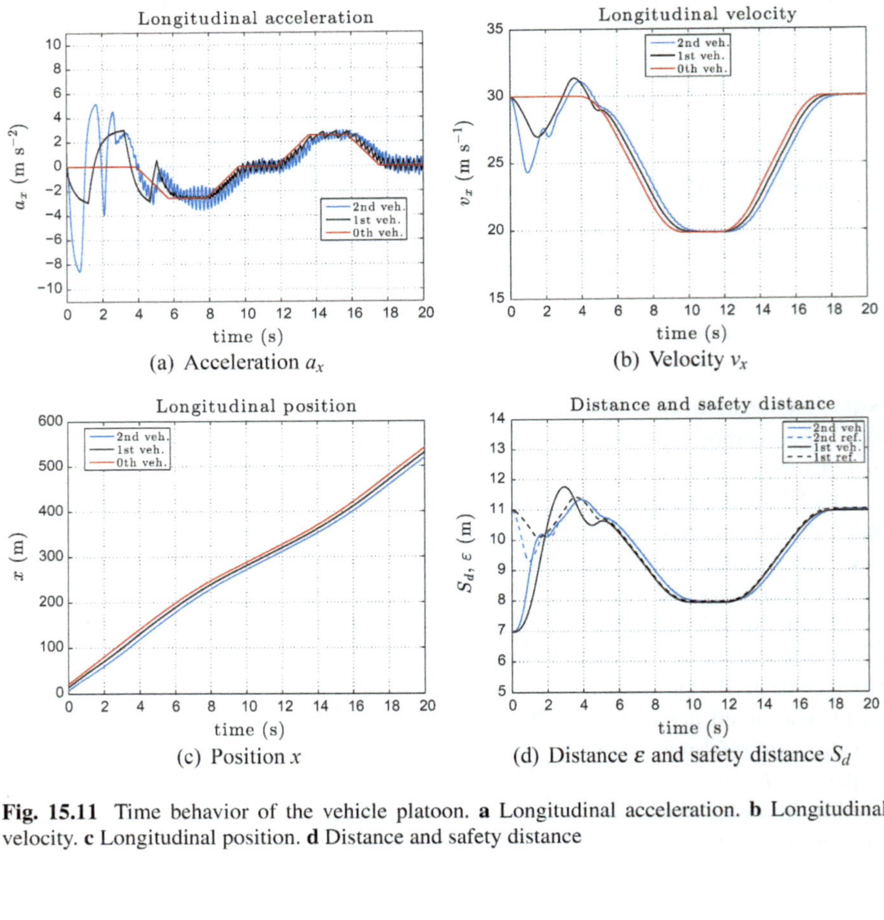

(a) Acceleration a_x (b) Velocity v_x

(c) Position x (d) Distance ε and safety distance S_d

Fig. 15.11 Time behavior of the vehicle platoon. **a** Longitudinal acceleration. **b** Longitudinal velocity. **c** Longitudinal position. **d** Distance and safety distance

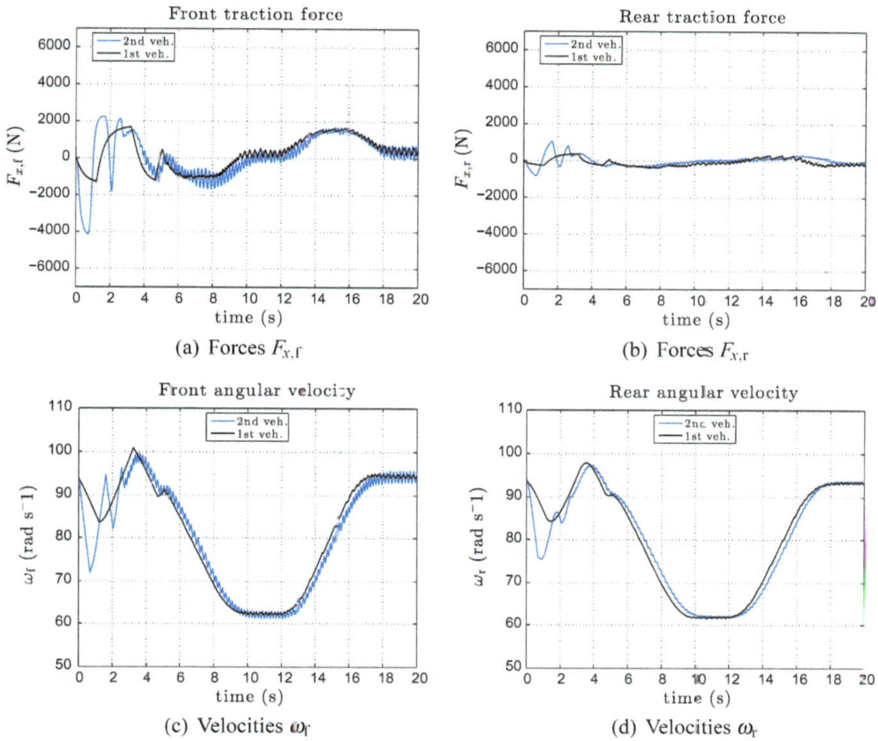

Fig. 15.12 Time behavior of the vehicle platoon. **a** Front traction forces. **b** Rear traction forces. **c** Front angular velocities. **d** Rear angular velocities

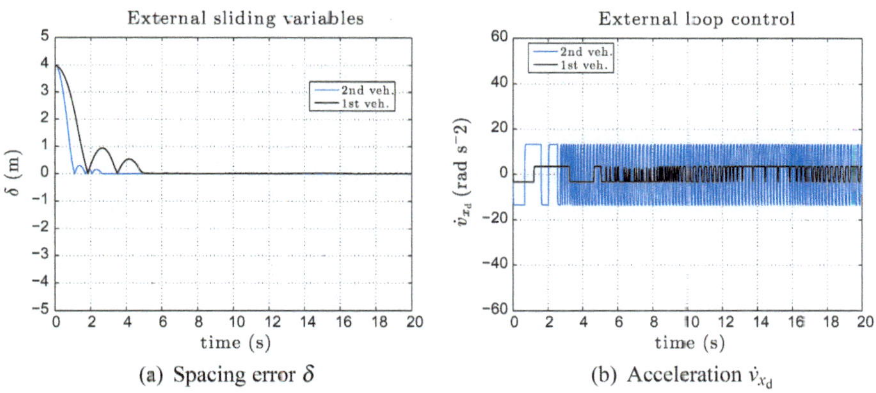

Fig. 15.13 External control loop. **a** Time evolution of the spacing error. **b** Time evolution of the desired acceleration

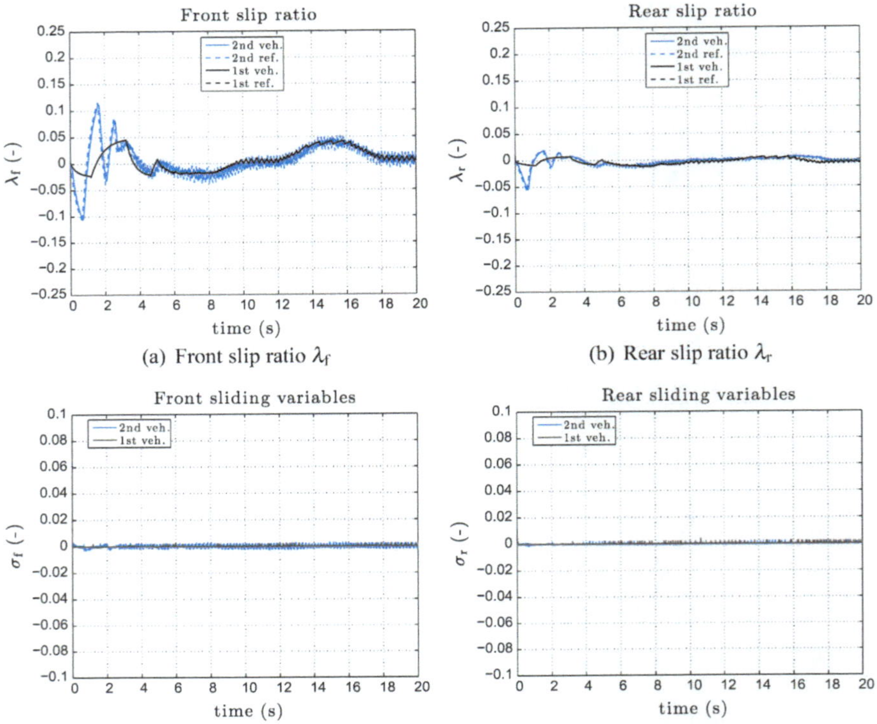

(a) Front slip ratio λ_f

(b) Rear slip ratio λ_r

(c) Front slip rate error σ_f

(d) Rear slip rate error σ_r

Fig. 15.14 Inner control loop. **a** Time evolution of the front slip ratio. **b** Time evolution of the rear slip ratio. **c** Time evolution of the front slip rate error. **d** Time evolution of the rear slip rate error

15.9 Simulation Results on Vehicles Platooning

In this section the vehicle platoon control scheme previously presented is evaluated in a test, where three front-wheel-driven cars (FWD) are involved, one leader and two followers. For the sake of simplicity, we assume that all the vehicles are identical and the parameters of the model used in simulation are the same reported in Table 15.2. Moreover, the subscript i will be omitted when obvious. The initial velocity is equal to 30 ms^{-1} for all the vehicles, while the acceleration profile of the leader is reported in Fig. 15.11 on the top left. The initial distance among vehicles is 7 m and the safety distance is computed according to the CTG spacing policy reported in (15.26), with $L = 2$ m and the headway time equal to $h = 0.3$ s. The parameter μ is assumed to be known and equal to 0.85, that corresponds to a dry asphalt condition. Note that, this parameter, in practical cases, should be correctly estimated, for instance by using the approach described in [56].

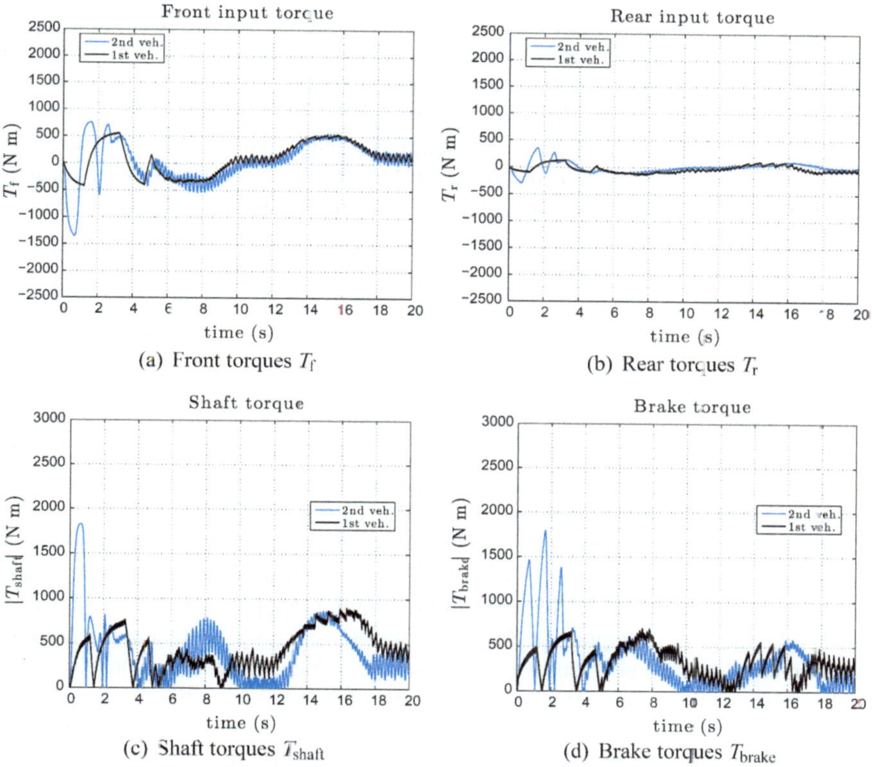

(a) Front torques T_f

(b) Rear torques T_r

(c) Shaft torques T_{shaft}

(d) Brake torques T_{brake}

Fig. 15.15 Inner control loop. **a** Time evolution of the front torques. **b** Time evolution of the rear torques. **c** Time evolution of the shaft torques. **d** Time evolution of the brake torques

Figure 15.11 on the top left shows the acceleration of the leader and the followers. It is possible to see that this profile is more oscillatory for the last follower. This is due to the effect of the higher control gain used for the last vehicle of the string. On the top right the longitudinal velocity is illustrated, while on the bottom from the left, the run distance and the evolution in time of the distance among the leader and the followers are illustrated. One can note that the safety distance among vehicles is reached in a finite time (about 5 s), as expected. Figure 15.12 shows the time evolution of the front and rear traction forces and the corresponding angular velocities for all the vehicles. The traction forces are not discontinuous since a SSOSM control is used. This strategy allows one to steer the spacing error to zero in a finite time, reaching the platooning condition as shown in Fig. 15.13 on the left, and generating the reference for acceleration profile, as shown on the right. Note that this last signal is however oscillating, but this does not represent a critical aspect since it is not directly fed into the plant but it is used to find the desired value of the slip ratio for the inner loop of the proposed control scheme.

Figure 15.14 reports, on the top, the time evolution of the slip ratio with respect to the reference values generated by the external loop, and on the bottom, the sliding variable for the front and rear axle, corresponding to the difference between the slip ratio and its reference value. Also in this case one can note that the sliding variables are steered to zero in a finite time, as expected. Finally, Fig. 15.15 shows the time evolution of the front and rear torques and the computed shaft and brake torques. Note that, as desired, the torque profiles are smooth so as to make the application of the proposed algorithm really feasible in field implementation.

15.10 Conclusions

In this chapter, the application of the sliding mode methodology to vehicle dynamics control is discussed, focusing, in particular, on traction control problems. Two approaches, a first order and a second order SMC (the latter belonging to the class of suboptimal algorithms), are presented, indicated in the chapter with the acronyms FOSM and SSOSM, respectively. The advantage of SSOSM over FOSM control is its capability of solving the control problem through the generation of a continuous control action, which makes this control methodology appropriate to be applied to a real vehicle. In this chapter the SSOSM control previously designed as a solution to the traction control problem is used to solve a vehicles platooning problem. The control objective is to make each vehicle of the platoon travel maintaining a desired safety distance with respect to the preceding vehicle, according to a predefined spacing policy. Simulation results both on traction control and platooning control have demonstrated the effectiveness of the proposals. Future works will be devoted to implement the proposed algorithm even on real vehicles in order to assess their performance. Moreover, the complexity of a real vehicle will require the introduction of advanced observers and control schemes capable of detecting the effective relative degree of the plant, thus suitably switching the control algorithm in order to guarantee better performance. Furthermore, the presence of a communication network makes worth further investigating the design of networked control schemes for this kind of systems. Some of these perspectives are among the main targets of the ITEAM project, in which the University of Pavia is involved. The project has the aim to train strong specialists skilled in research and development of novel technologies for multi-actuated ground vehicles.

Acknowledgements This work was supported by the EU Project ITEAM (project reference: 675999).

References

1. Acarman, T., Özgüner, Ü., Hatipoglu, C., Igusky, A.M.: Pneumatic brake system modeling for systems analysis. SAE 2000 Trans. J. Comm. Veh. - V109-2 (2000)
2. Ackermann, J., Guldner, J., Sienel, W., Steinhauser, R., Utkin, V.I.: Linear and nonlinear controller design for robust automatic steering. IEEE Trans. Control Syst. Technol. **3**(1), 132–143 (1995)
3. Amodeo, M., Ferrara, A., Terzaghi, R., Vecchio, C.: Wheel slip control via second order sliding modes generation. In: Proceedings of the 46th IEEE Conference on Decision and Control, pp. 3889–3894 (2007)
4. Amodeo, M., Ferrara, A., Terzaghi, R., Vecchio, C.: Wheel slip control via second-order sliding-mode generation. IEEE Trans. Intell. Transp. Syst. **11**(1), 122–131 (2010)
5. Bartolini, G., Ferrara, A., Usai, E.: Output tracking control of uncertain nonlinear second-order systems. Automatica **33**(12), 2203–2212 (1997)
6. Bartolini, G., Ferrara, A., Usai, E.: Chattering avoidance by second-order sliding mode control. IEEE Trans. Autom. Control **43**(2), 241–246 (1998)
7. Bartolini, G., Ferrara, A., Levant, A., Usai, E.: On second order sliding mode controllers. Variable Structure Systems, Sliding Mode and Nonlinear Control, pp. 329–350. Springer, London (1999)
8. Bartolini, G., Ferrara, A., Usai, E., Utkin, V.I.: On multi-input chattering-free second-order sliding mode control. IEEE Trans. Autom. Control **45**(9), 1711–1717 (2000)
9. Bartolini, G., Pisano, A., Punta, E., Usai, E.: A survey of applications of second-order sliding mode control to mechanical systems. Int. J. Control **76**(9–10), 875–892 (2003)
10. Bartolini, G., Punta, E., Zolezzi, T.: Approximability properties for second-order sliding mode control systems. IEEE Trans. Autom. Control **52**(10), 1813–1825 (2007)
11. Boiko, I., Fridman, L., Pisano, A., Usai, E.: Analysis of chattering in systems with second-order sliding modes. IEEE Trans. Autom. Control **52**(11), 2085–2102 (2007)
12. Choi, S.B., Choi, Y.T., Park, D.W.: A sliding mode control of a full-car electrorheological suspension system via hardware in-the-loop simulation. J. Dyn. Syst. Meas. Control **122**(1), 114–121 (1998)
13. Delle Monache, M.L., Goatin, P.: A numerical scheme for moving bottlenecks in traffic flow. Bull. Braz. Math. Soc. **47**(2), 605–617 (2016)
14. Dinuzzo, F., Ferrara, A.: Higher order sliding mode controllers with optimal reaching. IEEE Trans. Autom. Control **54**(9), 2126–2136 (2009)
15. Drakunov, S., Özgüner, Ü., Dix, P., Ashrafi, B.: ABS control using optimum search via sliding modes. IEEE Trans. Control Syst. Technol. **3**(1), 79–85 (1995)
16. Edwards, C., Spurgeon, S.: Sliding Mode Control: Theory and Applications. CRC Press, London (1998)
17. Edwards, C., Hebden, R.G., Spurgeon, S.K.: Sliding mode observers for vehicle mode detection. Veh. Syst. Dyn. **43**(11), 823–843 (2005)
18. Ferrara, A., Pisu, P.: Minimum sensor second-order sliding mode longitudinal control of passenger vehicles. IEEE Trans. Intell. Transp. Syst. **5**(1), 20–32 (2004)
19. Ferrara, A., Vecchio, C.: Low vibration vehicle traction control to solve fastest acceleration/deceleration problems via second order sliding modes. In: Proceedings of American Control Conference, pp. 5236–5241 (2007)
20. Ferrara, A., Vecchio, C.: Second order sliding mode control of vehicles with distributed collision avoidance capabilities. Mechatronics **19**(4), 471–477 (2009)
21. Ferrara, A., Librino, R., Massola, A., Miglietta, M., Vecchio, C.: Sliding mode control for urban vehicles platooning. In: Proceedings of the Intelligent Vehicles Symposium, pp. 877–882 (2008)
22. Fridman, L.: Singularly perturbed analysis of chattering in relay control systems. IEEE Trans. Autom. Control **47**(12), 2079–2084 (2002)
23. Fu, L., Özgüner, Ü., Haskara, İ.: Automotive applications of sliding mode control. IFAC Proc. Vol. **44**(1), 1898–1903 (2011)

24. Genta, G.: Motor Vehicle Dynamics: Modeling and Simulation. World Scientific, Singapore (1997)
25. Goggia, T., Sorniotti, A., De Novellis, L., Ferrara, A., Gruber, P., Theunissen, J., Steenbeke, D., Knauder, B., Zehetner, J.: Integral sliding mode for the torque-vectoring control of fully electric vehicles: theoretical design and experimental assessment. IEEE Trans. Veh. Technol. **64**(5), 1701–1715 (2014)
26. Goggia, T., Sorniotti, A., Novellis, L.D., Ferrara, A.: Torque-vectoring control in fully electric vehicles via integral sliding modes. In: Proceedings of American Control Conference, pp. 3918–3923 (2014)
27. Haskara, I., Hatipoglu, C., Özgüner, Ü.: Sliding mode compensation, estimation and optimization methods in automotive control. Variable Structure Systems: Towards the 21st Century, pp. 155–174. Springer, Berlin (2002)
28. Haskara, I., Kokotovic, V.V., Mianzo, L.A.: Control of an electro-mechanical valve actuator for a camless engine. Int. J. Robust Nonlinear Control **14**(6), 561–579 (2004)
29. Hingwe, P., Tomizuka, M.: Experimental evaluation of a chatter free sliding mode control for lateral control in AHS. In: Proceedings of American Control Conference, pp. 3365–3369 (1997)
30. Imine, H., Fridman, L., Shraim, H., Djemai, M.: Sliding Mode Based Analysis and Identification of Vehicle Dynamics. Springer, Berlin (2011)
31. Kiencke, U., Nielsen, L.: Automotive control systems: for engine, driveline, and vehicle. Meas. Sci. Technol. **11**(12), 1828 (2000)
32. Kim, Y.W., Rizzoni, G., Utkin, V.I.: Developing a fault tolerant power-train control system by integrating design of control and diagnostics. J. Robust Nonlinear Control **11**(11), 1095–1114 (2001)
33. Leclercq, L., Chanut, S., Lesort, J.B.: Moving bottlenecks in Lighthill–Whitham–Richards model: a unified theory. Transp. Res. Rec. J. Transp. Res. Board **1883**, 3–13 (2004)
34. Lee, H., Tomizuka, M.: Adaptive vehicle traction force control for intelligent vehicle highway systems (IVHSs). IEEE Trans. Ind. Electron. **50**(1), 37–47 (2003)
35. Lee, H., Love, D.W., Tomizuka, M.: Longitudinal maneuvering control for automated highway systems based on a magnetic reference/sensing system. In: Proceedings of American Control Conference, pp. 150–154 (1995)
36. Levant, A.: Chattering analysis. IEEE Trans. Autom. Control **55**(6), 1380–1389 (2010)
37. Li, L., Wang, F.Y., Zhou, Q.: Integrated longitudinal and lateral tire/road friction modeling and monitoring for vehicle motion control. IEEE Trans. Intell. Transp. Syst. **7**(1), 1–19 (2006)
38. Li, P., Shrivastava, A.: Traffic flow stability induced by constant time headway policy for adaptive cruise control vehicles. Transp. Res. Part C **10**(4), 273–301 (2002)
39. Liang, K.Y.: Fuel-efficient heavy-duty-vehicle platoon formation. Doctoral dissertation, KTH Royal Institute of Technology (2016)
40. M'sirdi, N.K., Rabhi, A., Fridman, L., Davila, J., Delanne, Y.: Second order sliding mode observer for estimation of velocities, wheel sleep, radius and stiffness. In: Proceedings of American Control Conference, pp. 14–16 (2006)
41. Newell, G.: A moving bottleneck. Transp. Res. Part B Methodol. **32**(8), 531–537 (1998)
42. Pacejka, H.B.: Tyre and Vehicle Dynamics. Butterworth-Heinemann, Oxford (2006)
43. Patel, N., Edwards, C., Spurgeon, S.K.: Optimal braking and estimation of tyre friction in automotive vehicles using sliding modes. Int. J. Syst. Sci. **38**(11), 901–912 (2007)
44. Reichhartinger, M., Horn, M.: Application of higher order sliding-mode concepts to a throttle actuator for gasoline engines. IEEE Trans. Ind. Electron. **56**(9), 3322–3329 (2009)
45. Reichhartinger, M., Spurgeon, S.K., Weyrer, M.: Design of an unknown input observer to enhance driver experience of electric power steering systems. In: Proceedings of European Control Conference, pp. 269–274 (2016)
46. Sànchez-Torres, J.D., Loukianov, A.G., Galicia, M.I., Dominguez, R.: Robust nested sliding mode integral control for anti-lock brake system. Int. J. Veh. Des. **62**(2–4), 188–205 (2013)
47. Santhanakrishnan, K., Rajamani, R.: On spacing policies for highway vehicle automation. IEEE Trans. Intell. Transp. Syst. **4**(4), 198–204 (2003)

48. Shladover, S.: Operation of automated guideway transit vehicles in dynamically reconfigured trains and platoons. Technical report, UMTA-MA-060085-79 (1979)
49. Shladover, S.E.: Longitudinal control of automotive vehicles in close formation platoons. J Dyn. Syst. Meas. Control **113**(2), 231–241 (1991)
50. Slotine, J.J.E., Li, W.: Applied Nonlinear Control. Prentice Hall, Upper Saddle River (1991)
51. Swaroop, D., Rajagopal, K.R.: Intelligent cruise control systems and traffic flow stability. Transp. Res. Part C **7**(6), 329–352 (1999)
52. Swikir, A., Utkin, V.I.: Chattering analysis of conventional and super twisting sliding mode control algorithm. In: Proceedings of the 14th International Workshop on Variable Structure Systems, pp. 98–102 (2016)
53. Tafner, R., Reichhartinger, M., Horn, M.: Robust online roll dynamics identification of a vehicle using sliding mode concepts. Contr. Eng. Pract. **29**, 235–246 (2014)
54. Tafner, R., Horn, M., Ferrara, A.: Experimental evaluation of nonlinear unknown input observers applied to an EPS system. In: Proceedings of American Control Conference, pp. 2409–2414 (2016)
55. Tanelli, M., Ferrara, A., Vecchio, C.: Switched second order sliding mode for wheel slip control of road vehicles. In: Proceedings of the 11th International Workshop on Variable Structure Systems, pp. 469–474 (2010)
56. Tanelli, M., Ferrara, A., Giani, P.: Combined vehicle velocity and tire-road friction estimation via sliding mode observers. In: Proceedings of IEEE International Conference on Control Applications, pp. 130–135 (2012)
57. Ulsoy, A.G., Peng, H., Cakmakci, M.: Automotive Control Systems. Cambridge University Press, New York (2012)
58. Utkin, V.I.: Sliding Modes in Control and Optimization. Springer, Berlin (1992)
59. Utkin, V.I., Guldner, J., Shi, J.: Sliding Model Control in Electromechanical Systems. CRC Press, London (1999)
60. Wang, J., Rajamani, R.: Should adaptive cruise-control systems be designed to maintain a constant time gap between vehicles? IEEE Trans. Veh. Technol. **53**(5), 1480–1490 (2004)
61. Yokoyama, M., Hedrick, J.K., Toyama, S.: A model following sliding mode controller for semi-active suspension systems with MR dampers. In: Proceedings of American Control Conference, pp. 2652–2657 (2001)
62. Zhao, J., Oya, M., Kamel, A.E.: A safety spacing policy and its impact on highway traffic flow. In: Proceedings of IEEE Intelligent Vehicles Symposium, pp. 960–965 (2009)

Chapter 16
Sliding Mode Control of Power Converters with Switching Frequency Regulation

Víctor Repecho, Domingo Biel, Josep M. Olm and Enric Fossas

16.1 Introduction

Sliding mode control (SMC) constitutes a natural control tool for variable structure systems (VSS), such as power converters, which are nonlinear systems where the control inputs are inherently discontinuous functions of time. Several first order SMC applications for linear and nonlinear systems can be found in the literature [21].

In most cases SMC designs assume an infinite switching frequency of the control action in accordance with the sign of a certain function, but this entails issues when implemented in real systems. In the field of power converters, the first realistic SMC implementations are reported in [2, 22]. In these works, the sign function is replaced by a hysteresis comparator, and the control action is enforced to switch at finite frequency, but variable and system dependent [3, 4]. However, power converters require a fixed switching frequency operation since the design of their reactive components is highly dependent on the switching frequency of the system.

V. Repecho · D. Biel (✉) · J.M. Olm · E. Fossas
Institute of Industrial and Control Engineering,
Universitat Politècnica de Catalunya, 08028 Barcelona, Spain
e-mail: domingo.biel@upc.edu

V. Repecho
e-mail: victor.repecho.del@upc.edu

J.M. Olm
e-mail: josep.olm@upc.edu

E. Fossas
e-mail: enric.fossas@upc.edu

D. Biel
Department of Electronic Engineering,
Universitat Politècnica de Catalunya, Barcelona, Spain

S. Li et al. (eds.), *Advances in Variable Structure Systems and Sliding Mode Control—Theory and Applications*, Studies in Systems, Decision and Control 115, DOI 10.1007/978-3-319-62896-7_16

Several different approaches have been proposed to regulate the SMC switching frequency to a fixed value. Some of them adapt the comparator hysteresis band, adjusting its level in accordance with the system state [5, 7, 8, 10, 12, 17]. The procedure provides good results, but requires perfect knowledge of the plant, and it is not robust in the face of parametric variations. Additional sensors and/or observers can be included to get a proper adaptation of the hysteresis band amplitude but, in this case, the system reliability decreases and the cost raises.

Fixed switching frequency can also be achieved by using an external signal to force the switching instants [11, 18]. This approach needs some additional hardware on the controller and requires the switching frequency to be low enough with respect to the system time constants, otherwise the state dynamics drifts away from the ideal sliding mode and an unexpected steady-state error appears.

The Zero Averaged Dynamics (ZAD) concept was presented in [6]. The method computes a duty cycle that guarantees zero T-periodic mean value of the switching function, with T denoting the switching period. Therefore, fixed switching frequency is reached in the steady-state, and the averaged behaviour is close to the ideal sliding mode one. The ZAD strategy has been successfully implemented in [13]. The results presented therein show a good performance of the ZAD, but also point out the requirement of a fast digital processor to solve the complex calculations involved in the duty cycle computation, which in the end constitute the main drawbacks of ZAD-based SMC fixed frequency implementations.

Pulse Width Modulators (PWM) at fixed frequency have been used to implement the so-called PWM-SMC. Initially proposed in [9, 19], the method implements directly the equivalent control and obtains the switching instants comparing the equivalent control with the fixed frequency saw-tooth waveform at the PWM. The results presented in [20] show overall good performance, but it should be noted that the same solution can also be derived by calculating the duty cycle required to obtain the desired system dynamics. Moreover, some sliding mode properties, such as order reduction or robustness in the face of disturbances, could be lost.

Alternatively, a simple hysteresis band controller in charge of fixing the switching frequency of a sliding mode controller is presented in the next sections. The controller is based on a variable hysteresis band comparator which regulates the switching frequency to a desired constant value. The analysis allows to develop a large signal model for the frequency control loop, and the controller parameters design guarantees stability and asymptotic tendency to a fixed switching frequency when the system is on the sliding surface. Furthermore, in order to cover the case of tracking time-varying references, the switching frequency controller design has also been extended with the addition of a feedforward term which, once properly designed following the guidelines presented here, is able to provide the desired switching frequency in the steady state.

16.2 Hysteresis Band Controller for Switching Frequency Regulation

Let us consider a single input single output (SISO) system, with dynamics given by

$$\dot{x} = f(x) + g(x)u, \tag{16.1}$$

where x denotes the state vector, $f(x)$, $g(x)$ are smooth nonlinear functions, and $u \in \{u^+, u^-\}$ is the control input. According to [3, 21], a system with the structure presented in (16.1), where a sliding motion is enforced over a switching surface $\sigma(x) = 0$ in a comparator with a fixed hysteresis band value $\Delta > 0$, see Fig. 16.1, as

$$u = \begin{cases} u^+ \text{ if } \sigma < -\Delta, \text{ or } (|\sigma| < \Delta \;\&\; \dot{\sigma} > 0) \\ u^- \text{ if } \sigma > \Delta, \text{ or } (|\sigma| < \Delta \;\&\; \dot{\sigma} < 0) \end{cases} \tag{16.2}$$

produces a series of consecutive kth switching periods ($k > 0$), corresponding to

$$T_k = T_k^+ + T_k^- = 2\Delta\left(\rho_k^+ - \rho_k^-\right), \tag{16.3}$$

where ρ_k^+, ρ_k^- are defined as the inverses of $\dot{\sigma}$ for each control input state:

$$\rho_k^+ = \frac{1}{\dot{\sigma}_{k_{u=u^+}}}, \quad \rho_k^- = \frac{1}{\dot{\sigma}_{k_{u=u^-}}}$$

The obtaining of (16.3) relies on the assumption of piecewise linear behavior for σ, which implies that ρ_k^+, ρ_k^- are constant during the switching interval. This is a standard hypothesis in the SMC literature [3, 21] which holds if the switching frequency is high enough with respect to the system dynamics.

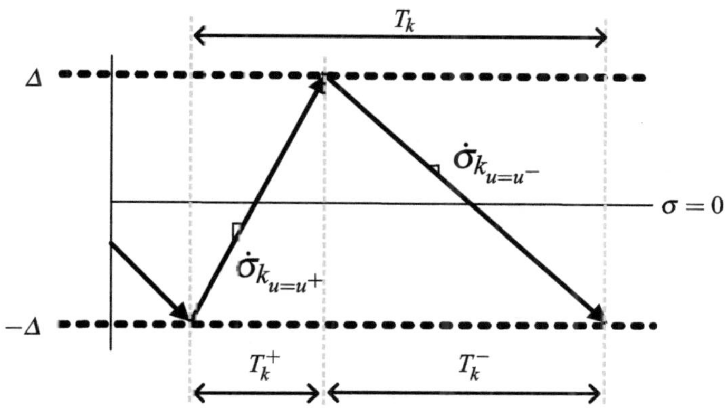

Fig. 16.1 Behavior of σ within a constant amplitude boundary layer

Notice that the expected switching period depends on ρ_k^+, ρ_k^-, which are inversely proportional to the switching function slopes, and this implies that the switching period varies as the state vector does. This phenomenon is sometimes disadvantageous for specific systems, as happens with power converters. Hence, a solution is provided hereafter.

16.2.1 Control Architecture

The proposed structure, already presented in [14–16], includes a control loop that regulates the switching period of the control action under sliding motion, thus achieving a fixed switching frequency in the steady state. The idea is sketched in Fig. 16.2. The control loop measures each switching period of the control action and compares it with the desired switching period, T^*. The difference is processed by the switching frequency controller (SFC), which will update the hysteresis band value of the hysteretic comparator in such a way that $T_k \to T^*$.

16.2.2 Discrete-Time Modelling of the Control Loop

It is assumed that the hysteresis band amplitude can be updated at the beginning of each switching interval by the SFC, keeping it constant up to the next switching interval. The behavior of σ when confined in a time-varying boundary layer is represented in Fig. 16.3. Therefore, the expression of the switching period needs to be revisited. Following an analogue procedure to the derivation of (16.3), the kth switching period in the time-varying case is now given by:

$$T_k = T_k^+ + T_k^- = \rho_k^+ (\Delta_k + \Delta_{k-1}) - 2\rho_k^- \Delta_k = \hat{\rho}_k \Delta_k + \left(\tilde{\rho}_k - \hat{\rho}_k\right) \Delta_{k-1}, \quad (16.4)$$

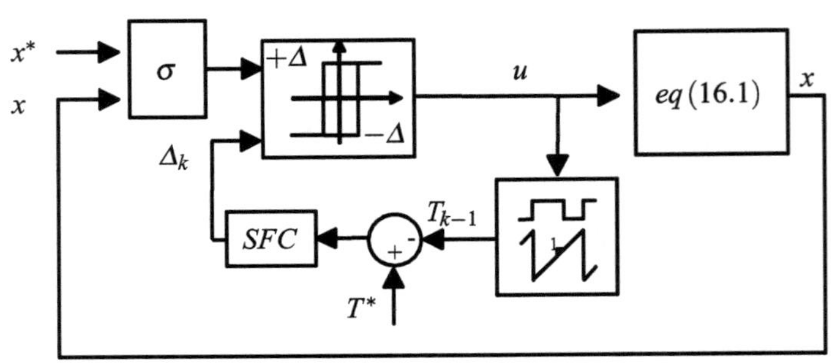

Fig. 16.2 Overall controller architecture

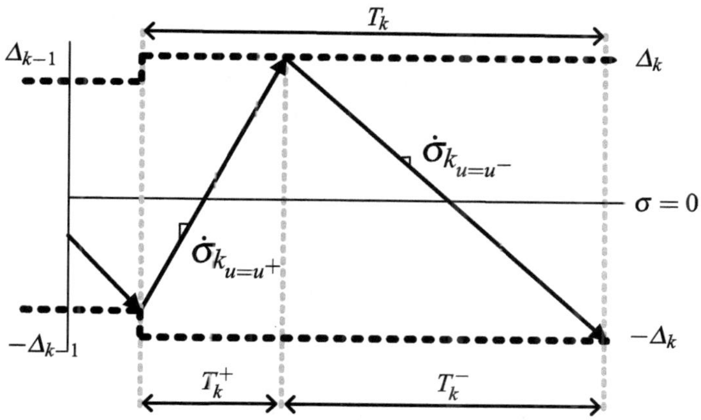

Fig. 16.3 Behavior of σ within a time-varying amplitude boundary layer

with

$$\hat{\rho}_k = \rho_k^+ - 2\rho_k^-,$$
$$\tilde{\rho}_k = 2\left(\rho_k^+ - \rho_k^-\right).$$

Let us define the switching period error as $e := T^* - T$. Therefore, using (16.4) one easily finds out that

$$e_k - e_{k-1} = \hat{\rho}_k\left(\Delta_{k-1} - \Delta_k\right) + \rho_{k-1}^+\left(\Delta_{k-2} - \Delta_{k-1}\right) + \left(\tilde{\rho}_{k-1} - \tilde{\rho}_k\right)\Delta_{k-1}. \quad (16.5)$$

Next subsections will particularize expression (16.5) in two different working conditions, namely: the regulation case and the tracking case.

16.2.2.1 The Regulation Case

In regulation tasks the state vector reference, x^*, is constant. Assuming that the amplitude of the ripple, 2Δ, of σ in the vicinity of $\sigma = 0$ is small, the steady state vector can be considered also constant, and hence $x = x^*$. As a consequence, the switching function derivatives and their inverses are constant in the steady state as well. Therefore, from a certain discrete-time instant k_0 it results that:

$$\rho_k^\pm = \rho\left(x^*, u^\pm\right) := \rho_*^\pm, \quad \hat{\rho}_k := \hat{\rho}^*, \quad \tilde{\rho}_k := \tilde{\rho}^*, \ \forall k \geq k_0. \quad (16.6)$$

With these approximations, (16.5) can be simplified up to the following expression:

$$e_k - e_{k-1} = \hat{\rho}^*\left(\Delta_{k-1} - \Delta_k\right) + \rho_*^+\left(\Delta_{k-2} - \Delta_{k-1}\right). \quad (16.7)$$

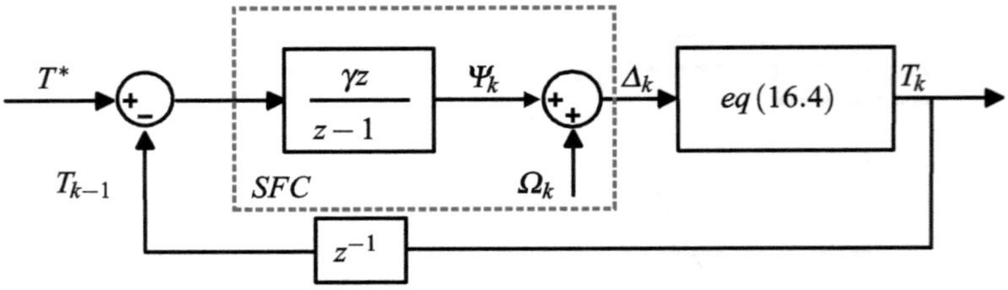

Fig. 16.4 Switching frequency regulation control loop with feedforward action. The inherent time delay due to the switching period measurement is represented by z^{-1}, see [14] for details

The control law proposed for the hysteresis band amplitude in the regulation case is of integral type and answers to the following difference equation:

$$\Delta_k = \Delta_{k-1} + \gamma e_{k-1}, \tag{16.8}$$

with $\gamma > 0$ denoting the integral constant. Notice that taking (16.8) to (16.7) results in the following linear homogeneous difference equation with constant coefficients:

$$e_k = \left(1 - \gamma \hat{\rho}^*\right) e_{k-1} - \gamma \rho_*^+ e_{k-2}. \tag{16.9}$$

The stability of the zero solution of (16.9), which means $T_k \to T^*$, is studied in Sect. 16.2.3.

16.2.2.2 The Tracking Case

When the system tracks a time-varying reference $x^* = x^*(t)$, the time derivatives of the switching functions can not be considered constant values, i.e. $\rho_k^+ \neq \rho_{k-1}^+$, $\rho_k^- \neq \rho_{k-1}^-$. Hence, when using the integral action (16.8) as SFC, the corresponding closed-loop response given by (16.5) results in:

$$e_k = \left(1 - \gamma \hat{\rho}_k\right) e_{k-1} - \gamma \rho_{k-1}^+ e_{k-2} + \Delta_{k-1}\left(\tilde{\rho}_{k-1} - \tilde{\rho}_k\right). \tag{16.10}$$

Notice that (16.10) is non-homogeneous, and does not have $e_k = 0$ as an equilibrium solution. In order to overcome this drawback the proposal presented here adds a feedforward loop that compensates the undesirable effect of the last term of (16.10). Therefore, the new SFC structure for systems under tracking tasks is shown in Fig. 16.4 and consists of setting

$$\Delta_k = \Psi_k + \Omega_k, \tag{16.11}$$

where Ψ_k is the integral control action

$$\Psi_k = \Psi_{k-1} + \gamma e_{k-1}, \tag{16.12}$$

while the feedforward term Ω_k responds to:

$$\Omega_k = \frac{\hat{\rho}_{k-1} - \rho_k^+}{\hat{\rho}_k}\Omega_{k-1} + \frac{\rho_{k-1}^+}{\hat{\rho}_k}\Omega_{k-2} + \frac{\tilde{\rho}_{k-1} - \tilde{\rho}_k}{\hat{\rho}_k}\Psi_{k-1}. \tag{16.13}$$

Merging (16.11)–(16.13) the new closed-loop error dynamics is given by

$$e_k = \left(1 - \gamma\hat{\rho}_k\right) e_{k-1} - \gamma\rho_{k-1}^+ e_{k-2}. \tag{16.14}$$

Now the equation of the switching period error boils down to a homogeneous time-varying discrete-time linear system recovering $e_k = 0$ as the desired equilibrium solution. Under sliding motion and in the steady state, the state vector profile $x^*(t)$ will produce time-varying values for ρ_k:

$$\begin{aligned} \rho_k^+ &= \rho_k\left(x^*(t), u^+\right) := \rho_{*k}^+, \\ \rho_k^- &= \rho_k\left(x^*(t), u^-\right) := \rho_{*k}^-, \\ \hat{\rho}_k &= \hat{\rho}_k\left(x^*(t)\right) := \hat{\rho}_k^*, \\ \tilde{\rho}_k &= \tilde{\rho}_k\left(x^*(t)\right) := \tilde{\rho}_k^*, \end{aligned} \tag{16.15}$$

$\forall k \geq k_0$, and the preceding error equation becomes

$$e_k = \left(1 - \gamma\hat{\rho}_k^*\right) e_{k-1} - \gamma\rho_{*k-1}^+ e_{k-2}. \tag{16.16}$$

The stability analysis of the zero solution of (16.16) is conducted in Sect. 16.2.3.

16.2.3 Stability Analysis and Design Criteria

The obtained results rely upon the hypotheses established in the above analysis. These can be summarized as follows:

Assumption 16.1 The control law (16.2) induces system (16.1) to evolve within a boundary layer defined by $|\sigma\left(x, x^*(t)\right)| < \Delta$. Moreover, sliding motion exists on the switching hyperplane $\sigma\left(x, x^*(t)\right) = 0$ for $\Delta \to 0$, with $x^*(t) \in \mathbb{R}^n$ being the steady state of the ideal sliding dynamics. Finally, $\sigma\left(x, x^*(t)\right)$ shows constant time derivatives at either sides of the switching hyperplane during a complete switching period within the boundary layer.

16.2.3.1 The Regulation Case

Theorem 16.1 *Let Assumption 16.1 be fulfilled, with x^* being a constant regulation point, and let the hysteresis band amplitude, Δ, be updated according to* (16.8). *If the integral gain γ is selected as*

$$0 < \gamma < \min\left\{ \left(\rho_*^+\right)^{-1}, \left|\rho_*^-\right|^{-1} \right\},$$

with $\rho_*^\pm$ defined in (16.6), *then the switching period, T_k, converges asymptotically to its reference value, T^*, in the steady state.*

Proof It follows applying Jury stability criterion to the characteristic polynomial associated to the difference equation (16.5), see [14] for details.

16.2.3.2 The Tracking Case

Theorem 16.2 *Let Assumption 16.1 be fulfilled, with $x^* = x(t)$ being a time-varying reference signal, and let the hysteresis band amplitude, Δ, be updated according to* (16.11)–(16.13). *If the integral gain γ is selected as*

$$\gamma_m := \max \left\{ \frac{\hat{\rho}_k^* - \sqrt{\frac{1}{2}\left(\hat{\rho}_k^{*^2} - \rho_{*k}^{+^2}\right)}}{\hat{\rho}_k^{*^2} + \rho_{*k}^{+^2}}, \ \forall k \geq 0 \right\},$$

$$\gamma_M := \min \left\{ \frac{\hat{\rho}_k^* + \sqrt{\frac{1}{2}\left(\hat{\rho}_k^{*^2} - \rho_{*k}^{+^2}\right)}}{\hat{\rho}_k^{*^2} + \rho_{*k}^{+^2}}, \ \forall k \geq 0 \right\},$$

with $\hat{\rho}_k^*$ and ρ_{*k}^+ defined in (16.15), *then the switching period, T_k, converges asymptotically to its reference value, T^*, in the steady state.*

Proof It follows using a Lyapunov-based discrete time approach, see [15] for details.

16.3 Application to Power Electronics

In this section, the previously proposed structures for switching frequency regulation in SMC are designed for several power converters. Specifically, three different cases are considered: a SMC in a regulation task for a buck converter, a SMC in a regulation case for a boost converter, and a SMC in a tracking task for a voltage source inverter (VSI).

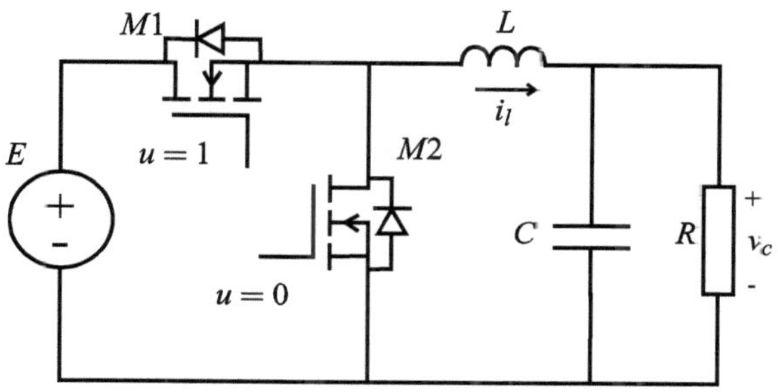

Fig. 16.5 Buck converter

16.3.1 *Output Regulation of a Linear System: The Buck Converter*

A buck converter circuit scheme is shown in Fig. 16.5, and the values of its parameters are listed in Table 16.1.

The converter state space equations are:

$$C \frac{d\,v_c}{dt} = i_l - \frac{v_c}{R}$$
$$L \frac{d\,i_l}{dt} = E\,u - v_c,$$

where u is the control signal and takes values in the set $\{0, 1\}$. The power switches $M1$ and $M2$ work in a complementary way, remaining closed when u takes the values showed in Fig. 16.5.

16.3.1.1 Sliding Mode Control

Taking into account that the relative degree of the buck converter with respect to the output voltage is two, the chosen switching surface for output voltage regulation is:

Table 16.1 Buck converter parameters

Parameter	Symbol	Value
Input voltage	E	48 V
Desired output voltage	v_c^*	12–24 V
Inductor	L	22 μH
Output capacitor	C	50 μF
Load resistance	R	2 Ω
Switching period reference	T^*	10 μs

$$\sigma(v_c, i_l) := \lambda_1 e_v + \lambda_2 \dot{e}_v = 0, \quad \lambda_{1,2} > 0,$$

where $v_c{}^*$ and $e_v = v_c{}^* - v_c$ are the output voltage reference and the voltage error, respectively. The switching function derivative becomes:

$$\dot{\sigma}(v_c, i_l) = f_1(v_c, i_l) - \frac{\lambda_2}{LC} E\, u \tag{16.17}$$

where

$$f_1(v_c, i_l) = i_l \left(\frac{\lambda_2}{R\, C^2} - \frac{\lambda_1}{C} \right) + v_c \left(\frac{\lambda_1}{RC} + \frac{\lambda_2}{LC} - \frac{\lambda_2}{R^2\, C^2} \right).$$

From (16.17), it is clear that sliding motion exists if $\frac{\lambda_2 E}{LC} > f_1 > 0$. In turn, the equivalent control results in:

$$u_{eq} = \frac{LC}{E\,\lambda_2} f_1(v_c, i_l),$$

and the control law that enforces a real sliding motion in the vicinity of $|\sigma(v_c, i_l)| < \Delta$ is:

$$u = \begin{cases} 0 \text{ if } \; \sigma < -\Delta_k \; \text{ or } \; (|\sigma| < \Delta_k \;\&\; \dot{\sigma} > 0) \\ 1 \text{ if } \; \sigma > \Delta_k \; \text{ or } \; (|\sigma| < \Delta_k \;\&\; \dot{\sigma} < 0)\,. \end{cases}$$

Under sliding motion the system dynamics are governed by:

$$\frac{\mathrm{d}\, v_c}{\mathrm{d}t} = -\frac{\lambda_1}{\lambda_2} v_c + \frac{\lambda_1}{\lambda_2} v_c^* + \dot{v}_c^* \tag{16.18}$$

$$\frac{\mathrm{d}\, i_l}{\mathrm{d}t} = \left(\frac{1}{RC} - \frac{\lambda_1}{\lambda_2} \right) \left(i_l - \frac{v_c}{R} \right), \tag{16.19}$$

which is a linear system with equilibrium point $v_c = v_c^*$, $i_l = \frac{v_c^*}{R}$. From (16.18), (16.19) it is evident that system will be asymptotically stable if $\frac{\lambda_1}{\lambda_2} > \frac{1}{RC}$. According to Table 16.1, the selected values for the sliding coefficients are: $\lambda_1 = 0.2$, $\lambda_2 = 1.9 \cdot 10^{-5}$, which ensures stability and delivers a good transient response.

16.3.1.2 Switching Frequency Regulation

In order to select γ for the SFC, ρ_k^+ and ρ_k^- have to be evaluated. This requires (16.17) to be particularized for the ideal steady-state sliding mode dynamics, namely $v_c = v_c^*$, $i_l = \frac{v_c^*}{R}$:

$$\dot{\sigma}(v_c^*, i_l^*) = \frac{\lambda_2}{LC} \left(v_c^* - E\, u \right),$$

which yields

$$\rho_{*k}^{+} = \left[\dot{\sigma}(v_c^*, i_l^*)_{u=u^-}\right]^{-1} = \frac{LC}{\lambda_2 v_c^*}$$
$$\rho_{*k}^{-} = \left[\dot{\sigma}(v_c^*, i_l^*)_{u=u^+}\right]^{-1} = \frac{LC}{\lambda_2 \left(v_c^* - E\right)}.$$

Then, with the data given in Table 1.1, one gets:

$$v_c^* = 12V \rightarrow \rho_{*k}^{+} = 4.82e^{-6};\ \rho_{*k}^{-} = -1.61e^{-6},$$
$$v_c^* = 24V \rightarrow \rho_{*k}^{+} = 2.41e^{-6};\ \rho_{*k}^{-} = -2.41e^{-6}.$$

According to Theorem 16.1, the values of γ within the range (0, 207470) provide stability for the SFC. It should be noted that this range corresponds to 12 V at the output, corresponding to the worst case for the SFC stability. Consequently, the chosen value is $\gamma = 2 \cdot 10^4$.

16.3.1.3 Simulation Results

The simulations are performed using Matlab Simulink, with the data shown in Table 16.1 and with the previously selected control parameters, namely $\lambda_1 = 0.2$, $\lambda_2 = 1.9 \cdot 10^{-5}$, and $\gamma = 2 \cdot 10^4$.

Figure 16.6 shows the response of the system with different initial conditions, Δ_{ini}, for the hysteresis value. From the top plots it can be seen how the system always reaches the desired steady state, Δ_{ss}, i.e. when $\Delta_{ini} < \Delta_{ss}$ and also when $\Delta_{ini} > \Delta_{ss}$. The second and third plots show the evolution of the hysteresis band and the corresponding switching period, respectively, confirming a good regulation to the desired value, 10^{-5} s i.e. 100 kHz, in both cases.

In Fig. 16.7 the system response to a variation of the voltage reference between 24 and 12 V is plotted. Besides a correct regulation of the output voltage, it is possible to confirm how, after the sliding transient, the desired switching frequency is reached in both cases.

The results shown in Fig. 16.8 correspond to the variation of the switching period reference when the value of γ brings the system close to the unstable region. Such tests are performed in order to numerically verify the theoretical values that ensure stable behaviour of the SFC. Specifically, γ is set to $2 \cdot 10^5$. In the test, the switching period reference is step varied from 14 to 12 μs and from 10 to 12 μs, respectively. From the results, it is clear that this value of γ is close to the ones which would produce instability, as Theorem 16.1 claims.

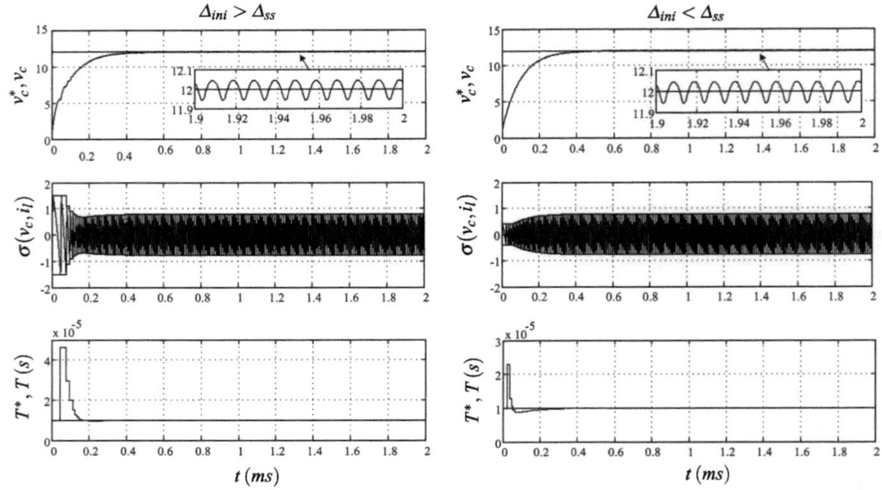

Fig. 16.6 Buck Converter: start-up with different initial values for Δ. From *top to bottom*. 1- Output voltage, v_c, and reference voltage, v_c^*. 2- Switching function σ. 3- Desired and real switching period (T^*, T)

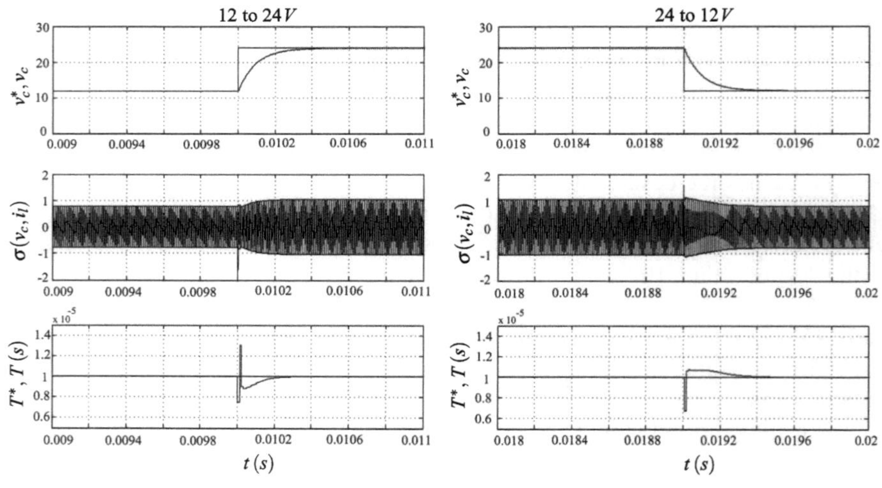

Fig. 16.7 Buck Converter: output voltage response to a step-changing reference. From *top to bottom*. 1- Output voltage, v_c, and reference voltage, v_c^*. 2- Switching function σ. 3- Desired and real switching period (T^*, T)

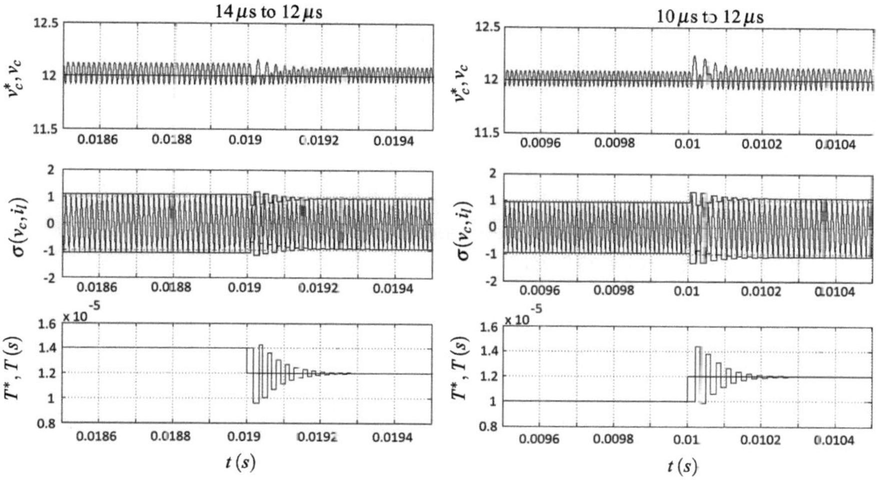

Fig. 16.8 Buck Converter: switching period regulation. From *top to bottom*. 1- Output voltage, v_c, and reference voltage, v_c^*. 2- Switching function σ. 3- Desired and real switching period (T^*, T)

Fig. 16.9 Boost converter

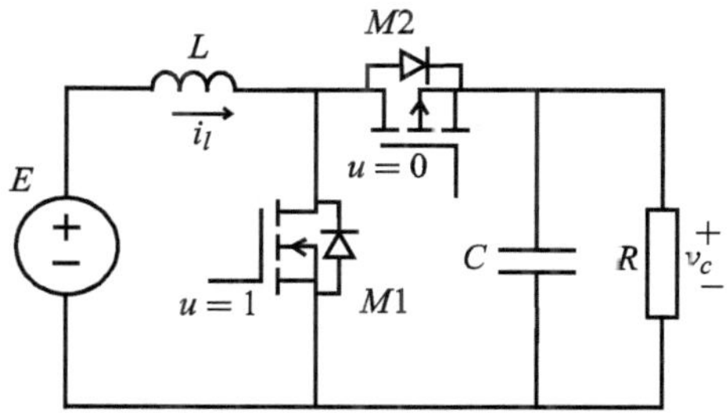

Table 16.2 Boost Converter Parameter

Parameter	Symbol	Value
Input voltage	E	12 V
Desired output voltage range	v_c^*	36–48 V
Output capacitor	C	50 μF
Inductance	L	22 μH
Load Resistance	R	20 Ω
Switching period reference	T^*	10 μs

16.3.2 Output Regulation of a Nonlinear System: The Boost Converter

A boost converter circuit scheme is shown in Fig. 16.9, and the values of its parameters are listed in Table 16.2.

The nonlinear state space equations of the converter are:

$$L\frac{\mathrm{d}i_l}{\mathrm{d}t} = E - v_c(1-u)$$
$$C\frac{\mathrm{d}v_c}{\mathrm{d}t} = i_l(1-u) - \frac{v_c}{R},$$

where u is the control signal and takes values in $\{0, 1\}$. The power switches $M1$ and $M2$ work in a complementary way, as in the Buck converter case.

16.3.2.1 Sliding Mode Control

The relative degree between the output voltage and the control input is one. However, imposing a sliding dynamics directly over the output voltage results in an unstable behaviour of the inductor current, which prevents its practical use [1]. An alternative solution in order to regulate the output voltage, v_c, is to consider the following switching function:

$$\sigma(v_c, i_l) := \kappa_1 e_v + \kappa_2 \int e_v \mathrm{d}t - \kappa_3 i_l, \quad \kappa_{1,2,3} > 0$$

where $e_v = v_c^* - v_c$.

The switching function derivative results in

$$\dot{\sigma}(v_c, i_l) = -\psi_1(v_c, i_l) + (1-u)\,\psi_2(v_c, i_l), \tag{16.20}$$

where

$$\psi_1(v_c, i_l) = \frac{\kappa_3 E}{L} - \frac{\kappa_1}{RC}v_c - \kappa_2 e_v, \quad \psi_2(v_c, i_l) = \frac{\kappa_3}{L}v_c - \frac{\kappa_1}{C}i_l. \tag{16.21}$$

Notice from (16.20) that sliding motion can be enforced on $\sigma(v_c, i_l) = 0$ if $1 > \frac{\psi_1(v_c,i_l)}{\psi_2(v_c,i_l)} > 0$. Using the last expression, the equivalent control is easily derived:

$$u_{eq} = \frac{\psi_2(v_c, i_l) - \psi_1(v_c, i_l)}{\psi_2(v_c, i_l)}. \tag{16.22}$$

Therefore, the equivalent system in sliding mode is:

$$L\frac{\mathrm{d}i_l}{\mathrm{d}t} = E - v_c\frac{\psi_1(v_c, i_l)}{\psi_2(v_c, i_l)} \tag{16.23}$$

$$C\frac{\mathrm{d}v_c}{\mathrm{d}t} = i_l\frac{\psi_1(v_c, i_l)}{\psi_2(v_c, i_l)} - \frac{v_c}{R}, \tag{16.24}$$

which is highly nonlinear. It is straightforward to check that (i_l^*, v_c^*), with

$$i_l^* = \frac{v_c^{*2}}{ER},$$

is an equilibrium point for this system. In the following, conditions will be obtained to guarantee local asymptotic stability of such equilibrium.

Indeed, defining the error variables $e_1 = i_l - i_l^*$, $e_2 = v_c - v_c^*$, the linearized model of the error system corresponding to (16.23), (16.24) reads as:

$$\begin{aligned} L\frac{\mathrm{d}e_1}{\mathrm{d}t} &= -\frac{E^2\kappa_1}{Cv_c^*\psi_1(v_c^*, i_l^*)}e_1 - \frac{E}{\psi_1(v_c^*, i_l^*)}\left(\kappa_2 - \frac{2\kappa_1}{RC}\right)e_2 \\ C\frac{\mathrm{d}e_2}{\mathrm{d}t} &= \frac{E^2\kappa_3}{Lv_c^*\psi_1(v_c^*, i_l^*)}e_1 + \frac{1}{R\psi_1(v_c^*, i_l^*)}\left(v_c^*\kappa_2 - \frac{2E\kappa_3}{L}\right)e_2, \end{aligned} \tag{16.25}$$

where it follows from (16.21) that

$$\psi_1(v_c^*, i_l^*) = \frac{E\kappa_3}{L} - \frac{\kappa_1 v_c^*}{RC}.$$

The characteristic polynomial of (16.25) is given by:

$$P(\lambda) = \lambda^2 + \frac{1}{\psi_1(v_c^*, i_l^*)}\left(\frac{E^2\kappa_1}{LCv_c^*} - \frac{\kappa_2 v_c^*}{RC} + \frac{2E\kappa_3}{RLC}\right)\lambda + \frac{E^2\kappa_2}{LCv_c^*\psi_1(v_c^*, i_l^*)}$$

Hence, under the current hypotheses $(\kappa_1, \kappa_2, \kappa_3 > 0)$, the origin of (16.25) will be locally asymptotically if and only if

$$\frac{Ev_c^*\kappa_3}{L} - \frac{\kappa_1 v_c^{*2}}{RC} > 0, \quad \text{and} \quad \frac{E^2\kappa_1}{L} - \frac{\kappa_2 v_c^{*2}}{R} + \frac{2Ev_c^*\kappa_3}{RL} > 0.$$

In this simulation case, the chosen values are: $\kappa_1 = 0.8$, $\kappa_2 = 4500$, $\kappa_3 = 0.6$, which deliver a good transient response for the output voltage. Finally, using (16.20), the hysteretic control law that confines the switching function within the space region $|\sigma(v_c, i_l)| < \Delta_k$ is:

$$u = \begin{cases} 0 \text{ if } \ \sigma \cdot \text{sign}\,(\psi_2) < -\Delta_k \ \text{ or } \ (|\sigma| < \Delta_k \ \& \ \dot{\sigma} > 0) \\ 1 \text{ if } \ \sigma \cdot \text{sign}\,(\psi_2) > \Delta_k \ \text{ or } \ (|\sigma| < \Delta_k \ \& \ \dot{\sigma} < 0)\,. \end{cases}$$

16.3.2.2 Switching Frequency Regulation

In order to select γ for the SFC, ρ_k^+ and ρ_k^- have to be evaluated. This requires (16.22) to be particularized for the steady state sliding mode, i.e. assuming $v_c = v_c^*$ and $i_l = \frac{v_c^{*2}}{RE}$,

$$\dot{\sigma}(v_c^*, i_l^*) = \psi_2(v_c^*, i_l^*)\left(1 - \frac{E}{v_c^*} - u\right)$$

where

$$\psi_2(v_c^*, i_l^*) = \frac{\kappa_3 v_c^*}{L} - \frac{\kappa_1 v_c^{*2}}{ERC}.$$

Using the data given in Table 16.2, $\psi_2(v_c^*, i_l^*)$ results positive; therefore, the expected switching function slopes become:

$$\rho_{*k}^{+} = \left[\dot{\sigma}(v_c^*, i_l^*)_{u=0}\right]^{-1} = \psi_2(v_c^*, i_l^*)^{-1}\left(1 - \frac{E}{v_c^*}\right)^{-1}$$

$$\rho_{*k}^{-} = \left[\dot{\sigma}(v_c^*, i_l^*)_{u=1}\right]^{-1} = -\psi_2(v_c^*, i_l^*)^{-1}\frac{v_c^*}{E}.$$

Replacing the values of the parameter shown in the Table 16.2 one gets:

$$\begin{array}{l} v_c^* = 36V \rightarrow \rho_{*k}^{+} = 1.67 \cdot 10^{-5};\ \rho_{*k}^{-} = -3.35 \cdot 10^{-5}; \\ v_c^* = 48V \rightarrow \rho_{*k}^{+} = 1.15 \cdot 10^{-5};\ \rho_{*k}^{-} = -3.46 \cdot 10^{-5}; \end{array}$$

According to Theorem 16.1, the closed-loop system is stable for $\gamma \in (0, 2.89 \cdot 10^5)$. Hence, we choose $\gamma = 2 \cdot 10^4$.

16.3.2.3 Simulation Results

The simulations are performed using Matlab Simulink with the data shown in Table 16.2 and the control parameters κ_1 =0.8, $\kappa_2 = 4500$, $\kappa_3 = 0.6$, and $\gamma = 2 \cdot 10^4$.

Figure 16.10 shows the response of the system with different initial conditions, Δ_{ini}, for the hysteresis value Δ. Both voltage and frequency regulation are confirmed from the results.

The simulation shown in Fig. 16.11 plots the system response when the voltage reference is step changed between 48 and 36 V (see the top plots). Once the sliding motion is recovered, the switching period reaches the desired value in around 500μs. Notice from the mid plots of the figure how the SFC adjusts the hysteresis value in order to keep the switching period at the desired value.

The results presented in Fig. 16.12 show the switching period response when $\gamma = 2.75 \cdot 10^5$, which is close to the maximum value that guarantees stability, i.e. $\gamma = 2.88 \cdot 10^5$. The underdamped response illustrates the validity of the stability range.

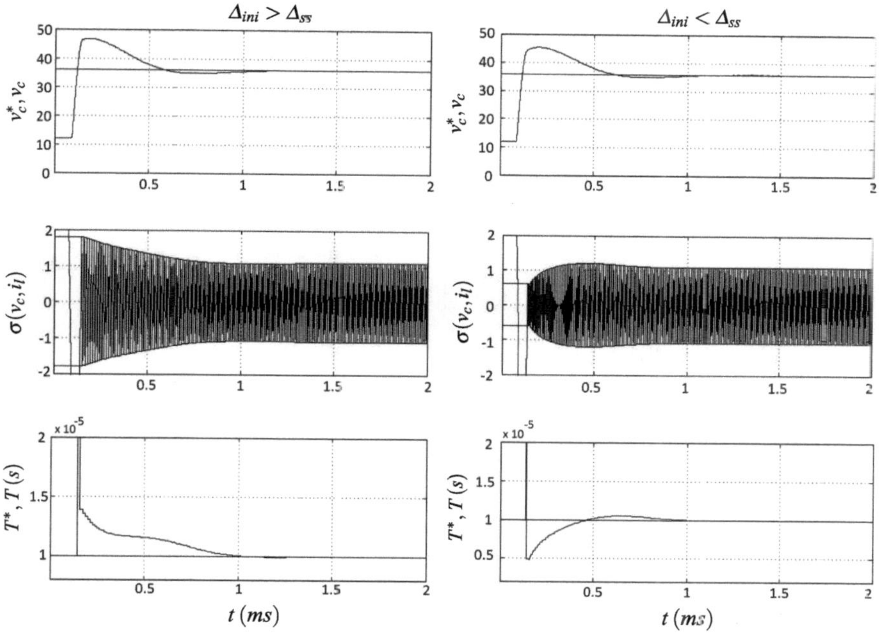

Fig. 16.10 Boost Converter: start-up with different initial values for Δ. From *top to bottom*. 1- Output voltage, v_c, and reference voltage, v_c^*. 2- Switching function σ. 3- Desired and real switching period (T^*, T)

16.3.3 Output Tracking: The Voltage Source Inverter

The voltage source inverter (VSI) circuit scheme is depicted in Fig. 16.13. This circuit is commonly employed to generate a sinusoidal signal at its output and is classified as DC/AC converter.

The VSI dynamics are governed by:

$$C\frac{\mathrm{d}v_c}{\mathrm{d}t} = -\frac{v_c}{R} + i_L, \tag{16.26}$$

$$L\frac{\mathrm{d}i_L}{\mathrm{d}t} = -v_c + E\,u, \tag{16.27}$$

where i_L is the inductor current, v_c is the output voltage, R is the resistive load, L is the inductance, C is the capacitor and E is the input voltage. The control action u takes values in $\{-1, 1\}$. The power switches are represented by M_1, M_2, M_3, and M_4. As it is shown in Fig. 16.13, M_1 and M_4 are short circuited when $u = 1$, and remain open when $u = -1$, whereas M_2 and M_3 work in a complementary way. Table 16.3 presents the specific values of the converter parameters used in the simulation.

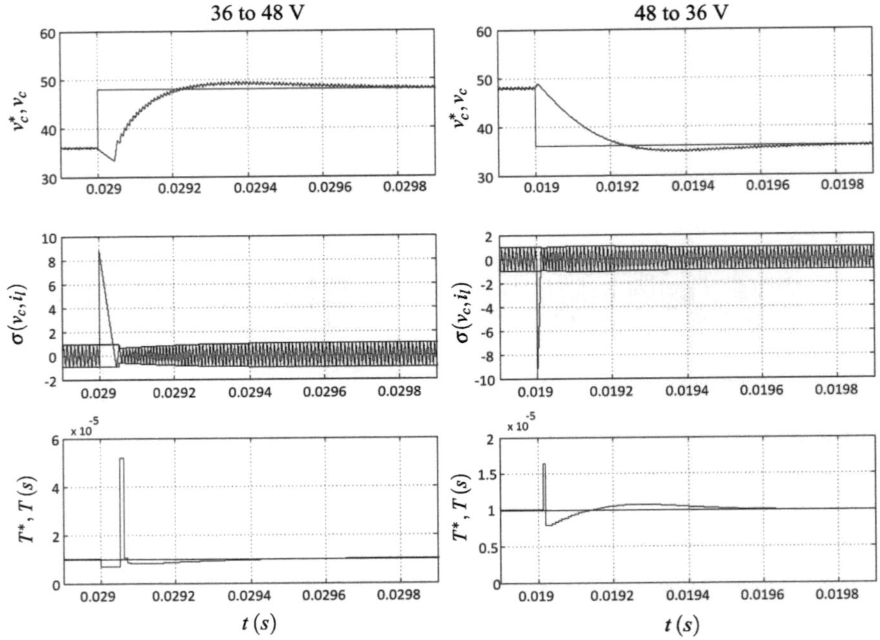

Fig. 16.11 Boost Converter: step-changing output voltage reference. From *top to bottom*. 1- Output voltage, v_c, and reference voltage, v_c^*. 2- Switching function σ. 3- Desired and real switching period (T^*, T)

16.3.3.1 Sliding Mode Control

In this case, the control objective is to track a time-varying reference at the output. The signal to be tracked is defined as:

$$v_c^*(t) = A \sin \omega t.$$

Since the relative degree of the output voltage with respect to the control is two, the following first order linear switching surface is used [4]:

$$\sigma\,(v_c, \dot{v}_c) = \phi_1 e_v + \phi_2 C \dot{e}_v = 0, \quad \phi_{1,2} > 0, \tag{16.28}$$

where $e_v = v_c^* - v_c$.

The switching function derivative becomes:

$$\dot{\sigma}\,(v_c, \dot{v}_c) = f_{vsi} - \frac{\phi_2 E}{L} u \tag{16.29}$$

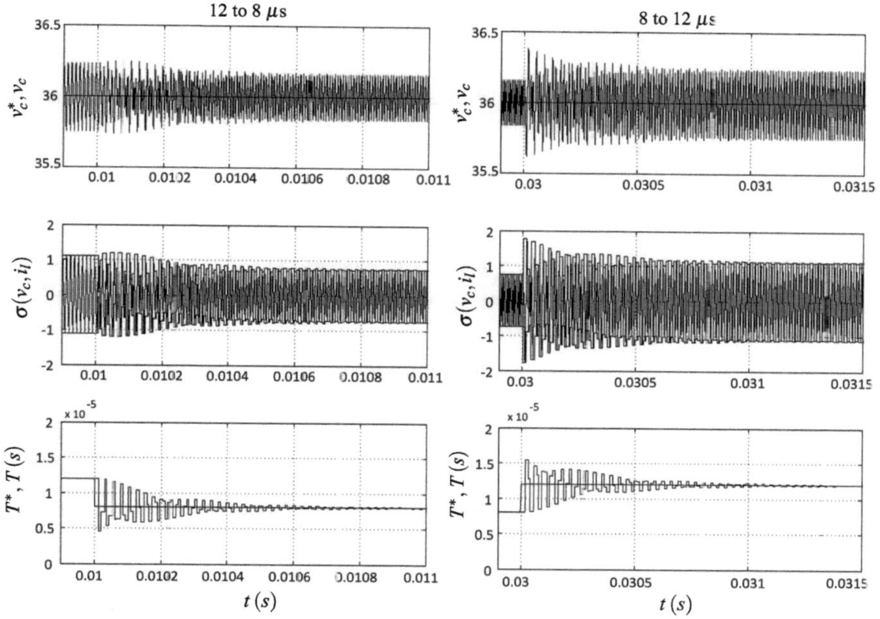

Fig. 16.12 Boost Converter: switching period regulation with $\gamma = 275000$. From *top to bottom*. 1- Output voltage, v_c, and reference voltage, v_c^*. 2- Switching function σ. 3- Desired and real switching period T^*, T

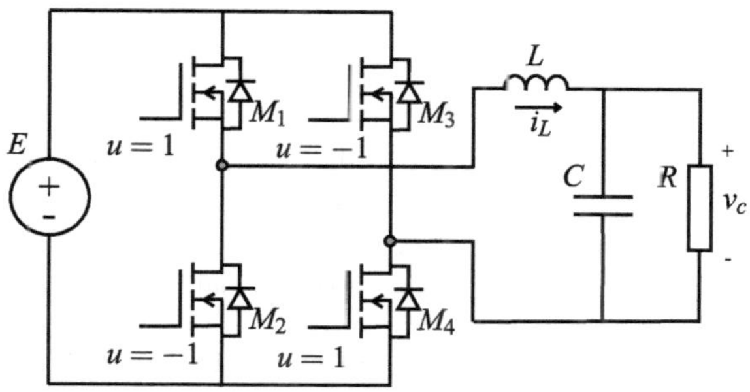

Fig. 16.13 Voltage source inverter structure

where

$$f_{vsi} = \phi_1 \dot{v}_c^* + \phi_2 C \ddot{v}_c^* + v_c \left(\frac{\phi_1}{RC} - \frac{\phi_2}{R^2 C} + \frac{\phi_2}{L} \right) + i_l \left(\frac{\phi_2}{RC} - \frac{\phi_1}{C} \right).$$

It is clear that sliding motion exists if $\frac{\phi_2 E}{L} > |f_{vsi}|$. The equivalent control results in:

$$u_{eq} = \frac{L}{\phi_2 E} f_{vsi}.$$

Table 16.3 Voltage Source Inverter parameters

Parameter	Symbol	Value
Input voltage	E	400 V
Desired output voltage amplitude	A	$230\sqrt{2}$ V
Output voltage frequency range	f	50–200 Hz
Inductor	L	450 μH
Output capacitor	C	100 μF
Load range	R	1 kΩ–10 Ω
Switching period reference	T^*	50 μs

According to the definition of the equivalent control, the expression (16.29) can be redefined as:

$$\dot{\sigma}\,(v_c, \dot{v}_c) = \frac{\phi_2\, E}{L}(u_{eq} - u) \tag{16.30}$$

The corresponding ideal sliding behavior is given by the linear time-varying system:

$$\begin{aligned} C\frac{\mathrm{d}v_C}{\mathrm{d}t} &= -\frac{\phi_1}{\phi_2}v_C + \frac{\phi_1}{\phi_2}v_C^* + C\dot{v}_C^*, \\ L\frac{\mathrm{d}i_L}{\mathrm{d}t} &= \frac{\alpha\phi_1}{R\phi_2}v_C - \frac{\alpha\phi_1}{\phi_2}i_L + Lh(t), \end{aligned}$$

where

$$\alpha := \frac{L}{C}\left(1 - \frac{\phi_2}{R\phi_1}\right) \quad \text{and} \quad h(t) := \frac{\phi_1}{\phi_2}\dot{v}_C^* + C\ddot{v}_C^*.$$

It is then immediate that the system is asymptotically stable if $R > \phi_2\phi_1^{-1} > 0$. According to the VSI parameters values defined in Table 16.3, the sliding coefficients are selected as: $\phi_1 = 0.1$, and $\phi_2 = C$.

Finally, the hysteretic control law that confines σ within a boundary layer of width $2\Delta_k$, is:

$$u = \begin{cases} -1 \text{ if } & \sigma < -\Delta_k \text{ or } (|\sigma| < \Delta_k \,\&\, \dot{\sigma} > 0) \\ 1 \text{ if } & \sigma > \Delta_k \text{ or } (|\sigma| < \Delta_k \,\&\, \dot{\sigma} < 0)\,. \end{cases}$$

16.3.3.2 Switching Frequency Regulation

In order to select γ for the SFC, the values of ρ_{*k}^+ and ρ_{*k}^- have to be evaluated. The switching function slopes can be obtained from (16.30). The equivalent control in the steady sliding motion can be derived from (16.26), (16.27) imposing that $v_c = v_c^*$.

Therefore, (16.30) becomes:

$$\dot{\sigma}\left(v_c^*, \dot{v}_c^*\right) = \phi_2\left[\frac{v_c^* - Eu}{L} + \frac{\dot{v}_c^*}{R} + C\ddot{v}_c^*\right],$$

and ρ_{*k}^+ and ρ_{*k}^- are finally given by:

$$\rho_{*k}^+ = \left[\dot{\sigma}\left(v_c^*, \dot{v}_c^*\right)_{u=-1}\right]^{-1} = \phi_2^{-1}\left[\frac{v_c^* + E}{L} + \frac{\dot{v}_c^*}{R} + C\ddot{v}_c^*\right]^{-1}$$
$$\rho_{*k}^- = \left[\dot{\sigma}\left(v_c^*, \dot{v}_c^*\right)_{u=1}\right]^{-1} = \phi_2^{-1}\left[\frac{v_c^* - E}{L} + \frac{\dot{v}_c^*}{R} + C\ddot{v}_c^*\right]^{-1}.$$

According to Theorem 16.2, the previous expressions allow to select the value of γ which ensures stability. In the top plot, Fig. 16.14 shows the desired output voltage, v_c^*, and the dynamic evolution of ρ_{*k}^+, ρ_{*k}^- in the mid plot. Finally, the set of

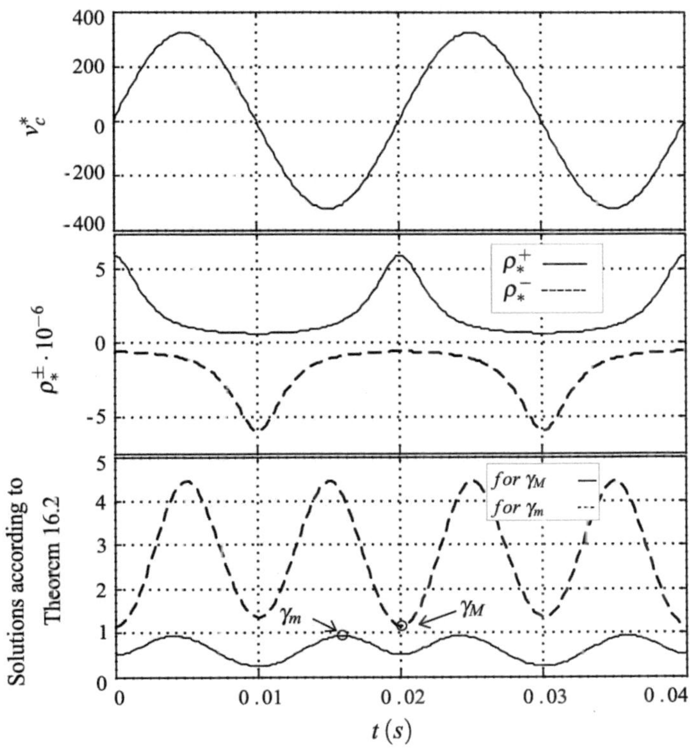

Fig. 16.14 VSI Converter. From *top to bottom*. 1- Desired output voltage, v_c^*. 2- Dynamic evolution of ρ_{*k}^+ and ρ_{*k}^-. 3- Groups of roots produced by the conditions given in Theorem 16.2

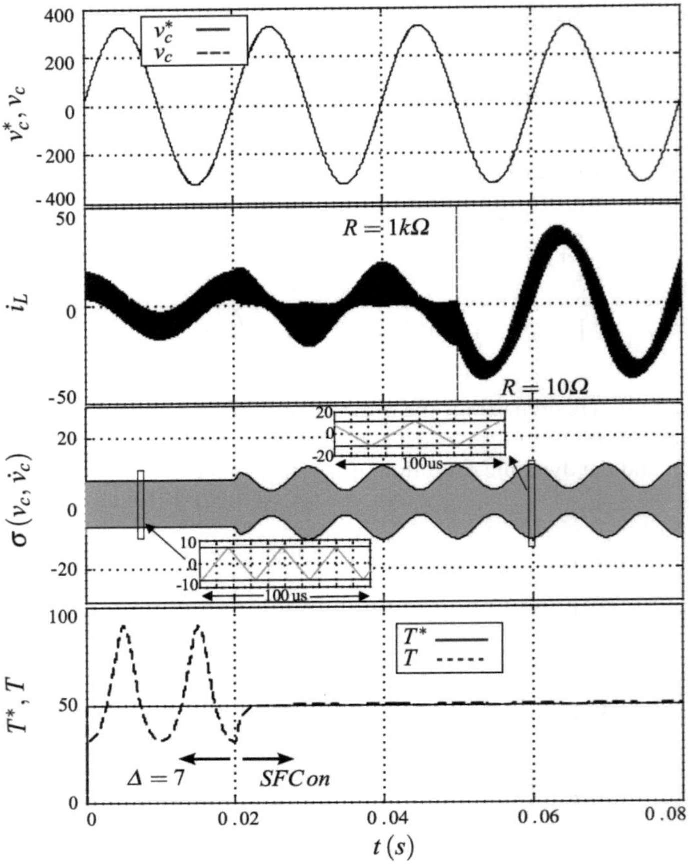

Fig. 16.15 VSI Converter. From *top to bottom*. 1- Desired and real output voltage (v_c^*, v_c). 2- Inductor current, i_l. 3- Switching surface, σ. 4- Desired and real switching period of the control action (T^*, T)

solutions of the condition stated at Theorem 16.2 for the resulting values of ρ^+_{*k}, ρ^-_{*k} are presented in the bottom plot. With such signals, it is straightforward to find the maximum and minimum values which guarantee stability of the SFC. Specifically, the exact values that define the stability margin are $\gamma_M = 1.14 \cdot 10^5$ and $\gamma_m = 9.3 \cdot 10^4$, i.e. $9.3 \cdot 10^4 < \gamma < 1.14 \cdot 10^5$. The chosen value for the simulations is $\gamma = 1 \cdot 10^5$.

16.3.3.3 Simulation Results

The simulations are performed with Matlab-Simulink. Figure 16.15 shows the response of the system under sliding motion when some variations are introduced. On the one hand, at the beginning of the simulations a fixed value for the hysteresis band is used, which leads to an expected time-varying switching period. At time instant $t = 0.02$ s the proposed SFC structure is enabled. In the bottom plot in Fig. 16.15 one can observe how the switching period converges to the desired value, confirming a proper performance of the SFC. Additionally, a load transient is introduced at time $t = 0.05$ s, from $R = 1\,\mathrm{k}\Omega$ to $R = 10\,\Omega$. Notice how the output voltage v_c tracks perfectly the desired voltage v_c^* during the entire test. The use of the feedforward signal Ω (see (16.13)) implies knowledge of $\rho_k^{\pm}$, but when the SFC is implemented the information to calculate Δ_k is related to the last interval measured $k-1$, since $\rho_k^{\pm}$ is not available until the kth interval ends. Specifically, $\rho_k^{\pm}$ are approximated by the immediately preceding values:

$$\rho_{k-1}^{+} = \frac{T_{k-1}^{+}}{\Delta_{k-1} + \Delta_{k-2}}, \quad \rho_{k-1}^{-} = \frac{T_{k-1}^{-}}{2\Delta_{k-1}}.$$

The last test, shown in Fig. 16.16, presents the switching period response when some parameters are varied, as the amplitude and frequency of the time-varying reference signal, and the desired switching frequency. An overall good performance of the system is confirmed. However, it is worthwhile commenting on the switching period oscillation that appears when the desired frequency, ω, of the tracking signal is set to 200 Hz (see second plot in Fig. 16.16 at $t = 0.07$ s). When the frequency signal increases, the values of $\rho_k^{\pm}$ have a higher time variation, and the assumption

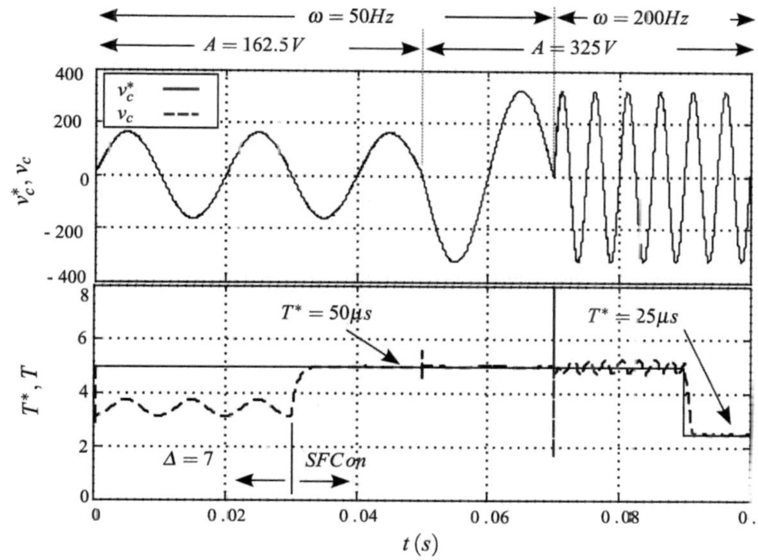

Fig. 16.16 VSI Converter. From *top to bottom*. 1- Desired and real output voltage (v_c^*, v_c). 2- Desired and real switching period of the control action (T^*, T)

of constant slopes during the switching interval is not completely fulfilled. As a consequence, the variable hysteresis band provided by the SFC does not perfectly reject the period oscillations. In the same way, notice that when the desired switching frequency is increased ($t = 0.09\,\mathrm{s}$), the assumption is newly met, and the switching period recovers the desired fixed value.

16.4 Conclusions

Fixing the switching frequency is a key issue in sliding mode control implementations when it is applied in inherently switched systems. This chapter presented a hysteresis band controller capable of setting a constant value for the steady-state switching frequency of a sliding mode controller in regulation and tracking tasks. Problem statement, practical assumptions, stability proofs and control parameters design criteria were also provided. The proposal was numerically validated through a set of simulations in power converters such as a Buck converter, a Boost converter, and a voltage source inverter.

Acknowledgements This work was partially supported by the Spanish projects DPI2013-41224-P (Ministerio de Educación) and 2014 SGR 267 (AGAUR).

References

1. Biel, D., Fossas, E.: SMC applications in power electronics. In: Variable Structure Systems: From Principles to Implementation, pp. 265–293. Institution of Electrical Engineers, London (2004)
2. Bilalovic, F., Music, O., Sabanovic, A.: Buck converter regulator operating in the sliding mode. In: Proceedings of the VII International PCI Conference, pp. 331–340 (1983)
3. Bühler, H.: Réglage par mode de glissement. Presses polytechniques et universitaires romandes (1986)
4. Carpita, M., Marchesoni, M.: Experimental study of a power conditioning system using sliding mode control. IEEE Trans. Power Electron. **11**(5), 731–742 (1996)
5. Chiarelli, C., Malesani, L., Pirondini, S., Tomasin, P.: Single-phase, three-level, constant frequency current hysteresis control for UPS applications. In: Proceedings of the 15th European Conference on Power Electronics and Applications, pp. 180–185 (1993)
6. Fossas, E., Griñó, R., Biel, D.: Quasi-sliding control based on pulse width modulation, zero averaged dynamics and the L_2 norm. In: Proceedings of the 6th IEEE International Workshop on Variable Structure Systems, Analysis, Integration and Applications, pp. 335–344 (2001)
7. Guzman, R., de Vicuña, L.G., Morales, J., Castilla, M., Matas, J.: Sliding-mode control for a three-phase unity power factor rectifier operating at fixed switching frequency. IEEE Trans. Power Electron. **31**(1), 758–769 (2016)
8. Holmes, D.G., Davoodnezhad, R., McGrath, B.P.: An improved three-phase variable-band hysteresis current regulator. IEEE Trans. Power Electron. **28**(1), 441–450 (2013)
9. Mahdavi, J., Emadi, A., Toliyat, H.: Application of state space averaging method to sliding mode control of PWM DC/DC converters. In: Proceedings of the 32nd Anual Conference on Industry Applications, pp. 820–827 (1997)

10. Malesani, L., Rossetto, L., Spiazzi, G., Zuccato, A.: An AC power supply with sliding-mode control. IEEE Ind. Appl. Mag. **2**(5), 32–38 (1996)
11. Mattavelli, P., Rossetto, L., Spiazzi, G., Tenti, P.: General-purpose sliding-mode controller for DC/DC converter applications. In: Proceedings of the 24th Annual IEEE Power Electronics Specialists Conference, pp. 609–615 (1993)
12. Mattavelli, P., Rossetto, L., Spiazzi, G., Tenti, P.: Sliding mode control of Sepic converters. In: Proceedings of European Space Power Conference, pp. 173–178 (1993)
13. Ramos, R.R., Biel, D., Fossas, E., Guinjoan, F.: A fixed-frequency quasi-sliding control algorithm: application to power inverters design by means of FPGA implementation. IEEE Trans. Power Electron. **18**(1), 344–355 (2003)
14. Repecho, V., Biel, D., Fossas, E.: Fixed switching frequency sliding mode control using a hysteresis band controller. In: Proceedings of the 13th International Workshop on Variable Structure Systems, pp. 1–6 (2014)
15. Repecho, V., Biel, D., Olm, J.M., Colet, E.F.: Switching frequency regulation in sliding mode control by a hysteresis band controller. IEEE Trans. Power Electron. **32**(2), 1557–1569 (2017)
16. Repecho, V., Biel, D., Olm, J.M., Fossas, E.: Sliding mode control of a voltage source inverter with switching frequency regulation. In: Proceedings of the 14th International Workshop on Variable Structure Systems, pp. 296–301 (2016)
17. Ruiz, J., Lorenzo, S., Lobo, I., Amigo, J.: Minimal UPS structure with sliding mode control and adaptive hysteresis band. In: Proceedings of the 16th Annual Conference of IEEE Industrial Electronics Society, pp. 1063–1067 (1990)
18. Silva, J.F., Paulo, S.S.: Fixed frequency sliding mode modulator for current mode PWM inverters. In: Proceedings of the 24th Annual IEEE Power Electronics Specialists Conference, pp. 623–629 (1993)
19. Tan, S.C., Lai, Y., Tse, C.K., Cheung, M.K.: A fixed-frequency pulse width modulation based quasi-sliding-mode controller for buck converters. IEEE Trans. Power Electron. **20**(6), 1379–1392 (2005)
20. Tan, S.C., Lai, Y.M., Chi, K.T.: General design issues of sliding-mode controllers in DC-DC converters. IEEE Trans. Ind. Electron. **55**(3), 1160–1174 (2008)
21. Utkin, V., Guldner, J., Shi, J.: Sliding Mode Control in Electro-Mechanical Systems. CRC Press, Boca Raton (2009)
22. Venkataramanan, R.: Sliding mode control of power converters. Ph.D. thesis, California Institute of Technology (1986)

MIX
Papier aus verantwortungsvollen Quellen
Paper from responsible sources
FSC® C105338

If you have any concerns about our products,
you can contact us on
ProductSafety@springernature.com

In case Publisher is established outside the EU,
the EU authorized representative is:
Springer Nature Customer Service Center GmbH
Europaplatz 3, 69115 Heidelberg, Germany

Printed by Libri Plureos GmbH
in Hamburg, Germany